NEUROPSYCHOPHARMACOLOGY OF MONOAMINES AND THEIR REGULATORY ENZYMES

Advances in Biochemical Psychopharmacology
Volume 12

Advances in Biochemical Psychopharmacology

Series Editors:

Erminio Costa, M.D.

*Chief, Laboratory of Preclinical Pharmacology
National Institute of Mental Health
Washington, D.C., U.S.A.*

Paul Greengard, Ph.D.

*Professor of Pharmacology
Yale University School of Medicine
New Haven, Connecticut, U.S.A.*

NEUROPSYCHOPHARMACOLOGY OF MONOAMINES AND THEIR REGULATORY ENZYMES

Advances in Biochemical Psychopharmacology Volume 12

Editor

Earl Usdin, Ph.D.

*Psychopharmacology Research Branch
National Institute of Mental Health
Rockville, Maryland*

Raven Press ■ New York

© 1974 by Raven Press Books, Ltd. All rights reserved. This book is protected by copyright. No part of it may be duplicated or reproduced in any manner without written permission from the publisher.

Made in the United States of America

International Standard Book Number 0-911216-77-4
Library of Congress Catalog Card Number 74-77231

ISBN outside North and South America only:
0-7204-7519-8

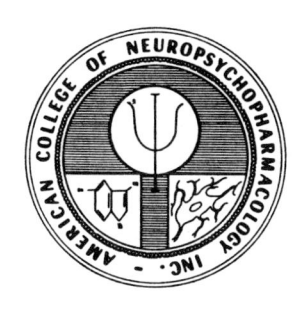

Officers for 1973
American College of Neuropsychopharmacology Inc.

J. R. Wittenborn, Ph.D., *President*
Leo E. Hollister, M.D., *President-Elect*
Philip R. A. May, M.D., *Vice-President*
Alberto DiMascio, Ph.D., *Secretary-Treasurer*
Jonathan O. Cole, M.D., *Assistant Secretary-Treasurer*
Alfred Freedman, M.D., *Past President*

Preface

The area of neuropsychopharmacology is expanding so rapidly that it has become impossible for the American College of Neuropsychopharmacology (ACNP) to allow time at its Annual Meeting for each of the Study Groups proposed by members of the ACNP. Thus at the December 4–7, 1973 meeting of the ACNP, four proposed study groups, all with topics related to the neuropsychopharmacology of monoamines and their regulatory enzymes, were combined into one. The texts of the manuscripts for this combined group are included here as well as a summary of the open discussions which were held after the individual papers were presented.

As Dr. Wyatt points out in his introduction, one of the principal purposes in holding the MAO session was to determine whether multiple molecular forms of this enzyme are of clinical psychiatric significance. This question is considered from such aspects as diagnosis and prognosis and even from the point of view of designing selective MAO inhibitors.

It is probably more symptomatic of our knowledge than of their significance that all the monoaminergic enzymes except MAO are treated in a chapter of about the same length as MAO. The current controversies with regard to the significance of these enzymes, even with regard to the most appropriate assay techniques, are covered in the papers and in the lively discussions which they engendered.

Finally, current thinking on dopamine and its behavioral effects is summarized in the next chapter. Some of the papers are factual, others are more speculative. Hopefully the reader will be able to distinguish between fact and speculation and be able to come to his own conclusions.

The helpful assistance of the ACNP Secretary-Treasurer, Dr. Alberto DiMascio, and his secretary, Ms. Lorraine Josof, is most gratefully acknowledged, as well as the long and loyal help in transcribing tapes, etc., of my secretaries, Ms. Dorothy Eisel and Ms. Karen Deel.

Earl Usdin

This book was edited by Earl Usdin in his private capacity. No official support or endorsement by the NIMH is intended or should be inferred.

Contents

MULTIPLE MOLECULAR FORMS OF MONOAMINE OXIDASE

1   Introduction: The Multiple Monoamine Oxidases
R. J. Wyatt

3   Multiple Forms of Monoamine Oxidase: Some *In Vivo* Correlations
M. Sandler, S. Bonham Carter, B. L. Goodwin, C. R. J. Ruthven, M. B. H. Youdim, E. Hanington, M. F. Cuthbert, and C. M. B. Pare

11   Physicochemical Properties, Development, and Regulation of Central and Peripheral Monoamine Oxidase Activity
M. B. H. Youdim, M. Holzbauer, and H. F. Woods

29   Some Interrelated Properties of Brain Monoamine Oxidase
J. C. Shih and S. Eiduson

37   Some Considerations in the Design of Substrate and Tissue-Specific Inhibitors of Monoamine Oxidase
Y. C. Martin and J. H. Biel

49   The Use of Selective Monoamine Oxidase Inhibitor Drugs to Modify Amine Metabolism in Brain
N. H. Neff, H.-Y. T. Tang, and J. A. Fuentes

59   Comparison of Monoamine Oxidase, Substrate Activities in Twins, Schizophrenics, Depressives, and Controls
A. Nies, D. S. Robinson, L. S. Harris, and K. R. Lamborn

71   Monoamine Oxidase in Man: Enzyme Characteristics in Platelets, Plasma, and Other Human Tissues
D. L. Murphy and C. H. Donnelly

87   *Discussion*

MONOAMINERGIC ENZYMES OTHER THAN MONOAMINE OXIDASE

93   Introduction: Monoaminergic Enzymes Other Than Monoamine Oxidase
M. Goldstein

95	Dopamine-β-hydroxylase and the Regulation of the Noradrenergic Neuron *P. B. Molinoff, D. L. Nelson, and J. C. Orcutt*
105	Human Serum Dopamine-β-hydroxylase: Relationship to Sympathetic Activity in Physiological and Pathological States *M. Goldstein, L. S. Freedman, R. P. Ebstein, D. H. Park, and T. Kashimoto*
121	Evaluation of Serum Dopamine-β-hydroxylase Activity as an Index of Sympathetic Nervous Activity in Man *W. Lovenberg, E. A. Bruckwick, R. W. Alexander, D. Horwitz, and H. R. Keiser*
129	Serum Dopamine-β-hydroxylase: A Possible Aid in the Evaluation of Hypertension *S. M. Schanberg, R. A. Stone, N. Kirshner, J. C. Gunnells, and R. R. Robinson*
135	Short-Term Control of Tyrosine Hydroxylase *A. Carlsson, W. Kehr, and M. Lindqvist*
143	A Critical Assessment of Methods for the Determination of Monoamine Synthesis and Turnover Rates *In Vivo* *N. Weiner*
161	Short- and Long-Term Regulation of Tyrosine Hydroxylase *E. Costa, A. Guidotti, and B. Zivkovic*
177	Regulation of Function of Tryptophan Hydroxylase *A. J. Mandell and S. Knapp*
189	Comparative Properties of Soluble and Particulate Catechol-O-methyl Transferases from Rat Red Blood Cells: Preliminary Observations *M. Roffman, T. G. Reigle, P. J. Orsulak, and J. J. Schildkraut*
195	Genetic Aspects of Monoamine Mechanisms *J. D. Barchas, R. D. Ciaranello, J. A. Dominic, T. Deguchi, E. Orenberg, J. Renson, and S. Kessler*
205	*Discussion*

DOPAMINE

211	Dopamine and Psychotic States: Preliminary Remarks *B. Angrist and S. Gershon*

221	Effect of Chronic Treatment with Central Stimulants on Brain Monoamines and Some Behavioral and Physiological Functions in Rats, Guinea Pigs, and Rabbits *T. Lewander*
241	Drug-Induced Stereotyped Behavior: Similarities and Differences *M. B. Wallach*
261	Central Dopamine Function in Affective Illness: Evidence from Precursors, Enzyme Inhibitors, and Studies of Central Dopamine Turnover *F. K. Goodwin and R. L. Sack*
281	Behavioral and EEG Changes in the Amphetamine Model of Psychosis *E. H. Ellinwood, Jr.*
299	Serum Prolactin Levels in Newly Admitted Psychiatric Patients *H. Y. Meltzer, E. J. Sachar, and A. G. Frantz*
317	Dopamine, Psychomotor Stimulants, and Schizophrenia: Effects of Methylphenidate and the Stereoisomers of Amphetamine in Schizophrenics *D. S. Janowsky and J. M. Davis*
325	Catecholamines, Drugs, and Behavior: Mutual Interactions *L. S. Seiden and A. B. Campbell*
339	The Effect of Catechol-O-methyl Transferase Inhibitors on Behavior and Dopamine Metabolism *G. M. McKenzie*
353	Relationship of Dopamine Neural Systems to the Behavioral Alterations Produced by 6-Hydroxydopamine Administration into Brain *B. R. Cooper and G. R. Breese*
369	Effects of Alterations in Impulse Flow on Transmitter Metabolism in Central Dopaminergic Neurons *R. H. Roth, J. R. Walters, and V. H. Morgenroth III*
385	Ingestive Behavior Following Damage to Central Dopamine Neurons: Implications for Homeostasis and Recovery of Function *M. J. Zigmond and E. M. Stricker*
403	Behavioral Effects of Direct- and Indirect-Acting Dopaminergic Agonists *K. E. Moore*

415	The Use of a Dopaminergic Receptor Stimulating Agent (Piribedil, ET 495) in Parkinson's Disease *A. Lieberman, Y. Le Brun, D. Boal, and M. Zolfaghari*
427	Clinical Studies of Dopaminergic Mechanisms *T. N. Chase*
435	*Discussion*
443	Discussion and Summary *Morris A. Lipton*
455	*Index*

Participants

Burton Angrist
New York University Medical Center
New York, New York

Jack D. Barchas
Stanford University
Palo Alto, California

Robert H. Belmaker
National Institute of Mental Health
Bethesda, Maryland

John H. Biel
Abbott Laboratories
North Chicago, Illinois

George Breese
University of North Carolina
Chapel Hill, North Carolina

Wagner H. Bridger
Yeshiva University
Bronx, New York

Bernard B. Brodie
University of Arizona Medical Center
Tucson, Arizona

Benjamin S. Bunney
Yale University School of Medicine
New Haven, Connecticut

Arvid Carlsson
University of Göteborg
Göteborg, Sweden

Thomas N. Chase
National Institute of Mental Health
Bethesda, Maryland

William G. Clark
Veterans Administration Hospital
Sepulveda, California

Barrett R. Cooper
University of North Carolina School
  of Medicine
Chapel Hill, North Carolina

Erminio Costa
National Institute of Mental Health
St. Elizabeth's Hospital
Washington, D.C.

John M. Davis
Illinois State Psychiatric Institute
Chicago, Illinois

Samuel Eiduson
University of California
Los Angeles, California

Everett E. Ellinwood, Jr.
Duke University School of Medicine
Durham, North Carolina

Ronald S. Fieve
New York State Psychiatric Institute
New York, New York

Sabit Gabay
Veterans Administration Hospital
Brockton, Massachusetts

Menek Goldstein
New York University Medical Center
New York, New York

Frederick Goodwin
National Institute of Mental Health
Bethesda, Maryland

Ingeborg Hanbauer
National Institute of Mental Health
Bethesda, Maryland

Ernest Hartmann
Boston State Hospital
Boston, Massachusetts

David S. Janowsky
University of California School
  of Medicine
La Jolla, California

Murray E. Jarvik
University of California
Los Angeles, California

PARTICIPANTS

Eva K. Killam
University of California
Davis, California

Tommy Lewander
Ulleraker Hospital
Uppsala, Sweden

A. Lieberman
New York University Medical Center
New York, New York

Morris A. Lipton
University of North Carolina
Chapel Hill, North Carolina

Walter Lovenberg
National Heart and Lung Institute
Bethesda, Maryland

L. R. Mandel
Merck & Co.
Rahway, New Jersey

Arnold J. Mandell
University of California School
of Medicine
La Jolla, California

William R. Martin
NIMH Addiction Research Center
Lexington, Kentucky

Y. C. Martin
Abbott Laboratories
North Chicago, Illinois

James McGaugh
University of California
Irvine, California

Gerald M. McKenzie
Wellcome Research Laboratories
Research Triangle Park, North Carolina

Herbert Y. Meltzer
University of Chicago School of Medicine
Chicago, Illinois

Perry B. Molinoff
University of Colorado School of Medicine
Denver, Colorado

Kenneth E. Moore
Michigan State University
East Lansing, Michigan

Dennis L. Murphy
National Institute of Mental Health
Bethesda, Maryland

Norton H. Neff
National Institute of Mental Health
St. Elizabeth's Hospital
Washington, D.C.

Alexander Nies
Dartmouth Medical School
Hanover, New Hampshire

Richard H. Rech
Michigan State University
East Lansing, Michigan

Donald S. Robinson
University of Vermont
College of Medicine
Burlington, Vermont

Mark Roffman
Massachusetts Mental Health Center
Boston, Massachusetts

Robert H. Roth
Yale University School of Medicine
New Haven, Connecticut

Merton Sandler
Queen Charlotte's Maternity Hospital
London, England

Alfred T. Sapse
Rom-Amer Pharmaceuticals
Beverly Hills, California

Saul M. Schanberg
Duke University School of Medicine
Durham, North Carolina

Joseph J. Schildkraut
Harvard Medical School
Boston, Massachusetts

Lewis S. Seiden
University of Chicago
Chicago, Illinois

Jean Chen Shih
University of California School
of Medicine
Los Angeles, California

PARTICIPANTS

Sydney Spector
Roche Institute for Molecular Biology
Nutley, New Jersey

Larry Stein
Wyeth Laboratories
Philadelphia, Pennsylvania

Edward M. Stricker
University of Pittsburgh
Pittsburgh, Pennsylvania

Earl Usdin
National Institute of Mental Health
Rockville, Maryland

Marshall B. Wallach
Syntex Research
Palo Alto, California

Norman Weiner
University of Colorado School
  of Medicine
Denver, Colorado

Richard Wyatt
National Institute of Mental Health
St. Elizabeths Hospital
Washington, D.C.

Moussa B. H. Youdim
MRC Clinical Pharmacology Unit
Oxford, England

Michael J. Zigmond
University of Pittsburgh
Pittsburgh, Pennsylvania

Abbreviations

AIM = abnormal involuntary movement
AMMT = α-methyl-*m*-tyrosine
AMPT = α-methyl-*p*-tyrosine
BOL = brom-LSD
CA = catecholamine(s)
cAMP = cyclic AMP
CAR = conditioned avoidance response
cGMP = cyclic GMP
CNS = central nervous system
COMT = catechol-O-methyl transferase
CSF = cerebrospinal fluid
DA = dopamine
DBA = 3,4-dihydroxybenzoic acid
DBH = dopamine-β-hydroxylase
DBP = diastolic blood pressure
2-DG = 2-deoxy-D-glucose
DMCI = 2,9-dimethyl-β-carboline iodide
DMI = desmethylimipramine
DMPEA = dimethoxyphenylethylamine
$DMPH_4$ = 2-amino-4-hydroxy-6,7-dimethyl-5,6,7,8-tetrahydropteridine
DOM = 2,5-dimethoxy-4-methylamphetamine
DOPA = 3,4-dihydroxyphenylalanine
DOPAC = dihydroxyphenylacetic acid
DTT = dithiothreitol
DZ = dizygotic
EGTA = ethylene glycol bis(β-aminoethyl ether) tetracetic acid
EPS = extrapyramidal side effects
F.D. = familial dysautonomia
FI = fixed interval
FR = fixed ratio
GBL = γ-butyrolactone
GDH = glutamate dehydrogenase
GHB = γ-hydroxybutyrate
GLC = gas/liquid chromatography
6-HDA = 6-hydroxydopamine
5-HIAA = 5-hydroxyindole acetic acid
HMPG = 4-hydroxy-3-methoxyphenyl glycol

ABBREVIATIONS

5-HT = 5-hydroxytryptamine = serotonin
HVA = homovanillic acid
IR = immunoreactive
IRT = interresponse time(s)
LH = lateral hypothalamus
MAO = monoamine oxidase
MAOI = monoamine oxidase inhibitor(s)
MHPG = 3-methoxy-4-hydroxyphenyl glycol
MMPI = Minnesota Multiphasic Personality Inventory
6-MPH$_4$ = 2-amino-4-hydroxy-6-methyl-5,6,7,8-tetrahydropteridine
α-MPT = α-methyl-p-tyrosine
MS = mass spectroscopy
αMT = α-methyl-p-tyrosine
MZ = monozygotic
NA = noradrenaline = norepinephrine
NE = norepinephrine
6-OHD = 6-hydroxydopamine
PAA = phenylacetic acid
PCPA = p-chlorophenylalanine
PIF = prolactin inhibitory factor
PNMT = phenylethanol N-methyl transferase
PRF = prolactin releasing factor
SAR = structure-activity relationship(s)
TH = tyrosine hydroxylase
VI = variable interval

Neuropsychopharmacology of Monoamines and Their Regulatory Enzymes, edited by E. Usdin.
Raven Press, New York © 1974.

Introduction: The Multiple Monoamine Oxidases

Richard J. Wyatt

National Institute of Mental Health, Saint Elizabeths Hospital, Washington, D.C. 20032

The enzyme monoamine oxidase (MAO) is strategically placed in biogenic amine metabolism. An abnormality in its activity would greatly jeopardize the function of most of the putative neurotransmitters. As a cornerstone of our knowledge, this enzyme is important to most biochemical theories of brain function and dysfunction, and researches in such disparate fields as temperature regulation, eating, and schizophrenia have an interest in it. We now know that there are multiple forms of this enzyme, and it is likely that any or all of them are capable of alteration.

One of the principal purposes of holding a session on multiple forms of MAO was to determine if any of these enzyme forms might be of significance to the etiology, diagnosis, or treatment of medical and psychiatric disorders. Recent reports that MAO activity is abnormal in some patients with affective disorders and schizophrenia and that, in the latter, an apparent decrease in activity may be genetically determined, suggest the possibility of a relationship between the etiology and diagnosis of these illnesses.

Many enzymes, as well as other proteins, exhibit genetic polymorphism and occur more frequently in a population than can be expected on the basis of recurrent mutation. Some of the more rare variants produce clinical abnormalities associated with "inborn errors of metabolism." Others are, under normal conditions, entirely silent—becoming apparent only under unusual circumstances. Teleologically, we might expect to see at least two different kinds of enzyme variants. The first would be a variant of high specificity toward a certain substrate or group of substrates. These enzymes would be located in the central nervous system and perhaps other areas of the body where the particular substrate was found in high concentrations. In the brain they would serve to regulate the neurotransmitters, but in the gastrointestinal system they might prevent exogenous amines from entering the body.

The second type of variant would be an alteration in one of the functional MAOs. This alteration could be either silent or cause a decrease in the activity of that MAO. In the latter case we might well see a buildup in a certain substrate or a fall in an important catabolite. In relation to current theories regarding schizophrenia, if either tryptamine or dopamine (both regulated by MAO) was substantially increased, an aberrant metabolite such as dimethyltryptamine (in man, DMT is a short-acting hallucinogen) or perhaps dopamine itself could be causative of the psychosis. Further alterations are also possible which might manifest themselves in altered enzyme activity. These include deviations in cofactor activity, protein binding, and mechanisms of enzyme destruction. Furthermore, natural enzyme inhibitors may be present.

While the clinical material studied has been limited to plasma, platelets, and on some occasions autopsied brains, for the most part examinations of these materials have used only one substrate and may not have been of sufficient specificity to pick up a specific abnormality if it should occur. Recently Meltzer and Stahl *(personal communication)* measured platelet MAO activity in 12 chronic and 10 acute schizophrenics as well as in 15 normal controls using four substrates. The MAO activity of the chronic schizophrenics was lower than that of the controls for all four substrates. Enzyme activity for the acute schizophrenics was lower than that of controls only when *meta*-iodobenzylamine and tyramine were used as substrates. On the other hand, Schwartz, Neff, and Wyatt *(unpublished data)* studied MAO activity in the brains of eight controls and eight schizophrenics. There was no difference between the groups in either the type A or B enzyme.

Inhibitors of MAO are at times useful in treating hypertension, angina, affective disorders, and narcolepsy. The clinician, however, is reluctant to treat patients with MAO inhibitors because they occasionally interact with dietary tyramine to produce a hypertensive crisis. It is hoped that by understanding the multiple isozymes of MAO, MAO inhibitors which are effective in the treatment of one or more of these illnesses, but do not inhibit the destruction of tyramine, might be synthesized.

Following a similar line of reasoning, if some affective disorders should prove to be caused by a functional deficit in an indoleamine or catecholamine, an MAO inhibitor which selectively blocked the destruction of one but not the other might be able to correct this deficit without raising the concentration of both substrates. This may be significant if the balance of these substances is equally or more important than their absolute concentrations. Realizing this, drug companies are actively searching for drugs with greater specificity, and when they are found, a resurgence of interest in these agents will follow.

Neuropsychopharmacology of Monoamines and Their Regulatory Enzymes, edited by E. Usdin. Raven Press, New York © 1974.

Multiple Forms of Monoamine Oxidase: Some In Vivo Correlations

M. Sandler, Susan Bonham Carter, B. L. Goodwin, C. R. J. Ruthven, *M. B. H. Youdim, **Edda Hanington, †M. F. Cuthbert, and ††C. M. B. Pare

*Bernhard Baron Memorial Research Laboratories, and Institute of Obstetrics and Gynaecology, Queen Charlotte's Maternity Hospital, London W6 OXG; *MRC Clinical Pharmacology Unit, Radcliffe Infirmary, Oxford OX2 6HE; **Wellcome Trust, London W1M 9LA; †Committee on Safety of Medicines, London EC2; and ††Department of Psychological Medicine, St. Bartholomew's Hospital, London EC1, England*

I. INTRODUCTION

Available information relating to the electrophoretic separation of multiple forms of solubilized monoamine oxidase (MAO; EC 1.4.3.4.) and to their physicochemical properties has recently been reviewed in detail (Sandler and Youdim, 1972). Although enzyme multiplicity has been observed directly in a number of separate laboratories, the finding has not met with universal acceptance. The view has been put forward that the electrophoretically separated bands of activity may be an artifact of the solubilization process, a possibility never completely ruled out by the present authors. However, the likelihood is probably remote, since a characteristic pattern of multiple forms is obtained from a particular tissue whatever method of solubilization is employed; an infinite electrophoretic gradation of activity might have been predicted had the disintegration pattern of the outer mitochondrial membrane, where the enzyme is located, been a preparation artifact (Sandler and Youdim, 1974).

Houslay and Tipton (1973) have stressed the importance of phospholipid for the existence of multiple forms of enzyme activity which they consider to be more apparent than real. They treated mitochondrial preparations from rat liver with a so-called chaotropic agent, sodium perchlorate, to remove phospholipid which is normally tightly bound to the enzyme. The resulting MAO preparation was homogeneous, manifesting a single band of electrophoretic activity. However, it is important to emphasize that even though these experiments may provide some insight into the nature of the enzyme protein, they take no account of its tertiary structure in the *in vivo* situation: It seems more than likely, in view of the known high affinity of enzyme from certain sources for phospholipid (Olivecrona and Oreland, 1971), that MAO exists within the cell strongly bound to different amounts of phospholipid. This association may in turn account for observed differences in substrate and inhibitor specificity. Indeed, we have made some sighting observations that indicate that enzyme multiplicity is a true *in vivo* phenomenon (Youdim, Collins, Sandler, Bevan Jones, Pare, and Nicholson, 1972; Collins, Youdim, and Sandler, 1972).

II. STUDIES ON MAO A AND B

Latterly, further direct evidence of MAO multiplicity has come from another source. Hidaka, Hartman, and Udenfriend (1971) have established the existence of two distinct immunochemical forms of the enzyme. Very recently, McCauley and Racker (1973) have achieved a complete physical separation of these two forms, which appear to possess many of the properties of MAO types A and B as originally identified by Johnson (1968).

Using the MAO inhibitor clorgyline, Johnson presented indirect evidence of differential inhibition compatible with the existence of two different forms of the enzyme which he termed A and B. Neff and Goridis (1972) expanded this approach. They showed that type A enzyme is specific for the oxidation of norepinephrine and 5-hydroxytryptamine, whereas type B preferentially oxidizes phenylethylamine. Tyramine is a substrate for both forms. Clorgyline specifically inhibits type A, while deprenyl inhibits type B. Deprenyl is chemically related to pargyline, which also appears to inhibit type B enzyme preferentially (McCauley and Racker, 1973).

Neff, Yang, and Goridis (1973) have provided evidence to suggest that such differential inhibition may similarly occur *in vivo*. After groups of rats had been pretreated with either clorgyline or deprenyl, a proportionately greater impairment of ability to oxidize norepinephrine and 5-hydroxytryptamine compared with phenylethylamine was noted in the former group and the reverse situation was noted in the latter.

We ourselves have administered clorgyline or deprenyl, in the dosage used by Neff et al. (1973), to two different groups of rats in an attempt to demonstrate whether the *in vivo* existence of type A and B enzymes, as indicated by differential inhibition with the drugs, is a reality or merely a drug artifact (Sandler, Goodwin, and Ruthven, 1974a). Twenty-four-hour urine samples were collected over 6 N HCl and subjected to gas chromatographic analysis for endogenously produced 4-hydroxy-3-methoxyphenylglycol (HMPG), 5-hydroxyindoleacetic acid (5-HIAA), and phenylacetic acid, the major metabolites in the rat of norepinephrine, 5-hydroxytryptamine, and phenylethylamine, respectively. The results (Sandler et al., 1974a) (Fig. 1), however, were inconclusive. Whereas a significantly smaller output of HMPG and 5-HIAA was observed after clorgyline compared with deprenyl administration, the two groups did not differ with respect to phenylacetic acid administration. It is quite possible, however, that large dietary contributions of phenylacetic acid masked small fluctuations in excretion value resulting from changes in endogenous phenylethylamine metabolism.

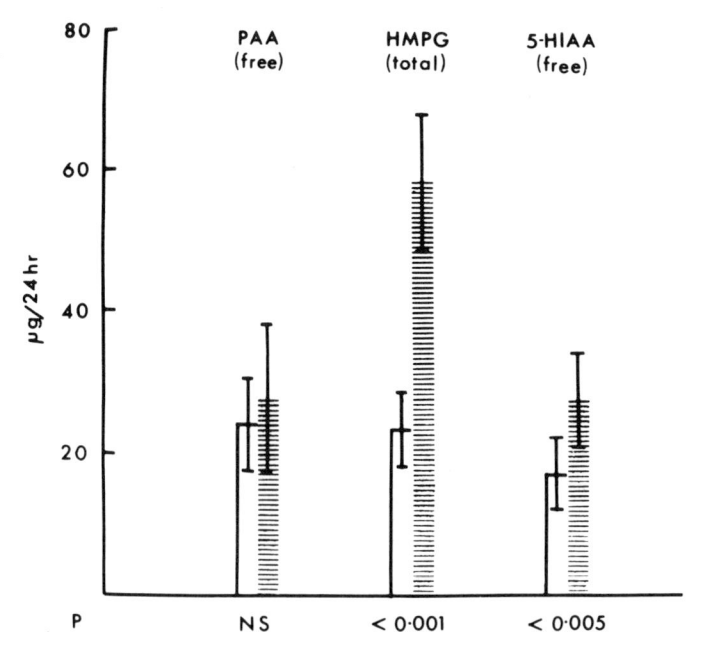

FIG. 1. Comparing the differences in 24-hr urinary excretion values (mean ± SD) of phenylacetic acid (PAA), 4-hydroxy-3-methoxyphenylglycol (HMPG), and 5-hydroxyindoleacetic acid (5-HIAA) in two groups of rats given clorgyline (1 mg/kg) (open columns) or deprenyl (4 mg/kg) (hatched columns) intravenously.

III. PHENYLETHYLAMINE, MAO B DEFICIENCY, AND MIGRAINE

There has been a considerable resurgence of interest in phenylethylamine economy in man and animals. Very recently, evidence has emerged which points to a possible role of this amine in the pathogenesis of migraine. Although tyramine has been implicated as a trigger in some cases of dietary migraine (Hanington, 1967), the most common article of diet blamed by patients for their attacks tends to be chocolate (Hanington, Horn, and Wilkinson, 1970), which does not appear to contain any tyramine. However, Dr. M. J. Saxby (British Food Manufacturing Industries Research Association, Leatherhead, UK) was able to detect large amounts of phenylethylamine in every sample of plain chocolate he examined, about 3 mg per 2-ounce bar. Many, but not all, cheeses and some red wines also contained phenylethylamine.

A double-blind trial of oral phenylethylamine compared with lactose in patients with chocolate-sensitive migraine revealed that attacks were triggered in a considerably higher proportion after phenylethylamine ingestion than after placebo (Sandler, Youdim, and Hanington, 1974b).

Certain monoamines, such as tryptamine, 5-hydroxytryptamine (Alabaster and Bakhle, 1970), dopamine, and tyramine (Bakhle and Smith,

1972), share the ability to release from the lung into the systemic circulation a number of different pharmacologically active substances—prostaglandins, "slow-reacting substance," rabbit aorta-contracting substance, and so on—when introduced into the pulmonary circulation. Sandler (1972) has utilized this observation as the basis for his hypothesis of the pathogenesis of migraine, suggesting that either abnormal concentrations of monoamine reaching the lung or an abnormal sensitivity to the releasing effect of normal concentrations may result in an abnormal liberation of spasmogens into the systemic arterial circulation.

Although it has not been practicable to study systematically the spasmogen-liberating effect of various monoamines in human lung, it has recently been shown that phenylethylamine acts strongly in this capacity in some animal species (Y. S. Bakhle, *personal communication*). One possible spasmogen, prostaglandin E_1, is known to initiate a headache similar to migraine when injected intravenously (Carlson, 1967). It therefore seemed possible that if, for any reason, abnormal concentrations of phenylethylamine were able to reach the proximal pulmonary circulation, spasmogen release into the systemic circulation and, perhaps, migrainous headache might follow. One circumstance which might result in such a state of affairs would be a defect in phenylethylamine-oxidizing ability, that is, a specific

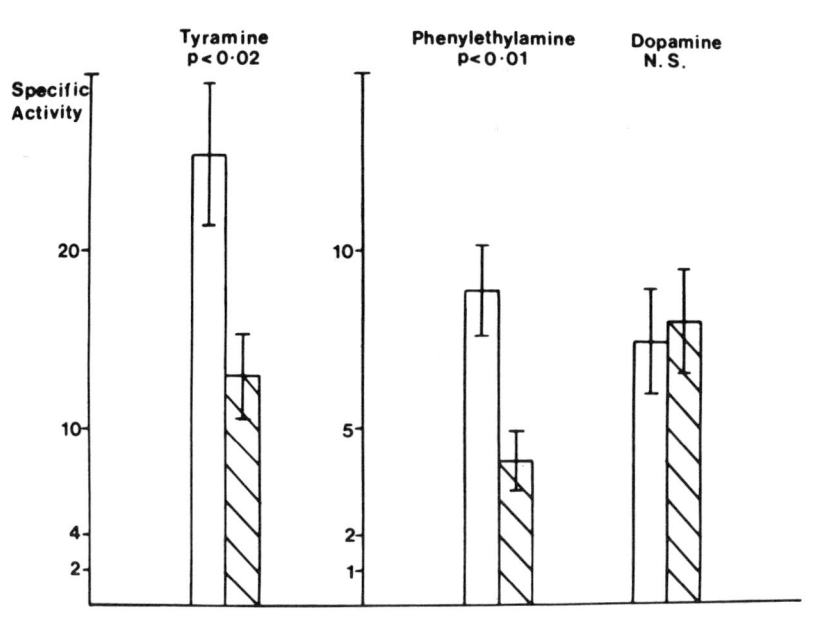

FIG. 2. Specific activity (mean ± SEM, nanomoles of product formed per milligram of platelet protein per 30 min) of platelet MAO activity toward ^{14}C-tyramine, ^{14}C-phenylethylamine, and ^{14}C-dopamine in a migrainous (hatched columns, n = 13) and normal (open columns, n = 12) group.

decrease in MAO B. We have obtained *prima facie* evidence for the existence of such a defect (Sandler, et al., 1974b) (Fig. 2).

Platelet samples were obtained during an attack-free period from a group of migrainous patients and a normal control group. MAO activity was measured using ^{14}C-labeled dopamine, tyramine, and phenylethylamine as substrates. Whereas there was no significant difference in dopamine-oxidizing ability between the two groups, significantly lower activity was present in the migrainous group, both for phenylethylamine and tyramine (Fig. 2). While final interpretation must await similar experiments employing norepinephrine and/or 5-hydroxytryptamine as substrates, the present data are compatible with the existence of a specific deficit of MAO B in migrainous patients. They agree with our earlier preliminary results, which gave the first indication of decreased MAO activity in this clinical group (Sandler, Youdim, Southgate, and Hanington, 1970), data which later received support from Sicuteri, Buffoni, Anselmi, and Del Bianco (1972).

There are many important implications of these findings, not the least of which is the possible origin of the attacks of hypertension and headache which occasionally supervene in patients on MAO-inhibitor therapy after the ingestion of cheese or other dietary trigger. They may, in fact, stem from a secondary pulmonary release of spasmogens in response to proximal challenge with such monoamines as tyramine or phenylethylamine.

IV. SULFATE CONJUGATION AND TYRAMINE SENSITIVITY IN RELATION TO MAO-INHIBITOR THERAPY

The findings in migrainous subjects described above are somewhat difficult to reconcile with earlier observations from these laboratories: Youdim, Bonham Carter, Sandler, Hanington, and Wilkinson (1971) observed that a group of patients with tyramine-sensitive migraine were not able to conjugate oral tyramine as well as normal controls. Despite the present finding of an MAO deficit, output of *p*-hydroxyphenylacetic acid, the major oxidative deamination metabolite of tyramine, was unaffected in the migraine group. Sulfate conjugation of tyramine (Mullen and Smith, 1971) appears to account for about 10% of an oral tyramine load; Youdim et al. (1971) suggested that this pathway might act as a "safety valve" mechanism and considered that subjects with a deficient conjugating mechanism might be at special risk if they were to be treated with MAO inhibitors. It is well known that by no means all patients ingesting tyramine-containing foods during MAO-inhibitor therapy undergo an adverse reaction. It seemed possible, therefore, that those who did were lacking in this "safety valve" mechanism. By assessing the conjugation response to an oral tyramine load, it might be possible to identify, in advance, patients especially at risk.

With the help of the Committee on Safety of Medicines, it proved possible to trace and investigate a number of patients who had undergone a severe adverse reaction to certain foods while undergoing MAO-inhibitor treatment. Preliminary results (Sandler and Youdim, 1972) seemed to point to a highly significant decrease in conjugation ability in this group compared with normal controls. However, more recent data (Sandler, Bonham Carter, Cuthbert, and Pare, 1974c) (Fig. 3) have made it necessary to reevaluate this finding. The control group initially used for the investigation consisted of normal nondepressed subjects. It has now been possible to obtain a group of depressed patients who were able to eat cheese with impunity while on MAO-inhibitor therapy. One regularly ingested at least $3\frac{1}{2}$ lb per week. It was therefore surprising to find that a deficit in conjugating ability existed in this group which was similar to that in the patients who had undergone an adverse reaction. A long-term depressant action of MAO inhibitors on con-

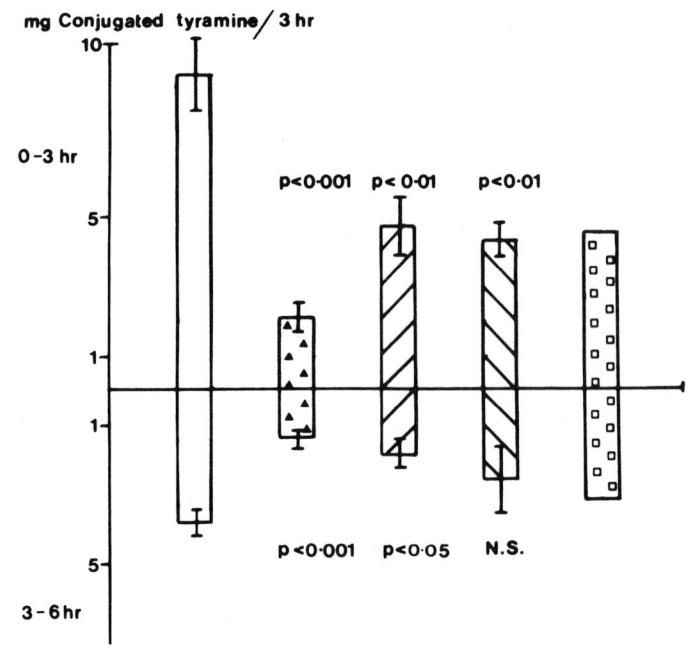

FIG. 3. Urinary excretion of conjugated tyramine (milligrams per 3 hr, mean ± SEM) 3 hr (above line) and 6 hr (below line) after ingestion of tyramine hydrochloride (equivalent to 100 mg of free base) by ☐ normal subjects (n = 12 for 0 to 3 hr period; n = 4 for 3 to 6 hr period), ▲ subjects with dietary migraine (n = 8 for 0 to 3 hr period; n = 6 for 3 to 6 hr period), ▨ subjects who suffered an adverse reaction while on MAO-inhibitor therapy (n = 12), ◩ subjects who suffered no adverse reaction while on MAO-inhibitor therapy (n = 7), ▫ subjects who had never been on MAO-inhibitor therapy but who were considered to be suitable cases for it (n = 2).

jugation, although theoretically possible (Hargreaves, 1968), can probably be ruled out, since two subjects who were deemed suitable candidates for MAO-inhibitor therapy but had not yet received any drug had a similar decrease in sulfate conjugation ability. Very recently, preliminary data (not seen in Fig. 3) have shown that patients who had been depressed and were treated with MAO inhibitors but were free from depression at the time of testing do not have any conjugation deficit. Thus it may well be that, in common with certain other enzyme systems, sulfate-conjugating mechanisms are damped down in depressive illness. Indeed, perhaps the members of the original dietary migraine group of Youdim et al. (1971), who presented themselves in response to a questionnaire, were in reality masked depressives.

Before another enzyme deficit in depression can definitively be added to the canon, however, rather more direct measurements of activity are necessary—perhaps on gut biopsy material. Such experiments should be supplemented by tests in which oral isoprenaline is substituted for tyramine. The former is predominantly metabolized by conjugation (Morgan, Ruthven, and Sandler, 1969). It may even be that the apparent decrease in tyramine conjugation is brought about by an increase in oxidative deaminating ability. There are precedents for such a situation. MAO activity is known to increase with age (Robinson, Davis, Nies, Colburn, Davis, Bourne, Bunney, Shaw, and Coppen, 1972) and under the action of certain hormonal influences (Southgate, Grant, Pollard, Pryse-Davies, and Sandler, 1968; Collins, Pryse-Davies, Sandler, and Southgate, 1970). It would be fascinating if the wheel were to have come full circle and MAO inhibitors turn out to be a rational therapy for depression illness because the immediate cause is an increase in MAO. It is even possible that such an increase, if it exists, is specific for one of the multiple forms of MAO, so that 5-hydroxytryptamine or tyramine (Diaz Borges and D'Iorio, 1972) is perhaps preferentially inactivated to a greater extent than usual. Thus the monoamine hypothesis of depression, as defined in terms of a single amine rather than a group of them, might yet be vindicated.

REFERENCES

Alabaster, V. A., and Bakhle, Y. S. (1970): The release of biologically active substances from isolated lungs by 5-hydroxytryptamine and tryptamine. *British Journal of Pharmacology*, 40:582P–583P.

Bakhle, Y. S., and Smith, T. W. (1972): Release of spasmogenic substances induced by vasoactive amines from isolated lungs. *British Journal of Pharmacology*, 46:543P–544P.

Carlson, L. A. (1967): Metabolic and cardio-vascular effects *in vivo* of prostaglandins. In: *Prostaglandins. Nobel Symposium 2*, edited by S. Bergström and B. Samuelsson. Wiley-Interscience, New York, pp. 123–132.

Collins, G. G. S., Pryse-Davies, J., Sandler, M., and Southgate, J. (1970): Effect of pretreatment with oestradiol, progesterone and DOPA on monoamine oxidase activity in the rat. *Nature*, 226:642–643.

Collins, G. G. S., Youdim, M. B. H., and Sandler, M. (1972): Multiple forms of monoamine oxidase: Comparison of *in vitro* and *in vivo* inhibition patterns. *Biochemical Pharmacology*, 21:1995–1998.

Diaz Borges, J. M., and D'Iorio, A. (1972): Multiple forms of mitochondrial monoamine oxidase. *Advances in Biochemical Psychopharmacology,* 5:79–89.
Hanington, E. (1967): Preliminary report on tyramine headache. *British Medical Journal,* 2:550–551.
Hanington, E., Horn, M., and Wilkinson, M. (1970): Further observations on the effects of tyramine. In: *Background to Migraine. Third Migraine Symposium,* edited by A. L. Cochrane. Heinemann, London, pp. 113–119.
Hargreaves, T. (1968): The effect of monoamine oxidase inhibitors on conjugation. *Experientia,* 24:157–158.
Hidaka, H., Hartman, B. K., and Udenfriend, S. (1971): Comparison of mitochondrial monoamine oxidase from bovine brain and liver using antibody to purified liver monoamine oxidase. *Archives of Biochemistry and Biophysics,* 147:805–809.
Houslay, M. D., and Tipton, K. F. (1973): The nature of the electrophoretically separable multiple forms of rat liver monoamine oxidase. *Biochemical Journal,* 135:173–186.
Johnson, J. P. (1968): Some observations upon a new inhibitor of monoamine oxidase in brain tissue. *Biochemical Pharmacology,* 17:1285–1297.
McCauley, R., and Racker, E. (1973): Separation of two monoamine oxidases from bovine brain. *Molecular and Cellular Biochemistry,* 1:73–81.
Morgan, C. D., Ruthven, C. R. J., and Sandler, M. (1969): The quantitative assessment of isoprenaline metabolism in man. *Clinica Chimica Acta,* 26:381–386.
Mullen, P. E., and Smith, I. (1971): Tyramine metabolism and migraine: A metabolic defect. *British Journal of Pharmacology,* 41:413P–414P.
Neff, N. H., and Goridis, C. (1972): Neuronal monoamine oxidase: Specific enzyme types and their rates of formation. *Advances in Biochemical Psychopharmacology,* 5:307–323.
Neff, N. H., Yang, H.-Y. T., and Goridis, C. (1973): Degradation of the transmitter amines by specific types of monoamine oxidase. *Abstracts, 3rd International Catecholamine Symposium, Strasbourg.*
Olivecrona, T., and Oreland, L. (1971): Reassociation of soluble monoamine oxidase with lipid-depleted mitochondria in the presence of phospholipid. *Biochemistry,* 10:332–340.
Robinson, D. S., Davis, J. M., Nies, A., Colburn, R. W., Davis, J. N., Bourne, H. R., Bunney, W. E., Shaw, D. M., and Coppen, A. J. (1972): Aging, monoamines, and monoamine oxidase levels. *Lancet,* i:290–291.
Sandler, M. (1972): Migraine: A pulmonary disease? *Lancet,* i:618–619.
Sandler, M., Bonham Carter, S. M., Cuthbert, M. F., and Pare, C. M. B. (1974c): The role of tyramine conjugation in adverse reactions during monoamine oxidase inhibitor therapy. *(Submitted for publication.)*
Sandler, M., Goodwin, B. L., and Ruthven, C. R. J. (1974a): Monoamine metabolism and monoamine oxidase multiplicity. *(Submitted for publication.)*
Sandler, M., and Youdim, M. B. H. (1972): Multiple forms of monoamine oxidase: Functional significance. *Pharmacology Reviews,* 24:331–348.
Sandler, M., and Youdim, M. B. H. (1974): Monoamine oxidases: The present status. *International Pharmacopsychiatry,* 9:27–34.
Sandler, M., Youdim, M. B. H., and Hanington, E. (1974b): A phenylethylamine-oxidising defect in migraine. *Nature (in press).*
Sandler, M., Youdim, M. B. H., Southgate, J., and Hanington, E. (1970): The role of tyramine in migraine: Some possible biochemical mechanisms. In: *Background to Migraine. Third Migraine Symposium,* edited by A. L. Cochrane. Heinemann, London, pp. 103–112.
Sicuteri, F., Buffoni, F., Anselmi, B., and Del Bianco, P. L. (1972): An enzyme (MAO) defect on the platelets in migraine. *Research and Clinical Studies on Headache,* 3:245–251.
Southgate, J., Grant, E. C. G., Pollard, W., Pryse-Davies, J., and Sandler, M. (1968): Cyclical variation in endometrial monoamine oxidase: Correlation of histochemical and quantitative biochemical assays. *Biochemical Pharmacology,* 17:721–726.
Youdim, M. B. H., Bonham Carter, S., Sandler, M., Hanington, E., and Wilkinson, M. (1971): Conjugation deficit in tyramine-sensitive migraine. *Nature,* 230:127–128.
Youdim, M. B. H., Collins, G. G. S., Sandler, M., Bevan Jones, A. B., Pare, C. M. B., and Nicholson, W. J. (1972): Human brain monoamine oxidase: Multiple forms and selective inhibitors. *Nature,* 236:225–228.

Neuropsychopharmacology of Monoamines and Their Regulatory Enzymes, edited by E. Usdin. Raven Press, New York © 1974.

Physicochemical Properties, Development, and Regulation of Central and Peripheral Monoamine Oxidase Activity

Moussa B. H. Youdim, Margarethe Holzbauer, and H. Frank Woods

MRC Unit and Department of Clinical Pharmacology, Radcliffe Infirmary, Oxford; and ARC Laboratory of Animal Physiology, Babraham, Cambridge, England

I. INTRODUCTION

The enzyme monoamine oxidase (MAO) [monoamine:oxygen-oxidoreductase (deaminating) EC 1.4.3.4.] is strongly bound to the outer mitochondrial membrane. It is difficult to purify, and as a consequence, until very recently we lacked basic information as to the nature, composition, and cofactor requirements of the enzyme. Soluble and highly purified MAO is now considered to be an —SH-containing protein; flavin is a cofactor and iron is now thought to play a role in its activity (see Youdim, 1973*a*, for review).

There have been numerous studies on the biochemical and physical properties of MAO, but relatively little is known about the control and development of its activity. Although fluctuations in the activity of MAO during the menstrual cycle are well documented (Southgate, 1972), no adequate explanation has been put forward for the mechanism of action. Changes in the rate of synthesis or degradation of the enzyme protein may occur or, alternatively, the hormones may influence enzyme activity by causing changes in the physical properties of the enzyme molecule, thus acting as allosteric effectors. Estrogen pretreatment leads to inhibition and progesterone pretreatment to stimulation of MAO activity (Collins, Pryse-Davies, Sandler, and Southgate, 1970*a*; Holzbauer and Youdim, 1973). Pretreatment with these steroids leads also to a change in substrate specificity.

Evidence has accumulated to suggest that MAO may exist in more than one form, each form having different substrate and inhibitor specificities. The exact molecular nature and physiological significance of the multiple forms have yet to be clarified. Their natural existence has been challenged recently, and the chemical findings have been attributed to the effects of preparative procedures (Collins, 1972; Houslay and Tipton, 1973). Thus the suggestion has been made that a study of the enzyme in its natural environment should be made which would furnish more valid information concerning the role of MAO activity in the control of tissue monoamine metabolism *in vivo* (Youdim 1973b).

These are some of the experimental approaches we have made in the study of this enzyme. The results of these investigations have given some idea of the control and function of MAO activity in tissues and indicate experimental approaches which may yield further information.

II. MULTIPLE FORMS – PHYSICOCHEMICAL PROPERTIES

There is good evidence that MAO exists in more than one form. Rat liver MAO can be separated into five forms by polyacrylamide gel electrophoresis. Multiple forms from other tissues have been reported in a number of species, but their presence is not universal (see Sandler and Youdim, 1972, for review). Comparison of the results reported by the various investigators is impossible, since each research group used different methods of solubilization and electrophoretic separation.

Although multiple forms of MAO differ in substrate and inhibitor specificities and in thermal stability, no detailed study of their kinetic properties has until now been carried out (Youdim and Collins, 1973). The Michaelis constant (K_m values) of form 5 for dopamine (3.6 μM) is considerably lower than that of the other forms (K_m values between 19 and 210 μM). Similarly, the K_m of form 5 for the other substrates is very much higher than for dopamine, ranging from 46 to 465 μM. Form 1 has a greater affinity for tyramine, tryptamine, and kynuramine (K_m values 12 to 24 μM). Over the concentration range of the substrates used (10 to 100 μM), double reciprocal plots were linear. However, when the concentration exceeds 100 μM (kynuramine, dopamine, and phenylethylamine), the results suggest that, with the exception of form 5, either substrate or product inhibition occurred. Our recent studies have shown that the inhibition of the deamination of substrates by form 1 to 4 was due to substrate rather than to product inhibition (Youdim and Collins, 1973).

A. Inhibition of Rat Liver MAO

Phenylethylhydrazine is an active site-directed irreversible inhibitor of MAO (Clineschmidt and Horita, 1969a, b). At a concentration of 5 μM, incubation at 37°C for 40 min resulted in a 95% inhibition of activity. The inhibition was a function of the amount of inhibitor bound to the enzyme protein (Youdim and Collins, 1973), and inclusion of the substrate benzylamine in the incubation mixture blocked the inhibitory action of phenylethylhydrazine on the enzyme activity.

Incubation of (^{14}C)-phenylethylhydrazine (5 μM) for 40 min with the multiple forms of MAO resulted in complete inhibition except with form 5.

A similar result has been reported by Tipton, Youdim, and Spires (1972). After prolonged dialysis against a buffer containing 100 μM of nonradioactive phenylethylhydrazine, 1.03 moles of the drug was bound to each 150,000 g of the enzyme forms 1 to 4, respectively. When the concentration of (^{14}C)-phenylethylhydrazine was increased to 100 μM, the activity of form 5 was completely inhibited and 0.91 mole of labeled inhibitor was bound to each 150,000 g of enzyme.

MAO inhibitors have been used as antidepressants in the chemotherapy of affective disorders. It is generally accepted that they owe this property to the inhibition of MAO rather than to some other effect. It is not known why some MAO inhibitors are more potent antidepressants than others. The suggestion has been made that the answer lies in the inhibition of multiple forms of MAO at specific sites in the central nervous system (Youdim, Collins, and Sandler, 1969; Collins, Sandler, Williams, and Youdim, 1970b; Youdim, Collins, Sandler, Bevan Jones, Pare, and Nicholson, 1972), thus raising the level of a particular monoamine at that site.

A number of new inhibitors have been studied for possible selective inhibition of the different forms of MAO. Of these, clorgyline [M8B9302, N-methyl-N-propargyl-3-(2,4-dichlorophenoxy) propylamine hydrochloride] (Johnson, 1968; Hall and Logan, 1969; Hall, Logan, and Parsons, 1969), deprenyl (E.250) (Knoll and Magyar, 1972), and Lilly 51641 (N-cyclopropyl-o-chloro-phenoxyethylamine) (Fuller, 1968, 1972) have achieved prominence. These inhibitors have one property in common: The plot of percentage inhibition against concentration of the inhibitor does not show a simple sigmoid curve but consists of a pair of sigmoid curves joined by a horizontal section where the inhibition is invariant. The two sigmoid curves have been attributed to the presence of two monoamine oxidases termed types A and B (Johnson, 1968). Type A is said to be sensitive to clorgyline and oxidatively deaminates tyramine and 5-hydroxytryptamine. The type B enzyme is more resistant to clorgyline and oxidizes tyramine but not 5-hydroxytryptamine. Deprenyl has been reported to give a picture almost opposite to that of clorgyline (Neff and Yang, 1973; Yang and Neff, 1973). It has been suggested that the relative proportions of these two enzymes vary in various tissues (Neff and Goridis, 1972; Coquil, Goridis, and Neff, 1973). The multiple forms of MAO as separated by gel electrophoresis have been shown to have different substrate and inhibitor specificities. An attempt was made, therefore, to see whether separated forms of the enzyme would show different substrate-dependent inhibitions with clorgyline. Furthermore, if types A and B exist, can they be separated by present electrophoresis? Multiple forms of MAO were isolated according to the method of Youdim, Collins, and Sandler (1970). The effect of varying concentration of clorgyline on activity of the five forms of rat liver is shown in Fig. 1. With the possible exception of MAO-1, all other forms show a single sigmoid curve, whether kynuramine or dopamine is used as substrate. We have

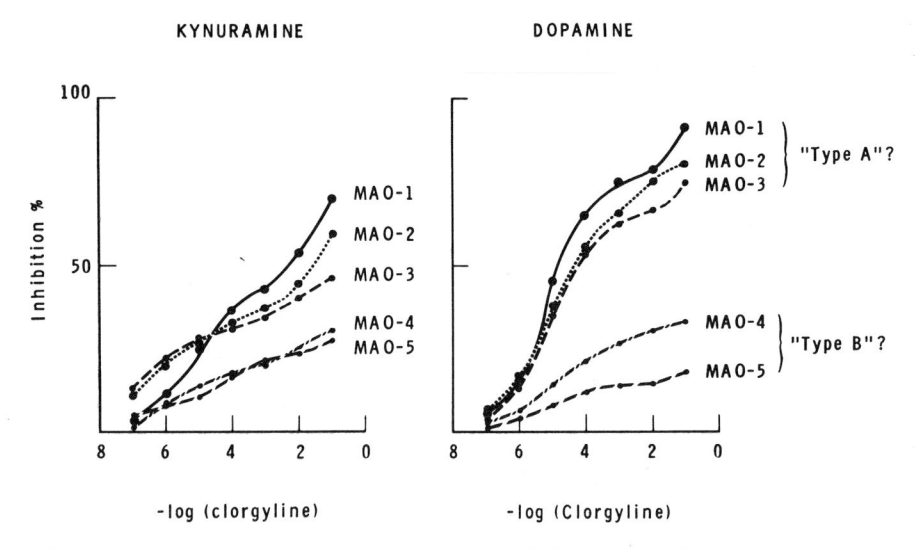

FIG. 1. Effect of clorgyline on the activity of multiple forms of rat liver MAO. Samples of electrophoretically separated monoamine oxidase (Youdim, Collins, and Sandler, 1970) were incubated at 37°C for 15 min with 50 µl of a clorgyline solution in water to give the final concentration indicated. The activities of enzyme used were determined using either kynuramine (Kraml, 1965) or ^{14}C-dopamine (Robinson, Lovenberg, Keiser, and Sjoerdsma, 1968) as substrate.

previously suggested (Youdim and Collins, 1971) that MAO-1 could represent a polymerized form of the enzyme that does not enter the gel. A more important fact is that MAO-1, MAO-2, and MAO-3 are more sensitive to inhibition by clorgyline than MAO-4 and MAO-5. One other point of interest is that the degree of inhibition is variable with the two substrates. These results indicate that we may have separated the type A and type B enzymes. However, this does not prove that two different enzyme proteins exist within the mitochondrion. Houslay and Tipton (1973) have suggested that the degree of inhibition by clorgyline and also the double sigmoid curve obtained by this inhibitor depend solely on the associated phospholipid membraneous material. Removal of phospholipid membraneous material by chatotropic reagents results in the disappearance of substrate-selective inhibition and the double sigmoid curve. Our present results support the findings of Houslay and Tipton (1973) in that the degree of inhibition of multiple forms of MAO appears to depend on the phospholipid-phosphorus content. Multiple forms have already been shown to have varying amounts of phospholipid (Tipton, 1972; Tipton et al., 1972; Youdim and Collins, 1973). The enzymes MAO-4 and MAO-5, with lowest phospholipid, are more resistant to inhibition by clorgyline and phenelzine. A very similar result has been obtained for the whole enzyme after partial removal of phospholipid (Houslay and Tipton, 1973).

These studies suggest that enzyme form 5 (cathodic form) is not a classical MAO. However, in addition to the differences mentioned, this enzyme also possesses some similarities with other enzymic forms. For example, all forms of rat liver mitochondrial MAO have the same absorption spectra characteristic of flavin moiety (Youdim, 1972) and, in addition, both gel filtration (Youdim and Collins, 1971; Tipton et al., 1972) and ultracentrifugation studies provide evidence that the multiple forms have similar molecular weight. The resistance of form 5 to loss of activity in the presence of substrate, phenylethyl hydrazine, urea, and guanidine hydrochloride may be overcome by changes in the experimental conditions (Youdim and Collins, 1973). However, the anomalous properties of the cathodally migrating forms of MAO described suggest that the enzyme differs fundamentally from the other enzyme forms.

The differences in the substrate and inhibitor specificity of multiple forms may of course be attributed to the phospholipid content (Tipton, 1972; Youdim, 1973b) or to attachment of membraneous material (Houslay and Tipton, 1973). This in no way diminishes the importance of the enzyme environment and its physiological implications. Indeed, it is possible that membrane-embedded MAOs depend on bound phospholipid or on the fluid paraffin-like nature of their surroundings (Youdim, 1973b). Collins (1972) has already pointed out that exclusion of lipid and membraneous material from enzyme preparations will not necessarily clarify the situation, since such factors may be essential to stabilize the enzyme forms. In this connection, the report that mitochondria from rat brain (Kroon and Veldstra, 1972; Youdim, 1973c), liver (Youdim, 1973d), and chick brain (Barberis and Gayet, 1973) are heterogenous and can be separated into a number of subcategories each having different relative MAO activity with the use of more than one substrate is of interest. It suggests that multiple forms could arise from either different mitochondria or mitochondria having varying amounts and types of phospholipid membraneous material. It is noteworthy that mitochondria, as separated by sucrose and ficoll density gradient (see Cascarno and Wilson, 1972, for references), show relative differences in other enzyme activities, including cytochromes, succinate dehydrogenase, NADH-dehydrogenase, and α-glycerol phosphate dehydrogenase.

The problem that still exists is whether multiple forms of MAO are different enzyme proteins or one enzyme existing in different environments (Tipton et al., 1972; Youdim, 1973b). Hartman, Yasunobu, and Udenfriend (1969) were the first to induce antibody to whole beef liver MAO and its multiple forms. They obtained a single immuno precipitant and came to the conclusion that liver enzyme was a single entity; however, using the same antibody, not all the beef brain MAO could be precipitated (Hartman, 1972). A brain-specific MAO different from that of the liver was suggested. Recently, McCauley and Racker (1973) and Youdim (1973c) have separated two enzymes from bovine and rat brain, and liver, respectively, which are immunologically different and may be distinct forms.

III. INFLUENCE OF ENDOCRINE GLANDS ON MAO ACTIVITY

Fluctuations of MAO activity during the female sexual cycle have been reported by a number of workers. In the human uterine endometrium, up to 10-fold increases were observed between days 19 and 21 of the menstrual cycle (Southgate, Grant, Pollard, Pryse-Davies, and Sandler, 1968), a period during which the plasma progesterone content reaches peak values. In the rat, progesterone was found to stimulate (Collins and Southgate, 1970) and estradiol to inhibit (Collins et al., 1970a) the enzyme activity. Variations in the MAO activity of different tissues were also reported during the estrous cycle of the guinea pig and the rat (for references see Youdim, 1973a). Since it was difficult to draw general conclusions from the reported observations, a more detailed study was carried out recently (Holzbauer and Youdim, 1973) in which MAO activity was studied in rat tissues during different phases of the cycle. The rats were not only killed in daylight hours when they were usually asleep, but also in dark periods when they were active. Enzyme activity was measured in four brain regions known to be rich in monoamines (hypothalamus, septum, caudate nucleus, and part of the posterior telencephalon), in the uterus, the ovaries, and the adrenal glands (Fig. 2). MAO activity in these tissues was highest during di-estrus

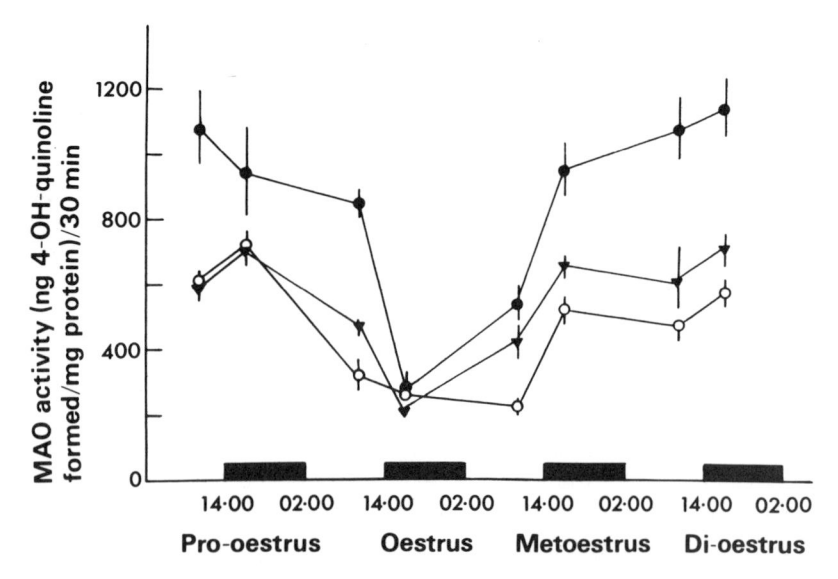

FIG. 2. MAO activity in the hypothalamus (●——●), the caudate nucleus (▼——▼), and the septum (○——○) of rats during different phases of the estrous cycle. A similar pattern for the posterior telencephalon was observed but not shown for the sake of clarity. Dark bars on abscissas indicate periods of red light (dark period). Rats were killed either after 8 hr in white light (light period) or 3 hr in red light (mean values and standard errors of the mean; n = 5 or 6).

and pro-estrus and fell during estrus, reaching lowest values in late estrus. Only uterine MAO activity showed a distinct peak in metestrus, fell in di-estrus and rose again to a second peak in late pro-estrus. The fact that we also made observations during the dark periods of the day enabled us to detect the very steep fall in MAO activity in late estrus which was especially pronounced in the hypothalamus and the uterus. There were no variations in MAO activity of the liver.

Because of the suggested link between MAO activity and ovarian steroid production rates (Southgate et al., 1968), we attempted an indirect assessment of the changes in progesterone production during the cycle by measuring the progesterone content in ovaries and adrenal glands. In previous experiments it had been established that in steroid-synthesizing glands the tissue content is a valid indicator for the rates at which steroids are secreted (Fajer, Holzbauer, Newport, 1971; Holzbauer, 1971). The highest ovarian progesterone contents were found in metestrus, the period when MAO activity was especially high in the uterus, the main target organ for progesterone. Progesterone secretion into ovarian venous blood was also found to be very high during this phase of the cycle (Fajer et al., 1971). In the adrenal gland, changes in MAO activity and in progesterone content ran nearly parallel (Fig. 3). In this context, it may be mentioned that in the

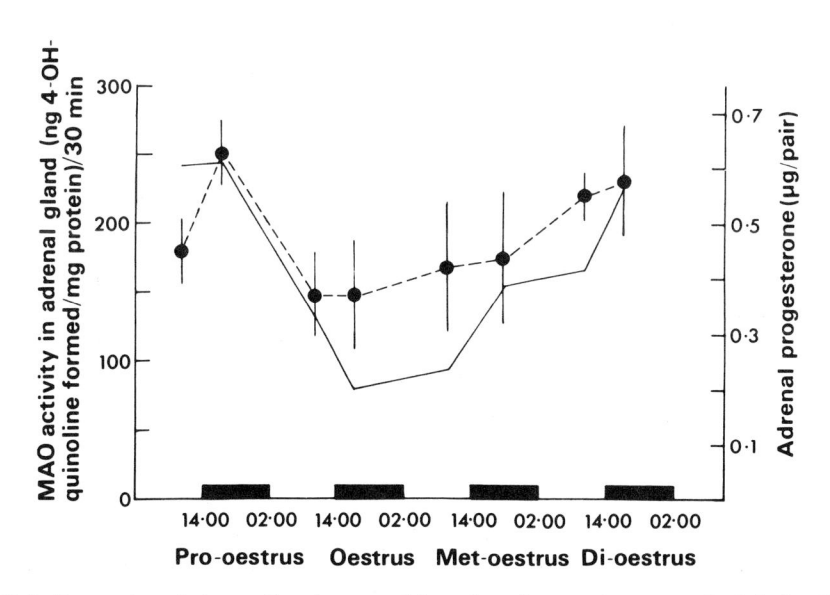

FIG. 3. Comparison between the changes of the adrenal progesterone content during the estrous cycle and the changes of the MAO activity in the adrenal gland. Progesterone (●-----●): mean values and standard errors of the mean; pro-estrus: n = 12; early: n = 10; late estrus: n = 8; metestrus and di-estrus: n = 4 to 8. Continuous line: adrenal monoamine oxidase. Dark bars indicate periods of red light. Killing schedule as described for Fig. 2.

adrenocortical tissue of the dog the largest proportion of progesterone was found in the mitochondrial fraction, which also contained the highest MAO activity, the enzyme being located on the outer mitochondrial membrane (Schnaitman, Erwin, and Greenwalt, 1967). When this fraction was further subjected to ultracentrifugation on a sucrose density gradient, the true mitochondrial subfraction contained the bulk of progesterone in addition to the highest MAO activity (Holzbauer, Bull, Youdim, Wooding, and Godden, 1972).

Since MAO probably exists in multiple forms, further experiments were carried out on the effect of progesterone and estrogens on tissue MAO activity in rats in which MAO activity was measured using four different substrates. The results are summarized in Table 1. The adrenal gland showed the highest substrate specificity in its reaction to steroid pretreatment. Estradiol decreased the oxidative deamination of all substrates with the exception of dopamine. In contrast, the largest stimulatory effect of progesterone was seen when dopamine was used as substrate (+265%).

TABLE 1. *The in vivo effect of steroids on adrenal gland MAO activity*

SPECIFIC ACTIVITY*

	DA	TRY	5-HT	TYR	KYN
CONTROL	3.55±0.30 (100)	1.20±0.15 (100)	2.80±0.30 (100)	4.80±0.55 (100)	2.40±0.15 (100)
PROGESTERONE (1 MG/KG)	9.40±1.20 (265)**	1.90±0.10 (158)	3.50±0.22 (135)	8.75±1.10 (182)	4.30±0.29 (179)
OESTRADIOL (0.5 MG/KG)	3.25±0.25 (92)	0.55±0.07 (45)	2.20±0.19 (71)	2.45±0.20 (51)	1.40±0.15 (58)

* μ MOLE DEAMINATED PRODUCT FORMED/30 MIN/MG PROTEIN
** PERCENTAGE OF CONTROL VALUES

Other evidence which points toward a control of sex steroids on MAO was observed in experiments on ovariectomized rats. This operation decreased MAO activity in all tissues tested to values similar to the lowest found during the estrous cycle (Holzbauer and Youdim, 1973).

We therefore suggest that the observed changes in MAO activity during the estrous cycle are the resultant of simultaneous changes in the blood concentration of progesterone (stimulator) and estrogens (inhibitor). In addition, the ovaries secrete several centrally active progesterone derivatives at varying rates during the cycle (Holzbauer, 1971), and a possible effect of these compounds on MAO is at present under investigation. Corticosteroids have also been reported to decrease MAO activity (Parvez and Parvez, 1973a,b), and cyclic variations in their secretion rates are probable.

The mechanism of action by which steroids alter MAO activity is not fully understood. There are a number of possible ways in which these effects could be achieved: Steroids could act

1. as coenzyme or apoenzyme,
2. as allosteric effectors,
3. on cyclic-AMP,
4. by inducing new enzyme formation at the transcriptional or post-transcriptional level,
5. by decreasing enzyme degradation, or
6. by changing the permeability of the outer mitochondrial membrane, the site of MAO (Schnaitman et al., 1967).

The first suggestion has so far not received much support, and stimulation of adenyl cyclase does not seem to play an important role in steroid hormone action (Lang, 1971).

The second alternative, that hormones may influence enzyme activities through bringing about changes in the physical properties of enzyme molecules, thus acting as allosteric effectors, is exemplified by the observations of Yielding and Tomkins (1960) on glutamate dehydrogenase (GDH). They found that estrogens can inhibit crystalline GDH by disaggregating the enzyme molecule from a polymeric into a monomeric form, thus leading to a change in substrate specificity. Similar results have been reported by Collins and Southgate (1970) for uterine MAO. The same might hold true for the action of steroids on MAO activity in other tissues, as shown in Table 1.

The formation of new enzymes at the transcriptional level seems at present an unlikely cause for the relatively fast increases in MAO activity occurring during the cycle because of the slow rate of enzyme synthesis. The half-life of MAO has been reported to be 14 to 17 days (Callingham and Della Corte, 1972), and when the enzyme is inhibited by an irreversible inhibitor it takes 6 to 8 weeks for MAO activity to return (Horita, 1968). Studies with inhibitors of protein synthesis are necessary to clarify this point and are in progress.

An action of steroids on the outer mitochondrial membrane which leads to conformational modifications of its structure, thus affecting MAO which is part of this membrane, seems to be the most likely site of steroid action. The lack of an immediate response to progesterone *in vitro (unpublished results)* may be due to the fact that the initial effect on membrane permeability is too small to be detected and that the duration of the exposure to progesterone is too short.

The physiological significance of steroid-induced changes in MAO activity is not yet fully understood. There is, however, little doubt that this effect represents an important additional action of these steroid hormones, since it may play a role in the regulation of the catabolism of monoamines.

IV. DEVELOPMENT OF MAO

During a study of the effect of steroid hormones on the activity of MAO, a number of observations have been made on untreated male rats aged 1 to 4 months. A retrospective analysis of these observations has confirmed that the activity of the enzyme increased with age in liver, heart, and brain tissue (for references see, e.g., Vaccari, Mura, Marchi, and Cugurra, 1972; Eiduson, 1972).

In the present experiments the enzyme activity in the brain was measured in well-defined, small regions, and differences became apparent in the rates at which the enzyme activity increased in the various regions (Fig. 4)

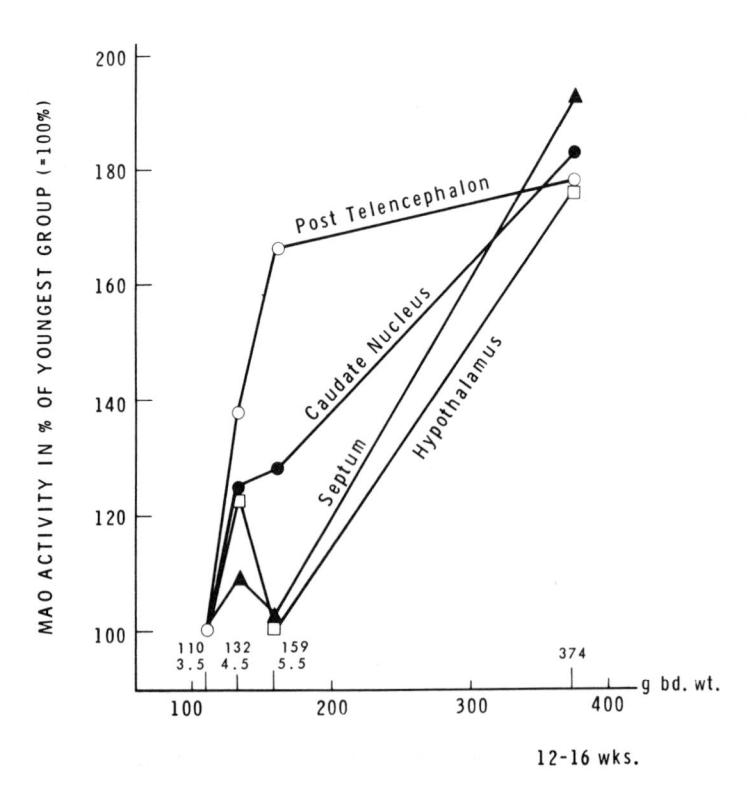

FIG. 4. Increase in MAO activity with age in different brain regions of young rats expressed as percentage of the MAO activity in 3- to 4-week-old rats (group I, see Table 2). The substrate was kynuramine (Kraml, 1965).

(Holzbauer, Youdim, and Blatchford, 1973). Three of these regions contained especially high concentrations of monoamines (hypothalamus, septum, and caudate nucleus). The fourth was the "posterior telencephalon," a complex part of the brain containing many functionally quite different structures. Not enough information on the developing brain is as yet available to attempt an interpretation of these observations. It is possible that the enzyme present in the mitochondria of monoaminergic neurons develops at a rate different from that in the mitochondria in other neurons or in glia cells. Evidence for the heterogeneity of MAO in brain mitochondria has recently been obtained (Youdim, 1973c) in experiments in which these particles were isolated from a total rat brain homogenate and distributed on a sucrose-ficoll gradient. When the MAO activity in different layers was measured using four substrates, a different distribution of MAO was observed depending on the substrate.

The increase in the activity of MAO in the whole brain of the rat with increasing age is by now a well-established phenomenon. It was reported by different groups of workers using different substrates. For example, Karki, Kuntzman, and Brodie (1962) and Bennet and Giarman (1965) found that the MAO activity toward 5-hydroxytryptamine increased severalfold between birth and maturity. Similar observations were made by Kuznya and Nagatsu (1969) and by Vaccari et al. (1972), using kynuramine as substrate, as well as by Ghosh and Guha (1972), who used tyramine. Since these studies were carried out on homogenates of whole rat brains, a direct comparison with our results is not possible. Studies on subdivided rat brains were reported by Porcher and Heller (1972). However, the regions used by these authors were much larger than those used by us, with the exception of the telencephalon. The region defined by Porcher and Heller as "telencephalon" (the entire neocortex, hippocampus, and amygdala) did not show any significant increase in MAO activity after 3 weeks of age, using tryptamine as substrate. Similar results were obtained with the regions called "mesencephalon," "septum-caudate," and "pons-medulla." Only in the "diencephalon" was the MAO activity significantly higher in $6\frac{1}{2}$-week-old rats than in 3-week-old rats.

In contrast to the enzyme activity in the brain, we found that the activity in the adrenal glands increased nearly linearly with age (Table 2).

There was a large difference in the rate of increase in enzyme activity between liver and heart. In the liver, maximal values were already reached in 5- to 6-week-old rats, whereas the activity in the hearts of 12- to 16-week-old rats was six times higher than that in 5- to 6-week-old rats (Table 2). Similar observations were reported by Vaccari et al. (1972).

The occurrence of significant increases in the MAO activity of many tissues in the young rat (100 to 150 g) during periods as short as 1 week emphasizes the necessity of using rats of exactly the same age and possibly the same litter when factors which influence the activity of the enzyme are being studied.

TABLE 2. *The effect of age on MAO activity in rat tissues*

MAO ACTIVITY
µG 4-OH-QUINOLINE FORMED/30 MIN/MG PROTEIN

GROUP NO. AND AGE		BODY WEIGHT (G)	LIVER	HEART	ADRENAL GLAND
I 3.5 WEEKS	MEAN	110	577	73	91
	S.E.:	±2.7	±29	±6	±6
	N.:	(21)	(19)	(21)	(20)
	% OF GROUP I	100%	100%	100%	100%
II 4.5 WEEKS	MEAN	132	715	101	106
	S.E.:	±1.9	±48	±6	±5
	N.:	(23)	(23)	(23)	(23)
	% OF GROUP I	120%	124%	138%	116%
III 5.5 WEEKS	MEAN	159	867	112	118
	S.E.:	±1.8	±48	±7	±6
	N.:	(23)	(21)	(22)	(23)
	% OF GROUP I	145%	150%	153%	130%
IV 12-16 WEEKS	MEAN	374	991	656	188
	S.E.:	±11	±55	±80	±17
	N.:	(11)	(11)	(11)	(10)
	% OF GROUP I	340%	172%	899%	206%

V. INTACT-ORGAN MAO ACTIVITY

If conclusions concerning the properties of MAO *in vivo* are to be drawn from the properties of the enzyme as determined *in vitro*, it is important to ascertain whether the properties of MAO in an intact organ differ from those displayed by the enzyme in tissue homogenates or in even further purified preparations *in vitro*. There is evidence that the "apparent K_m" of an enzyme determined from rates of reaction measured in the isolated perfused liver may differ from the K_m determined *in vitro*. For example, the initial concentration of xylitol giving half-maximal rates of removal in perfused rat livers is 3.2 mM (Woods and Krebs, 1973). This value is higher than the K_m for the NAD-linked purified cytoplasmic polyol dehydrogenase of sheep liver (0.18 mM at pH 9.6; Smith, 1962) and that of guinea pig liver (0.6 mM at pH 8.1; Hollmann, 1969).

To investigate this further, we have examined MAO activity in the isolated perfused rat liver using kynuramine as substrate. Livers from well-fed female Wistar rats weighing about 200 g were perfused using the method of Hems, Ross, Berry, and Krebs (1966) with a semisynthetic perfusion medium whose composition has been described previously (Woods, Eggleston, and Krebs, 1970). The medium contained 5 mM glucose in all experiments and varying initial concentrations of kynuramine (0.01 mM to 4.0 mM). During perfusion, 0.5-ml aliquots of the medium were withdrawn at intervals and the protein precipitated with 6% v/v perchloric acid. The MAO reaction was followed by determining the quinoline content of supernatant using the method of Kraml (1965). The initial rate of metabolic change was calculated from the gradient of the plot of total 4-hydroxyquinoline in the medium versus time and expressed as 4-hydroxyquinoline formed per minute per gram of wet liver, the liver being weighed wet at the end of each perfusion.

The rate of 4-hydroxyquinoline formation rose with increasing kynuramine concentration, being 1.4 γ/min/g of wet liver when the initial kynuramine concentration was 0.01 mM and rising to 15.4 γ/min/g of wet liver when the initial substrate concentration was 0.25 mM (Fig. 5). On increasing the concentration, the rate fell a little, being 12.33 γ/min/g of wet liver at 1 mM kynuramine and rising to 16.8 at 4 mM kynuramine. This result is in contrast to the observations of Gaby and Valcourt (1968) and Youdim (1973a), who found substrate inhibition *in vitro* with kynuramine concentrations above 0.1 mM, using purified MAO from rabbit and rat liver. The initial concentra-

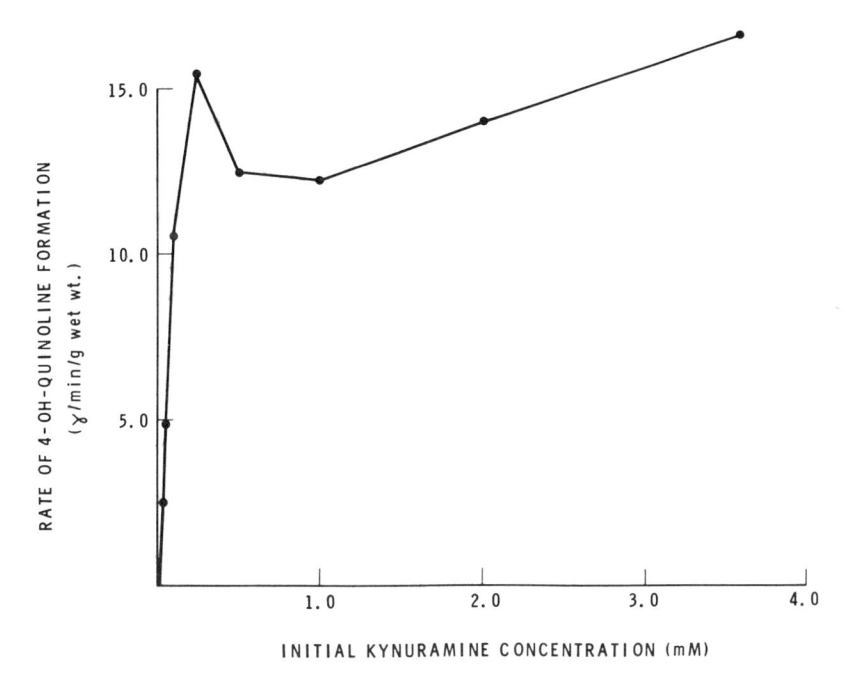

FIG. 5. The effect of kynuramine concentration on the rate of 4-hydroxyquinoline formation in the isolated perfused rat liver.

tion of kynuramine giving half-maximal rates of 4-OH-quinoline formation in the perfused rat liver was 6.5 × 10^{-5} mM, which is very close to the K_m reported for 4-OH-quinoline found in rat liver and other tissue preparations *in vitro* (see Tipton, 1973, for review). A reciprocal plot of initial reaction rate against the reciprocal of substrate concentration (Dixon and Webb, 1958) gives a very similar value (7.5 × 10^{-5} mM) (Fig. 6). It is important to

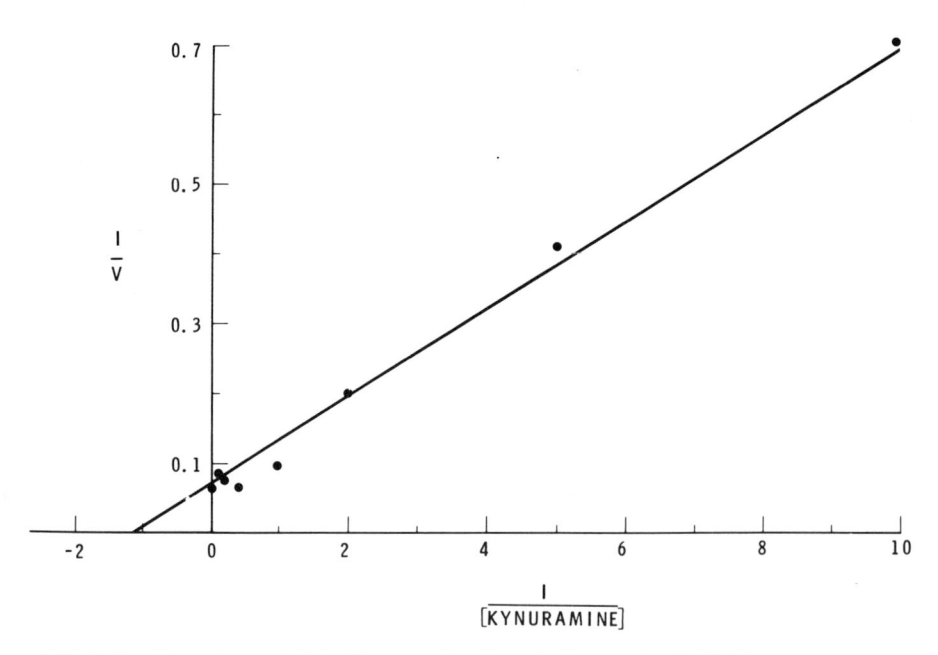

FIG. 6. Double reciprocal plot of kynuramine deamination against initial kynuramine concentration as observed in the isolated perfused rat liver.

remember that these are "apparent" intact-organ concentrations for half-maximal rates of reaction, but the similarity between these figures and the K_m for the purified enzyme is striking. In the intact organ the enzyme has not been subjected to methods used in purification and is retained in an environment which is closer to the physiological situation. However, it appears from the results that one of the conclusions drawn from *in vitro* studies may not be applicable to intact organs or the *in vivo* situation, viz. the lack of substrate inhibition.

VI. CONCLUSION

This chapter concerns some of the factors which control the activity of monoamine oxidase. The results of studies using purified enzymes, tissue extracts, and slices must be interpreted with caution, because when these techniques are used to study metabolic pathways the resultant system is artificial (Ross, 1972). For example, it has been pointed out that in the liver slice, the rates of metabolic activities such as gluconeogenesis, ketogenesis, and glycolysis are lower than those found in the intact, isolated, perfused organ (Woods and Krebs, 1971). In this chapter we report evidence which shows that the properties of an enzyme may differ depending on the type of preparation studied. The substrate inhibition of MAO with kynuramine as substrate, which occurs when purified enzyme preparations are used, does not occur in the isolated intact perfused organ, although the K_m may be the same in the two systems.

These observations go some way toward supporting the concept that the preservation of the structural integrity of an organ at a gross and microscopical level, together with such features as permeability barriers, is of great importance in the control of enzyme activity.

We suggest that studies of the pathways of monoamine synthesis and degradation in intact organs will yield information more directly applicable to the *in vivo* situation than that gained from experiments using tissue extracts and fractions.

REFERENCES

Barberis, C., and Gayet, J. (1973): Cytochrome oxidase and monoamine oxidase activities in mitochondrial populations isolated from the developing chick brain. *Biochemical Society Transactions (in press).*

Bennet, D. S., and Giarman, N. J. (1965): Schedule of appearance of 5-hydroxytryptamine (serotonin) and associated enzymes in the developing rat brain. *Journal of Neurochemistry,* 11:911–918.

Callingham, B. A., and Della Corte, L. (1972): The influence of growth and of adrenalectomy upon some rat heart enzymes. *British Journal of Pharmacology,* 46:530P–531P.

Clineschmidt, B. V., and Horita, A. (1969a): The monoamine oxidase catalyzed degradation of phenelzine-1-^{14}C, an irreversible inhibitor of monoamine oxidase. I. Studies *in vitro. Biochemical Pharmacology,* 18:1011–1021.

Clineschmidt, B. V., and Horita, A. (1969b): The monoamine oxidase catalyzed degradation of phenelzine-1-^{14}C, an irreversible inhibitor of monoamine oxidase. II. Studies *in vivo. Biochemical Pharmacology,* 18:1021–1029.

Collins, G. G. S. (1972): Summary of Section I. *Advances in Biochemical Psychopharmacology,* 5:129–132.

Collins, G. G. S., Pryse-Davies, J., Sandler, M., and Southgate, J. (1970a): Effect of pretreatment with estradiol, progesterone and DOPA on monoamine oxidase activity in the rat. *Nature,* 226:642–643.

Collins, G. G. S., Sandler, M., Williams, E. D., and Youdim, M. B. H. (1970b): Multiple forms of human brain mitochondrial monoamine oxidase. *Nature,* 225:817–820.

Collins, G. G. S., and Southgate, J. (1970): The effect of progesterone and oestradiol on rat uterine monoamine oxidase activity. *Biochemical Journal,* 117:38P.

Coquil, J. F., Goridis, C., and Neff, N. H. (1973): Monoamine oxidase in rat arteries: Evidence for different forms and selective localization. *British Journal of Pharmacology,* 48:590–599.

Dixon, M., and Webb, E. C. (1958): In *Enzymes.* Longmans, London, pp. 19–22.

Eiduson, S. (1972): Ontogenetic development of monoamine oxidase. *Advances in Biochemical Psychopharmacology,* 5:271–289.

Fajer, A. B., Holzbauer, M., and Newport, H. M. (1971): The contribution of the adrenal gland to the total amount of progesterone produced in the female rat. *Journal of Physiology* (London), 214:115–126.

Fuller, R. W. (1968): Kinetic studies and *in vivo* effects of a new monoamine oxidase inhibitor, N-2(o-chlorophenoxy)-ethylcyclopropylamine. *Biochemical Pharmacology,* 17:2097–2106.
Fuller, R. W. (1972): Selective inhibition of monoamine oxidase. *Advances in Biochemical Psychopharmacology,* 5:339–354.
Gabay, S., and Valcourt, A. J. (1968): Studies on monoamine oxidases. 1. Purification and properties of the rabbit liver mitochondrial enzyme. *Biochimica Biophysica Acta,* 159:440–450.
Ghosh, S. K., and Guha, S. R. (1972): Oxidation of monoamines in developing rat and guinea-pig brain. *Journal of Neurochemistry,* 19:229–231.
Hall, D. W. R., and Logan, B. W. (1969): Further studies on the inhibition of monoamine oxidase by M&B 9302 (chlorgyline)–II: Comparison of M&B 9302 inhibition with that of iproniazid. *Biochemical Pharmacology,* 18:1955–1959.
Hall, D. W. R., Logan, B. V., and Parsons, G. H. (1969): Further studies on the inhibition of monoamine oxidase by M&B 9302 (chlorgyline)–II: Substrate specificity in various mammalian species. *Biochemical Pharmacology,* 18:1447–1454.
Hartman, B. K. (1972): The discovery and isolation of a new monoamine oxidase from brain. *Biological Psychiatry,* 4:147–155.
Hartman, B. K., and Udenfriend, S. (1972): The application of immunological techniques to the study of enzymes regulating catecholamine synthesis and degradation. *Pharmacological Reviews,* 24:311–330.
Hartman, B., Yasunobu, K. T., and Udenfriend, S. (1969): Immunological identity of the multiple forms of beef liver mitochondrial monoamine oxidase. *Archives of Biochemistry and Biophysics,* 147:797–804.
Hems, R., Ross, B. D., Berry, M. N., and Krebs, H. A. (1966): Gluconeogenesis in the perfused rat liver. *Biochemical Journal,* 101:284–292.
Hollmann, S. (1969): In: *Pentoses and Pentitols,* edited by B. L. Harecker, D. Lang, and Y. Takagi. Springer Verlag, Berlin, pp. 97–108.
Holzbauer, M. (1971): *In vivo* production of steroids with central depressant actions by the ovary of the rat. *British Journal of Pharmacology,* 43:560–569.
Holzbauer, M., Bull, G., Youdim, M. B. H., Wooding, F. B. P., and Godden, U. (1972): Subcellular distribution of steroids in the adrenal gland. *Nature New Biology,* 242:117–119.
Holzbauer, M., and Youdim, M. B. H. (1973): Monoamine oxidase and oestrous cycle. *British Journal of Pharmacology,* 48:600–608.
Holzbauer, M., Youdim, M. B. H., and Blatchford, D. (1973): Regional differences in the rate of increase of brain monoamine oxidase with age. *(In preparation.)*
Horita, A. (1968): The influence of age on the recovery of cardiac monoamine oxidase after irreversible inhibition. *Biochemical Pharmacology,* 17:2091–2096.
Houslay, M. D., and Tipton, K. F. (1973): The nature of the electrophoretically separable multiple forms of rat liver monoamine oxidase. *Biochemical Journal,* 135:173–186.
Johnson, J. P. (1968): Some observations upon a new inhibitor of monoamine oxidase in brain tissue. *Biochemical Pharmacology,* 17:1285–1297.
Karki, N., Kuntzman, R., and Brodie, B. B. (1962): Storage, synthesis and metabolism of monoamines in the developing brain. *Journal of Neurochemistry,* 9:53–58.
Knoll, J., and Magyar, K. (1972): Some puzzling pharmacological effects of monoamine oxidase inhibitors. *Advances in Biochemical Psychopharmacology,* 5:393–408.
Kraml, M. (1965): A rapid microfluorometric determination of monoamine oxidase. *Biochemical Pharmacology,* 14:1684–1686.
Kroon, M. C., and Veldstra, H. (1972): Multiple forms of rat brain mitochondrial monoamine oxidase: Subcellular localization. *FEBS Letters,* 24:173–177.
Kuzuya, H., and Nagastu, T. (1969): Flavins and monoamine oxidase activity in the brain, liver and kidney of developing rat. *Journal of Neurochemistry,* 16:123–126.
Lang, N. (1971): Steroid hormones and enzyme induction. In: *The Biochemistry of Steroid Hormone Action,* edited by R. M. S. Smellie. Academic Press, London and New York.
McCauley, R., and Racker, E. (1973): Separation of two monoamine oxidases from bovine brain. *Molecular and Cellular Biochemistry,* 1:73–81.
Neff, N. H., and Goridis, C. (1972): Neuronal monoamine oxidase. Specific enzyme types and their rates of formation. *Advances in Biochemical Psychopharmacology,* 5:307–323.
Neff, N. H., and Yang, H. Y. T. (1973): Monoamine oxidase: II. Evaluation of the physiological role of type A and B enzyme of brain. *Federation Proceedings,* 31:797.

Parvez, H., and Parvez, S. (1973a): The effects of metopirone and adrenalectomy on the regulation of the enzymes monoamine oxidase and catechol-O-methyl transferase in different brain regions. *Journal of Neurochemistry,* 20:1011-1020.
Parvez, H., and Parvez, S. (1973b): The regulation of monoamine oxidase activity by adrenal corticoid steroids. *Acta Endocrinologia,* 73:509-517.
Percher, W., and Heller, A. (1972): Regional development of catecholamine biosynthesis in rat brain. *Journal of Neurochemistry,* 19:1917-1931.
Robinson, D. S., Lovenberg, W., Keiser, H., and Sjoerdsma, A. (1968): Effect of drugs on human platelet and plasma amine oxidase activity *in vitro* and *in vivo. Biochemical Pharmacology,* 17:109-119.
Ross, B. D. (1972): In: *Perfusion Techniques in Biochemistry.* Clarendon Press, Oxford, p. 4.
Sandler, M., and Youdim, M. B. H. (1972): Multiple forms of monoamine oxidase: Functional significance. *Pharmacological Reviews,* 24:331-348.
Schnaitman, C., Erwin, V. G., and Greenawalt, J. W. (1967): Submitochondrial localization of monoamine oxidase. *Journal of Cell Biology,* 34:719-735.
Smith, M. G. (1962): Polyol dehydrogenases. 4. Crystallization of the L-iditol dehydrogenase of sheep liver. *Biochemical Journal,* 83:135-144.
Southgate, J. (1972): Endometrial monoamine oxidase: The effect of sex steroids. *Advances in Biochemical Psychopharmacology,* 5:263-270.
Southgate, J., Grant, E. C. G., Pollard, W., Pryse-Davies, J., and Sandler, M. (1968): Cyclic variation in endometrial monoamine oxidase: Correlation of histochemical and quantitative biochemical assays. *Biochemical Pharmacology,* 17:721-726.
Tipton, K. F. (1972): Some properties of monoamine oxidase. *Advances in Biochemical Psychopharmacology,* 5:11-24.
Tipton, K. F. (1973): The adrenal gland. In: *Handbook of Physiology,* section 7: Endocrinology, edited by H. Blaschko and A. D. Smith. The American Physiological Society, Washington, D.C. *(in press).*
Tipton, K. F., Youdim, M. B. H., and Spires, I. P. C. (1972): Beef adrenal gland monoamine oxidase. *Biochemical Pharmacology,* 21:2197-2204.
Vaccari, A., Maura, M., Marchi, M., and Cugurra, F. (1972): Development of monoamine oxidase in several tissues in the rat. *Journal of Neurochemistry,* 19:2453-2457.
Wilson, M. A., and Cascarno, J. (1972): Biochemical heterogeneity of rat liver mitochondria separated by rate zonal centrifugation. *Biochemical Journal,* 129:209-218.
Woods, H. F., Eggleston, L. V., and Krebs, H. A. (1970): The cause of hepatic accumulation of fructose 1-phosphate on fructose loading. *Biochemical Journal,* 119:501-510.
Woods, H. F., and Krebs, H. A. (1971): Lactate production in the perfused rat liver. *Biochemical Journal,* 125:129-139.
Woods, H. F., and Krebs, H. A. (1973): Xylitol metabolism in the isolated perfused rat liver. *Biochemical Journal,* 134:437-443.
Yang, H. Y. T., and Neff, N. H. (1973): Monoamine oxidase: I. A natural substrate for type B enzyme. *Federation Proceedings,* 31:797.
Yielding, K. L., and Tomkins, G. M. (1960): An effect of enzymic reduction of steroids on triphosphopyridine nucleotide-dependent glucose 6-phosphate oxidation. *Biochimica Biophysica Acta,* 39:348-357.
Youdim, M. B. H. (1972): Multiple forms of monoamine oxidase and their properties. *Advances in Biochemical Psychopharmacology,* 5:67-78.
Youdim, M. B. H. (1973a): Monoamine deaminating system in mammalian tissues. In: *MTP International Review of Science,* edited by H. Blaschko. Butterworths, London *(in press).*
Youdim, M. B. H. (1973b): Multiple forms of mitochondrial monoamine oxidase. *British Medical Bulletin,* 29:120-122.
Youdim, M. B. H. (1973c): Heterogeneity of rat brain mitochondrial monoamine oxidase. *Advances in Biochemical Psychopharmacology,* 11:59-63.
Youdim, M. B. H. (1973d): Heterogeneity of rat liver and brain mitochondrial monoamine oxidase: Subcellular fractionation. *Biochemical Society Transactions,* 1:1126-1127.
Youdim, M. B. H., and Collins, G. G. S. (1971): The disassociation and reassociation of rat liver mitochondrial monoamine oxidase. *European Journal of Biochemistry,* 18:73-78.
Youdim, M. B. H., and Collins, G. G. S. (1973): Some physico-chemical properties of the multiple forms of rat liver mitochondrial monoamine oxidase. *(Submitted for publication.)*
Youdim, M. B. H., Collins, G. G. S., and Sandler, M. (1969): Multiple forms of rat brain monoamine oxidase. *Nature,* 223:626-628.

Youdim, M. B. H., Collins, G. G. S., and Sandler, M. (1970): Isoenzymes of soluble mitochondrial monoamine oxidase. In: FEBS symposium 18, *Enzymes and Isoenzymes,* edited by D. Shugar. Academic Press, London.

Youdim, M. B. H., Collins, G. G. S., Sandler, M., Bevan Jones, A. B., Pare, C. M. B., and Nicholson, W. J. (1972): Human brain monoamine oxidase; Multiple forms and selective inhibitors. *Nature,* 236:225–228.

Neuropsychopharmacology of Monoamines and
Their Regulatory Enzymes, edited by E. Usdin.
Raven Press, New York © 1974.

Some Interrelated Properties of Brain Monoamine Oxidase

Jean C. Shih and Samuel Eiduson

Departments of Psychiatry and Biological Chemistry, The Neuropsychiatric and Brain Research Institutes, UCLA Medical School, University of California, Los Angeles, California 90024

I. INTRODUCTION

Although several groups of investigators (Gorkin, 1963; Youdim and Sourkes, 1966; Kim and D'Iorio, 1968; Shih and Eiduson, 1969, 1971; Youdim, Collins, and Sandler, 1969) have suggested that monoamine oxidase (MAO) [monoamine:O_2 oxidoreductase (deaminating) EC 1.4.3.4.] may exist in multiple forms, the nature of the multiple forms and the relationship among these forms are still obscure. Our previous work (Shih and Eiduson, 1973) demonstrated that rat brain mitochondrial monoamine oxidase can be solubilized and separated into several fractions by either agarose column chromatography or by Sephadex electrophoresis. Accordingly, we have attempted to characterize further these multiple forms; in this chapter we will describe the relationship among these fractions and speculate about the physiological roles they may play *in vivo*.

II. THE RELATIONSHIP BETWEEN ISOLATED FRACTIONS ON AN AGAROSE COLUMN: EFFECT OF CONCENTRATIONS AND UREA

In these studies the rat brain mitochondrial MAO was solubilized by Triton X-100 (Shih and Eiduson, 1973), followed by ammonium sulfate fractionation (20 to 45%), and then absorbed on calcium phosphate gel. The MAO active fraction eluted from the gel (by 0.2 M phosphate buffer pH 7.4, containing 0.2% Triton X-100) was applied to an agarose column (Bio-Gel A 1.5 m, 2.2 × 115 cm). Essentially two major protein peaks were obtained (Fig. 1). The first protein peak (A) came down with the void volume of the column. The second peak, B, appeared in the later fractions. Both A and B forms had similar enzymic activity when benzylamine was used as substrate. However, when serotonin (5-HT) was used as substrate for the enzyme, fraction A was considerably more active than fraction B. The molecular weight of the A form of the enzyme as estimated by gel filtration was approximately equal to or greater than 1.5×10^6 daltons. The B form appeared

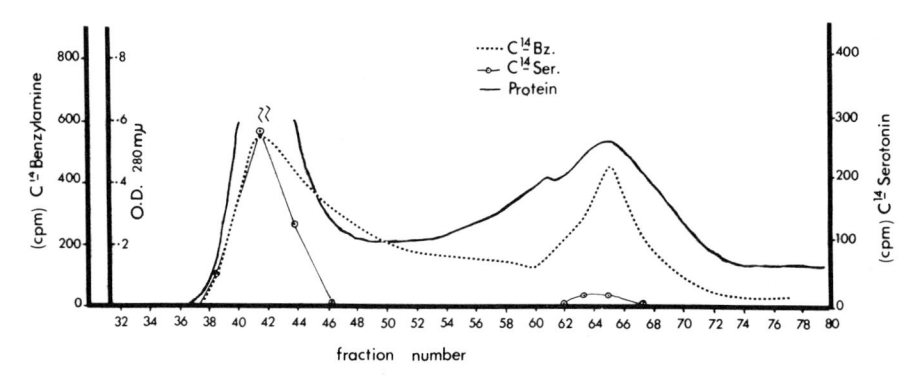

FIG. 1. Separation of MAO forms on agarose (2.2 × 115 cm column). The c.p.m. have been corrected for background. Solid line = protein content; dotted line = MAO activity measured with ¹⁴C-benzylamine; circles = MAO activity measured with ¹⁴C-serotonin. Buffer used was 0.05 M phosphate, pH 7.4 (Shih and Eiduson, 1973).

to be approximately 400,000 daltons. In order to ascertain the relationship between A and B, more specifically whether A is an aggregate of B, we subjected the mitochondrial MAO preparation to agarose column chromatography in the presence of 8 M urea or 6 M guanidine-HCl.

It was hypothesized that the A form might disappear under the influence of the urea or guanidine hydrochloride if the A form is an aggregate of B.

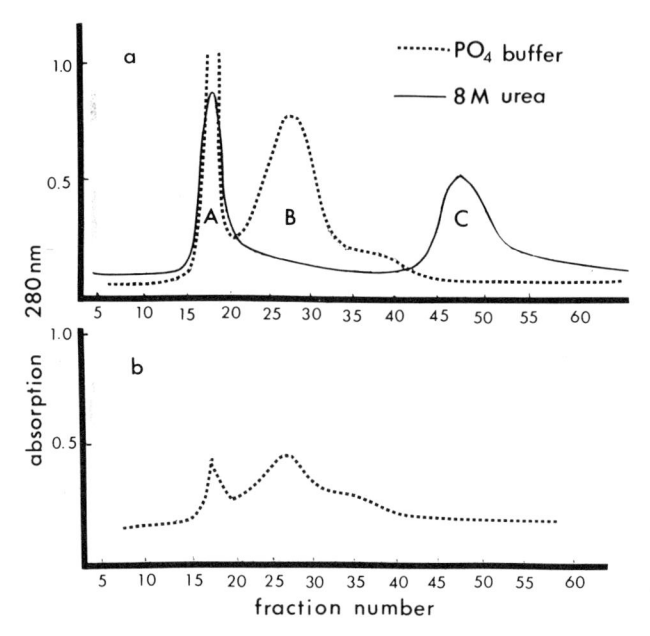

FIG. 2. a. Separation of MAO forms on agarose (2.5 × 45 cm column) in the presence (solid line) and in the absence (dotted line) of 8 M urea. The same buffer was used as in Fig. 1. b. Pattern of peak B rechromatographed on agarose column in the absence of urea. Column size is the same as in a. The same buffer was used as in Fig. 1.

The results are indicated in Fig. 2. It is clear that considerable protein came down in the void volume of the column as before. A similar pattern was obtained when the column was equilibrated with guanidine hydrochloride instead of urea. Interestingly, the protein peak eluted from the column (labeled C in Fig. 2) did not appear in the same set of fractions as did B previously in the absence of 8 M urea. It is tempting to suggest, therefore, that B may be an aggregate of C since it disappeared under the influence of the urea while the C form appeared. Equally interesting is the observation that not all of the protein peak in A was disassociated (Fig. 2a). When the fractions in each peak were pooled and dialyzed to eliminate the urea and subsequently concentrated, it was found that A and C possessed MAO activity when tryptamine was used as substrate for the enzyme.

The fraction A from the urea column had about 43% of the enzyme activity toward tryptamine compared to the fraction A obtained from a column without urea.

When we rechromatographed the original peak B (without 8 M urea), we obtained the results shown in Fig. 2b. It is clear that the B peak was still present, but in addition an appreciable aggregation occurred to form material which comes through the void volume (as did the original A peak). Therefore, we surmise that B indeed aggregates to A.

There is one additional important observation we would like to mention. The B form appeared to be concentration dependent, since at low concentrations peak B was converted to peak C. (Fig. 3). This C form corresponds

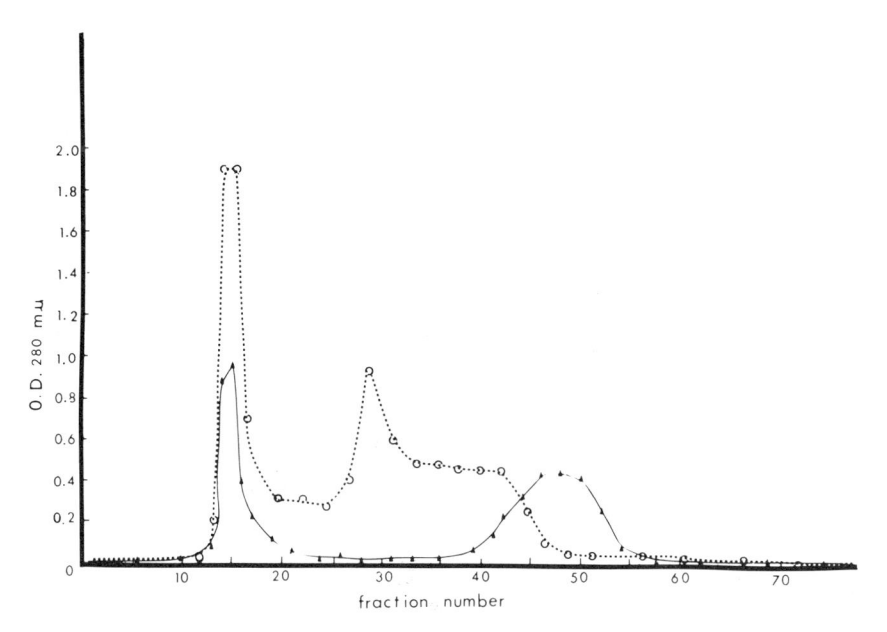

FIG. 3. Separation of MAO forms on agarose (2.2×115 cm column) when different amounts of enzyme are applied to column. Solid line = 10 to 12 mg of protein applied; dotted line = 5 to 6 mg of protein applied. The same buffer was used as in Fig. 1.

to the C form obtained using an agarose column equilibrated with 8 M urea (cf. Fig. 2*a*).

These results suggest that the B form obtained with agarose column chromatography is an intermediate form of the enzyme which may dissociate into C either at low concentration of the enzyme or in the presence of urea or guanidine-HCl. In addition, the B form may also aggregate to A. The A form is a stable form because of the observation that when peak A was rechromatographed, no dissociation of A was observed. The previously mentioned observation that the protein peak which came down in the void volume in the presence of 8 M urea and which still contained MAO activity after dialysis suggests that there may be at least one form of MAO in peak A which is not an aggregate of B.

It is clear that the different MAO forms in A, B, and C have different substrate affinities (Table 1).

TABLE 1. *Specific activities of various forms of MAO with different substrates*

MAO form	5-HT*	Tryptamine*	Benzylamine*	Phenylethylamine*
A	1 (0.45)	1 (13.25)	1 (12.04)	1 (14.50)
B	5.5	1.8	5.3	2.8
C	21.4	36.9	703.0	9.3

*Nanomoles of substrate oxidized/mg protein/30 min = values for specific activities referred to A form as 1.

Figures in parentheses are the actual specific activities of the A form for each of the substrates.

III. THE RELATIONSHIP BETWEEN THE FRACTIONS ISOLATED BY MEANS OF SEPHADEX-ELECTROPHORESIS AND BY AGAROSE COLUMN CHROMATOGRAPHY

When the mitochondrial MAO preparation was subjected to Sephadex-electrophoresis, two proteins fractions, I and II, were obtained (Fig. 4). The enzyme in I appeared to be more active when 5-HT was used as substrate than was that in II. In order to clarify further the relationship between these two fractions, they were rechromatographed using an agarose column.

When fraction I was rechromatographed on agarose (independent of the concentration of the fraction), a major peak was obtained in the void volume of the column, similar to peak A of Fig. 1. Fraction I was found to have a molecular weight of approximately 1.5×10^6 daltons (Fig. 5).

When fraction II was rechromatographed, the protein separation profile was observed to be dependent on the concentration. When a diluted fraction was chromatographed, fractions II_A and II_B were obtained (Fig. 6) which had the same size as fractions A and C from the agarose column (cf. Fig. 3).

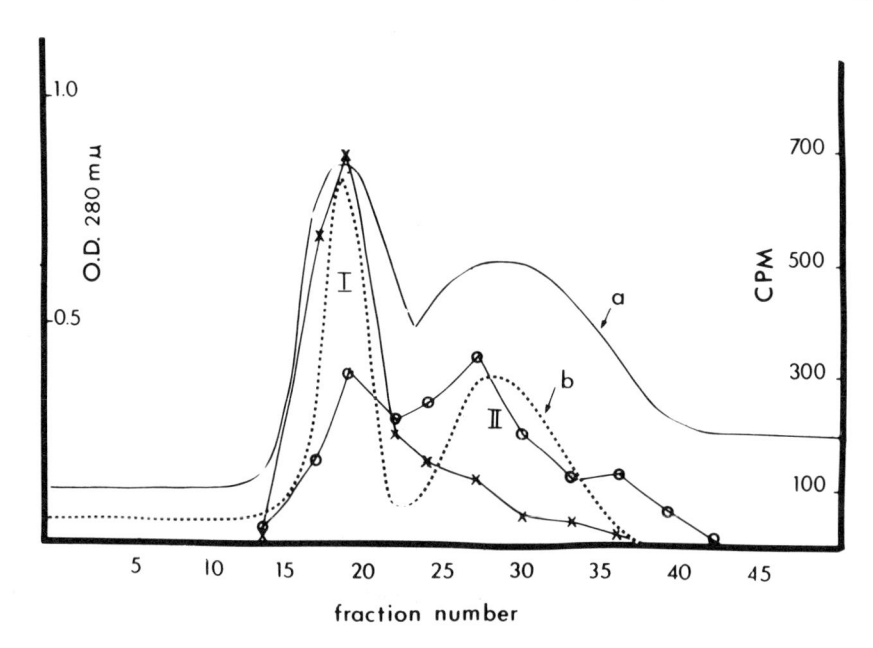

FIG. 4. Separation of MAO by Sephadex-electrophoresis and MAO activity determined with ^{14}C-serotonin or ^{14}C-benzylamine as substrates. The c.p.m. have been corrected for background. Key to plots: curves a and b = patterns of protein content derived from two different experiments: O———O = MAO activity measured with ^{14}C-benzylamine; ×———× = MAO activity measured with ^{14}C-serotonin; enzyme activity was assayed on samples taken from curve a. Buffer used was 0.038 M Tris-glycine, pH 8.3. (Shih and Eiduson, 1973.)

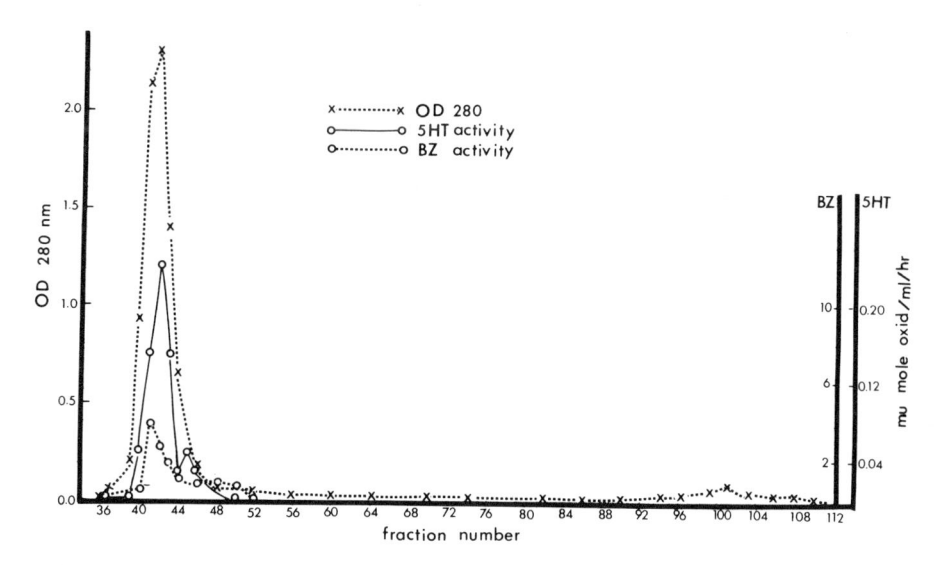

FIG. 5. MAO pattern of peak I from Sephadex-electrophoresis on agarose (2.2 × 115 cm column). (×----×) O.D. at 280 μ; (O———O) activity using 5-HT as substrate; (O----O) activity using benzylamine as substrate.

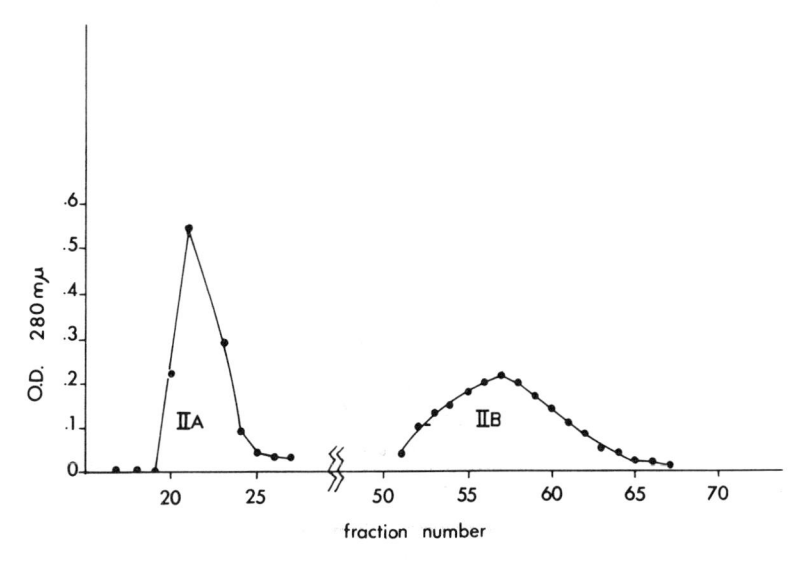

FIG. 6. MAO pattern of peak II from Sephadex-electrophoresis on agarose (2.2 × 45 cm column). 1.0 mg of protein applied to the column.

If a more concentrated peak II was put on agarose column, two peaks were obtained (Fig. 7) which had the same pattern as fractions A and B from agarose column (cf. Fig. 3). This apparent aggregation and dissociation phenomenon exhibited by peak II is very similar to the B peak from the agarose column. From their sensitivity toward serotonin and benzylamine

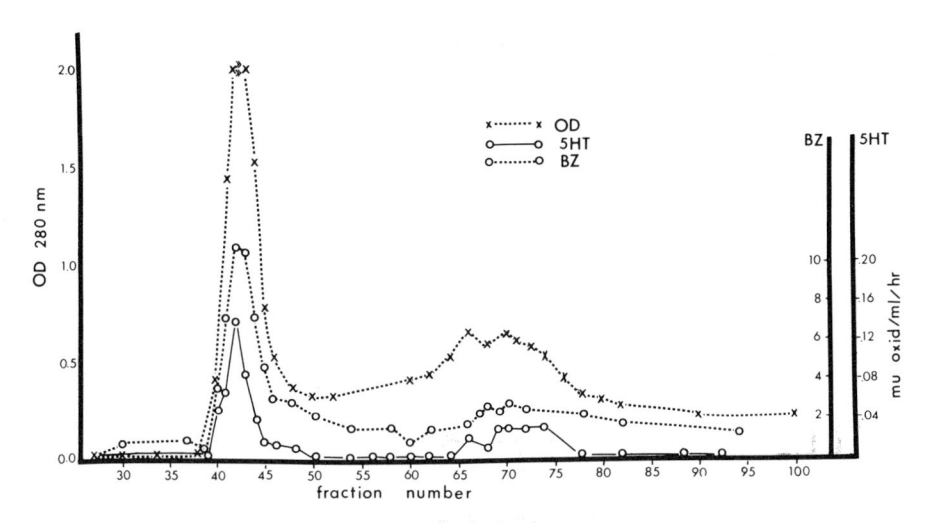

FIG. 7. MAO pattern of peak II from Sephadex-electrophoresis on agarose (2.2 × 115 column). 8.3 mg of protein applied to the column.

(Figs. 1 and 2) and from their aggregation and dissociation properties, we concluded that peak I from the Sephadex-electrophoresis (SE) column was very similar to the A peak from the agarose separation. Peak II from SE was very similar to the B peak derived from the agarose column.

IV. SUBSTRATE SPECIFICITIES OF MAO FORMS

It is apparent that each of the various forms of MAO has different specific activities depending upon the substrate used. Further, and importantly, as the MAO is converted to forms B or C, there is a differential increase in specific activity of the enzyme, again depending upon the substrate used. The data suggest that as disaggregation of the MAO takes place from A to B to C, there is a corresponding conformational change of the enzyme such that there is a change in the affinity of the enzyme for a particular substrate.

V. SUMMARY AND DISCUSSION

It seems clear that, based on substrate sensitivity and characteristics of aggregation, the MAO forms separated by means of Sephadex-electrophoresis are the same or very similar to those separated using an agarose column. The fact that the protein fraction with a molecular weight of 1.5×10^6 was still apparent in the presence of 8 M urea (or 6 M guanidine-HCl) and that the fraction possessed MAO activity indicated that there are at least two forms of MAO in this fraction. These data suggest that one of the forms may be an aggregate of the B form while the other is probably another and separate form of MAO. Certainly the possibility of this form being a partially solubilized MAO is not excluded by these data.

The results obtained from the rechromatographing of the B form indicate that B does indeed aggregate to A regardless of the concentration. (Tipton, 1968, has suggested that the aggregation of MAO forms may be due to their content of phospholipid.) At diluted concentrations, however, B appeared dissociated to the C form as well. The possibility that these various forms may be interdependent and concentration dependent suggests that perhaps one form of regulation of the turnover of the biogenic amines may be this interconversion of the various MAO forms with different substrate affinities.

Although these different forms have been observed *in vitro*, there is as yet no compelling evidence that they exist *in vivo*. They may, of course, exist *in situ* and, if they do, the suggestion that they may play a regulatory role *in vivo* is a tenable hypothesis.

ACKNOWLEDGMENTS

Support from the Grant Foundation is gratefully acknowledged. This work was also supported in part by U.S. Public Health Service grant MH 19734.

REFERENCES

Gorkin, V. Z. (1963): Partial separation of rat liver mitochondrial amine oxidases. *Nature,* 200:77.

Kim, H. C., and D'Iorio, A. (1968): Possible isoenzymes of monoamine oxidase in rat tissues. *Canadian Journal of Biochemistry,* 46:295–297.

Shih, J. C., and Eiduson, S. (1969): Multiple forms of monoamine oxidase in developing brain. *Nature,* 224:1309–1310.

Shih, J. C., and Eiduson, S. (1971): Multiple forms of monoamine oxidase in developing brain: Tissue and substrate specificities. *Journal of Neurochemistry,* 18:1221–1227.

Shih, J. C., and Eiduson, S. (1973): Monoamine oxidase (EC 1.4.3.4): Isolation and characterization of multiple forms of the brain enzyme. *Journal of Neurochemistry,* 21:41–49.

Tipton, K. F. (1968): The purification of pig brain mitochondrial monoamine oxidase. *European Journal of Biochemistry,* 4:103–107.

Youdim, M. B. H., Collins, G. G. S., and Sandler, M. (1969): Multiple forms of rat brain monoamine oxidase. *Nature,* 223:626–628.

Youdim, M. B. H., and Sourkes, T. L. (1966): Properties of purified, soluble monoamine oxidase. *Journal of Biochemistry,* 44:1397–1400.

Neuropsychopharmacology of Monoamines and Their Regulatory Enzymes, edited by E. Usdin. Raven Press, New York © 1974.

Some Considerations in the Design of Substrate and Tissue-Specific Inhibitors of Monoamine Oxidase

Yvonne C. Martin and John H. Biel

Abbott Laboratories, North Chicago, Illinois 60064

I. INTRODUCTION

The clinical literature during the past 15 years abounds with sufficient evidence for the therapeutic utility of monoamine oxidase inhibitors (MAOIs) in the treatment of certain types of mental depression. The main restriction for the widespread clinical use of these agents has been their propensity for potentiating the central and peripheral pressor responses of tyramine following the ingestion of tyramine-rich foods. From both a theoretical and a practical standpoint, it would be highly desirable to design MAOIs which could be both tissue and substrate specific. For instance, an MAOI which would have special affinity for brain tissue and inhibit selectively the oxidation of either norepinephrine, serotonin, or dopamine would go far in elucidating the implied role of these neurotransmitters in many types of psychiatric disorders: the depressions, psychoses, neuroses, and anxieties.

Rather than arrive at such selectively acting MAOIs purely through the empirical route (via the random synthesis of thousands of compounds), we present here a drug design technique which could yield such an agent by determining the physical-chemical characteristics required of a drug to fulfill a certain set of biochemical criteria. This approach may be a more effective means of producing the optimally acting compound within a shorter period of time and with a substantially lessened requirement for heavy investment in chemical synthesis of empirically designed structures.

While this technique is not purported to be the ultimate in drug design, it nevertheless offers definite opportunities for "short-cutting" the laborious process of achieving optimal drug action through systematic structural modification of organic molecules.

II. PHYSICOCHEMICAL BASIS OF DRUG DESIGN

The underlying philosophy of the drug design examples to be discussed is that the pharmacological and biochemical activity and potency of a molecule can be entirely explained on the basis of its physicochemical properties.

If we knew enough of the physical chemistry of the receptor or enzyme, the effector molecule, the competing processes, and the compartments through which it passes, we could explain the observed properties of a drug without reference to such concepts as "the β-blocker side chain." Obviously we are a long way from such understanding. At present we must start with certain active chemical series, usually arrived at empirically, to help us characterize the physical-chemical nature of a receptor.

There are certain simplifications possible when one studies structure-activity (or more precisely, structure-potency) relationships (SAR) within a series of related drug molecules tested in the same biological system (Hansch, 1971). In this case one focuses on changes, and one need not examine those physical-chemical events which are constant within the series. The problem is therefore considerably reduced in complexity.

Intuitively, one expects that changes in drug structure can result in changes in any or all of the three primary physical properties: lipophilicity, distribution of electrons, and arrangement of atoms in three-dimensional space. It is convenient that physical organic chemists have measured the effects of common substituents on these physical properties. The experimentally derived quantities Π, σ, and E_s^c are used as measures of changes in the logarithm of the octanol-water partition coefficient (P) (lipophilic effect), pK_a (electronic effect), and rate of reaction in a sterically controlled situation (steric effect), respectively. These quantities are general; that is, although they were measured on one set of molecules and in one experimental situation, they may be applied with suitable precautions and modification to other sets of molecules or experimental situations. In the case of drug analogues, one would therefore expect the potency to be a function of Π, σ, and E_s^c. The functional relationship intuitively may be thought of as deviations from an optimum. Mathematically, this is expressed as a coefficient times the parameter minus a second coefficient times the parameter squared. We thus come to the classic Hansch (1971) equation:

1. $$\log \Delta A = a + b\Pi - c\Pi^2 + d\sigma - e\sigma^2 + fE_s^c - g(E_s^c)^2$$

A reasonable approach to the discovery of a new drug is to plan one's work so that ultimately one can design new analogues on the basis of specific equations of the above type. This program involves four steps. *The first step* is to choose and prepare a series of analogues in which the lipophilic, electronic, and steric factors are varied as much as possible and in which these factors are not correlated with each other. For example, if the "lead" compound has an aromatic ring, one would expect substitution on this ring to affect potency. A typical series of 10 analogues which it would be reasonable to synthesize is listed in Table 1. The substituents chosen are based on a cluster analysis of all possible substituents (Hansch, Unger, and Forsythe, 1973a). In actual practice one would select from the choices generated by cluster analysis that member of each cluster which was easiest to synthesize and which seemed least likely to be metabolized. This approach is satisfactory when the analogues are easy to synthesize and test. Typical clusters of substituents from which one would choose one analogue each are shown in Table 2.

TABLE 1. *Rational series of phenyl analogues*

	Π^a	σ_p^a	MR^a
H	0.00	0.00	1.03
CH=CHCO$_2$H	0.00	0.90	17.91
Cl	0.71	0.23	6.03
SO$_2$NH$_2$	−1.82	0.57	12.28
OCH$_3$	−0.02	−0.27	7.87
CF$_3$	0.88	0.54	5.02
SO$_2$C$_6$H$_5$	0.27	0.70	33.20
COC$_6$H$_5$	1.05	0.43	30.33
C$_6$H$_5$	1.96	−0.01	25.36
Adamantyl	3.30	−0.13	40.63

[a] Parameter values from Hansch, Leo, Unger, Kim, Nikaitani, and Lien (1973b). MR is the molar refractivity of the substituent, a measure of its volume.

TABLE 2. *Typical clusters of aromatic substituents with similar physical properties (in order of increasing distance from each other)*

Cluster no.	Members
1	B(OH)$_2$; CH$_3$; CH$_2$CH$_3$; CH$_2$OH; H
2	CH=CHCOOH
3	CN; NO$_2$; CO$_2$H; COCH$_3$; C≡CH; Cl; SH
4	CONH$_2$; CONHCH$_3$; SO$_2$NH$_2$; SO$_2$CH$_3$; NHCONH$_2$
5	F; OCH$_3$; NH$_2$; OH; N(CH$_3$)$_2$
6	Br; CF$_3$; I; SO$_2$CF$_3$
7	CH$_2$Br; NHCO$_2$CH$_2$CH$_3$; SO$_2$C$_6$H$_5$; OSO$_2$Me
8	NHCOC$_6$H$_5$; N=CHC$_6$H$_5$; NHSO$_2$C$_6$H$_5$; COC$_6$H$_5$
9	3,4-(CH$_2$)$_3$; Pr; 3,4-(CH)$_4$; NHC$_4$H$_9$; C$_6$H$_5$
10	Ferrocenyl; adamantyl

Topliss (1972) has suggested that when analogues are difficult to synthesize, the relative potency of each new molecule should be used as a guide for further synthesis. Two synthetic schemes or decision trees have been proposed, one for aromatic substitution, the other for side chains. In the aromatic substitution case one first compares the —H and —Cl analogues. If the —Cl is more potent, the —CF$_3$ analogue is prepared on the basis that increases in Π or σ are most likely the cause of the increased potency. If the —Cl is less potent, one makes the —OCH$_3$ analogue in which both Π and σ are negative. The schemes are followed for four to six analogues, at which point inspection of the results will suggest further "follow-up." An abbreviated scheme is shown in Fig. 1.

The second step in the rational design of a drug analogue is biological testing. The test must be relevant to the problem. For example, if the goal is to design a tissue-selective inhibitor of serotonin oxidation, each analogue must be tested as an inhibitor of serotonin oxidation in each tissue of interest. The actual experiments could be administration of the compounds *in vivo* followed by some biochemical measure of the inhibition of MAO in each tissue of interest. Alternatively, the same problem could be studied by establishing *in vitro* the pI_{50} of each analogue on each tissue of interest and determining the tissue concentration of each analogue in an independent experiment. In a similar way, if the goal is to find a substrate-selective MAOI for CNS use, each analogue must be tested against the substrates of interest. The ability of each analogue to reach the CNS after oral administration must also be measured by biochemical or pharmacological methods.

If it is known that certain side effects, tolerance, or toxicity will be an important factor in the evaluation of a compound, quantitative values for this activity should be evaluated for each analogue. The quantitative drug design is only as relevant or precise as the biological data on which it is based.

The third step in the analysis is the calculation, by standard statistical techniques, of the relationship between potency in each biological test and physical properties. Multiple regression analysis has been the standard calculation method, but other techniques may also be useful (Martin, Holland, Jarboe, and Plotnikoff, 1974). One must, of course, be careful that the calculated relationships are true and not artifacts of the data.

The fourth step is to design a new "optimal" analogue on the basis of a

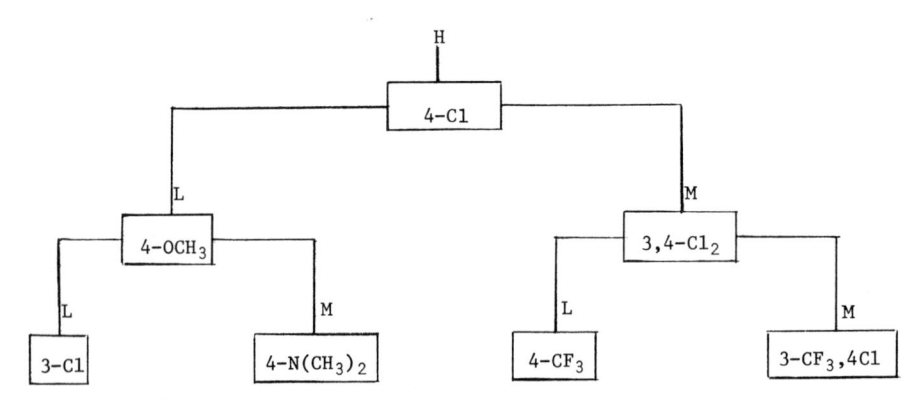

FIG. 1. Abbreviated operational scheme for aromatic substitution (Topliss, 1972). L indicates the analogue to make if the latest analogue is *less* potent than the one prior to it; M indicates the analogue to make if the latest analogue is *more* potent than the one prior to it.

comparison of the relationships for each activity. If different physical properties control the different activities, the task is straightforward. For example, suppose that *in vitro* inhibition were electronically controlled ($+\sigma$) and that brain penetration has an optimum log P of 2.3. If the goal were a CNS-active compound, one would make one with a log P of 2.3 and as high a σ as possible. If the same physical property controls both potencies, a graph is often the best way to visualize the ideal compound. An example of this latter case is the β-carboline MAOIs to be discussed later.

The type of drug design described above has been applied to MAOIs. Several years ago Fuller, Marsh, and Mills (1968) reported a Hansch analysis and successful prediction of the MAOI potency of a series of N-(phenoxyethyl)cyclopropylamines. Their equations suggested a large positive coefficient of Π and σ. Two compounds were synthesized on the basis of their equation; the potencies of both were as predicted. One analogue was more potent than any of the original series.

The above approach to the design of a new analogue may seem tedious and unimaginative as compared with the "stroke of genius" methods often used. However, it offers the promise that when an initial series has been tested and the SAR calculated, one knows how to get to the optimally acting members of a series with the least number of compounds. (It might be that no solution is possible and the effort should be abandoned.) On the other hand, in the more traditional approach one often ends up with a variety of weakly active analogues not suited for further SAR evaluation.

III. RATIONALE FOR TISSUE SELECTIVITY OF 2,9-DIMETHYL-β-CARBOLINIUM IODIDE

The earliest attempts at describing the physical-chemical basis for drug activity were involved with studies of CNS-active drugs. In the 1890s Meyer (1899) and Overton (1897) proposed that narcotic action parallels oil/water partition coefficients. Thus it seems logical to all of us 70 years later that a less lipophilic analogue of a drug of interest would not enter brain as well but would exert its activity, if any, peripherally.

The quantitative relationship between potency on the CNS and lipophilicity has been studied extensively by Hansch and Clayton (1973). As an example, Fig. 2 contains a plot of the equation which describes the extent of penetration into brain of various phenyl-substituted benzene boronic acids (Hansch, Steward, and Iwasa, 1965). The optimum log P is 2.3. In this case the log P of the neutral form of the drug is plotted. The neutral form is chosen on the basis of a whole body of scientific information which supports the concept that it is the neutral form of an ionizable drug that penetrates most easily into the CNS (Goldstein, Aronow, and Kalman, 1968). The other sets of data studied by Hansch and Clayton also suggest an optimum log P of approximately 2.0 for penetration into brain (1973). From Fig. 2

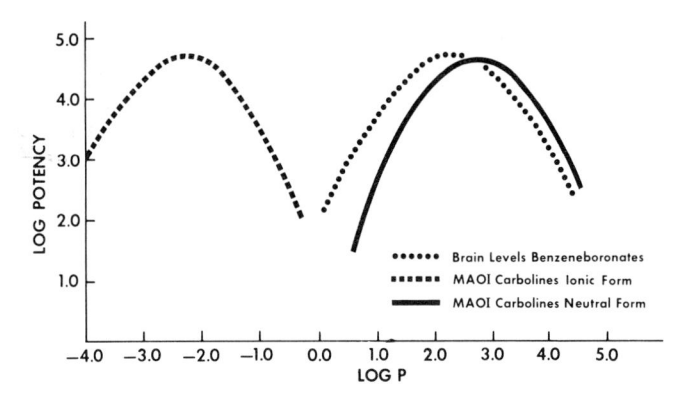

FIG. 2. Dependence of MAOI and CNS penetration on log P.

it can be seen that if an analogue which does not penetrate the brain is desired, one with a log P below 0 or above 4 or 5 should be made.

Lien, Hussain, and Tong (1970) studied the quantitative structure-activity relationships of a series of β-carbolines (Fig. 3) which had been prepared and tested as MAOI versus tyramine by Ho, McIsaac, Tansley, and Walker (1969). The parabola with an optimum log P of 2.7 (Fig. 2) is that reported by Lien. However, in this case it is possible that it is the protonated form of the β-carboline which inhibits the MAO. The pK_a of the pyridyl nitrogen of these compounds is in the biological range. The log P-potency curve for this situation is the parabola with an optimum potency at -2.25 in Fig. 2.

FIG. 3. β-Carbolines.

The design of a new β-carboline can now be considered. If it is the neutral form of the carboline which reacts with the MAO, no analogue which does not enter the brain can be designed. The log P-activity and log P-brain penetration curves are either parallel or else brain penetration is favored. On the other hand, if it is the ionized form of the carboline which reacts with the MAO, selectivity can be accomplished by the synthesis of a permanently charged analogue. This was done by Ho, Gardner, Pong, and Walker (1973), who prepared and tested 2,9-dimethyl-β-carbolinium iodide (DMCI) (Table 3). The potency of this compound (log $P = -2.2$) is as expected if one assumes that the ionizable carbolines react with monoamine oxidase in their protonated form. From Fig. 2, DMCI is obviously expected not to penetrate the brain. The data in Table 3 confirm the lack of *in vivo* effect of this drug on brain MAO.

TABLE 3. *Inhibition of rat MAO by DMCI*

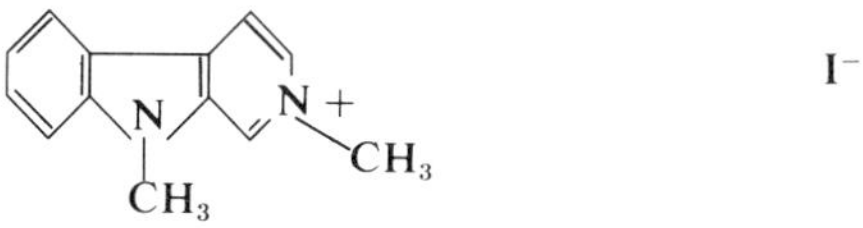

	in vitro		2 hr *in vivo* post 15 mg/kg, i.v. level	
Tissue	pI$_{50}$	% dose/g	% I	% increase 5-HT
Liver	5.15	0.6	16	4
Heart	5.85	2.4	57	37
Brain	5.40	0	0	0

B. T. Ho et al. (1973).

The selective action of DMCI on heart and not liver can be explained by the observation that it is rapidly secreted by liver into bile. Eighteen percent of an intravenous dose is recovered in the bile within 2 hr after an intravenous dose. Thus, although DMCI reaches the liver, it is apparently rather effectively removed from this organ. As a result, no significant increase in amine levels is observed in liver (Ho et al., 1973). This biliary secretion was not expected, but such a biological phenomenon could obviously be one of the activities considered in the quantitative structure-activity analysis.

IV. INHIBITION OF SEROTONIN OXIDATION BY PROPARGYL AMINES

The interest of Abbott Laboratories in MAOIs dates back to the program which resulted in the synthesis of pargyline (Fig. 4) in 1959 (Taylor, Wykes, Gladish, and Martin 1960). The pI$_{50}$'s versus serotonin and pK_a's of approximately 60 analogues were measured in 1960 (Swett, Martin, Taylor, Everett, Wykes, and Gladish 1963). It is only very recently, however, that the complex structure-activity relationships within this series have been discerned (Y. C. Martin, W. B. Martin, and J. D. Taylor, *in preparation*).

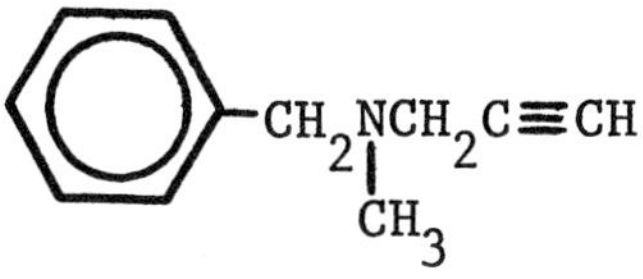

FIG. 4. Pargyline.

Preliminary evaluation of the data suggested that analogues in which the smallest substituent on the nitrogen is larger than methyl are essentially inactive. Variation of the propargyl portion of the molecule also destroys activity. Therefore, a multiple regression analysis of the relationship between pI_{50} and physical properties was calculated for those 47 analogues in which variations are made in the benzyl portion of pargyline. The compounds include analogues substituted in the benzylamine portion, pyridyl methyl-, phenethyl-, and alkyl amine analogues. The following relationship was observed:

2. $\quad pI_{50} = -7.48(4.66) + 4.38(1.38)pK_a - 0.35(0.10)pK_a^2 + 0.25(0.19)\Pi$
$\hspace{10em} + 1.02(0.45)D_2$

The numbers in parentheses are the 95% confidence intervals of the coefficients. For this equation the multiple correlation coefficient, R, is 0.87; the standard deviation of estimate is 0.58. The parameter D_2 is a dummy parameter which is set equal to 1.0 if the analogue is substituted in the ortho position and 0.0 if the ortho position is free.

This equation provides us with a great deal of information about the correlation of potency variation within the series. Potency is parabolically related to pK_a. The optimum value is 6.2 with a 95% confidence interval of 5.8 to 6.5. This parabolic dependence indicates that two pK_a-dependent processes occur. One increases potency as pK_a is increased; the second decreases potency as pK_a is increased. One may postulate that, starting with those compounds with the lowest pK_a as the analogues become more basic, they would bind more strongly to the amine-binding site in the active center of the enzyme. Ultimately, however, further increases in basicity perhaps retard the rate of shift of electrons that produces the irreversibly bound inhibitor.

The equation suggests a very slight hydrophobic effect.

There is a 10-fold increase in the potency of 2-substituted benzylamine analogues. There is too much correlation between the steric, hydrophobic, and electronic parameters for substituents at this position to allow one to determine which is the physical property responsible for the enhanced potency. The correlation between the parameters for ortho substitution is illustrated by the following equation:

$\quad MR = 1.05(0.73) + 5.49(1.92)\Pi - 7.03(1.47)PP \quad\quad R = 0.93, n = 22$

MR is the molar refractivity, Π is $\Delta \log P$, and PP is an electronic parameter. Thus steric, lipophilic, and electronic effects cannot be separated unless one makes additional analogues in which the above chance correlation is removed. This problem emphasizes one of the advantages of consulting a drug design expert before beginning synthesis in a series. A different selection of substituents might allow establishment of which property increases potency.

The compounds that were excluded from the calculation may now be reevaluated. All analogues in which the third substituent on the nitrogen is larger than methyl are at least 10 times less potent than calculated from Eq. 2. We conclude that this is a specific steric effect, since overall log P and pK_a are included in the calculation of the expected activities. Variations in the propargyl portion of the molecule produced analogues 1000 times less potent than predicted by Eq. 2. This deviation indicates that the special electronic or steric effects of the propargyl portion of pargyline are in addition to the pK_a and Π effects included in Eq. 2.

We are now determining the pI_{50}'s of these same molecules versus tyramine. It is known that pargyline is approximately 10-fold better as an irreversible inhibitor of tyramine than of serotonin oxidation (Fuller, Warren, and Molloy, 1970). From our new data it may be possible to design an analogue for which the reverse substrate specificity is seen by comparison of the equations for the two activities.

V. DESIGN OF SUBSTRATE-SELECTIVE AMINOINDANE OR AMINOTETRALIN ANALOGUES

The final example concerns some simple MAOIs that we discovered in our search for antiparkinsonism agents (Martin, Jarboe, Krause, Lynn, Dunnigan, and Holland, 1973). In early 1970 we noticed that one could superimpose the heteroatoms of dopamine and oxotremorine when models of these molecules in their proposed preferred conformations are built. We therefore postulated that dopamine and oxotremorine interact with a common receptor. The concept is strengthened by the observation that atropine and the anticholinergic antiparkinsonism drugs also have atoms that could fit the proposed receptor. We decided to test the hypothesis by the synthesis of phenolic 1-aminoidanes and 1-aminotetralins (Fig. 5). The first compounds, 6-methoxy-1-aminotetralin and its N-methyl analogue, showed activity in the modified DOPA potentiation test. However, when they were tested in mice they were found to be MAOIs.

Since we wanted to evaluate this series as dopaminergics, the goal of the first analogues made was to *decrease* MAOI potency. Table 4 shows the strategy we followed. The two compounds on which the MAOI effect was

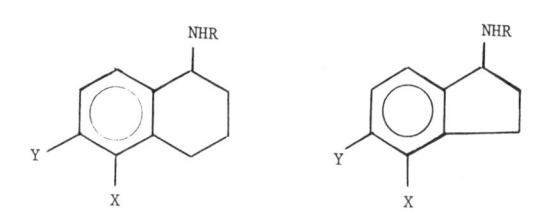

FIG. 5. Aminotetralins and aminoindanes. Y = H, OH, or OCH_3; X = H, OH, or OCH_3; Y ≠ X.

initially observed were those in which the substituents on the nitrogen were —H and —CH$_3$. The methyl analogue is more lipophilic and more sterically hindered; it is also less potent. Therefore, we reasoned that further increases in lipophilicity or size should further decrease potency. We also wanted to know which factor, lipophilicity or size, was responsible for activity. Hence, we designed a set of analogues of 6-methoxytetralins in which these factors were varied independently. The E_s^c or steric factor could not be varied through as wide a range as could lipophilicity. The ethyl group is of approximately the same size as the methyl; further substitutions on the terminal carbon atom of the ethyl group, whether additional methylenes or other functional groups, do not change E_s^c substantially. Branching the σ-methylene carbon of the ethyl side chain to form isopropyl does decrease E_s^c. The (larger) tertiary butyl analogue ($E_s^c = -2.46$) could not be made by the synthetic scheme used for the other analogues. Therefore, its synthesis was deferred and finally deemed unnecessary. The Π values of the analogues chosen varied from approximately that of the H analogue to +3.4.

TABLE 4. *Rational design of non-MAOI analogues of 6-methoxy-1-aminotetralin*

R	Π	E_s^c	in vivo activity
Initial compounds			
H	0.0	0.32	3+
CH$_3$	0.4	0.0	2+
Series designed			
CH$_2$CH$_3$	0.9	−0.07	2+
(CH$_2$)$_2$CN	0.1	−0.66	0
(CH$_2$)$_2$OH	0.3	−0.66	1+
(CH$_2$)$_2$OCH$_3$	0.4	−0.66	1+
CH(CH$_3$)$_2$	1.2	−1.08	0
(CH$_2$)$_2$CH$_3$	1.4	−0.66	1+
(CH$_2$)$_5$CH$_3$	2.9	−0.68	1+
(CH$_2$)$_6$CH$_3$	3.4	−0.68	0

The *in vivo* MAOI potency of the 6-methoxytetralins is also listed in Table 4. It can be seen that E_s^c and not Π is the primary determinant of potency. Other positions of methoxy substitution and various 1-aminoindanes were also synthesized. The compounds were tested at 100 mg/kg for their ability to potentiate the effects of DOPA (Martin et al., 1974). Since this is essentially an all-or-none response, the statistical method of discriminant analysis was used rather than regression analysis to study the relationship between physical properties and activity versus inactivity of the 20 analogues. This analysis showed that *in vivo* DOPA potentiation is independent of log P or the size of the saturated ring. The more bulky the substituent on nitrogen, the less likely it is to be active. Molecules with the methoxy or hydroxy substitution at the position adjacent to the ring junction (X in Fig. 5) are less likely to be active than those in which the substitution is at the Y position.

Our current interest in this series is to examine the possibility of finding a serotonin-selective inhibitor. For this purpose we are now in the process of analyzing the pI_{50}'s of these compounds versus tyramine and versus serotonin oxidation by rat liver mitochondria. We have already observed one interesting difference (Table 5). The type and position of aromatic substitution influences the selectivity of the inhibitor. It can be seen that there is apparently a correlation of selectivity with σ. We therefore plan to explore this relationship by the synthesis of analogues with even larger σ constants. A likely candidate is the 4-fluoro analogue, which has a larger σ constant than the 4-methoxy compound but which is essentially identical in other physical properties.

TABLE 5. Aminoindane MAOIs in vitro pI_{50}

Tyramine	4.42	3.77	3.37
Serotonin	3.25	3.10	3.40
Hammett σ	−0.27	0.0	0.12

VI. CONCLUSION

The above examples illustrate the physical-chemical approach to drug design. This approach promises to lead to the most potent analogue in fewer compounds than more traditional approaches. Whether this new analogue becomes a new drug is critically dependent on the relevance to the clinical situation of the biological properties used in its design. A close collaboration of chemist, biochemist, pharmacologist, clinician, and drug design person is required in order for the development of new drugs to be most efficient.

REFERENCES

Fuller, R. W., Marsh, M. M., and Mills, J. (1968): Inhibition of monoamine oxidase by N-(phenoxyethyl)cyclopropylamines. Correlation of inhibition with Hammett constants and partition coefficients. *Journal of Medicinal Chemistry*, 11:397-398.
Fuller, R. W., Warren, B. J., and Molloy, B. B. (1970): Selective inhibition of monoamine oxidase in rat brain mitochondria. *Biochemical Pharmacology*, 19:2934-2936.
Goldstein, A., Aronow, L., and Kalman, S. M. (1968): *Principles of Drug Action*, Harper & Row, New York, pp. 167-174.
Hansch, C. (1971): Quantitative structure-activity relationships in drug design. In: *Drug Design*, Vol. 1, edited by E. J. Ariëns. Academic Press, New York.
Hansch, C., and Clayton, J. M. (1973): Lipophilic character and biological activity of drugs II: The parabolic case. *Journal of Pharmaceutical Science*, 62:1-21.
Hansch, C., Leo, A., Unger, S. H., Kim, K. H., Nikaitani, D., and Lien, E. J. (1973b): "Aromatic" substituent constants for structure-activity correlations. *Journal of Medicinal Chemistry*, 16:1207-1216.
Hansch, C., Steward, A. R., and Iwasa, J. (1965): The correlation of localization rates of benzenboronic acids in brain and tumor tissue with substituent constants. *Molecular Pharmacology*, 1:87-92.
Hansch, C., Unger, S. H., and Forsythe, A. B. (1973a): Strategy in drug design. Cluster analysis as an aid in the selection of substituents. *Journal of Medicinal Chemistry*, 16:1217-1222.
Ho, B. T., Gardner, P. M., Pong, S. F., and Walker, K. E. (1973): A new peripheral monoamine oxidase inhibitor: 2,9-Dimethyl-β-carbolinium iodide. *Experientia*, 29:527-529.
Ho, B. T., McIsaac, W. M., Tansley, L. W., and Walker, K. E. (1969): Inhibition of monoamine oxidase III:9-Substituted β-carbolines. *Journal of Pharmaceutical Sciences*, 58:219-221.
Lien, E. J., Hussain, M., and Pong, G. L. (1970): Role of hydrophobic interactions in enzyme inhibition by drugs. *Journal of Pharmaceutical Science*, 59:865-868.
Martin, Y. C., Jarboe, C. H., Krause, R. A., Lynn, K. R., Dunnigan, D., and Holland, J. B. (1973): Potential anti-Parkinson drugs designed by receptor mapping. *Journal of Medicinal Chemistry*, 16:147-150.
Martin, Y. C., Holland, J. B., Jarboe, C. H., and Plotnikoff, N. (1974): Discriminant analysis of the relationship between physical properties and the inhibition of monoamine oxidase by aminotetralins and aminoindanes. *Journal of Medicinal Chemistry*, 17:409-413.
Meyer, H. (1899): Zur Theorie der Alkoholnarkose. Erste Mittheilung. Welche Eigenschaft der Anästhetica bedingt ihre narkotisch Wirkung? *Archives of Experimental Pathology and Pharmacology*, 42:109-118.
Overton, E. (1897): Osmotic properties of cells in their bearing on toxicology and pharmacy. *Zeitschrift für Physikalische Chemie*, 22:189-209.
Swett, L. R., Martin, W. B., Taylor, J. D., Everett, G. M., Wykes, A. A., and Gladish, Y. C. (1963): Structure-activity relations in the pargyline series. *Annals of the New York Academy of Sciences*, 107:891-898.
Taylor, J. D., Wykes, A. A., Gladish, Y. C., and Martin, W. B. (1960): New inhibitor of monoamine oxidase. *Nature*, 187:941-942.
Topliss, J. G. (1972): Utilization of operational schemes for analog synthesis in drug design. *Journal of Medicinal Chemistry*, 15:1006-1011.

Neuropsychopharmacology of Monoamines and
Their Regulatory Enzymes, edited by E. Usdin.
Raven Press, New York © 1974.

The Use of Selective Monoamine Oxidase Inhibitor Drugs to Modify Amine Metabolism in Brain

Norton H. Neff, Hsiu-Ying T. Yang, and Jose A. Fuentes

Laboratory of Preclinical Pharmacology, National Institute of Mental Health, Saint Elizabeths Hospital, Washington, D.C. 20032

I. INTRODUCTION

The existence of multiple forms of monoamine oxidase (MAO) in brain can be demonstrated by electrophoretic techniques (Youdim, 1973), with inhibitor drugs (Johnston, 1968), and with enzyme-specific substrates (Hall, Logan, and Parsons, 1969; Yang and Neff, 1973). We will present some of the characteristics of the enzymes that have been identified with drugs and specific substrates.

II. SUBSTRATES AND INHIBITORS OF MAO

Johnston (1968) found that the drug clorgyline could be used *in vitro* to identify two forms of MAO in rat brain. He called them enzyme A and B. We will refer to them as types A and B enzyme, since there is now evidence that they may not be single species (Yang and Neff, 1974). Type A enzyme was sensitive to clorgyline and it oxidatively deaminated tyramine and serotonin; in contrast, type B enzyme was insensitive to clorgyline and it oxidatively deaminated tyramine, but not serotonin (Johnston, 1968). Since Johnston's original observation, other specific drugs and substrates have been recognized. Table 1 lists some of these compounds. The drugs are listed as being specific or nonspecific. When one type of enzyme is inhibited at lower concentrations of drug than the other, the drug is designated as specific. When the same concentration of drug is required to inhibit both types of enzyme, the drug is designated as nonspecific. For example, deprenyl inhibits type B enzyme at a lower concentration than is required to inhibit type A enzyme. Substrates are listed as preferred or common substrates. Common substrates are metabolized by both types of enzyme, whereas preferred substrates are metabolized almost exclusively by one form of the enzyme. For example, tyramine is metabolized by both enzymes, whereas β-phenylethylamine is metabolized by type B enzyme.

There are several points to be noted in Table 1. Almost all of the drugs that have been used clinically apparently block both enzymes; they are

TABLE 1. *Some inhibitor drugs and substrates of types A and B MAO*

	MAO	
	Type A	Type B
Preferred substrates	Norepinephrine Serotonin Normetanephrine	Benzylamine β-Phenylethylamine
Specific inhibitor drugs	Clorgyline Lilly 51641 Harmaline	Deprenyl
Common substrates	Dopamine Tyramine Tryptamine	
Nonspecific inhibitor drugs	Pargyline[a] Isocarboxazid Phenelzine Iproniazid	Tranylcypromine Nialamide Pheniprazine

[a] May preferentially inhibit type B enzyme.
Modified from Johnston (1968); Hall, Logan, and Parsons (1969); Fuller (1972); Squires (1972); Yang, Goridis, and Neff (1972); and Yang and Neff (1973).

nonspecific drugs. The therapeutic effects associated with MAO-inhibitor drugs are usually ascribed to the interruption of the metabolism of the transmitter amines. Norepinephrine (Goridis and Neff, 1971) and serotonin (Johnston, 1968) are oxidatively deaminated by type A enzyme and not by type B enzyme. Dopamine is a substrate for both enzymes (Hall et al., 1969, Yang and Neff, 1974). Therefore a more rational therapeutic approach would be to administer a drug that specifically blocks type A enzyme rather than a drug that blocks all enzyme activity as has been done in the past.

III. PARTIAL SEPARATION OF THE MAO TYPES BY DENSITY GRADIENT CENTRIFUGATION

MAO is found on the mitochondrial outer membrane (Schnartman and Green, 1968). Mitochondria isolated from glial cells and neuronal cell bodies have different bouyant densities (Hamberger, Blomstrand, and Lehninger, 1972), and the activities of several enzymes vary in mitochondria of different bouyant densities (Blokhuis and Veldstra, 1970). Type A and B enzymes are also associated with different mitochondrial fractions (Yang and Neff, 1973). Two regions of activity were found on a continuous sucrose gradient when tyramine was the substrate (Fig. 1). Type B enzyme activity, as judged by the ratio of β-phenylethylamine (PEA) to serotonin deamination (Fig. 1), was enriched in the mitochondria from the more dense sucrose regions of the gradient. The differences of activity were not due to

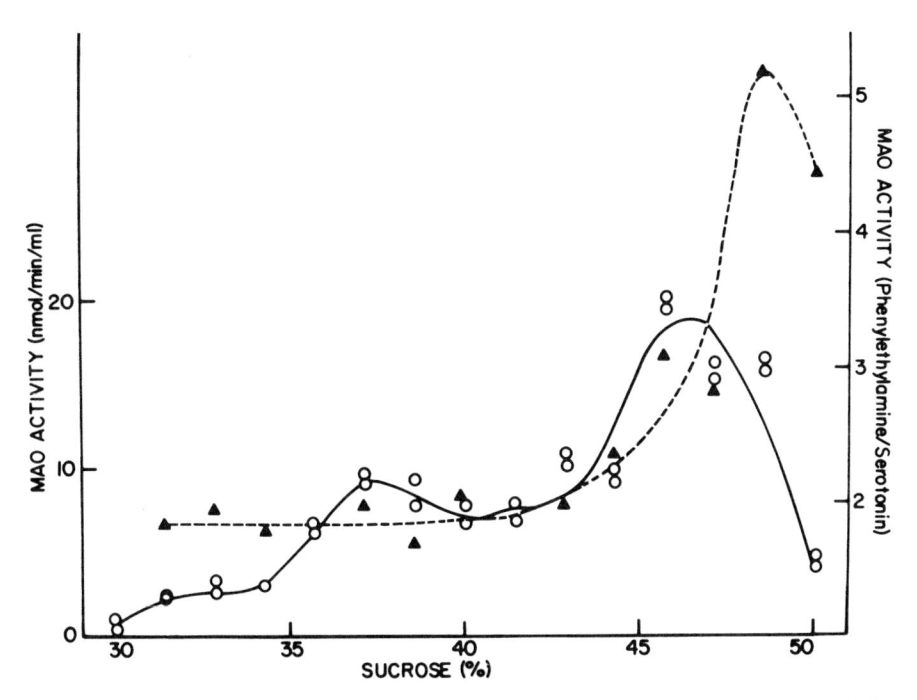

FIG. 1. Distribution of MAO activity in a continuous sucrose gradient. Open circles and the solid line represent activity using tyramine (2.1 mM) as substrate. Triangles and the broken line represent the ratio of activity of β-phenylethylamine (0.2 mM) to serotonin (0.2 mM).

the partial disruption of mitochondrial membranes nor to the presence of mitochondria in synaptosomes, because deliberately disrupting membranous structures by freezing and thawing had no effect on enzyme activity (Table 2). Apparently not all mitochondria are endowed with the same forms of MAO.

TABLE 2. *Effect of freezing and thawing on the substrate specificity of rat brain MAO*

Sample	Control (nmole/sample/30 min) Serotonin	PEA	Ratio	After freezing and thawing (nmole/sample/30 min) Serotonin	PEA	Ratio
Low-density mitochondria	3.5	1.8	0.51	3.2	1.7	0.53
High-density mitochondria	3.9	6.2	1.6	3.9	6.4	1.6

Data are presented as the mean for duplicate determinations on a single sample. Low-density mitochondria (from the 35 to 40% sucrose fraction) or high-density mitochondria (from the 45 to 50% sucrose fraction) were assayed for MAO activity using serotonin (0.2 mM) or PEA (0.2 mM) as substrate or frozen and thawed three times and then assayed for MAO activity.

IV. ADMINISTERING DRUGS THAT BLOCK SPECIFIC MAO ENZYMES

Clorgyline and deprenyl block MAO differentially when studied *in vitro* (Yang and Neff, 1973) or when injected into animals (Yang and Neff, 1974). Figure 2 shows the MAO activity of rat brain 2 hr after administering increasing doses of clorgyline or deprenyl. The deamination of sero-

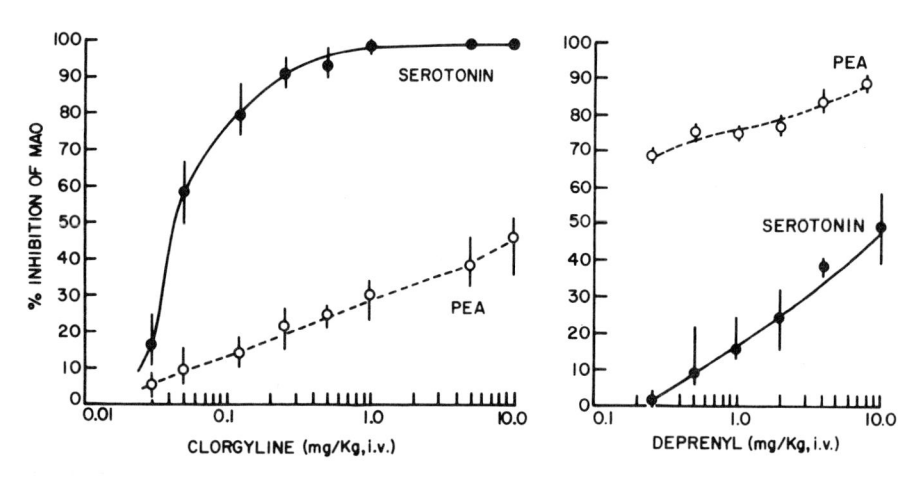

FIG. 2. MAO activity in rat brain 2 hr after the intravenous administration of increasing doses of clorgyline or deprenyl. Enzyme activity was measured using serotonin (0.2 mM) or β-phenylethylamine (0.2 mM) as substrate.

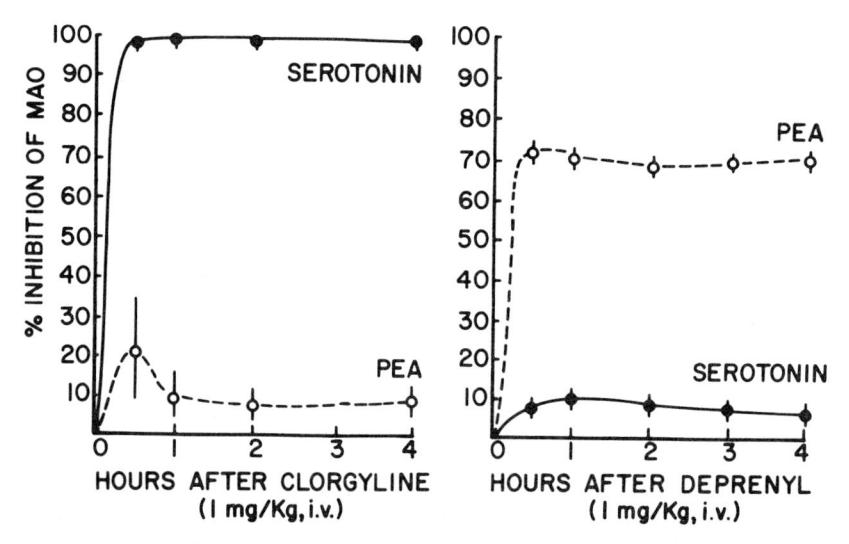

FIG. 3. MAO activity in rat brain during 4 hr after the administration of a single intravenous dose of clorgyline or deprenyl (1 mg/kg).

tonin (preferred substrate for type A enzyme) was inhibited more than the deamination of β-phenylethylamine (preferred substrate for type B enzyme) after administering clorgyline. The opposite was true after administering deprenyl. Moreover, the selectivity of the drugs decreased as the doses of the drugs increased, as would be predicted from *in vitro* studies (Yang and Neff, 1973).

When a single dose (1 mg/kg, i.v.) of clorgyline or deprenyl was administered, activity was inhibited within 30 min and did not return during the 4-hr period studied (Fig. 3). The prolonged inhibition of activity is consistent with the observation that these drugs act irreversibly (Yang and Neff, 1974).

V. AMINE METABOLISM IN BRAIN FOLLOWING THE ADMINISTRATION OF CLORGYLINE OR DEPRENYL

Serotonin and norepinephrine are deaminated by type A MAO, while β-phenylethylamine is deaminated by type B MAO when evaluated in mitochondrial preparations from rat brain. Dopamine is deaminated by both enzymes (Table 1). Therefore, by administering drugs that block one enzyme preferentially, we should selectively modify amine metabolism in brain. Following the administration of clorgyline, the transmitter amines—norepinephrine, serotonin, and dopamine—increased in brain (Fig. 4), but the metabolism of β-phenylethylamine remained essentially unaltered (Fig. 5) (Yang and Neff, 1974). Following the administration of deprenyl, only dopamine—and not norepinephrine or serotonin—increased in brain and

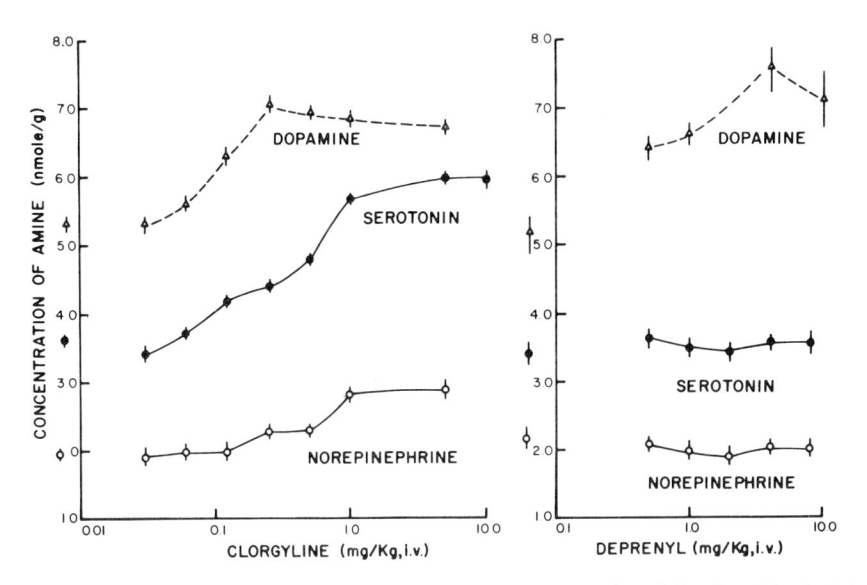

FIG. 4. The concentrations of dopamine, serotonin, and norepinephrine in rat brain 2 hr after the intravenous injection of increasing doses of clorgyline or deprenyl.

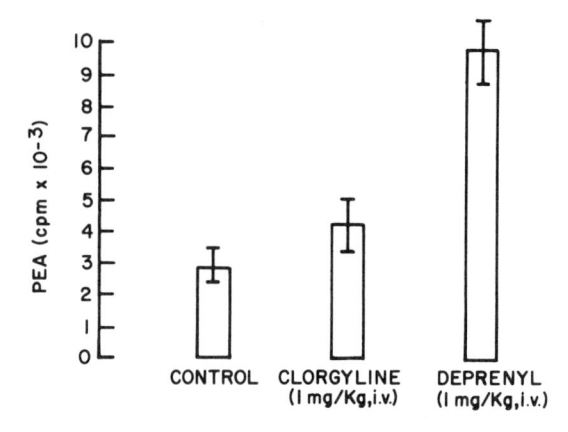

FIG. 5. Radioactive β-phenylethylamine remaining in brain 10 min after an intraventricular injection to clorgyline- or deprenyl-treated rats. β-phenylethylamine (15 nmole) was administered to rats that had been treated with the inhibitor drugs 2 hr earlier. Only deprenyl treatment significantly ($p < 0.01$) delayed the metabolism of the radioactive amine.

the metabolism of β-phenylethylamine was curtailed (Fig. 4). Dopamine apparently has access to both enzymes, because administering clorgyline in a dose that almost completely inhibited type A enzyme, together with deprenyl, induced an accumulation of dopamine that was greater than that following either drug alone (Table 3). Differential inhibition of amine me-

TABLE 3. *Dopamine concentrations in brain following the administration of clorgyline and deprenyl*

Group	Treatment	Dopamine nmole/g ± SEM (5)
A	Saline	5.1 ± 0.1
B	Deprenyl	5.9 ± 0.3
C	Clorgyline	6.5 ± 0.3
D	Deprenyl + clorgyline	7.8 ± 0.4

One mg/kg of deprenyl and/or clorgyline was administered intravenously and the rats were killed 2 hr later. Data were compared using a Student's *t* test. A vs. B, $p < 0.01$; B vs. C, NS; B vs. D, $p < 0.01$; C vs. D, $p < 0.01$.

tabolism was also evident from the concentrations of the acidic metabolites of serotonin and dopamine in brain. 5-Hydroxyindoleacetic acid decreased following clorgyline treatment, but not after deprenyl treatment. In contrast, 3,4-dihydroxyphenylacetic acid concentrations declined after deprenyl or clorgyline treatment (Table 4). These studies provide clear evidence that multiple forms of MAO exist *in vivo* and that they can be inhibited differentially with drugs (Yang and Neff, 1974).

TABLE 4. *5-Hydroxyindoleacetic acid and 3,4-dihydroxyphenylacetic acid concentrations in rat brain following the administration of clorgyline or deprenyl*

Treatment	5-Hydroxyindoleacetic acid nmole/g ± SE (4)	3,4-Dihydroxyphenylacetic acid nmole/g ± SE (4)
Saline	2.4 ± 0.2	4.9 ± 0.5
Deprenyl	2.4 ± 0.1	2.4 ± 0.3[a]
Clorgyline	0.93 ± 0.02[a]	1.3 ± 0.1[b]

Deprenyl or clorgyline was administered intravenously in a dose of 5 mg/kg and the rats were killed 2 hr later. 5-Hydroxyindoleacetic acid was measured in whole brain while 3,4-dihydroxyphenylacetic acid was assayed in striatum.
[a] $p < 0.01$, [b] $p < 0.001$ when compared with saline treatment

VI. REVERSAL OF THE RESERPINE SYNDROME BY SPECIFIC MAO-INHIBITOR DRUGS

Reserpine is capable of precipitating a state of mental depression in normal man (Achor, Hanson, and Gifford, 1955). Because it can, the reserpine syndrome is often used as a pharmacological model in animals for the naturally occurring disorders (Brodie, Spector, and Shore, 1959). MAO-inhibitor drugs prevent the pharmacological response to reserpine if they are administered before reserpine. Apparently, reserpine releases the transmitter amines from their storage sites, allowing MAO to degrade them, so they are not available for nerve transmission which results in central depression. After inhibiting MAO, the released transmitter amines are not metabolized; they supposedly overwhelm neuroreceptors, producing central excitation (Brodie et al., 1959).

With clorgyline and deprenyl, it is now possible to evaluate whether reversal of the reserpine syndrome is associated with amines that are metab-

TABLE 5. *Reversal of the reserpine syndrome by clorgyline and deprenyl*

Treatment[a]	Motor activity[b] 30 min	90 min	Ptosis[c] 30 min	90 min	Catatonia[d] 30 min	90 min
Reserpine	0	0	+	+	+	+
Clorgyline and reserpine	+	+	0	0	0	0
Deprenyl and reserpine	0	0	+	+	+	+

[a] Clorgyline or deprenyl was administered (1 mg/kg, i.v.) 1 hr before reserpine (2.5 mg/kg, i.v.) to groups of four rats each. Animals were evaluated 30 and 90 min following the injection of reserpine.
[b] Motor activity: 0, no activity; +, hyperactivity.
[c] Ptosis: 0, normal appearance; +, complete closure of the eyes.
[d] Catatonia: 0, normal appearance; +, marked catatonia.

olized by type A or B enzyme or by both enzymes. As a simple working hypothesis, we might assume that if clorgyline (inhibitor of type A MAO) and deprenyl (inhibitor of type B MAO) both reverse the reserpine syndrome, then amines that are common substrates are responsible for the reversal. If either clorgyline or deprenyl reverses the syndrome, then the amines involved must be preferred substrates for type A or B MAO. As shown in Table 5, only clorgyline was capable of preventing the reserpine syndrome in the doses tested. Thus, we might speculate that the amines that are preferred substrates for type A enzyme (norepinephrine and serotonin) might be responsible for the antidepressant activity of the MAO-inhibitor drugs used clinically. Our observation is consistent with many previous reports that norepinephrine and serotonin are probably responsible for the ability of MAO-inhibitor drugs to prevent the reserpine syndrome (Brodie et al., 1959). Apparently, endogenous dopamine plays only a minor role in this pharmacological test. Moreover, this study provides additional support for the premise that administering nonspecific drugs may increase the potential for side effects without producing additional therapeutic effects.

VII. CONCLUSIONS

Multiple forms of MAO are present in brain (Youdim, 1973). The nature of these enzymes is controversial (Houslay and Tipton, 1973); they may be separate proteins or they may be a single protein whose properties have been modified by attached phospholipid. These enzymes deaminate separate substrates as well as some of the same substrates. They can be partially separated by density gradient centrifugation. Moreover, they can be inhibited differentially with drugs. Because of these findings, it is now possible to modify amine metabolism in brain selectively. Although this chapter emphasizes studies in the rat, enzymes with similar properties have been identified in human brain *(unpublished observation)*.

The specific inhibitor drugs may be useful tools for studying the functional role of the amines of brain. For example, it appears that blocking type B MAO does not alter the pharmacology of reserpine, whereas blocking type A MAO reverses the effects of reserpine. Apparently, only the amines that are specific substrates for type A MAO, that is, norepinephrine and serotonin, but not dopamine, are responsible for reversing the reserpine syndrome.

The existence of multiple forms of MAO, the fact that they can be inhibited differentially, and the knowledge that these enzymes deaminate specific amines will probably alter the treatment of many disorders ascribed to abnormal amine metabolism. Perhaps, these observation will also renew interest by the pharmaceutical industry in developing more specific and less toxic MAO-inhibitor drugs.

REFERENCES

Achor, R. W., Hanson, N. O., and Gifford, R. W. (1955): Hypertension treated with *Rauwolfia serpentina* (whole root) and with reserpine; Controlled study disclosing occasional severe depression. *Journal of the American Medical Association,* 159:841-845.

Blokhuis, G. G. D., and Veldstra, H. (1970): Heterogeneity of mitochondria in rat brain. *FEBS Letters,* 11:197-199.

Brodie, B. B., Spector, S., and Shore, P. A. (1959): Interaction of drugs with norepinephrine in the brain. *Pharmacological Reviews* 11:548-564.

Fuller, R. W. (1972): Selective inhibition of monoamine oxidase. *Advances in Biochemical Psychopharmacology,* 5:339-354.

Goridis, C., and Neff, N. H. (1971): Monoamine oxidase in sympathetic nerves: A transmitter specific enzyme type. *British Journal of Pharmacology,* 43:814-818.

Hall, D. W. R., Logan, B. W., and Parsons, G. H. (1969): Further studies on the inhibition of monoamine oxidase by M & B 9302 (clorgyline)—I, Substrate specificity in various mammalian species. *Biochemical Pharmacology,* 18:1447-1454.

Hamberger, A., Blomstrand, C., and Lehninger, A. L. (1972): Comparative studies on mitochondria isolated from neuron-enriched and glia-enriched fractions of rabbit and beef brain. *Cellular Biology,* 45:221-234.

Houslay, M. D., and Tipton, K. F. (1973): The nature of the electrophoretically separable multiple forms of rat liver monoamine oxidase. *Biochemical Journal,* 135:173-186.

Johnston, J. P. (1968): Some observations upon a new inhibitor of monoamine oxidase in brain tissue. *Biochemical Pharmacology,* 17:1285-1297.

Schnartman, C., and Greenawalt, J. W. (1968): Enzymatic properties of the inner and outer membranes of rat liver mitochondria. *Journal of Cell Biology,* 38:158-175.

Squires, R. F. (1972): Multiple forms of monoamine oxidase of intact mitochondria as characterized by selective inhibitors and thermal sability: A comparison of eight mammalian species. *Advances in Biochemical Psychopharmacology,* 5:355-370.

Yang, H.-Y. T., Goridis, C., and Neff, N. H. (1972): Properties of monoamine oxidases in sympathetic nerve and pineal gland. *Journal of Neurochemistry,* 19:1241-1250.

Yang, H.-Y. T., and Neff, N. H. (1973): β-Phenylethylamine: A specific substrate for type B monoamine oxidase of brain. *Journal of Pharmacology and Experimental Therapeutics,* 187:365-371.

Yang, H.-Y. T., and Neff, N. H. (1974): The monoamine oxidases of brain: Selective inhibition with drugs and the consequences on the metabolism of the biogenic amines. *Journal of Pharmacology and Experimental Therapeutics (in press).*

Youdim, M. B. H. (1973): Multiple forms of mitochondrial monoamine oxidase. *British Medical Bulletin,* 29:120-122.

Neuropsychopharmacology of Monoamines and Their Regulatory Enzymes, edited by E. Usdin. Raven Press, New York © 1974.

Comparison of Monoamine Oxidase Substrate Activities in Twins, Schizophrenics, Depressives, and Controls

Alexander Nies,* Donald S. Robinson,** Lawrence S. Harris,** and Kathleen R. Lamborn**

*Veterans Administration Center, White River Junction, Vermont 05001 and Dartmouth Medical School, Hanover, New Hampshire 03755, and ** University of Vermont College of Medicine, Burlington, Vermont 05401*

I. INTRODUCTION

In the course of studying the application of an assay of monoamine oxidase (MAO) activity to the clinical problem of monitoring and predicting treatment with MAO inhibitors, we have also surveyed platelet MAO activity levels in several psychiatric and control populations. In 1971 we presented data based on the assay of platelet MAO with benzylamine in a series of depressed patients with a wide range of symptomatology compared to a control population selected for the absence of a personal or family history of psychiatric disorder (Nies, Robinson, Davis, and Ravaris, 1971). At that point our work was concerned mainly with the application of a relatively simple and rapid assay to the clinical situation, hence we used only benzylamine as substrate in the platelet MAO assay (Robinson, Lovenberg, Keiser, and Sjoerdsma, 1968).

It has subsequently been reported that bipolar depressed patients have reduced MAO activity in blood platelets (Murphy and Weiss, 1972), that schizophrenic patients have reduced platelet MAO activity (Murphy and Wyatt, 1972), and that a series of monozygotic twin pairs discordant for schizophrenia have reduced platelet MAO activity (Wyatt, Murphy, Belmaker, Cohen, Donnelly, and Pollin, 1973). These three studies employed tryptamine as substrate.

This chapter reports the results of a study of platelet MAO activity in monozygotic (MZ) and dizygotic (DZ) twin pairs compared to concurrent control pairs with both benzylamine and tryptamine as substrates. The purpose of this controlled study was to investigate genetic influences on MAO activity. In addition, we also compared MAO activities using both substrates in a carefully selected sample of nonhospitalized schizophrenic patients and in a series of depressed outpatients. Finally, the assay of MAO activity in human brain from three schizophrenic patients and eight controls is also reported.

II. METHODS

A. Subjects

Twin Study. Twin pairs were obtained by a call for volunteers in a letter sent to physicians and medical center employees. Age- and sex-matched control pairs were obtained in a similar manner. Subjects were chosen from among the respondents on the basis of good physical health, absence of psychiatric symptoms, and absence of drug treatment. Arrangements were made for the twin and matched control pair to appear at the laboratory at the same time so that blood specimens from both groups could be processed concurrently. Twin pairs were designated prospectively as MZ or DZ on the basis of established clinical criteria (Cederlof, Friberg, Jonsson, and Kaijg, 1961).

Schizophrenic Series. Schizophrenic subjects who met the following criteria were chosen: (1) individuals who had suffered at least two unequivocal episodes of a schizophrenic illness according to Schneiderian criteria (Taylor, 1972) resulting in hospitalization; (2) individuals who had been hospitalized for no longer than 4 months in any year; (3) individuals who had spent no more than a total of 12 months in hospital during the previous 5 years. When not hospitalized they were either employed, attending school, or in active rehabilitation programs, although they might still exhibit stigmata of schizophrenia such as minor and subtle incongruities of language, thought, and affect. Control subjects in this series were chosen from age- and sex-matched individuals without a history of psychiatric disorder. Blood samples were drawn on the same day as that of their matched schizophrenic subject, but in some instances in a different location. However, blood samples were processed concurrently on the same day.

Depressed Series. Two hundred and two patients with a clinical diagnosis of depression admitted to the inpatient service and outpatient psychopharmacology clinic of a university medical center from October 1968 through February 1971 had blood specimens drawn for MAO assay with benzylamine. A control group of 162 subjects for this series was collected in groups of from 8 to 16 at intervals during the course of these $2\frac{1}{2}$ years, and therefore are not strictly concurrent with the patient series. The controls were screened by a short interview and a screening of health records from their industrial health services. Subjects with a personal or family history of psychiatric disorder or who had at any time been treated with a major psychotropic drug were excluded.

A second series of depressed subjects were outpatients in a prospective double-blind study of the efficacy of an MAO inhibitor. They were referred from group and individual family practices as well as other primary-care subspecialty groups. They were chosen to be representative of a broad range of depressive symptoms which specifically included individuals with fatigue, hypochondriasis, and anxiety which would warrant a descriptive diagnosis of depressive neurosis or atypical depression. The pretreatment baseline platelet MAO activity was determined with both substrates.

Brain Series. Specimens of human hippocampus were obtained at postmortem by gross dissection following a standard protocol. The specimens were placed in vials, sealed in a nitrogen atmosphere, immediately frozen, and stored at −10°C for subsequent analysis.

Specimens were obtained from three subjects with a clinical diagnosis of schizophrenia who suffered acute medical deaths. One subject, aged 47, died suddenly following a bowel obstruction on the surgical service of the medical center; the other two, aged 65 and 83, died at the nearby State Hospital of sudden death associated with coronary artery disease. In these subjects the time from death to postmortem ranged from 2 to 18 hr, and the times from opening of the dura to the freezing of the specimens under nitrogen were $1\frac{1}{2}$, 3, and 5 hr. Identically dissected specimens were obtained at necropsy from eight accident victims who died suddenly from causes not involving head trauma or poisoning. In these subjects the times from death to postmortem ranged from 2 to 24 hr and the times from opening of the dura to freezing of the specimens were from 1 to 2 hr.

B. Laboratory

Blood samples were collected by venipuncture with nonwettable tubing into a 50-ml polypropylene centrifuge tube containing ethylenediaminetetraacetic acid. Suspensions of washed platelets were prepared as previously described and assayed in duplicate for MAO activity using ^{14}C-benzylamine and ^{14}C-tryptamine as substrates (Robinson et al., 1968). Brain tissue homogenates were prepared by homogenizing approximately 100 mg of tissue in 5 ml of 0.2 M phosphate buffer, pH 7.4. After freezing and thawing, an aliquot of this suspension was assayed in duplicate using both substrates by the method described above.

III. RESULTS

A. Twin Study

The analysis was based primarily on intrapair differences. Figure 1 shows the distributions of the intrapair differences in platelet MAO activity with benzylamine (upper half) and tryptamine (lower half) as substrates. It can be seen that the intrapair differences of the MZ pairs cluster more tightly near zero than those of the DZ and control pairs, although this seems to be somewhat less striking in the tryptamine data. Table 1 shows that for both substrates the sample means and variances of the intrapair differences follow the rank order MZ < DZ < controls. Two-sample Wilcoxon tests (be-

BENZYLAMINE SUBSTRATE

[Frequency histograms for MZ, DZ, and C groups on axis from 0 to 40]

TRYPTAMINE SUBSTRATE

[Frequency histograms for MZ, DZ, and C groups on axis from 0 to 4.0]

INTRAPAIR DIFFERENCE

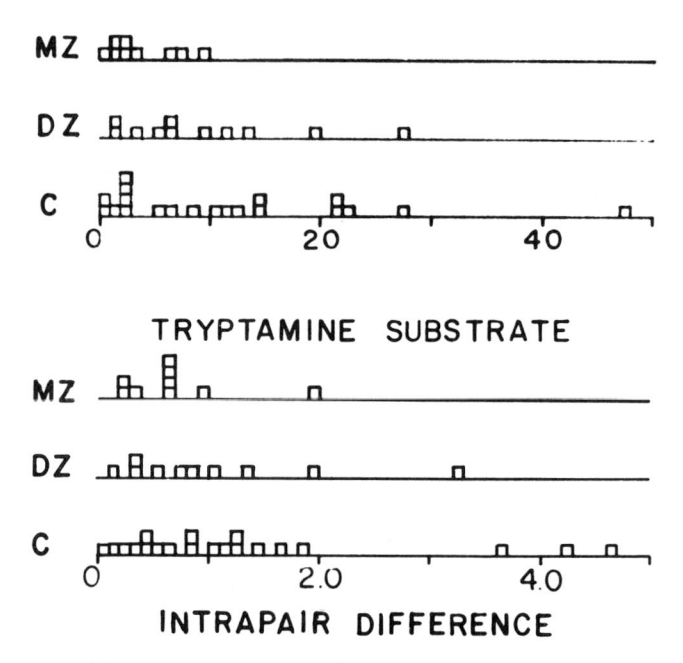

FIG. 1. Frequency histogram of intrapair differences in platelet MAO activity (nmoles/mg protein/hr) with benzylamine and tryptamine for nine pairs of monozyous (MZ) twins, 11 pairs of dizygous (DZ) twins, and 20 control (C) pairs.

TABLE 1. *Intrapair difference data in platelet MAO activity*

	Benzylamine		Tryptamine	
	Mean	Variance	Mean	Variance
MZ twins	3.9	10.8	0.69	0.27
DZ twins	9.5	64.3	1.11	1.14
Controls	11.5	138.1	1.32	1.70

cause of the non-normality of the data and the inequality of the variances) were performed to determine if the centers of the distribution (means) of intrapair differences with both substrates were identical. In addition, the Box-Anderson test for equality of the variance (standard tests such as the F-test are inappropriate for non-normal data) was applied in order to quantify the impression that the population variability increases from MZ to DZ to control groups. Using these two measures, it can be seen from Table

TABLE 2. *Statistical tests of intrapair difference for platelet MAO activity*

	Wilcoxon two-sample test	Box-Anderson test
Benzylamine		
MZ vs. controls	$p = 0.08$	$0.05 < p < 0.10$
DZ vs. controls	$p < 0.40$	$0.10 < p < 0.25$
MZ vs. DZ	$p = 0.02$	—
Tryptamine		
MZ vs. controls	$p = 0.15$	$p = 0.10$
DZ vs. controls	$p > 0.20$	$0.10 < p < 0.25$
MZ vs. DZ	$p > 0.20$	—

2 that for both substrates only the MZ twin pairs are more alike than the DZ and control pairs.

Two statistics, the heritability (h^2) (Osborne and DeGeorge, 1959) and the intraclass correlation coefficient (Neel and Schull, 1954), were calculated (Table 3). It can be seen that h^2 (the proportion of the DZ variance due to genetic variation) is large with both benzylamine ($h^2 = 0.83$) and tryptamine ($h^2 = 0.65$), and that the intraclass correlation with either substrate occurs in the order MZ > DZ > controls.

TABLE 3. *Intraclass correlation and heritability of platelet MAO activity*

	Intraclass correlation coefficient		Heritability (h^2)	
	Benzylamine	Tryptamine	Benzylamine	Tryptamine
MZ twins	0.76	0.47	—	—
DZ twins	0.39	0.31	0.83	0.65
Controls	−0.16	0.05	—	—

The distribution of enzyme activity was plotted because this can provide evidence for the existence of an underlying genetic heterogeneity. Frequency histograms for both substrates are shown in Fig. 2. The distribution of MAO activity with benzylamine in this data is very possibly nonunimodal, whereas that with tryptamine, while perhaps nonunimodal, is less clear cut.

B. Schizophrenic Patients

Platelet MAO with both substrates in the schizophrenic subjects is shown in Table 4. With benzylamine there is no statistically significant difference in MAO activity between the two groups, whereas with tryptamine platelet MAO activity is significantly lower in the schizophrenic group.

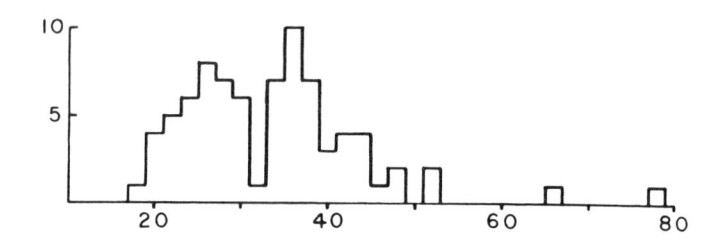

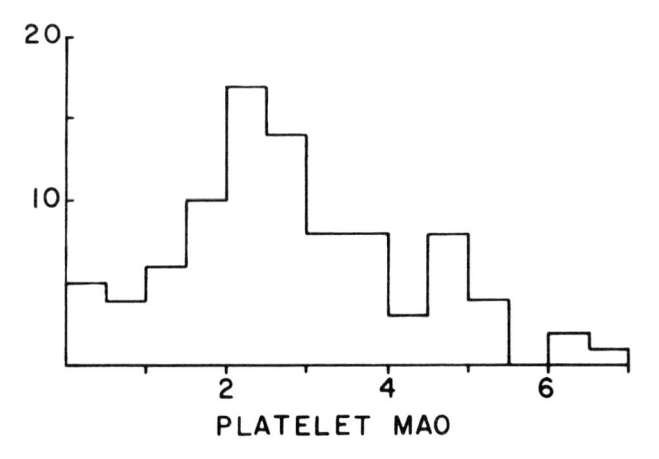

FIG. 2. Frequency histogram of platelet MAO activity with benzylamine (upper half) and tryptamine (lower half) for 80 subjects (40 twin and 40 control pairs).

TABLE 4. Mean platelet MAO activity[a] of schizophrenics and controls

	Benzylamine	Tryptamine
Schizophrenics (12)	19.1 ± 1.9	2.6 ± 0.2
Controls (12)	22.1 ± 2.8	3.6 ± 0.4
	$p > 0.25$	$p < 0.01$

[a] nmoles/mg protein/hr.

C. Depressed Patients

The mean platelet MAO activity with benzylamine by age decade for patients with depressive illnesses and for normal controls is shown in Fig. 3. Twenty-six of the 202 depressed patients could be classified as having unequivocal bipolar (manic-depressive) disorder. When age and sex adjustments are made, these bipolar individuals do not differ from the total sample of depressives with regard to platelet MAO activity with benzylamine.

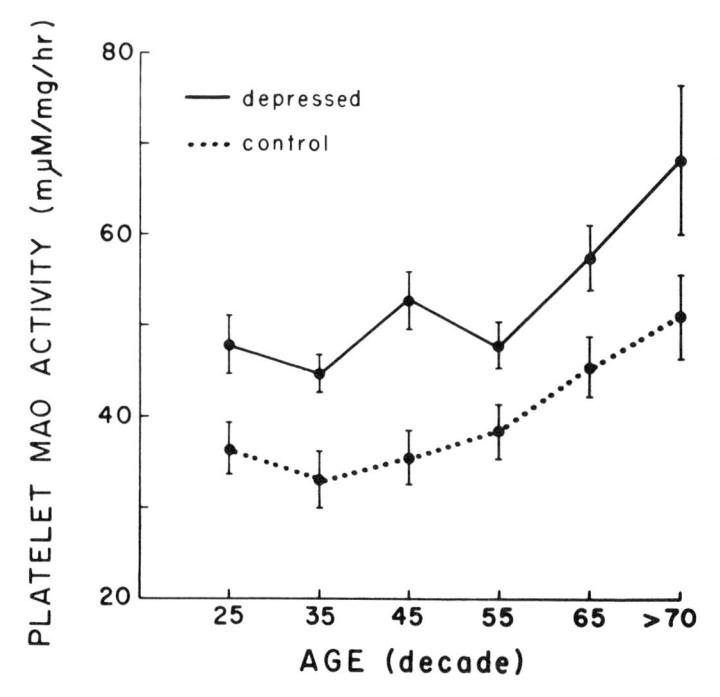

FIG. 3. Mean platelet MAO activity (nmoles/mg protein/hr) ± standard error for each age decade in 202 depressed patients (solid line) and 162 controls (dotted line).

D. Brain MAO Activity

The MAO activities in hippocampus of the eight control and the three schizophrenic subjects are given in Table 5. With both substrates the schizophrenic subjects have hippocampal MAO activity as great or greater than any of the controls.

TABLE 5. *Hippocampal MAO activity in schizophrenics and controls*

	MAO activity[a]									
Tryptamine										
Controls	6	12	12	14	14	18	19	20		
Schizophrenics								20	22	28
Benzylamine										
Controls	22	31	40	41	46	48	53	56		
Schizophrenics									64	82	116

[a] nmoles/mg protein/hr.

E. Comparison of Substrate Activities

An attempt was made to develop information bearing on the problem of the possible existence of multiple forms of MAO by examining our clinical assay data for correlations of MAO activity with benzylamine and tryptamine in groups of individuals in which both determinations had been done simultaneously. This was done by means of Kendall's tau (τ) because this statistic does not assume a bivariate normal population and is little influenced by isolated extreme values. The results in Table 6 show that the two subject groups show a weak positive and three correlation groups show no correlation.

TABLE 6. *Comparison of simultaneous determinations of MAO activity with benzylamine and tryptamine*

Group	N	τ^a
MZ twins	18	−0.059
DZ twins	22	0.029
All controls	40	0.401
Schizophrenics	12	−0.045
Depressed outpatients	56	0.440

[a] Kendall's tau.

IV. DISCUSSION

Significant age and sex effects on human platelet MAO activity (Robinson, Davis, Nies, Ravaris, and Sylwester, 1971) make family and sibship studies difficult because of the necessity for age and sex adjustments. We therefore chose the twin method to make an assessment of the genetic control of platelet MAO activity. The results of the present investigation are that, with respect to platelet MAO activities with both substrates, MZ twins are closer to each other than are DZ twins or control pairs (Fig. 1). As shown in Table 1, the means of the intrapair differences and the variances of the intrapair differences with both substrates follow the order MZ < DZ < controls.

The statistical tests of the intrapair difference data (Table 2) suggest that with benzylamine platelet MAO activity of MZ twins differs from both DZ twins and controls, or, put another way, that DZ twins (sibs) do not differ from randomly assigned control pairs from the general population. The results with tryptamine are similar except that in this instance the DZ intrapair difference does not differ significantly from that of the MZ group. Examination of the tests of the variances indicates that there is greater variability in the DZ data (with both substrates) and greater variability in the tryptamine than in the benzylamine data. The rank order of the means and variances of the intrapair differences (Table 1), with the DZ group being intermediate, suggests that in large-scale studies first-degree relatives would resemble each other more closely in MAO activities with both substrates than random pairs from the general population.

An overall distribution curve of enzyme activity can reveal nonunimodality indicative of genetic heterogeneity (Chen, Giblett, Anderson, and Fossum, 1972). The distribution of platelet MAO activities shown in Fig. 2 is compatible with the possible existence of heterogeneous forms of this enzyme. Previous work based on electrophoresis of solubilized enzyme indicates that human platelet MAO occurs as a single isoenzyme (Collins and Sandler, 1971). However, although it has been suggested that the human platelet contains primarily MAO_B (Squires, 1972), it has also been proposed that MAO_B is a class or group of enzymes (Neff and Goridis, 1972). Different results could occur depending on whether soluble or particulate forms of an enzyme are studied.

The qualitative statistics in Table 3 support the conclusion that genetic variation is responsible for the major portion of the phenotypic variation of human platelet MAO activity. This result is consistent with the knowledge that enzymes, like other proteins, have their structures and hence their functional specificities determined by chromosomal DNA, and is therefore not surprising. However, the possible occurrence of detectable phenotypic differences in MAO activity in the human could have implications for the study of neuropsychiatric conditions in which biogenic amine metabolism may be disordered (Davis, 1970).

Our results, which show that platelet MAO activity in schizophrenic patients is reduced with tryptamine as substrate when compared to normals, are consistent with those of Murphy and Wyatt (1972) and Wyatt et al. (1973) and include the additional finding that schizophrenics do not have reduced platelet MAO activity when benzylamine is used as the substrate. Our finding that bipolar depressed patients do not have reduced platelet MAO activity compared to other depressed patients when benzylamine is used as substrate is of interest in the light of the report by Murphy and Weiss (1972) that bipolar patients show reduced platelet MAO activity with tryptamine. Unfortunately, we have no data on bipolar patients with tryptamine as substrate. Discrepancies in results, depending on substrate, make it possible that either multiple forms and/or activities of MAO may have clinical significance.

The increased platelet MAO activity with benzylamine observed in the depressed patients must be subjected to methodologic and statistical reservations. First, the control series was selected for absence of a personal or family history of psychiatric disorder and is therefore not representative of the general population, and although control subjects were obtained throughout the same span of time as patients, the two series are not strictly concurrent. The result could therefore be an artifact, although we are unaware of any systematic differences in processing blood samples between the two groups. Second, as indicated by the standard errors for the mean of each age decade in Fig. 3, there is considerable overlap and the increased MAO activity is a trend perhaps produced by a subpopulation of depressed patients.

The limited results with brain show a contrast between hippocampal and platelet and controls. They suggest that in schizophrenics the substrate differences which occur in platelets very possibly do not hold for brain. The higher brain MAO activity in schizophrenics than in the normal subjects must be interpreted with caution because of the age discrepancies between the schizophrenics (range, 47 to 83 years) and the controls (range, 19 to 72 years), since it has been shown that MAO activity increases with age in the human (Robinson et al., 1971). A further qualification is that all assays of brain MAO were carried out on homogenates of sizeable brain volumes (100 mg), so that the relative contributions of intraneuronal and glial and other extraneuronal MAO cannot be taken into account.

The correlation data (Table 6) are an attempt at crude examination of the data for evidence of heterogeneity of MAO activity. In the two instances in which there are positive correlations, these are not as strong as one might expect if there were a direct linear relationship between the activity of the same enzyme on two substrates. A low correlation of MAO activity with the two substrates could be explained by the occurrence of mixtures of different forms of MAO activity. Intact mitochondria exhibit MAO activity with characteristics that have been divided into fractions A and B (Johnston, 1968). Since benzylamine is almost exclusively deaminated by fraction B and tryptamine by both fractions A and B (Squires, 1972), differences in proportions of these fractions could account for a lack of correlation. The possibility that each of these fractions of mitochondrial MAO represent groups of enzymes (Neff and Goritis, 1972) provides an additional explanation.

Discussion of the results of the determination of MAO activity in clinical investigations must be placed within the context of the uncertainty and controversy concerning the existence of multiple forms of MAO. Collins (1972) has pointed out that the "harsh processes" used in purifying and solubilizing MAO could produce artifacts, and Youdim (1973) has pointed out that such "drastic" methods may produce results not representative of conditions *in vivo*. There is excellent evidence that mitochondrial MAO activity is located in the outer membrane of the mitochondrion (Greenawalt, 1972). It is possible that an assay which relies on the freezing and thawing of tissue to disrupt membranes yields mixtures of whole mitochondria and fragments of mitochondrial membrane. Assay of MAO activity in such preparations could be more representative of conditions *in vivo* than methods which rely on extraction, solubilization, and purification. However, even if this were true, depending on the uniformity of such procedures, variability could result from differences in the distribution of membrane fragments. This in turn might also lead to apparent differences introduced by the procedure.

In view of results which thus far indicate that measures of MAO outside the nervous system do not necessarily parallel the findings in CNS, one might question the use of peripheral measures (such as the platelet). However, the suggestion that the local molecular environment of MAO enzyme protein could induce differences in substrate activity (Youdim, 1973) raises the possibility that although location (brain, platelet) modifies enzyme function, underlying genetic differences detectable in peripheral sites might still be appropriate genetic or biological markers for human disease.

V. SUMMARY

Human platelet activity with benzylamine and tryptamine as substrates seems to be under genetic control. Heterogeneity of the distributions and differences in substrate activities suggest the possibility of genetic heterogeneity. Our results, together with those of Murphy and Weiss (1972), Murphy and Wyatt (1972), and Wyatt et al. (1973), suggest that differences in platelet MAO activity may occur in schizophrenics and some depressed patients. Preliminary results comparing brain MAO activity in three schizophrenic subjects and eight psychiatrically normal controls are in contrast with studies of platelet MAO activity and indicate that studies of enzyme activity in neuropsychiatric populations based on peripheral tissues require cautious interpretation.

ACKNOWLEDGMENTS

The authors thank Dr. Irwin Kay for brain specimens from one schizophrenic patient, Dr. Roy Butler for specimens from two patients, and John Mullen for performing dissections of brain specimens. Mrs. Janet Lord provided secretarial assistance, and Mrs. Lorrainne Korson provided indispensable research assistance. This work was supported by grants from the Pharmaceutical Manufacturers Association, the Warner-Lambert Company, and grants PHS 5501 RR-05429-11 and PHS T02 05935-22 from the U.S. Public Health Service.

REFERENCES

Cederlof, R., Friberg, L., Jonsson, E., and Kaij, L. (1961): Studies on similarity diagnosis in twins with the aid of mailed questionnaires. *Acta Genetica,* 11:338–362.
Chen, S. H., Giblett, E. R., Anderson, J. E., and Fossum, B. L. G. (1972): Genetics of glutamic-pyruvic transaminase: Its inheritance, common and rare variants, population distribution, and differences in catalytic activity. *Annals of Human Genetics,* 35:401–409.
Collins, G. G. S. (1972): Summary of section I. In: *Monoamine Oxidases—New Vistas,* edited by E. Costa and M. Sandler. Raven Press, New York.

Collins, G. G. S., and Sandler, M. (1971): Human blood platelet monoamine oxidase. *Biochemical Pharmacology,* 20:289–296.

Davis, J. M. (1970): Theories of biologic etiology of affective disorders. *International Review of Neurobiology,* 12:145–175.

Greenawalt, J. W. (1972): Localization of monoamine oxidase in rat liver mitochondria. In: *Monoamine Oxidases—New Vistas,* edited by E. Costa and M. Sandler. Raven Press, New York.

Murphy, D. L., and Weiss, R. (1972): Reduced monoamine oxidase activity in blood platelets from bipolar depressed patients. *American Journal of Psychiatry,* 128:1351–1357.

Murphy, D. L., and Wyatt, R. J. (1972): Reduced monoamine oxidase activity in blood platelets from schizophrenic patients. *Nature,* 238:225–226.

Neel, J. V., and Schull, W. J. (1954): *Human Heredity.* University of Chicago Press, Chicago.

Neff, N. H., and Goridis (1972): Neuronal monoamine oxidase: Specific enzyme types and their rates of formation. In: *Monoamine Oxidases—New Vistas,* edited by E. Costa and M. Sandler. Raven Press, New York.

Nies, A., Robinson, D. S., Davis, J. M., and Ravaris, C. L. (1971): Amines and monoamine oxidase in relation to aging and depression in man. *Psychosomatic Medicine,* 33:470 (abstract).

Osborne, R. H., and DeGeorge, F. V. (1959): *Genetic Basis of Morphological Variation.* Harvard University Press, Cambridge, Mass.

Robinson, D. S., Davis, J. M., Nies, A., Ravaris, C. L., and Sylwester, D. (1971): Relation of sex and aging to monoamine oxidase activity of human brain, plasma and platelets. *Archives of General Psychiatry,* 24:536–539.

Robinson, D. S., Lovenberg, W., Keiser, H., and Sjoerdsma, A. (1968): The effects of drugs on human blood platelet and plasmas amine oxidase activity *in vitro* and *in vivo. Biochemical Pharmacology,* 17:109–119.

Squires, R. F. (1972): Multiple forms of monoamine oxidase in intact mitochondria as characterized by selective inhibitors and thermal stability: A comparison of eight mammalian species. In: *Monoamine Oxidases—New Vistas,* edited by E. Costa and M. Sandler. Raven Press, New York.

Taylor, M. W. (1972): Schneiderian first-rank symptoms and clinical prognostic features in schizophrenia. *Archives of General Psychiatry,* 26:64–67.

Wyatt, R. J., Murphy, D. L., Belmaker, R., Cohen, S., Donnelly, C. H., and Pollin, W. (1973): Reduced monoamine oxidase activity in platelets: A possible genetic marker for vulnerability to schizophrenia. *Science,* 179:916–918.

Youdim, M. B. H. (1973): Multiple forms of mitochondrial monoamine oxidase. *British Medical Bulletin,* 29:120–122.

Neuropsychopharmacology of Monoamines and Their Regulatory Enzymes, edited by E. Usdin. Raven Press, New York © 1974.

Monoamine Oxidase in Man: Enzyme Characteristics in Platelets, Plasma, and Other Human Tissues

Dennis L. Murphy and Cynthia H. Donnelly

Section on Clinical Neuropharmacology, NIMH Laboratory of Clinical Science, Bethesda, Maryland 20014 and the Department of Chemistry, University of Maryland, College Park, Maryland 20742

I. INTRODUCTION

Monoamine oxidase (MAO) was among the first of the monoamine-related enzymes discovered (Hare, 1928). Its major roles in deactivating putative neurotransmitter amines, in degrading other amines and amine metabolites, in regulating the concentration of free cellular amines, and in functioning via a feedback mechanism in regulating amine synthesis have been the subject of several recent reviews (Kopin, 1964; Costa and Greengard, 1972; Weiner and Bjur, 1972).

In man, most of the original evidence on the functional roles of MAO came as by-products of studies of the MAO-inhibiting drugs which are used as antidepressant and antihypertensive agents. More recently, direct studies of the enzyme in various tissues have suggested that there exist individual differences in the activity of the human enzyme which are primarily under genetic control (Murphy, 1973; Nies, Robinson, Lamborn, and Lampert, 1973; Wyatt, Murphy, Belmaker, Cohen, Donnelly, and Pollin, 1973; Murphy, Donnelly, and Buchsbaum, 1974*d*). Other studies have indicated that the enzyme in various human tissues is affected by drugs and some hormones (Levine, Oates, Vendsalu, and Sjoerdsma, 1962; Southgate, Grant, Pollard, Pryse-Davies, and Sandler, 1968; Tryding, Nilsson, Tufvesson, Berg, Carlstrom, Elmfors, and Nilsson, 1969; Belmaker, Murphy, Wyatt, and Loriaux, 1974), and that there exist age- and sex-related differences in the activity of the enzyme which may be hormone-mediated (Tryding et al., 1969; Robinson, Davis, Nies, Ravaris, and Sylwester, 1971; Robinson, Davis, Nies, Colburn, Davis, Bourne, Bunney, Shaw, and Coppen, 1972; Belmaker et al., 1974; Murphy, Donnelly, Belmaker, Carlson, Baker, and Wyatt, 1974*c*). Alterations in human MAO activity may also occur during some medical illnesses (Levine et al., 1962; McEwen and Harrison, 1965; McEwen and Castell, 1967; Tryding et al., 1969) and in association with some psychiatric disorders, including depression (Nies, Robinson, Ravaris, and Davis, 1971; Klaiber, Broverman, Vogel, Kobayashi, and Moriatry, 1972; Murphy and Weiss, 1972) and schizophrenia (Murphy and Wyatt, 1972; Wyatt et al., 1973; Murphy, Belmaker, and Wyatt, 1974a). Of additional neuropsychopharmacologic interest are the reports of significant correlations of MAO activity with electroencephalographic patterns elicited by visual stimuli (Vogel, Broverman, and Klaiber, 1971; Buchsbaum, Landau, Murphy, and Goodwin, 1974) and with personality characteristics derived from psychological tests including the MMPI and the Zuckerman sensation-seeking scale (Murphy, Belmaker, Wyatt, and Buchsbaum, 1974b).

Current questions concerning the apparent multiple molecular forms of MAO and their distribution in different tissues, with different cellular and subcellular locations (Sandler and Youdim, 1972; Squires, 1972), are of special pertinence to the studies of this enzyme in man. The majority of the investigations of MAO activity in man have utilized measurements of platelet or plasma MAO, with fewer studies of such other, more difficult-to-obtain tissues as brain samples collected at autopsy, liver and intestinal biopsy samples, and uterine cells collected by curetage. Although platelets have a number of characteristics in common with nerve-ending preparations from brain, including similarities in amine transport mechanisms, amine storage vesicle functions, receptor responses, and the possession of other enzymes involved in amine degradation besides MAO (Murphy, 1973), the relationship of the structure and function of the platelet MAO to that of MAOs in other tissues such as brain has been only partially elucidated (Robinson, Lovenberg, Keiser, and Sjoerdma, 1968; Collins and Sandler, 1973; Murphy et al., 1974c). While the soluble MAO in plasma is more easily studied than the platelet enzyme in man, it appears to differ in cofactor requirement and in substrate-inhibitor responses from the mitochondrial MAOs found in all other tissues, including the platelet (see below).

This chapter will review recent data and studies bearing on the questions of differences in substrate specificity, differences in susceptibility to inhibitors, and the direct evidence for the existence of multiple forms of MAO in man. The major focus will be on studies utilizing multiple substrates and inhibitors and on studies allowing comparisons of MAO activities in different human tissues, with special emphasis on the relationship of the platelet and plasma enzymes to each other and to the MAOs in other tissues. Although only limited information is available in comparison to the voluminous information on other species, this consideration will be limited to human tissues to permit a more comprehensive coverage of the studies of MAO in man.

II. HUMAN PLATELET MAO

Blood platelets provide the most readily accessible source of a mitochondrial MAO in man. The availability of these cells has permitted studies in outpatients and normal controls for the estimation of genetic contributions (Table 1) (Murphy, 1973; Nies et al., 1973; Wyatt et al., 1973; Murphy et al., 1974a,d) and of other factors such as age and sex (Tryding et al., 1969; Robinson et al., 1971, 1972; Belmaker et al., 1974; Murphy et al., 1974c) which influence the activity of MAO in different populations. Serial sam-

TABLE 1. *Genetic influences on platelet MAO activity*

	Intraclass correlation coefficients	
	Murphy et al. (1973, 1974d)	Nies et al. (1973)
Normals		
Monozygotic twins	0.88	0.76
Dizygotic twins	0.45	0.39
Sib pairs	0.28	—
Random pairs matched for age and sex	0.12	−0.16
Monozygotic twins discordant for schizophrenia (Wyatt et al., 1973)		0.65
Monozygotic twins concordant for bipolar manic-depressive illness (Murphy, 1973)		0.83

pling has also permitted studies of the *in vivo* effects of antidepressant and antihypertensive drugs on the enzyme, including studies of the time course of MAO inhibition and of the relationships between the magnitude of MAO inhibition and drug dose as well as clinical efficacy (Robinson et al., 1968; Zeller, Babu, Cavanaugh, and Stanich, 1969; Murphy et al., 1974c). The effects on *in vivo* MAO activity of other drugs, such as imipramine, methylphenidate, lithium carbonate, furizolidone, and chlorpromazine and other phenothiazines, have also been evaluated in man (Robinson et al., 1968; Murphy and Wyatt, 1972; Murphy et al., 1974c).

A. General Characteristics of Human Platelet MAO

Platelet MAO, like other tissue MAOs, is located in mitochondria (Paasonen and Solatunturi, 1965). Collins and Sandler (1971) have solubilized the enzyme, purifying it approximately 12-fold; its estimated molecular weight is 235,000, a value at the lower end of the range for other mitochondrial MAOs. Polyacrylamide gel electrophoresis has revealed only a single tetrazolium-stained band which migrates toward the anode, suggesting that the human platelet, unlike human brain and liver, possesses only one molecular form of MAO (Collins and Sandler, 1971).

B. Substrate-Related Characteristics of Platelet MAO

Tryptamine, tyramine, benzylamine, dopamine, and serotonin are among the substrates that have been studied with platelet MAO. Taking into account differences in preparation procedures, including partial purification measures, and differences in assay techniques, including pH and buffer differences, generally similar patterns of substrate specificity have been demonstrated (Robinson et al., 1968; Collins and Sandler, 1971). The nonphysiologic amine benzylamine is most actively deaminated, although its affinity for the enzyme (K_m) is 10 times less than that of most other substrates. The amine present in highest concentration in the platelet, serotonin, is least readily metabolized by the enzyme. Other physiologically occurring substrates such as dopamine, tyramine, and tryptamine are intermediate in affinity for the enzyme.

The possibility of heterogenous forms of the platelet MAO, with different responses to different substrates, was suggested by one report of the lack of a significant correlation between MAO activity measured by tryptamine deamination in comparison to that measured by benzylamine deamination in a small group of normal individuals (Robinson, Nies, and Lamborn, 1973). However, a more recent study (Donnelly and Murphy, 1974) of 75 normal controls indicated a very high correlation ($r = 0.89$, $p < 0.001$) between platelet MAO activities measured with these two substrates. In contrast, platelet MAO activities measured with either benzylamine or tryptamine as substrate did not significantly correlate with plasma MAO activities in the same blood sample from the same individuals (Table 2). In this study (Donnelly and Murphy, 1974), plasma MAO was determined with benzylamine as the substrate.

TABLE 2. *Correlations between human platelet and human plasma MAO activities with different substrates*

| | Correlation coefficients ||
	Platelet (tryptamine)	Platelet (benzylamine)
Plasma (benzylamine)	0.29 (N = 62)	0.16 (N = 62)
Platelet (benzylamine)	0.89[a] (N = 75)	—

[a] $p < 0.001$.

The substrate affinity pattern exemplified by platelet preparations is similar to that of the so-called B form of MAO identified in other tissues by substrate and inhibition characteristics (Neff and Goridis, 1972; Donnelly and Murphy, 1974). The platelet MAO's low affinity for serotonin and its apparent inability to deaminate norepinephrine, together with its relatively high affinity for phenylethylamine and for benzylamine (Donnelly and Murphy, 1974; Robinson et al., 1968), are characteristic of the B enzyme forms found, for example, after sympathetic denervation of rat or rabbit vas deferens (Knoll and Magyar, 1972). The contrasting A form of MAO demonstrated in other species has a higher affinity for serotonin and norepinephrine and has been suggested to represent an intraneuronal form of the enzyme (Neff and Goridis, 1972). In the superior cervical ganglion of the rat, for example, 90% of the MAO activity exhibits A form characteristics (Neff and Goridis, 1972).

C. Inhibitor-Related Characteristics of Human Platelet MAO

Most studies have reported that the platelet enzyme is most sensitive to inhibition by pargyline, moderately sensitive to tranylcypromine and pheniprazine, and relatively less sensitive to other inhibiting drugs such as iproniazid and harmaline (Robinson et al., 1968; Collins and Sandler, 1971; Donnelly and Murphy, 1974). While I_{50} and K_i values vary from study to study, this same rank order of inhibitor efficacy has been found in studies of whole brain and hypothalamus as well as in the platelet (Robinson et al., 1968; Collins et al., 1970; Collins and Sandler, 1971; Donnelly and Murphy, 1974). As with other mitochondrial MAOs, the platelet enzyme is resistant to inhibition by isoniazid, semicarbazide, and KCN (Robinson et al., 1968; Collins and Sandler, 1971; Donnelly and Murphy, 1974).

Since A and B forms of MAO in other tissues can be differentiated by their responses to two congeners of pargyline, deprenyl (± phenylisopropyl methyl propinylamine), which is a selective inhibitor of the B form, and clorgyline [N-methyl-N-propargyl-3-(2,4-dichlorphenoxy) propylamine], which is more effective in inhibiting the A form (Knoll and Magyar, 1972; Neff and Goridis, 1972; Squires, 1972), these two agents were recently studied in human platelet preparations. As indicated in Table 3, deprenyl was 200 to 500 times more effective than clorgyline in achieving 50% inhibition of the platelet enzyme with tryptamine, tyramine, and benzylamine as substrates (Donnelly and Murphy, 1974). Dose-response plots indicated simple sigmoidal responses to both inhibitors (Fig. 1); this type of curve contrasts with the more complex activity-inhibitor relationship found in other tissues possessing mixtures of both A and B MAO forms. This inhibi-

TABLE 3. *Inhibition of human platelet MAO by clorgyline and deprenyl*

Substrate	ID 50 (M) Clorgyline	Deprenyl
Tryptamine	5×10^{-6}	1×10^{-8}
Tyramine	2×10^{-6}	1×10^{-8}
Benzylamine	5×10^{-6}	1×10^{-8}

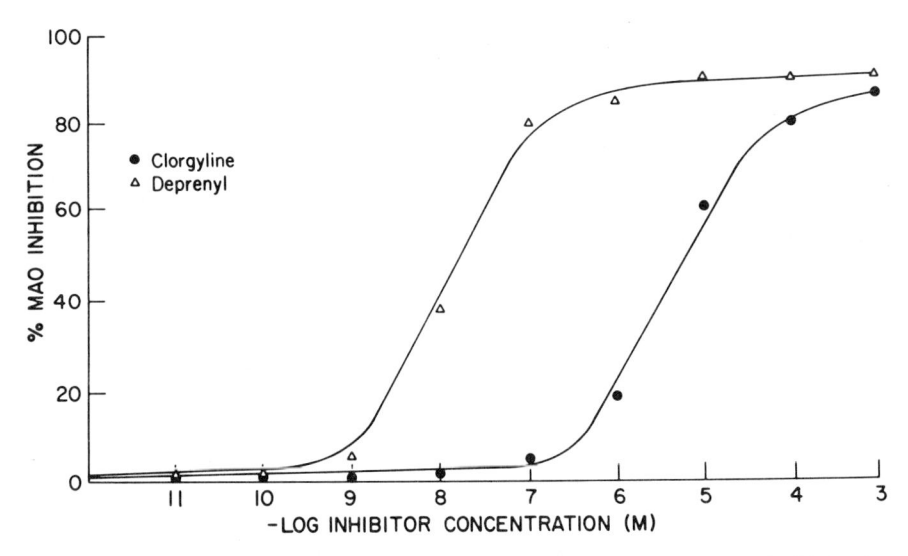

FIG. 1. Inhibition of human platelet MAO activity by clorgyline and deprenyl with tryptamine as substrate.

tor dose-response data, the relatively selective inhibition by deprenyl, the substrate characteristics described above (little or no affinity for serotonin or norepinephrine, high affinity for benzylamine and phenylethylamine), and the single tetrazolium-stained band found on electrophoresis all suggest that the human platelet enzyme is a homogenous MAO form, with characteristics like those of the B forms found in other tissues.

III. HUMAN PLASMA MAO

A soluble enzyme found in human plasma and serum which deaminates primary but not secondary or tertiary amines and which is distinct from diamine oxidase (histaminase) and ceruloplasmin has been most extensively studied by McEwen (1963, 1965a, 1965b, 1972). This enzyme appears to be distinctly different from mitochondrial MAOs in substrate and inhibitor responses and in its pyridoxal and copper cofactor requirements (Table 4).

Recent studies have identified age- and sex-related differences in plasma MAO activity levels (Tryding et al., 1969; Robinson et al., 1971). A twin study in normals indicated a large genetic component affecting the activity of this enzyme in plasma (Nies et al., 1973). However, plasma MAO levels are also affected by some disease states that alter liver function, including hepatic cirrhosis and metastatic tumors as well as congestive heart failure (McEwen and Harrison, 1965; McEwen and Castell, 1967; McEwen, 1972). Elevated plasma MAO levels have also been reported in diabetes mellitus and thyrotoxicosis (Tryding et al., 1969).

TABLE 4. Comparison of human platelet versus plasma MAO

	Human platelet MAO	Human plasma MAO
Localization:	Mitochondria	Soluble
Substrates:	Primary, secondary, and tertiary amines	Primary amines only
Cofactor:	Probably flavin adenine dinucleotide	Pyridoxal
K_m ($\times 10^{-4}$ M):		
Tyramine	0.5	1.8
Benzylamine	1.0	1.0–3.0
Inhibitors [ID_{50} (M)]:		
Pargyline	2×10^{-8}	$> 10^{-3}$
Tranylcypromine	1×10^{-6}	1×10^{-3}
Iproniazid	2×10^{-6}	1×10^{-3}
Isoniazid	$> 10^{-3}$	1×10^{-3}
KCN	$> 10^{-3}$	5×10^{-4}

Data from the studies of Collins and Sandler, 1971; Donnelly and Murphy, 1974; Knoll and Magyar, 1972; McEwen, 1965a, 1965b; McEwen and Cohen, 1963; Paasonen and Solatunturi, 1965; Robinson et al., 1968.

A. General Characteristics of the Human Plasma MAO

Plasma MAO is a soluble enzyme which has been purified 3000-fold by McEwen (1965). The K_m for benzylamine at pH 7.2 is 3×10^{-4} M; however, since benzylamine is ionizable and the active substrate for MAO is the nonionized amine (which has a pK_a of 9.37), the K_m has been observed to decrease with increasing pH of the medium such that at a pH of 9.0 the K_m is approximately 8×10^{-6} M (McEwen, 1972). Apparently, the molecular weight of the human plasma enzyme has not been determined, nor has purified plasma MAO been examined for electrophoretic homogeneity.

Recent studies of bovine plasma MAO have demonstrated that this enzyme has substrate and inhibitor-related characteristics which are similar to those of human plasma MAO. More complete physiochemical data indicate that bovine plasma MAO has a molecular weight of 170,000, and is composed of two subunits of 87,000 MW which are covalently linked by disulfide bridges; this enzyme contains 1.2 to 1.3 g-atoms of copper per molecule of the enzyme, two SH groups that are not involved in catalysis or needed for conformational stability, and one mole of pyridoxal 5-phosphate per mole of enzyme (Yasunobu, Achee, Chervenka, and Wang, 1968).

B. Substrate Characteristics of Human Plasma MAO

Benzylamine is the substrate most actively deaminated by this enzyme, with no other examined substrates manifesting more than 15% of the activity observed with this substrate (McEwen, 1965, 1972). Tyramine is the physiologically occurring amine that has been most frequently studied with plasma MAO. Under different assay conditions it is deaminated at only 1 to 10% of the rate of benzylamine (Tryding et al., 1969; Jarrott, 1971; McEwen, 1965, 1972).

Tyramine deamination by human plasma preparations has been shown to be altered in depressed patients compared to controls (Klaiber et al., 1972), and to vary during the menstrual cycle (Klaiber et al., 1972). Associations with certain personality features (Klaiber, Broverman, and Kobayashi, 1967) and with EEG responses (Vogel et al., 1971) have also been noted. In this series of studies (Otsuka and Kobayashi, 1964; Kobayashi, 1966; Klaiber et al., 1967, Vogel et al., 1971), plasma MAO apparently manifests some characteristics of mitochondrial MAOs that are difficult to reconcile with other studies of soluble plasma MAO (McEwen, 1965a,b, 1972). In a recently completed study comparing benzylamine and tyramine as substrates for plasma MAO, tyramine-based MAO activity in plasma was found to be markedly reduced by pargyline and other inhibitors of mitochondrial MAOs, but minimally affected by inhibitors of soluble plasma MAO only when routine clinical centrifuge preparations (1000 to 2000 g) of plasma were utilized (Donnelly and Murphy, 1974). Ultracentrifuged plasma preparations (80,000 g) manifested no change in MAO activity in comparison to the 1000-g preparations when benzylamine was the substrate, but manifested a 35-fold decrease when tyramine was the substrate (Table 5). Furthermore, the small residual activity in the ultracentrifuged preparations found with tyramine as the substrate was no longer inhibited by pargyline, but was inhibited by semicarbazide. These recent studies (Donnelly and Murphy, 1974) suggested that tyramine can indeed be deaminated by soluble plasma MAO, as McEwen (1965a,b) had demonstrated with purified plasma MAO preparations. However, it now appears that the anomalous properties of tyramine-related "plasma" MAO activity studied clinically with routine centrifugation procedures may result from the use of preparations containing varying amounts of incompletely separated MAO from platelets (Zeller et al., 1969; Donnelly and Murphy, 1974). Because of the much higher relative activity of the platelet enzyme compared to the

TABLE 5. *MAO activity remaining in human plasma as a function of substrate and preparative centrifuge g force*

Centrifugal force (g)	Percent MAO activity	
	Benzylamine	Tyramine
100	100	100
1,000	24.1	1.42
2,000	24.0	0.20
80,000	24.0	0.04

plasma enzyme when tyramine is used as the substrate, it is probable that the data from this series of clinical investigations (Otsuka and Kobayashi, 1964; Kobayashi, 1966; Klaiber et al., 1967, 1972; Vogel et al., 1971) predominantly reflect platelet, not plasma, enzyme activity.

C. Inhibitor-Related Characteristics of Human Plasma MAO

In *in vitro* studies, this enzyme has been demonstrated to be inhibited by KCN, isoniazid, semicarbazide, aminoguanidine, and other carbonyl reagents, as well as by the copper-chelating agent cuprizone; it is resistant to inhibition by pargyline, tranylcypromine, iproniazid, and some other clinically used MAO-inhibiting drugs (McEwen and Cohen, 1963; McEwen, 1965a,b, 1972; Robinson et al., 1971) (Table 4). Some hydrazine-type MAO-inhibiting drugs (e.g., phenelzine) affect the plasma and tissue MAOs equally (McEwen and Cohen, 1963; McEwen, 1965a,b, 1972; Robinson et al., 1971). *In vivo,* the greatest inhibition of plasma enzyme was obtained during treatment with isoniazid, furazolidone, and phenelzine, with lesser and more variable inhibition occurring during treatment with isocarboxazid, pargyline, nialamide, and tranylcypromine (Robinson et al., 1971). The enzyme has apparently not yet been studied with the selective A and B form MAO inhibitors, deprenyl, or clorgyline.

IV. MAO IN BRAIN, LIVER, AND OTHER HUMAN TISSUES

Monoamine oxidase has been detected in many human tissues, but the few detailed studies of substrate- and inhibitor-related characteristics of MAO have been accomplished in brain (McEwen and Cohen, 1963; Youdim, 1972; Youdim, Collins, Sandler, Jones, Pare, and Nicholson, 1972) and liver (Collins, Youdim, and Sandler, 1968; Youdim, 1972). Other human tissues with MAO activity include sympathetic ganglia, intestine, kidney, lung, uterus, placenta, thyroid gland, thymus, skeletal muscle, heart, and blood vessels (Franzen and Eysell, 1969).

A. Substrate-Related Characteristics of Human Tissue MAOs

Variations in the specific activities of brain tissue regions with different substrates have been most extensively studied by Collins, Sandler, Williams, and Youdim (1970). In general, MAO activity with tyramine, tryptamine, benzylamine, and dopamine is highest in the hypothalamus and lowest in the cerebral and cerebellar cortex, in agreement with other reports on human brain enzyme (Vogel, Orfei, and Century, 1969). In the hypothalamus, highest specific activities are found with benzylamine and dopamine as substrates, while tyramine and tryptamine are deaminated 7- to 10-fold less actively (Collins et al., 1970). Electrophoretic studies of solubilized and partially purified human brain samples revealed four separable bands of enzyme activity: MAO_{1-3} migrated toward the anode, while MAO_4 (also called MAO_R) migrated toward the cathode; the MAO_4 form was further distinguished by an extremely high specific activity with dopamine as the substrate, particularly in the hypothalamus and basal ganglia areas. MAO_1 had the highest specific activities for benzylamine, tyramine, and tryptamine. The K_m's also varied somewhat with the different substrates in the various brain regions, but less so than did the specific activities.

Human liver has also been demonstrated to exhibit four or five forms of MAO by polyacrylamide gel electrophoresis studies (Collins et al., 1968; Youdim, 1972). The MAO forms in liver have some similarities to the brain forms in their activity toward tyramine and tryptamine, but differ primarily in that the liver MAO forms manifest much lower specific activities with benzylamine as substrate (Collins et al., 1968; Youdim, 1972).

B. Inhibitor-Related Characteristics of Human Brain MAOs

MAO inhibition by different drugs appears to be substrate-dependent in human brain preparations. While the K_i values for pargyline, iproniazid, and harmaline studied with kynuramine as the substrate were generally similar for the different brain regions studied by Collins et al. (1970), a triphasic curve was found for inhibition of whole-brain MAO by clorgyline. This curve was most distinct with dopamine, less so with tyramine, and was replaced by a sigmoid curve when kynuramine was the substrate. These results agree with subsequent studies in other species in suggesting the existence of at least two MAO forms in brain that possess different inhibitor characteristics depending upon the substrate present (Neff and Goridis, 1972).

A study of *in vivo* treatment with isocarboxazid, tranylcypromine, and clorgyline compared the relative amounts of MAO activity inhibition produced by these three drugs in relation to MAO activities in control brain samples from nontreated patients (Youdim et al., 1972). Isocarboxazid produced greater effects on tyramine and kynuramine deamination than on tryptamine and dopamine deamination, while tranylcypromine led to relatively greater effects on dopamine deamination. The effects of clorgyline were more variable both in relation to the substrate used and to the different brain regions studied.

The effects of *in vivo* treatment on the MAO_{1-4} forms was also examined with these three inhibitors (Youdim et al., 1972). Tranylcypromine was the only drug to produce marked effects on the MAO_4 form, while tranylcypromine and isocarboxazid both inhibited the MAO_{1-2} forms. Clorgyline had more variable effects, again suggesting that this drug possesses more specific inhibitory effects which depend upon both substrate and brain region factors.

V. DISCUSSION

Immunological, electrophoretic, and substrate-inhibitor studies of MAO from nonhuman sources indicate that tissue MAO is composed of two or more active forms. The MAO antibody studies of Hartman, Hidaka, Udenfriend, and Yasunobu (1971, 1972) have demonstrated that beef liver possesses only one antigenically distinct MAO protein, while beef brain possesses both a MAO protein (representing 80% of total brain MAO activity) which cross-reacts with the liver MAO antibody and a second "brain-specific" form of MAO (representing 20% of total brain activity) which is antigenically distinct, and which exhibits some substrate and inhibitor-related characteristics different from the MAO form common to both liver and brain. Relative to kynuramine deamination by liver (set equal to 100), the "brain-specific" MAO exhibited rates for benzylamine and tyramine of 403 and 34, respectively. For whole brain, liver, and beef serum ratios of benzylamine to tyramine deamination were 78/27, 67/14, and 112/42, respectively. The "brain-specific" enzyme was relatively resistant to pargyline but was inhibited equally by iproniazid, in comparison to results obtained with these drugs in the whole-brain and liver preparations.

The evidence from electrophoretic studies of human platelet, brain, and liver MAOs, which revealed 1 to 5 MAO forms in the different tissues, would now seem to be better interpreted on the basis of recent data (Collins, 1972; Hartman and Udenfriend, 1972; Houslay and Tipton, 1973) as probably representing only two distinct MAO forms in the case of brain and liver, and one form in the case of platelets. The evidence from substrate-inhibitor interactions also suggests the existence of two MAO forms (A and B) which coexist in varying proportions in most tissues (Collins et al., 1970; Neff and Goridis, 1972; Sandler and Youdim, 1972; Squires, 1972).

Unfortunately, the presently published data from these three lines of approach were obtained under very different experimental conditions and in different species, and a combined experiment using the leverage of immunological, electrophoretic, and substrate-inhibitor approaches to human MAOs has yet to be reported. It would seem premature at the moment to attempt to integrate the available data from other species, and to predict its applicability to the studies of human MAO.

In considering the reported human tissue data, large gaps remain in the experimental evidence on hand, and caution continues to be required in generalizing from results with a limited number of substrates or inhibitors, or from one tissue to another. For example, although there are prominent similarities in the responses of platelet MAO and whole-brain MAO to various clinically used MAO-inhibiting drugs, and remarkable similarities in affinities for some substrates shared by these two tissues, there are also distinct differences in some other substrate and inhibitor-related characteristics. It is still unclear whether the platelet enzyme is identical to one of the brain forms or is a different molecular form of MAO. However, it does seem that methods are available which might define the relationship of the various putative MAO forms to each other, to their possible differences in mitochondrial (Blokhuis and Veldstra, 1970; Youdim, 1973) and tissue locations, to their interactions with different substrates and inhibitors, to their genetic and, perhaps, adaptive (Collins, 1972; Gorkin, 1972) regulatory mechanisms, and to their function in human biologic processes.

VI. SUMMARY

A comparison of available information on substrate, inhibitor, and electrophoretic characteristics of the monoamine oxidases in human platelet, plasma, brain, and liver tissue indicate that two or more multiple forms of MAO exist in human brain and liver. Human platelets apparently contain only one form of MAO, which manifests many substrate- and inhibitor-related characteristics similar to those of mitochondrial MAOs found in other tissues. In particular, the substrate specificities and the responses to deprenyl and clorgyline of platelet MAO suggest that this enzyme is closely similar to the B-type MAOs found in other tissues. In contrast, human plasma MAO appears to be a distinctly different enzyme, with different substrate and inhibitor-related responses. While plasma MAO activity measured with benzylamine as substrate and platelet MAO activity measured with either benzylamine or tryptamine as substrates were not significantly correlated in a large group of normal individuals, platelet MAO activity measured with benzylamine was highly correlated with platelet MAO activity with tryptamine as substrate.

ACKNOWLEDGMENTS

We thank Curtis Wright IV, Mrs. Anna Nichols, and Mrs. Irene Bellesky for their assistance; we also thank Dr. Richard Wyatt and Dr. William E. Bunney, Jr., for their support of some of these studies.

REFERENCES

Belmaker, R., Murphy, D. L., Wyatt, R. J., and Loriaux, L. (1974): Menstrual cycle-related changes in human platelet monoamine oxidase activity. *(In preparation.)*

Blokhuis, G. G. D., and Veldstra, H. (1970): Heterogeneity of mitochondria in rat brain. *FEBS Letters,* 11:197-199.

Buchsbaum, M., Landau, S., Murphy, D., and Goodwin, F. (1973): Average evoked response in bipolar and unipolar affective disorders: Relationship to sex, age of onset, and monoamine oxidase. *Biological Psychiatry,* 7:199-212.

Collins, G. G. S. (1972): Summary of Section I. In: *Advances in Biochemical Psychopharmacology,* Vol. 5, edited by E. Costa and P. Greengard. Raven Press, New York, pp. 129-132.

Collins, G. G. S., and Sandler, M. (1971): Human blood platelet monoamine oxidase. *Biochemical Pharmacology,* 20:289-296.

Collins, G. G. S., Sandler, M., Williams, E. D., and Youdim, M. B. H. (1970): Multiple forms of human brain mitochondrial monoamine oxidase. *Nature,* 225:817-820.

Collins, G. G. S., Youdim, M. B. H., and Sandler, M. (1968): Isoenzymes of human and rat liver monoamine oxidase. *FEBS Letters,* 1:215-218.

Costa, F., and Greengard, P., Eds. (1972): *Advances in Biochemical Psychopharmacology,* Vol. 5. Raven Press, New York.

Donnelly, C. H., and Murphy, D. L. (1974): Substrate- and inhibitor-related characteristics of human platelet monoamine oxidase. *(In preparation.)*

Franzen, F., and Eysell, K. (1969): *Biologically Active Amines Found in Man.* Pergamon Press, Oxford.

Gorkin, V. Z. (1972): Qualitative alterations in enzymatic properties of amine oxidases. In: *Advances in Biochemical Psychopharmacology,* Vol. 5, edited by E. Costa and P. Greengard. Raven Press, New York, pp. 55-65.

Hare, M. L. C. (1928): Tyramine oxidase I. A new enzyme system in liver. *Biochemical Journal,* 22:968-979.

Hartman, B. K., and Udenfriend, S. (1972): The use of immunological techniques for the characterization of bovine monoamine oxidase from liver and brain. In: *Advances in Biochemical Psychopharmacology,* Vol. 5, edited by E. Costa and P. Greengard. Raven Press, New York, pp. 119-128.

Hartman, B. K., Yasunobu, K. T., and Udenfriend, S. (1971): Immunological identity of the multiple forms of beef liver mitochondrial monoamine oxidase. *Archives of Biochemistry and Biophysics,* 147:797-804.

Hidaka, H., Hartman, B., and Udenfriend, S. (1971): Comparison of mitochondrial oxidases from bovine brain and liver using antibody to purified liver monoamine oxidase. *Archives of Biochemistry and Biophysics,* 147:805-809.

Houslay, M. D., and Tipton, K. F. (1973): The nature of the electrophoretically separable multiple forms of rat liver monoamine oxidase. *Biochemical Journal,* 135:173-186.

Jarrott, B. (1971): Occurrence and properties of monoamine oxidase in adrenergic neurons. *Journal of Neurochemistry,* 18:7-16.

Klaiber, E. L., Broverman, D. M., and Kobayashi, Y. (1967): The automatization cognitive style, androgens, and monoamine oxidase. *Psychopharmacologia,* 11:320-336.

Klaiber, E. L., Broverman, D. M., Vogel, W., Kobayashi, Y., and Moriatry, D. (1972): Effects of estrogen therapy on plasma MAO activity and EEG driving responses of depressed women. *American Journal of Psychiatry,* 128:1492-1498.

Knoll, J., and Magyar, K. (1972): Some puzzling pharmacological effects of monoamine oxidase inhibitors. In: *Advances in Biochemical Psychopharmacology,* Vol. 5, edited by E. Costa and P. Greengard. Raven Press, New York, pp. 393-408.

Kobayashi, Y. (1966): The effect of three monoamine oxidase inhibitors on human plasma monoamine oxidase activity. *Biochemical Pharmacology,* 15:1287-1294.

Kopin, I. J. (1964): Storage and metabolism of catecholamines: The role of monoamine oxidase. *Pharmacological Reviews,* 16:179-191.

Levine, R. J., Oates, J. A., Vendsalu, A., and Sjoerdsma, A. (1962): Studies on the metabolism of aromatic amines in relation to altered thyroid function in man. *Journal of Clinical Endocrinology,* 22:1242-1250.

McEwen, C. M., Jr. (1965*a*): Human plasma monoamine oxidase: I. Purification and identification. *Journal of Biological Chemistry,* 240:2003-2010.

McEwen, C. M., Jr. (1965*b*): Human plasma monoamine oxidase: II. Kinetic studies. *Journal of Biological Chemistry,* 240:2011-2017.

McEwen, C. M., Jr. (1972): The soluble monoamine oxidase of human plasma and sera. In: *Advances in Biochemical Psychopharmacology,* Vol. 5, edited by E. Costa and P. Greengard. Raven Press, New York, pp. 151-165.

McEwen, C. M., Jr., and Castell, D. O. (1967): Abnormalities of serum monoamine oxidase in chronic liver disease. *Journal of Laboratory and Clinical Medicine,* 70:36-47.

McEwen, C. M., Jr., and Cohen, J. D. (1963): An amine oxidase in normal human serum. *Journal of Laboratory and Clinical Medicine,* 62:766-776.

McEwen, C. M., Jr., and Harrison, D. C. (1965): Abnormalities of serum monoamine oxidase in chronic congestive heart failure. *Journal of Laboratory and Clinical Medicine,* 65:546-559.

Murphy, D. L. (1973): Technical strategies for the study of catecholamines in man. In: *Frontiers in *Catecholamines Research,* edited by E. Usdin and S. Snyder. Pergamon Press, Oxford, pp. 1077-1082.

Murphy, D. L., Belmaker, R., and Wyatt, R. J. (1974a): Monoamine oxidase in schizophrenia and other behavioral disorders. In: *Enzymes in the Neuropathology of Schizophrenia,* edited by S. Matthysse and S. S. Kety. Pergamon Press, Oxford.

Murphy, D. L., Belmaker, R., Wyatt, R. J., and Buchsbaum, M. (1974b): Personality variables and monoamine oxidase activity. *(In preparation.)*

Murphy, D. L., Donnelly, C. H., Belmaker, R., Carlson, H., Baker, M., and Wyatt, R. J. (1974c): Influence of age, sex and some drugs and hormones on human platelet monoamine oxidase activity. *(In preparation.)*

Murphy, D. L., Donnelly, C. H., and Buchsbaum, M. (1974d): Platelet monoamine oxidase activity in siblings and in monozygotic and dizygotic twins. *(In preparation.)*

Murphy, D. L., and Weiss, R. (1972): Reduced monoamine oxidase activity in blood platelets from bipolar depressed patients. *American Journal of Psychiatry,* 128:1351-1357.

Murphy, D. L., and Wyatt, R. J. (1972): Reduced monoamine oxidase activity in blood platelets from schizophrenic patients. *Nature,* 238:225-226.

Neff, N. H., and Goridis, C. (1972): Neuronal monoamine oxidase: Specific enzyme types and their rates of formation. In: *Advances in Biochemical Psychopharmacology,* Vol. 5, edited by E. Costa and P. Greengard. Raven Press, New York, pp. 307-323.

Nies, A., Robinson, D. S., Lamborn, K. R., and Lampert, R. P. (1973): Genetic control of platelet and plasma monoamine oxidase activity. *Archives of General Psychiatry,* 28:834-838.

Nies, A., Robinson, D. S., Ravaris, C. L., and Davis, J. M. (1971): Amines and monoamine oxidase in relation to aging and depression in man. *Psychosomatic Medicine,* 33:470.

Otsuka, A., and Kobayashi, Y. (1964): A radioisotopic assay for monoamine oxidase determination in human plasma. *Biochemical Pharmacology,* 13:995-1006.

Paasonen, M. K., and Solatunturi, E. (1965): Monoamine oxidase in mammalian blood platelets. *Annales Medicinae Experimentalis et Biologiae Fenniae,* 43:98-100.

Robinson, D. S., Davis, J. M., Nies, A., Colburn, R. W., Davis, J. N., Bourne, H. R., Bunney, W. E., Shaw, S. M., and Coppen, A. J. (1972): Ageing, monoamines, and monoamineoxidase levels. *Lancet,* 1:290-291.

Robinson, D. S., Davis, J. M., Nies, A., Ravaris, C. L., and Sylwester, D. (1971): Relation of sex and aging to monoamine oxidase activity of human brain, plasma, and platelets. *Archives of General Psychiatry,* 24:536-539.

Robinson, D. S., Lovenberg, W., Keiser, H., and Sjoerdma, J. (1968): Effects of drugs on human blood platelet and plasma amine oxidase activity *in vitro* and *in vivo. Biochemical Pharmacology,* 17:109-119.

Robinson, D. S., Nies, A., and Lamborn, K. (1973): Genetic control of blood MAO activity and substrate differences in platelet enzyme activity. *Clinical Pharmacological Therapeutics,* 14:144-145.

Sandler, M., and Youdim, M. B. H. (1972): Multiple forms of monoamine oxidase: Functional significance. *Pharmacological Reviews,* 24:331-348.

Southgate, J., Grant, E. C. G., Pollard, W., Pryse-Davies, J., and Sandler, M. (1968): Cyclical variations in endometrial monoamine oxidase: Correlation of histochemical and quantitative biochemical assays. *Biochemical Pharmacology,* 17:721-726.

Squires, R. F. (1972): Multiple forms of monoamine oxidase in intact mitochondria as characterized by selective inhibitors and thermal stability: A comparison of eight mammalian species. In: *Advances in Biochemical Psychopharmacology,* Vol. 5, edited by E. Costa and P. Greengard. Raven Press, New York, pp. 355-370.

Tryding, N., Nilsson, S. E., Tufvesson, G., Berg, B., Carlstrom, S., Elmfors, B., and Nilsson, J. E. (1969): Physiological and pathological influences on serum monoamine oxidase level. *Scandinavian Journal of Clinical and Laboratory Investigations,* 23:79-84.

Vogel, W., Broverman, D. M., and Klaiber, E. L. (1971): EEG responses in regularly menstruating women and in amenorrheic women treated with ovarian hormones. *Science,* 172:388-391.

Vogel, W. H., Orfei, V., and Century, B. (1969): Activities of enzymes involved in the formation and destruction of biogenic amines in various areas of human brain. *Journal of Pharmacology and Experimental Therapeutics,* 165:196-203.

Weiner, N., and Bjur, R. (1972): The role of intraneuronal monoamine oxidase in the regulation of norepinephrine synthesis. In: *Advances in Biochemical Psychopharmacology,* Vol. 5, edited by E. Costa and P. Greengard. Raven Press, New York, pp. 409–419.

Wyatt, R. J., Murphy, D. L., Belmaker, R., Cohen, S., Donnelly, C. H., and Pollin, W. (1973): Reduced monoamine oxidase activity in platelets: A possible genetic marker for vulnerability to schizophrenia. *Science,* 179:916–918.

Yasunobu, K. T., Achee, F., Chervenka, C., and Wang, T. (1968): Molecular properties of beef plasma amine oxidase. In: *Pyriodxal Enzymes,* edited by K. Yamada. Maruzen Co. Ltd., Tokyo, pp. 139–143.

Youdim, M. B. H. (1972): Multiple forms of monoamine oxidase and their properties. In: *Advances in Biochemical Psychopharmacology,* Vol. 5, edited by E. Costa and P. Greengard. Raven Press, New York, pp. 67–77.

Youdim, M. B. H. (1974): Heterogeneity of rat brain mitochondrial monoamine oxidase. In: *Advances in Biochemical Psychopharmacology,* Vol. 11, edited by E. Costa, G. L. Gessa, and M. Sandler. Raven Press, New York, pp. 59–63.

Youdim, M. B. H., Collins, G. G. S., Sandler, M., Jones, A. B. B., Pare, C. M. B., and Nicholson, W. J. (1972): Human brain monoamine oxidase: Multiple forms and selective inhibitors. *Nature,* 236:225–228.

Zeller, E. A., Babu, B. H., Cavanaugh, M. J., and Stanich, G. J. (1969): On the *in vivo* inhibition of human platelet monoamine oxidase (1.4.3.4) by pargyline. *Pharmacological Research Communications,* 1:20–24.

Neuropsychopharmacology of Monoamines and
Their Regulatory Enzymes, edited by E. Usdin.
Raven Press, New York © 1974.

Open Discussion: Multiple Molecular Forms of MAO

Reporter: Earl Usdin

Sandler. In reply to a series of questions by Dr. Spector, Dr. Sandler replied that sulfation had been observed for tyramine, MHPG, and other substrates, but that enzymatic mechanisms had not yet been studied in detail.

Youdim. Dr. Eiduson inquired about the molecular weights of Dr. Youdim's MAO forms and was told that Dr. Youdim had reported five forms, each with a molecular weight of about 300,000, in previously reported work with Dr. Collins on rat brain and liver. A similar size had been observed (with Dr. Tipton) for beef adrenal MAO. Dr. Gabay asked a number of questions about the mitochondria in Dr. Youdim's studies and was told that most of the present results concerned intact brain mitochondria but that Dr. Youdim had also had similar results with liver mitochondria. Dr. Youdim stated that in previous work (on adrenal mitochondria), he had obtained electron micrographs to demonstrate the integrity of the purified mitochondria. He agreed that different molecular forms might have altered pH optima, but that all present work was done at one pH.

In reply to a question from Dr. Schildkraut, Dr. Youdim mentioned the observation that MAO activity developed differentially in various areas of the brain. Dr. Youdim's studies have only been carried to 16 weeks of age, but he hopes to carry them to 1 year. Dr. Youdim told Dr. Spector that he did not know whether the MAO changes were intraneuronal or extraneuronal.

Dr. Neff objected that listeners might have received the impression that the corticoid treatment produced changes specific for MAO, whereas hormones influence mitochondrial protein synthesis and thereby change the activity of many enzymes. Dr. Neff also pointed out that the stated MAO half-life of 14 days was misleading, since this is shortened after corticoids. Dr. Youdim said that he had only referred to fluctuation of MAO during the estrous cycle. He also said that the *in vivo* effects of hormones (progesterone, estradiol, androgens) had not yet been really studied, although they do bring changes in MAO activity.

Finally, Dr. Gabay said that since freedom from contamination by other organelles is probably best determined by the absence of enzymes characteristic of those organelles, and in view of the numerous reports that there is a counterpart of MAO residing in the microsomes, mitochondrial preparations should be assayed for (1) microsomal contamination (monitored by glucose-6-phosphatase) and (2) lysosomes which constitute a major source of contamination in all mitochondrial fractions (monitored by N-acetyl-β-D-glucosaminidase and/or acid phosphatase). Dr. Youdim believes that his purification procedure had eliminated microsomes and he feels that previous work by others substantiates that his preparations were mitochondria.

Shih. To Dr. Lovenberg's query as to whether Dr. Shih had tested dopamine as a substrate, Dr. Shih replied that she had obtained similar results with dopamine as with serotonin. A number of questions were then asked by various members of the audience on the buffer used and the detergent (Triton X-100). Dr. Shih replied that she was able to show multiple MAO forms with and without Triton X-100. In amplification of Dr. Shih's talk, her co-worker, Dr. Eiduson, pointed out that Dr. Shih had repeated her work on multiple MAO forms using sodium perchlorate in Tris-HCl buffer, pH 8.2, and there was only a slight loss in the amounts of the A and C forms. He also mentioned some work in progress in his laboratory by Drs. Shih and Huang using an electron spin resonance (ESR) technique (labeling with hydroxy-amphetamine nitroxide), which measures rates for what they believe are two MAO forms. The ESR technique has the advantage that Triton X-100 does not affect assays.

Martin. After Dr. Martin's talk, Dr. Sapse described a Romanian drug, Gerovital H_3, which is presently being tested clinically in the United States. It is a weak MAO inhibitor; unlike pargyline, its inhibition is reversible. Further, it is a selective and fully competitive MAO inhibitor. Dr. Martin replied that her compounds were somewhat different; that the aminoindanes are reversible, competitive inhibitors, but she had no idea with which of the MAO forms. Dr. Martin thought there might be advantage to an MAO inhibitor whose effects would be over by 8 hr.

In reply to Dr. Youdim's comments that it would be better to synthesize MAO inhibitors on the basis of knowledge of the enzymes's active sites (pointing out that it is known from his own work that this includes an —SH as well as a pentapeptide group attached covalently to a flavin) rather than merely modifying marketed drugs, Dr. Martin said that she really did not yet know how to do this, although she was struggling to come up with concepts for wholly new structures. She also offered to try to come up with a new inhibitor if Dr. Youdim would furnish her with a three-dimensional structure of the active site of MAO.

Neff. Dr. Biel queried Dr. Neff about current usage of deprenyl and was told that in Hungary Dr. Knoll was testing this drug as an antiparkinsonian agent based on the fact that it increases dopamine in the brain specifically. Dr. Neff could not tell Dr. Meltzer whether octopamine had specificity for the A or B enzyme.

Several members of the audience presented questions on the pineal gland. Dr. Neff stated that the MAO concentration in pineal is very high and appears to be primarily the B type (the one that metabolizes phenethylamine). He elaborated this by stating that whereas MAO is B type in the gland itself, the enzyme is A in the nerves innervating the pineal gland. The NE and 5-HT of the pineal are thus metabolized in the sympathetic nerves, which contain the "correct MAO."

In reply to several of Dr. Brodie's questions, Dr. Neff stated that liver, like most tissues, contains both A and B enzymes, but that he did not know the specifics for the adrenal gland. He also pointed out that brain (about equal A and B) is not a good model for nerve, since brain is 90% glial. The superior cervical ganglion has predominantly A enzyme, which is essential for the metabolism of NE and 5-HT.

Dr. Martin wished to know about the MAO forms in cats and dogs. Dr. Neff stated that various workers have shown that all species examined except pig have been shown to have both A and B forms of MAO, although the ratio of A/B varies from species to species.

In reply to Dr. Gabay's questions concerning differences between the results reported by Dr. Sabelli and by Dr. Neff (data obtained by Drs. Willner, LeFevre, and Costa) for levels of phenethylamine, Dr. Neff stated that Sabelli's method is nonspecific and gives values about 1000 times higher than those obtained with either the method used by the group at the Laboratory of Preclinical Pharmacology, NIMH, or a new method recently described by Dr. Axelrod.

Dr. Abood questioned that freeze-thawing produced significant fragmentation of brain mitochondria, unless the freeze-thaw procedure was repeated many times. He also doubted the 90% glia figure for brain composition. He stated that the major constituent of brain tissue was comprised largely of axonal, dendritic, and glial processes. In reply, Dr. Neff said that electron micrographs (in collaboration with Dr. Bloom) had shown that the fractions studied did contain mitochondria and that they were broken after freeze thawing. In an attempt to localize enzyme activity (e.g., is A activity associated with nerves carrying biogenic amines?), he had severed the spinal cord of test animals and looked for change above and below the spinal cord 14 days later. No change in MAO activity was observed, possibly due to the fact that MAO was ubiquitous.

Dr. Youdim complained that all of Dr. Neff's data were expressed as ratio of phenethylamine activity to serotonin activity, that none of the actual data were presented. Dr. Neff said that he did this for the sake of clarity, since his object was to demonstrate either the presence of A or B enzyme or the ratio between them.

Finally, Dr. Murphy mentioned some work now in progress in collaboration with Dr. Richelson on MAO localization in gliablastoma and neuroblastoma cells. He said that preliminary work shows that it is not true that glial cells are pure B MAO.

Nies and Murphy (combined discussion). Dr. Sandler wondered whether

Dr. Nies had observed any differences in MAO values between unipolar depressive patients and cured patients in the nondepressed condition as compared to the depressed. Dr. Nies replied that no differences were detectable before and after the administration of tricyclic antidepressant drugs. There is no change in MAO activity with improvement or nonimprovement.

In reply to Dr. Schildkraut's question as to whether there was a conflict in the data on platelet MAO for subjects over 50 (Dr. Murphy had shown it going down in males over 50 while it was going up in females), Dr. Nies pointed out that his studies used benzylamine as a substrate whereas Dr. Murphy used tryptamine. Dr. Schildkraut concluded that tryptamine knew the difference between boys and girls over the age of 50 but that benzylamine did not.

Dr. Gabay questioned Dr. Murphy as to whether he had been using purified platelet preparation rather than platelet-enriched plasma and whether he had attempted to subfractionate platelet and obtain a platelet mitochondrial fraction. He also suggested that it might be profitable to use Dr. Virginia Davis' method for purification of platelet and subsequent fractionation. Using this method, Dr. Gabay found nearly complete platelet MAO inhibition (using kynurenine as substrate) within 1 hr after a single dose of 20 mg of Parnate to a 70-kg subject. Various samples were taken from 1 to 18 hr, although 5-HT was elevated by 95% at the end of 18 hr. Dr. Murphy said that some of his recent nonclinical work has been with purified platelet mitochondria; earlier work by Paasonen had indicated that platelet MAO activity is associated with the mitochondrial fraction.

Dr. Meltzer commented that he and Dr. Stahl had confirmed the results of Murphy and Wyatt that platelet MAO activity is low in chronic schizophrenics (with tryptamine as a substrate); he found that acute patients were not different from controls. Further, platelet MAO activity in acute schizophrenics is low with tyramine as substrate. Tyramine seems the best substrate of platelet MAO for distinguishing the largest number of schizophrenics from controls. In reply, Dr. Murphy agreed that recent data suggested other psychiatric patients than schizophrenics might have reduced MAO values. His results agree with those of Nies and Robinson: cycling manic depressives, improving depressed patients, as well as other patients, do not show changes in platelet MAO activity; this argues against the MAO changes resulting from stress or other nonspecific accompaniments of acute psychiatric disorders.

With regard to his twin study, Dr. Murphy thought that the fact that reduced MAO was found in both the schizophrenic and the nonschizophrenic twin indicated a greater likelihood of a genetic factor than a stress factor. Dr. Wyatt then mentioned that he had been informed by geneticists that the twin results were compatible with the inherited factor being related to the cause of schizophrenia rather than a genetically linked vulnerability. Their reasoning is that if the low MAO factor is on the same chromosome as the vulnerability factor, then crossovers would occur over a period of time, and this would eliminate the relationship. Thus, the geneticists assume that low MAO activity must be related to the cause of schizophrenia, either directly or through secondary or tertiary processes. Dr. Wyatt stated that he remains skeptical. Dr. Neff wondered whether the results could be explained on a hormonal basis, since Dr. Murphy had reported data showing male/female MAO differences, and Dr. Youdim had indicated that steroids influence MAO activity. Dr. Neff wondered whether the patient population might not include abnormal corticoid producers or steroid producers, so that Dr. Murphy might be measuring this phenomenon rather than something more directly involved with schizophrenia. Dr. Murphy said that he had not yet been able to define any differential characteristics (e.g., K_m) of the platelet enzymes of individuals with low activity, and dialysis studies had not disclosed the presence of any low-molecular-weight inhibitors. Recent studies of biologic alterations in schizophrenia have not found differences between patients and normals in plasma cortisol or thyroid hormone measures.

Dr. Youdim wondered whether attempts had been made to correlate low MAO in schizophrenics with anemia, since he has found the enzyme to be influenced by iron levels; iron affects both synthesis and activity of MAO. Dr. Youdim has found a highly significant reduction in platelet MAO among anemic patients. Dr. Murphy said that he had looked at routine hematologic measures in past and present patients and had not found a significant indication of anemia; in addition, serum iron levels in bipolar depressed patients who had reduced MAO activities were normal.

Dr. Fieve wondered about the stability of MAO over time in the same individual with respect to behavioral crises, going on and off drugs, etc. Dr. Murphy replied that a paper on this subject is in preparation, and that in general he finds the enzyme quite stable in the same individual over time. He cautioned that there are many variables in platelet preparation, and consistent procedures are a necessity. As a check, Dr. Murphy routinely uses internal controls of previously determined preparations in all assays of enzyme activity.

Finally, Dr. Murphy mentioned some work now in progress in collaboration with Dr. Richelson on MAO localization in tissue cultures of glial cells and neuroblastoma cells. He said that preliminary work shows that it is not true that glial cells contain pure B-type MAO.

In a comment on the data of both Dr. Nies and Dr. Murphy on MAO levels in older subjects, Dr. Sapse pointed out that MAO and NE levels run in parallel up to the age of 45 to 50, but after that, after menopause or andropause, MAO levels increase dramatically whereas NE levels decrease. He also mentioned that MAO seemed to be altered not only in depressives but also in schizophrenics; there are indications that parkinsonian patients have altered levels of MAO activity. Dr. Sapse wondered if it might not be of value to study MAO with regard to aging. Dr. Nies replied that it has been observed for some time that the prevalence of depression increases with aging and that both the frequency and duration of depressive episodes in bipolar patients increases with age. Dr. Sapse mentioned the work of Dr. Finch of the Gerontological Center, USC, on the relation of catecholamine levels to aging.

Neuropsychopharmacology of Monoamines and Their Regulatory Enzymes, edited by E. Usdin. Raven Press, New York © 1974.

Introduction: Monoaminergic Enzymes Other Than MAO

Menek Goldstein

Department of Psychiatry, Neurochemistry Laboratories, New York University Medical Center, New York, New York 10016

The question may be asked why a session on monoamines and their regulatory enzymes was organized at the Twelfth Annual Meeting of the American College of Neuropsychopharmacology. It became obvious in the last few years that the role of monoamines in determining mental states, emotions, and behavior can be understood only if the properties and the regulatory mechanisms of enzymes involved in their biosynthesis and degradation is explored. Indeed, remarkable advances in purification and characterization of enzymes involved in the biosynthesis and degradation of monoamines were recently achieved.

As a by-product of the advances in the enzymology of monoamines, new procedures for assay of human serum dopamine-β-hydroxylase (DBH) levels were developed. The participants of this session will discuss whether human serum DBH levels can serve as a reliable index of sympathetic activity in various physiological and pathological states. The remarkable variation in serum DBH activity in the normal population frustrates but also challenges investigation. Recent evidence suggests that not only sympathetic activity but also genetic factors influence the basal levels of human serum DBH activity (Ross et al., 1973; Weinshilboum et al., 1973). Thus, the interpretation of serum DBH data in clinical studies is complicated, and hopefully the discussions at this session will bring some clarification on this point.

Several chapters will discuss the mechanisms involved in the regulation of the adrenergic neurotransmitters. It will become apparent during this session that, in addition to end-product inhibition, other regulatory mechanisms are involved in the short- and long-term control of neurotransmitter biosynthesis. Recent data suggest that the activation or blockade of receptors for dopamine (Kehr et al., 1972; Goldstein et al., 1973*a*); and the neuronal concentrations of K^+ and Ca^{2+} (Goldstein et al., 1970; Harris and Roth, 1972; Roth et al., 1973) play a role in the short-term regulation of dopamine biosynthesis. In addition, evidence has now been accumulated that cyclic-AMP plays a role in the short-term (Goldstein et al., 1973*a,b*) and long-term (Guidotti et al., 1973; Thoenen et al., 1973) regulation of catecholamine biosynthesis. The new advances in this area will be presented at the end of this session.

Finally, I would like to express my gratitude to Dr. Earl Usdin for his advice and help in the organization of this session.

REFERENCES

Goldstein, M., Anagnoste, B., and Shirron, C. (1973a): The effect of trivastal, haloperidol and dibutyryl cyclic AMP on ^{14}C-dopamine synthesis in rat striatum. *Journal of Pharmacy and Pharmacology*, 25:348–351.

Goldstein, M., Anagnoste, B., Freedman, L. S., Roffman, M., Ebstein, R., Park, D., Fuxe, K., and Hokfelt, T. (1973b): Characterization, localization and regulation of dopamine-β-hydroxylase and of other catecholamine synthesizing enzymes. In: *Frontiers in Catecholamine Research*, edited by E. Usdin and S. H. Snyder. Pergamon Press, Oxford, pp. 69–78.

Goldstein, M., Backstrom, T., Ohi, Y., and Frenkel, R. (1970): The effects of Ca^{++} ions on C^{14}-catecholamine biosynthesis from C^{14}-tyrosine in slices from the striatum of rats. *Life Sciences*, 9:919–924.

Guidotti, A., Mao, C. C., and Costa, E. (1973): Transsynaptic regulation of tyrosine hydroxylase in adrenal medulla: Possible role of cyclic nucleotides. In: *Frontiers in Catecholamine Research*, edited by E. Usdin and S. H. Snyder. Pergamon Press, Oxford, pp. 231–236.

Harris, J. E., and Roth, R. H. (1972): Potassium-induced acceleration of catecholamine biosynthesis in brain slices. 1. A study on the mechanism of action. *Molecular Pharmacology*, 7:593–604.

Kehr, W., Carlsson, A., Lindqvist, M., Magnusson, T., and Atack, C. (1972): Evidence for a receptor-mediated feedback control of striatal tyrosine hydroxylase activity. *Journal of Pharmacy and Pharmacology*, 24:744–747.

Roth, R. H., Walters, J. R., and Morgenroth, V. H., III (1973): Effects of alterations in impulse flow on transmitter metabolism in central dopaminergic neurons. *This volume*.

Ross, S. B., Wetterberg, L., and Myrhed, M. (1973): Genetic control of plasma dopamine-β-hydroxylase. *Life Sciences*, 12:529–532.

Thoenen, H., Otten, U., and Oesch, F. (1973): Transsynaptic regulation of tyrosine hydroxylase. In: *Frontiers in Catecholamine Research*, edited by E. Usdin and S. H. Snyder. Pergamon Press, Oxford, pp. 179–185.

Dopamine-β-hydroxylase and the Regulation of the Noradrenergic Neuron

Perry B. Molinoff,* David L. Nelson, and James C. Orcutt

University of Colorado Medical Center, Department of Pharmacology, Denver, Colorado 80220

The enzyme dopamine-β-hydroxylase (EC 1.14.2.1; DBH) catalyzes the last step in the biosynthesis of norepinephrine. It is likely that the maximum rate of catecholamine synthesis is limited by the amount of tyrosine hydroxylase (EC 1.14.3; TH) in the cell. It is also true that stimulation of sympathetic nerves leads to increases in the rate of synthesis of catecholamines via an increase in the activity of TH. On the other hand, it is probably an oversimplification to suggest that TH is the only regulated step in this biosynthetic pathway (Molinoff and Orcutt, 1974). The central position of DBH in the biosynthetic pathway leading to norepinephrine and its unique subcellular distribution have led us to study this enzyme. Some of the established and potential factors which modulate either the amount or the activity of DBH in adrenergic nerves will be discussed below.

I. TRANS-SYNAPTIC INDUCTION OF DBH AND TH

A number of procedures which cause prolonged increases in the level of activity in adrenergic neurons lead to increases in the amounts of TH, DBH, and phenylethanolamine N-methyl transferase (EC 2.1.10; PNMT). The prolonged increase in nerve firing may be due to the administration of pharmacological agents (Mueller, Thoenen, and Axelrod, 1969a, b; Viveros, Arqueros, Connett, and Kirshner, 1969; Molinoff, Brimijoin, Weinshilboum, and Axelrod, 1970; Molinoff, Brimijoin, and Axelrod, 1972), to immobilization stress (Kvetňanský, Weise, and Kopin, 1970; Kvetňanský, Gewirtz, Weise, and Kopin, 1971), to psychological stress (Axelrod, Mueller, Henry, and Stephens, 1970), or to exposure to cold (Thoenen, Kettler, Burkard, and Saner, 1971). The administration of drugs that deplete catecholamines leads to a fall in blood pressure and a presumed reflex increase in activity in the sympathetic nervous system. Following reserpine, for example, an increase in TH and DBH activities is observed in sympathetic ganglia (Thoenen, Mueller, and Axelrod, 1969; Molinoff et al., 1970, 1972), in the

---
* Established Investigator of the American Heart Association.

adrenal gland, and in sympathetically innervated organs, such as the heart and salivary glands. In sympathetic ganglia the increase in the two enzyme activities requires about 20 hr. In the heart the increase in TH occurs only after a 3- to 4-day lag (Thoenen, Mueller, and Axelrod, 1970), while the increase in DBH activity is preceded by a small but significant decrease (Molinoff et al., 1970). PNMT activity in the adrenal also increases after reserpine or after chemical sympathectomy with 6-hydroxydopamine (Mueller et al., 1969b; Molinoff et al., 1970). The increase in PNMT activity after 6-hydroxydopamine is blocked by adrenal denervation.

In an attempt to duplicate the effect of nerve impulses on DBH activity, individual sympathetic ganglia were maintained in organ culture. The DBH activity of control ganglia remained unchanged for up to 4 days. When the ganglion cells were depolarized by increasing the potassium concentration of the medium to 50 mM, the DBH activity rose by 40% within 24 hr (Silberstein, Brimijoin, Molinoff, and Lemberger, 1972).

The enhancement of DBH activity after reserpine administration appears, from several types of experiments, to require the synthesis of new enzyme and does not reflect simply the activation of preexisting enzyme. The electrophoretic mobility of DBH from sympathetic ganglia (Ross, Weinshilboum, Molinoff, Vesell, and Axelrod, 1972) and the K_m of DBH for phenylethylamine (6×10^{-4} M; Molinoff et al., 1972) are the same for enzyme from control and reserpine-pretreated rats. The increase in DBH activity following the administration of reserpine was blocked by the protein synthesis inhibitor cycloheximide. After administering cycloheximide, the activity of DBH in sympathetic ganglia decreased ($t_{1/2}$ about 15 hr), both in control and in chronically reserpinized animals (Molinoff et al., 1972). The increase in DBH activity seen in ganglia cultured in the presence of increased potassium was also blocked by cycloheximide (Silberstein et al., 1972).

Direct evidence that increased DBH activity was due to an increase in the rate of synthesis of the enzyme was obtained by the use of a specific antibody against DBH. After administering a pulse of ³H-leucine to control and to reserpine-pretreated rats, DBH was isolated by immunoadsorption. The amount of radioactivity was a measure of the rate at which DBH was being synthesized. The rate of synthesis in the adrenals of reserpine-pretreated rats was approximately three times that seen in the adrenals of control animals (Hartman, Molinoff, and Udenfriend, 1970).

The specificity of the effect of reserpine on DBH and TH activities has been investigated in several ways. The activity of the enzymes monoamine oxidase and lactic dehydrogenase in rat hearts, sympathetic ganglia, or adrenal glands did not increase after reserpine treatment. This suggests that the effects observed are not simply trophic effects on protein synthesis (Molinoff et al., 1970). It appeared that the effect of reserpine was related to a depletion of catecholamines (Molinoff et al., 1972), since that depletion which occurred after the administration of the competitive inhibitor of TH activity, α-methyl-*p*-tyrosine, was also associated with an increase in DBH activity. In addition, several agents that diminished the reserpine-induced depletion of catecholamines also interfered with the inducing effect of reserpine. These agents included monoamine oxidase inhibitors, bretylium, DOPA, and dopamine. The administration of catecholamines had no effect on DBH activity in ganglia, but led to a decrease in its activity in homogenates of rat heart (Molinoff et al., 1972).

II. ENDOGENOUS INHIBITORS OF DBH

Homogenates of adrenergically innervated organs show little or no DBH activity, although activity is readily demonstrable after the homogenate has been fractionated (Levin, Levenberg, and Kaufman, 1960; Austin, Levitt, and Chubb, 1967). The marked increase in the units of enzyme activity which occurred during purification of the enzyme raised the possibility that endogenous inhibitors may be present (Creveling, 1962).

DBH has been purified and has been shown to be a copper-containing protein (Friedman and Kaufman, 1965). It appears that endogenous inhibitors of DBH, which are present in many tissues, act by binding to or interacting with the copper at the active site of the enzyme. DBH, purified from the bovine adrenal gland, is inhibited by sulfhydryl compounds such as cysteine, glutathione, and mercaptoethanol (Nagatsu, Kuzuya, and Hidaka, 1967). The effect of the endogenous inhibitors is reversed by Cu^{2+}, Hg^{2+}, *p*-hydroxymercuribenzoate, and N-ethylmaleimide (Duch, Viveros, and Kirshner, 1968). A partial purification of a DBH inhibitor from bovine heart has been achieved (Chubb, Preston, and Austin, 1969). This compound is heat-stable and is of low molecular weight. It apparently contains carbohydrate and organic phosphate. The purified inhibitor was neutralized by Cu^{2+}, but no reversal of the inhibition was observed with N-ethylmaleimide. This last experiment is particularly important in that the DBH-inhibitory activity in the bovine adrenal medulla has been reported to be reversed by N-ethylmaleimide (Duch et al., 1968). An endogenous inhibitor of DBH has been isolated from homogenates of bovine adrenal medulla (Duch and Kirshner, 1971). This inhibitor was purified using chromatography on Florisil, Sephadex G-15, and DEAE cellulose. The inhibitor was rapidly inactivated on storage, but could be reactivated by exposure to H_2S. After treatment with H_2S, two components were found, only one of which possessed significant inhibitory activity. Each of the two components was made up of the same three amino acids—glutamic acid, cysteine, and glycine. Although the inhibitors resembled glutathione in composition, they did not appear to be identical with it. It is important to note that both the bovine heart and bovine adrenal inhibitors that have been isolated were purified after heat denaturation of proteins at 95°C.

The assay for endogenous inhibitors of DBH that has been established in our laboratory involves incubating a sample of the tissue homogenate in question with an aliquot of DBH purified from the bovine adrenal medulla (Foldes, Jeffrey, Preston, and Austin, 1972). The curve that relates percent inhibition to amount of inhibitor-containing homogenate is extremely steep, and most experiments are performed at several dilutions of inhibitor so as to obtain one or more concentrations of inhibitor that produce approximately a 50% inhibition of the purified enzyme. Results are usually expressed in terms of the dilution required to obtain a 50% inhibition of a given amount of DBH.

Some of the basic properties of the endogenous inhibitors have been determined. Inhibitory activity was found to remain constant when whole organs were stored at −10°C for up to 4 days. Approximately half of the inhibitory activity in the adrenal gland, spleen, or brain was lost, however, if a homogenate was stored either at −10 or at 4°C for from 1 to 4 days. When homogenates of various organs were subjected to temperatures of 95°C for 5 min or longer, there was no effect on the inhibitory activity in the spleen. On the other hand, approximately 98% of the inhibitory activity in the heart was destroyed by this treatment. This type of experiment suggested that the nature of the inhibitory activity is at least partially organ-specific (Molinoff and Orcutt, 1974). However, when a homogenate of spleen is added to a homogenate of heart and the mixture is placed in a boiling water bath for 5 min, the inhibitory activity in the spleen is largely destroyed. We believe that divalent cations are responsible for the heat lability of the inhibitors. The inhibitor in the spleen is stable for at least 20 min at 95°C in the absence of such cations, but over 95% of the activity is lost if the splenic homogenate is boiled in the presence of 10^{-3} M Fe^{2+}, Ca^{2+}, or Mg^{2+}. The effect of several heme-containing proteins on the heat stability of the DBH inhibitor in splenic homogenates was examined. Homogenates were boiled in the presence of cytochrome c (10^{-3} M), lactic acid dehydrogenase (10^{-7} M), or hemoglobin (10^{-3} M). In all three cases there was a marked decrease in the level of inhibitory activity. The concentration of at least one of these proteins, cytochrome c, is about 10 times higher in heart than in spleen (Potter and Du Bois, 1942). This suggests that the apparent difference in the heat stability of heart and spleen inhibitors may be due to a higher concentration of metal-containing proteins in the heart.

A very striking property of the DBH inhibitors is the enormous amount of activity which is found in various organs. There is enough inhibitory activity in a 1:5000 dilution of a homogenate of rat spleen or adrenal gland to cause at least a 50% inhibition of the activity of purified bovine DBH. A 1:3000 dilution of heart or salivary gland will cause at least a 50% inhibition. The high concentration of inhibitory activity in many organs means that if even a small percentage of the inhibitors has access to DBH, they may be playing a significant physiological role. When DBH activity is determined in fractions of sucrose density gradients (see below), it is necessary to add Cu^{2+}

to inactivate inhibitors which are present in the same regions of the density gradients as are noradrenergic storage granules. Although most of the inhibitory activity is at the top of the gradients, a small percentage of it is found in seeming association with the storage granules. Consistent with this observation is the fact that DBH-inhibitory activity has been found in washed adrenal chromaffin granules (Duch et al., 1968).

III. USE OF DBH AS A MARKER FOR ADRENERGIC VESICLES

Von Euler and Hillarp (1956) were the first to observe that some of the norepinephrine present in homogenates of sympathetic nerves is contained within particles. Since that time, these particles have been extensively investigated utilizing centrifugation methods (for reviews see Geffen and Livett, 1971; Smith, 1972). Two types of norepinephrine-containing particles, distinguishable by a difference in density, exist in nerve terminals. In homogenates of spleen approximately 20% of the vesicles are of the dense variety (Bisby and Fillenz, 1971), while in the vas deferens only about 4% of the vesicles are of this type (Bisby and Fillenz, 1971; Fillenz, 1971).

It has been possible to correlate the data obtained by centrifugation with morphological observations of noradrenergic nerves. Three distinct types of membrane-limited vesicle have been distinguished in nerve terminals (Grillo, 1966). There are large, dense-cored vesicles approximately 800 Å in diameter; small, dense-cored vesicles, approximately 450 Å in diameter; and small, electron-lucent vesicles about 450 Å in diameter. Both the large and the small dense-cored vesicles probably contain catecholamines (Fillenz, 1971). Furthermore, it appears likely that the electron-lucent vesicles can take up and store exogenous amines (Tranzer, Thoenen, Snipes, and Richards, 1969). On the other hand, studies of noradrenergic storage vesicles in the bovine splenic nerve reveal only a single type, the so-called large, dense-cored vesicle, which appears to correspond to the "heavy" noradrenergic vesicle as identified by biochemical techniques in density gradients (Klein and Thureson-Klein, 1971; DePotter, Chubb, and De Schaepdryver, 1972).

In addition to DBH and norepinephrine, a number of other constituents are normally present in at least some types of adrenergic storage granules. These include a number of soluble proteins, called chromogranins (see Smith, 1972). There are also large amounts of adenine nucleotides in the storage granules. All of these constituents appear to be released on nerve stimulation. There is excellent evidence that catecholamines are released from the adrenal medulla by a process called exocytosis (see Smith and Winkler, 1972), and it appears likely that the same process occurs in sympathetic nerve terminals.

The biochemical nature of the different types of adrenergic storage vesicles is not yet completely settled. In the rat vas deferens, two populations of vesicle have been described (Bisby and Fillenz, 1971). When DBH activity was determined in density gradients of microsomal pellets of this organ, two peaks of approximately equal size were found at 0.5 to 0.6 and 0.9 to 1.0 M sucrose (Bisby, Fillenz, and Smith, 1973). However, only a single peak of norepinephrine was seen. This was at 0.6 M sucrose, which is the approximate position of the small, dense-cored vesicles (Bisby and Fillenz, 1971). Somewhat different results have been obtained after isopycnic gradient centrifugation of a particulate fraction from dog spleen (DePotter et al., 1972). In these experiments a bimodal distribution of norepinephrine was found, but DBH was associated only with the denser of the two types of particle.

In an attempt to resolve the apparent discrepancy with regard to the content of DBH and norepinephrine of the dense and light vesicles, we have carried out sucrose density gradient centrifugation of homogenates of rat organs. We have found that there are two peaks of DBH in both spleen (Fig. 1B) and heart. Both of the peaks of DBH activity are associated with peaks of endogenous norepinephrine (Fig. 1A). These results suggest that in both the rat heart and spleen there are two types of adrenergic vesicles, and that both of these types of vesicles contain DBH and are able to store and presumably synthesize norepinephrine. In most experiments, between 60 and 80% of the DBH activity is found at a density of approximately 1.05 g/ml, with the remainder being at a density of 1.14 to 1.16 g/ml. In some experiments 50 to 100 μC of ^{3}H-norepinephrine is injected intravenously 1 hr before the animal is killed. This type of experiment has shown that both the dense and light vesicles are able to take up exogenous norepinephrine. In most experiments a shoulder (Fig. 1B) or subpeak (Fig. 1A) is seen at a density of about 1.08 g/ml. This may suggest that the light vesicle peak is actually a mixture of two or more types of vesicle with differing densities perhaps due to differing relative amounts of DBH and norepinephrine.*

As one measure of the purity of the vesicle fractions, we have measured the protein throughout the gradient and have calculated the amount of norepinephrine per milligram of protein. The denser vesicles have a ratio of 32 ng of norepinephrine/mg of protein, while the less dense vesicles have a ratio of 12 ng of norepinephrine/mg of protein. These values are considerably lower than the highest values that have been reported for the extensively purified, large, dense-core vesicles of bovine splenic nerve (11 to 12 μg norepinephrine/mg of protein; Yen, Klein, and Chen-Yen, 1973).

In several preliminary experiments, we have investigated the effect of the detergent Triton X100 on DBH activity. Purified DBH is neither activated nor inhibited by low concentrations of Triton (0.1 to 0.4%). When the effects of the detergent on DBH activity in the large granular, small granular, and

---

* Similar findings have been obtained by R. Weinshilboum in experiments with the vas deferens *(personal communication)*.

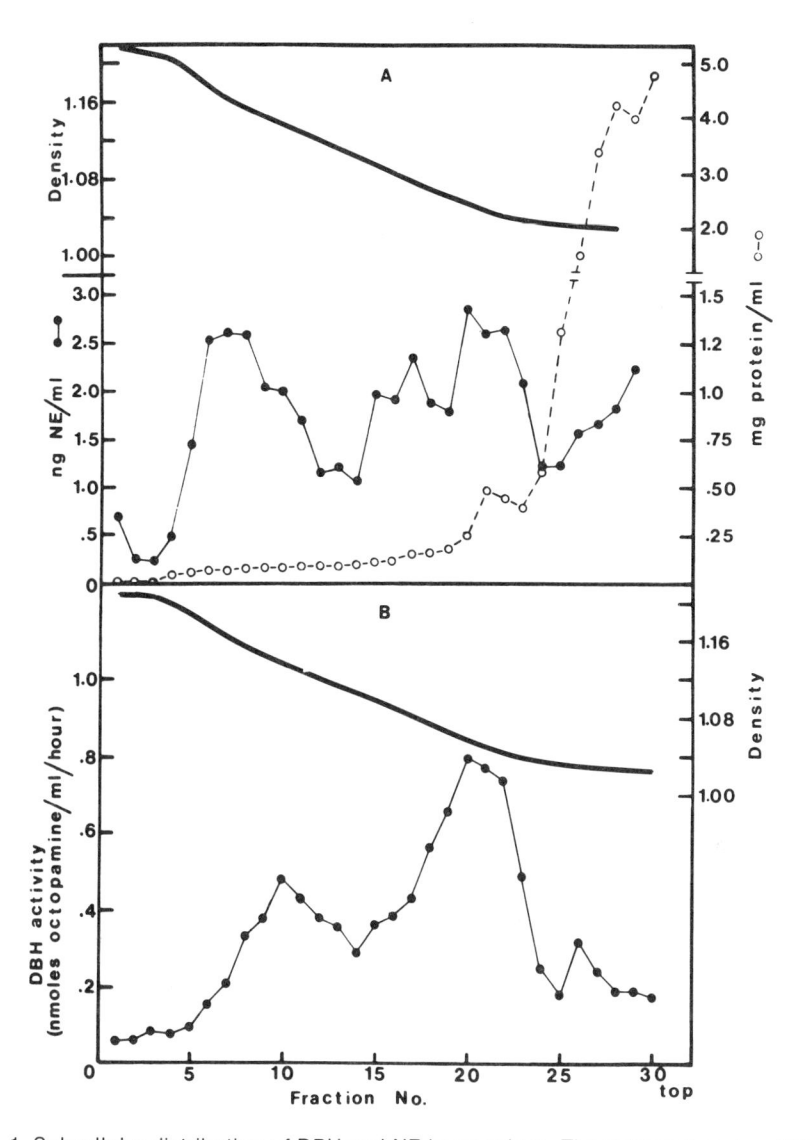

FIG. 1. Subcellular distribution of DBH and NE in rat spleen. The spleens from two 150-g male Sprague-Dawley rats were used for each part of this experiment (total weight in A, 1.45 g, in B, 1.90 g). The spleens were homogenized in 10 volumes of 5.05% glucose (w/v) containing tris-(hydroxymethyl) aminomethane (Tris) buffer, 0.005 M, pH 7.5. The homogenates were centrifuged for 10 min at 10,000 × g. The gradients were prepared 12 hr before being used. A 9.5-ml cushion of 53.5% (w/v) sucrose was placed in the gradient centrifuge tubes. A Beckman gradient generator was used to prepare a linear gradient which ran from 44.5% to 9% sucrose (w/v). A 10-ml buffer layer of 8% sucrose was layered above each gradient and a 5-ml aliquot of the 10,000 × g supernatant was placed above the buffer layer. All sugar solutions were buffered to pH 7.5 with 0.005 M Tris. Centrifugation was for 3 hr at 70,000 × g in a Beckman SW 25.2 rotor. A needle introduced through the gradient was used to collect 30 2-ml fractions from each gradient. DBH activity was determined on 100-μl aliquots of each fraction. In order to inactivate endogenous inhibitors of DBH, the activity in each fraction was determined in the presence of six different concentrations of $CuSO_4$. The assay was carried out by the method of Molinoff, Weinshilboum, and Axelrod (1971), using tyramine as substrate. The assay volume was 170 μl. Norepinephrine was determined by a method modified from that of Coyle and Henry (1973) and protein by the method of Lowry, Rosebrough, Farr, and Randall (1951). The refractive index was determined using a B & L refractometer.

supernatant regions of density gradients were examined, we found that in the first two cases there was a marked stimulation of DBH activity while the DBH at the top of the gradient, which is probably derived from vesicles ruptured during the homogenization procedure, was not affected by the detergent. These results suggest that at least with near-saturating concentrations of tyramine (the substrate used), the access of the substrate to the enzyme is a limiting step. This result does not imply that access is restricted at the very low concentrations of dopamine which are normally present in tissues, but it is certainly a possibility that should be investigated.

IV. CONCLUSIONS

The catecholamine biosynthetic pathway is extremely complex. A few of the established and potential regulatory mechanisms which bear on one aspect of this pathway have been discussed above. A great deal of work remains to be done, however, to understand the mechanisms and role of these and other of the many regulatory processes which have been and are being implicated in the control of the synthesis of the biogenic amines.

We would like to know the exact role of nerve impulses in the trans-synaptic induction of DBH and TH. The rate of synthesis of DBH and of TH may be a function of the frequency (or pattern) of nerve impulses, or the preganglionic nerve may be playing only a permissive role. If nerve impulses are required, they may be affecting the postganglionic cell via a change in its membrane potential; or the effect may be a trophic effect of the presynaptically released acetylcholine. Work in several laboratories has implicated cyclic nucleotides in the control of the synthesis of TH and DBH. This possibility should be extensively investigated over the next few years.

Current studies of the endogenous inhibitors of DBH are directed toward an examination of the possibility that some or all of these compounds have access to adrenergic storage vesicles and thus to DBH. If these compounds are playing a significant physiological role in regulating DBH activity, then the concentration of inhibitor in the nerve will probably be found to change in response to an appropriate exogenous stimulus.

The fact that several types of adrenergic storage vesicles exist in nerve terminals has been known for a number of years. It is only now, however, that their biochemical composition is being systematically defined. In several laboratories studies are now under way that are designed to investigate the specific roles that the different vesicle types play in the synthesis, uptake, storage, and release of norepinephrine and to investigate the relationship between the small and large granular vesicles.

ACKNOWLEDGMENTS

This work was supported by U.S. Public Health Service grant NS-10206.

REFERENCES

Axelrod, J., Mueller, R. A., Henry, J. P., and Stephens, P. M. (1970): Changes in enzymes involved in the biosynthesis and metabolism of noradrenaline and adrenaline after psychosocial stimulation. *Nature,* 225:1059.
Austin, L., Livett, B. J., and Chubb, I. W. (1967): Biosynthesis of noradrenaline in sympathetic nervous tissue. *Circulation Research,* Supp. III, 10:111-117.
Bisby, M. A., and Fillenz, M. (1971): The storage of endogenous noradrenaline in sympathetic nerve terminals. *Journal of Physiology,* 215:163.
Bisby, M. A., Fillenz, M., and Smith, A. D. (1973): Evidence for the presence of dopamine-β-hydroxylase in both populations of noradrenaline storage vesicles in sympathetic nerve terminals of the rat vas deferens. *Journal of Neurochemistry,* 20:245-248.
Chubb, I. W., Preston, B. N., and Austin, L. (1969): Partial characterization of a naturally occurring inhibitor of dopamine-β-hydroxylase. *Biochemical Journal,* 111:245-246.
Coyle, J. T., and Henry, D. (1973): Catecholamines in fetal and newborn rat brain. *Journal of Neurochemistry,* 21:61-68.
Creveling, C. R. (1962): Studies on dopamine-β-oxidase. Doctoral Thesis, George Washington University, Washington, D.C.
DePotter, W. P., Chubb, I. W., and De Schaepdryver, A. F. (1972): Pharmacological aspects of peripheral noradrenergic transmission. *Archives of International Pharmacodynamics,* 196:Supp., 258.
Duch, D. S., and Kirshner, N. (1971): Isolation and partial characterization of an endogenous inhibitor of dopamine-β-hydroxylase. *Biochimica Biophysica Acta,* 236:628-638.
Duch, D. S., Viveros, O. H., and Kirshner, N. (1968): Endogenous inhibitor(s) in adrenal medulla of dopamine-β-hydroxylase. *Biochemical Pharmacology,* 17:255.
Fillenz, M. (1971): Fine structure of noradrenaline storage vesicles in nerve terminals of the rat vas deferens. *Philosophical Transactions of the Royal Society of London,* Ser. B, 261:319.
Foldes, A., Jeffrey, P. L., Preston, B. N., and Austin, L. (1972): Dopamine-β-hydroxylase of bovine adrenal medulla: A rapid purification procedure. *Biochemical Journal,* 126:1209-1217.
Friedman, S., and Kaufman, S. (1965): 3,4-Dihydroxyphenylethylamine β-hydroxylase: Physical properties, copper content and role of copper in the catalytic activity. *Journal of Biological Chemistry,* 240:4763-4773.
Geffen, L. B., and Livett, B. G. (1971): Synaptic vesicles in sympathetic neurons. *Physiology Reviews,* 51:98.
Grillo, M. A. (1966): Electron microscopy of sympathetic tissues. *Pharmacological Reviews,* 18:387.
Hartman, B., Molinoff, P. B., and Udenfriend, S. (1970): Increased rate of synthesis of dopamine-β-hydroxylase in adrenals of reserpinized rats. *Pharmacologist,* 12:Ab. 470.
Klein, R. L., and Thureson-Klein, A. (1971): An electron microscopic study of noradrenaline storage vesicles isolated from bovine splenic nerve trunk. *Journal of Ultrastructure Research,* 34:473.
Kvetňanský, R., Gewirtz, G. P., Weise, V. K., and Kopin, I. J. (1971): Enhanced synthesis of adrenal dopamine-β-hydroxylase induced by repeated immobilization in rats. *Molecular Pharmacology,* 7:81.
Kvetňanský, R., Weise, V. K., and Kopin, I. J. (1970): Elevation of adrenal tyrosine hydroxylase and phenylethanolamine-N-methyl transferase by repeated immobilization of rats. *Endocrinology,* 87:744.
Levin, E. Y., Levenberg, B., and Kaufman, S. (1960): The enzymatic conversion of 3,4-dihydroxyphenylethylamine to norepinephrine. *Journal of Biological Chemistry,* 235:2080-2086.
Lowry, O. H., Rosebrough, N. J., Farr, A. L., and Randall, R. J. (1951): Protein measurement with the Folin phenol reagent. *Journal of Biological Chemistry,* 193:265-275.
Molinoff, P. B., Brimijoin, S., and Axelrod, J. (1972): Induction of dopamine-β-hydroxylase in rat hearts and sympathetic ganglia. *Journal of Pharmacology and Experimental Therapeutics,* 182:116.
Molinoff, P. B., Brimijoin, S., Weinshilboum, R., and Axelrod, J. (1970): Neurally mediated

increase in dopamine-β-hydroxylase activity. *Proceedings of the National Academy of Sciences,* 66:453.

Molinoff, P. B., and Orcutt, J. C. (1973): The trans-synaptic regulation of dopamine-β-hydroxylase. In: *Frontiers in Catecholamine Research,* edited by E. Usdin and S. H. Snyder. Pergamon Press, Elmsford, N.Y.

Molinoff, P. B., Weinshilboum, R., and Axelrod, J. (1971): A sensitive enzymatic assay for dopamine-β-hydroxylase. *Journal of Pharmacology and Experimental Therapeutics,* 178:425-432.

Mueller, R. A., Thoenen, H., and Axelrod, J. (1969a): Increase in tyrosine hydroxylase activity after reserpine administration. *Journal of Pharmacology and Experimental Therapeutics,* 169:74.

Mueller, R. A., Thoenen, H., and Axelrod, J. (1969b): Adrenal tyrosine hydroxylase: Compensatory increase in activity after chemical sympathectomy. *Science,* 158:468.

Nagatsu, T., Kuzuya, H., and Hidaka, H. (1967): Inhibition of dopamine-β-hydroxylase sulfhydryl compounds and the nature of the natural inhibitors. *Biochimica Biophysica Acta,* 139:319-327.

Potter, V. R., and Du Bois, K. P. (1942): The quantitative determination of cytochrome c. *Journal of Biological Chemistry,* 142:417-426.

Ross, S., Weinshilboum, R., Molinoff, P. B., Vessel, E., and Axelrod, J. (1972): Electrophoretic characteristics of dopamine-β-hydroxylase. *Molecular Pharmacology,* 8:50-59.

Silberstein, S. D., Brimijoin, S., Molinoff, P. B., and Lemberger, L. (1972): Induction of DBH in rat superior cervical ganglia in organ culture. *Journal of Neurochemistry,* 19:919-921.

Smith, A. D. (1972): Subcellular localization of noradrenaline in sympathetic neurons. *Pharmacological Reviews,* 24:435.

Smith, A. D., and Winkler, H. (1972): Fundamental mechanisms in the release of catecholamines. In: *Handbook of Experimental Pharmacology Catecholamines,* edited by H. Blaschko and E. Muscholl. Springer-Verlag, Berlin, p. 538.

Thoenen, H., Kettler, R., Burkard, W., and Saner, A. (1971): Neurally mediated control of enzymes involved in the synthesis of norepinephrine; are they regulated as an operational unit? *Naunyn Schmiedeberg's Archives of Pharmacology,* 270:146.

Thoenen, H., Mueller, R. A., and Axelrod, J. (1969): Trans-synaptic induction of adrenal tyrosine hydroxylase. *Journal of Pharmacology and Experimental Therapeutics,* 169:249.

Thoenen, H., Mueller, R. A., and Axelrod, J. (1970): Phase difference in the induction of tyrosine hydroxylase in cell body and nerve terminals of sympathetic neurones. *Proceedings of the National Academy of Sciences,* 65:58.

Tranzer, J. P., Thoenen, H., Snipes, R. L., and Richards, J. G. (1969): Recent developments on the ultrastructural aspect of adrenergic nerve endings in various experimental conditions. *Progress in Brain Research,* 31:33.

Viveros, O. H., Arqueros, L., Connett, R. J., and Kirshner, N. (1969): Mechanism of secretion from the adrenal medulla: IV. The fate of the storage vesicle following insulin and reserpine administration. *Molecular Pharmacology,* 5:69.

Von Euler, U. S., and Hillarp, N.-A. (1956): Evidence for the presence of noradrenaline in submicroscopic structures of adrenergic axons. *Nature,* 177:44.

Yen, S.-S., Klein, R. L., and Chen-Yen, S.-H. (1973): Highly purified splenic nerve vesicles: Early post mortem effects on norepinephrine content and pools. *Journal of Neurocytology,* 2:1-12.

Neuropsychopharmacology of Monoamines and Their Regulatory Enzymes, edited by E. Usdin.
Raven Press, New York © 1974.

Human Serum Dopamine-β-Hydroxylase: Relationship to Sympathetic Activity in Physiological and Pathological States

Menek Goldstein, Lewis S. Freedman, Richard P. Ebstein, Dong H. Park, and Takeshi Kashimoto

Department of Psychiatry, Neurochemistry Laboratories, New York University Medical Center, New York, New York 10016

I. INTRODUCTION

The question whether serum dopamine-β-hydroxylase (DBH) activity serves as a reliable index of sympathetic function in man is still unresolved. The following observations made in studies with experimental animals support the idea that serum DBH activity levels might serve as an index of sympathetic activity. First, DBH is released together with catecholamines during the stimulation of adrenal medulla (Viveros, Arqueros, and Kirshner, 1968) or of the sympathetic nerves (Smith, 1971; Geffen and Livett, 1971). Furthermore, a proportional release of norepinephrine (NE) and DBH from the peripheral nerves has been demonstrated (Weinshilboum, Thoa, Johnson, Kopin, and Axelrod, 1971*a*). Second, serum DBH activity has been shown to be increased by sympathetic stimulation such as forced immobilization (Weinshilboum, Kvetňansky, Axelrod, and Kopin, 1971*b*) or acute swim stress (Roffman, Freedman, and Goldstein, 1973). Third, destruction of sympathetic nerves by administration of 6-hydroxydopamine causes a significant reduction of serum DBH activity levels (Weinshilboum and Axelrod, 1971*a*). However, serum DBH activity levels in humans correlate poorly with sympathetic function. Although serum DBH activity levels responded to gross changes in sympathetic function produced by exposure of humans to various types of physical stress (i.e., cold pressor test and exercise) (Wooten and Cardon, 1973; Planz and Palm, 1973), no relationship was found between prevailing blood pressure and serum DBH activity (Freedman, Roffman, and Goldstein, 1973*a*). Furthermore, when compared to the striking alterations that occur in circulatory dynamics during the cold pressor test (Greene, Boltax, Lustig, and Rogow, 1965), the changes in serum DBH activity were rather small (Wooten and Cardon, 1973).

The great variation between individuals with normal or seemingly comparable sympathetic nervous system function in serum DBH activity levels indicates that other factors besides sympathetic activity are involved in determining the DBH levels in serum. The factors which influence the steady state levels of circulatory DBH are depicted in Fig. 1. The levels of circulatory DBH reflect a balance between rates of release of the enzyme from sympathetic nerve terminals and its clearance from the blood. The enzyme is inactivated to some extent prior to its removal from the circulation, and therefore total DBH levels (active plus inactive enzyme) as measured by the radioimmunoassay might rather reflect changes in sympathetic function than serum enzyme activity levels. Recently it was reported that hereditary factors influence the basal serum DBH activity levels (Ross, Wetterberg, and Myrhed, 1973; Weinshilboum, Raymond, Elveback, and Weidman, 1973). These observations led us to reexamine some previous studies on serum DBH activity levels in various physiological and pathological states, and we have interpreted our results in the light of these new findings.

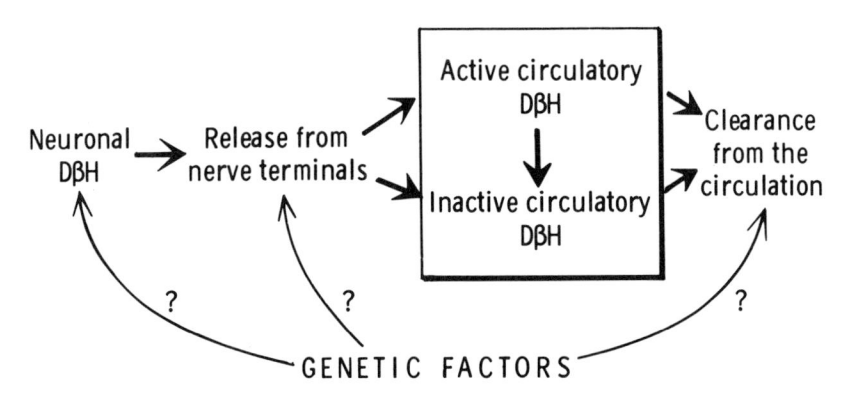

FIG. 1. Factors influencing the basal levels of serum DBH.

II. SERUM DBH LEVELS IN THE NORMAL POPULATION: ENZYMATIC ASSAY VERSUS RADIOIMMUNOASSAY

A solid state radioimmunoassay utilizing antibodies to sheep adrenal DBH and ^{125}I-labeled sheep adrenal DBH was described (Rush and Geffen, 1972). There is a considerable cross-species loss in immunoreactivity (Ohuchi, Joh, Freedman, and Goldstein, 1972), and therefore we found that the use of antibodies directed toward bovine DBH were not suitable for measuring human serum DBH levels. We have modified the radioimmunoassay for determination of circulatory human DBH by using ^{125}I-labeled human DBH to inhibit competitively the binding of human serum DBH to antibody directed toward human enzyme (Ebstein, Park, Freedman, Levitz, Ohuchi, and Goldstein, 1973). The results of our study have shown that serum immunoreactive (IR)-DBH levels vary among normal individuals, but the range of values is narrower than that of enzymatic activities. There is a preponderance of sera with low IR-DBH levels in the low enzyme activity group and a preponderance of sera with high IR-DBH levels in the high enzyme activity group. A plot of enzyme activity levels versus IR enzyme DBH levels reveals that there is a significant correlation between enzyme activity levels and IR-DBH levels (Fig. 2).

The simultaneous determination of circulatory DBH activity and of IR-DBH levels could yield information as to whether or not the alterations in circulatory enzyme levels are attributed to changes in the rate of enzyme inactivation prior to its elimination from the circulation. We have therefore measured the IR-DBH levels in various physiological and pathological states by the radioimmunoassay (Ebstein et al., 1973) and the enzyme activity levels by the enzymatic assay (Goldstein, Freedman, and Bonnay, 1971).*

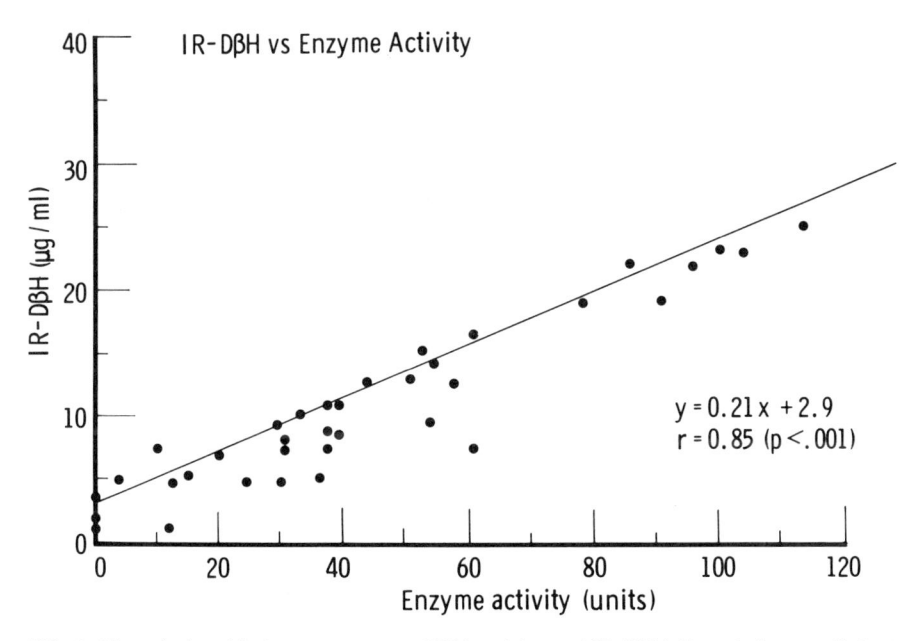

FIG. 2. The relationship between serum DBH activity and IR-DBH. Correlation coefficient $(r) = 0.85$ ($p < 0.001$). DBH activity units: nmole product/ml serum/20 min.

III. ELEVATION OF SERUM DBH LEVELS DURING COLD PRESSOR TEST

To determine whether serum DBH levels are sensitive to changes in sympathetic activity, we have investigated the effects of cold pressor test on serum IR-DBH levels and on DBH activity levels in seven healthy

---

* If not otherwise stated, the DBH activity values are expressed in units (1 nmole of product formed per 1 ml serum per hr) and IR-DBH is expressed as µg/ml.

108   HUMAN SERUM DBH

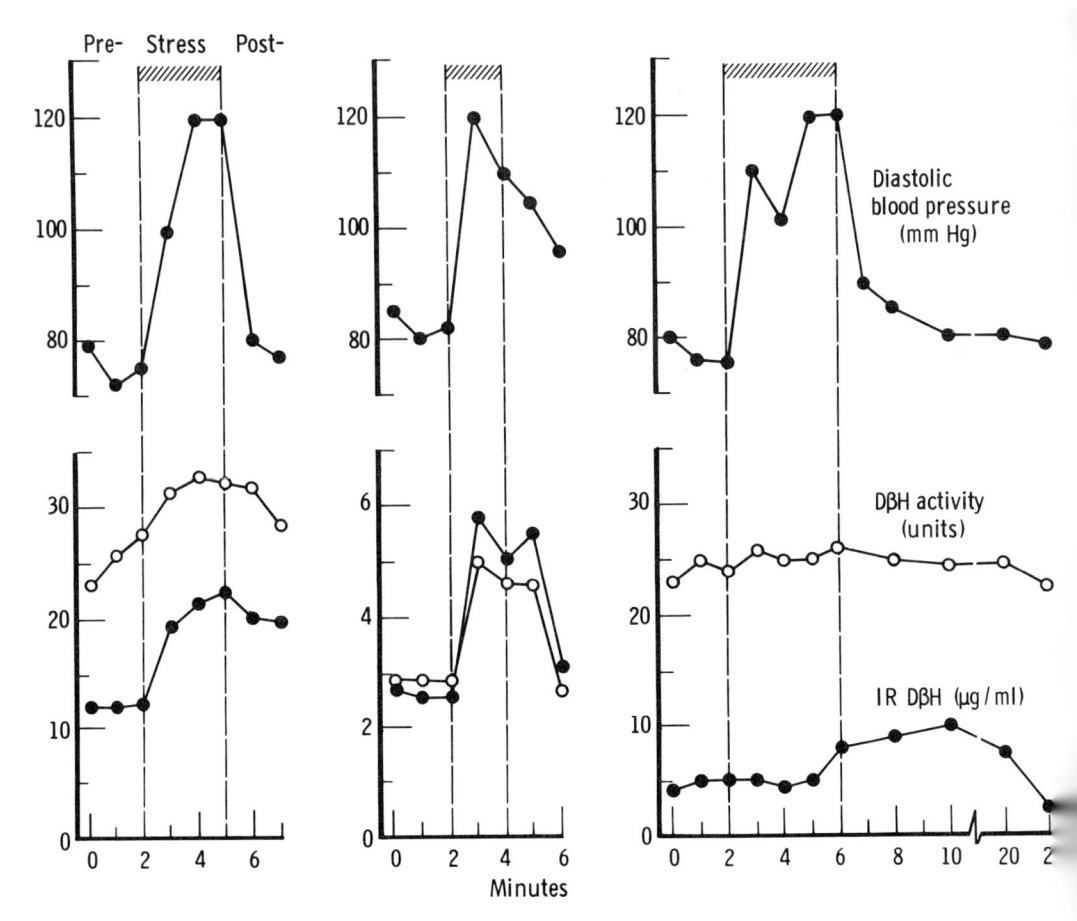

FIG. 3. The responses of three healthy individuals to changes in serum IR-DBH, DBH activity, and diastolic blood pressure during and following cold pressor test. DBH activity units: 1 nmole product/ml serum/20 min.

individuals (Freedman, Ebstein, Park, Levitz, and Goldstein, 1973b). Figure 3 depicts typical responses of three healthy individuals to changes in serum DBH levels during and following the cold pressor test. In two individuals there was a parallel increase in diastolic blood pressure (DBP), serum IR-DBH, and DBH activity. In the third individual there was no rise in serum DBH activity, and the increase in serum IR-DBH did not parallel the rise in DBP. It is of interest to note that the time required for restoration of serum IR-DBH and of DBH activity levels to prestress levels varies in each individual. The DBH levels were restored more rapidly in the individual with low basal DBH levels than in the two individuals with higher basal DBH levels. Based on these data, we are tempted to postulate that differences in basal DBH levels among normal individuals might be due to a different rate of enzyme elimination from the circulation.

The findings that in some individuals serum IR-DBH levels rise following cold pressor test while serum DBH activity remains unaltered indicate that DBH released into the circulation might be inactivated in various individuals at a different rate. Since radioimmunoassay measures the total DBH in the circulation rather than the variable activity of the enzymes, it seems that serum IR-DBH levels may reflect more sensitively changes in sympathetic function than serum DBH activity levels.

IV. SERUM DBH LEVELS IN SOME NEUROPSYCHIATRIC DISORDERS

A. Familial Dysautonomia

Familial dysautonomia (F.D.) is an inherited autosomal recessive disease with protean neurological manifestations, dysfunctions of the autonomic nervous system as indicated by vasomotor instability with hypertensive episodes, postural hypotension, and exaggerated responses to norepinephrine (NE). Autopsy examination of a child with F.D. revealed a deficiency in neuronal elements of the sympathetic ganglia (Pearson, Feingold, and Butzilovich, 1970). Since aberrations in catecholamine metabolism have been reported in F.D. (Smith, Taylor, and Wortis, 1963; Gitlow, Bertani and Wilk, 1970), we and others (Weinshilboum and Axelrod, 1971b; Freedman, Ohuchi, Goldstein, Axelrod, Fish, and Dancis, 1972) have investigated serum DBH activities in patients with this disorder. We have extended our studies to a larger population, and the results are summarized in Fig. 4.

Patients with F.D. show a similar variability in serum enzyme activity levels as the control population. There is no significant difference in serum DBH activity levels as compared to control subjects in the age group 1 to 5 years. In the 6-and-older age group the mean level in F.D. is slightly but significantly lower. Some F.D. patients have serum DBH activity levels one or more standard deviations below the mean of the control subjects, but some have enzyme activity levels higher than one standard deviation above the mean. The parents of the dysautonomia patients had a mean value similar to the age-matched control population. Enzyme activity levels in patients with F.D. correlate significantly with the levels of their mothers (correlation coefficient = 0.45; $p < 0.05$). A very good correlation exists in patients with F.D. between enzyme activity levels and IR-DBH levels (correlation coefficient = 0.80; $p < 0.01$).

The low values in serum DBH of some patients with F.D. might result from the diminution in the number of neurons in the sympathetic and sensory ganglia (Pearson et al., 1970). However, over half of the analyzed patients with F.D. failed to have a low level of serum DBH. It seems, therefore, that the disease may have some effect on the circulating levels of DBH, but not a decisive one. This point is perhaps best illustrated by the results obtained with one family. Three sons are affected with the disease; two have low levels and one has a normal serum DBH level. The healthy parents have normal serum DBH levels. F.D. is an autosomal recessive disease affecting one ethnic group, the Ashkenazi Jew, suggesting that the disease is genetically homogeneous. Thus, the findings that two affected siblings have low

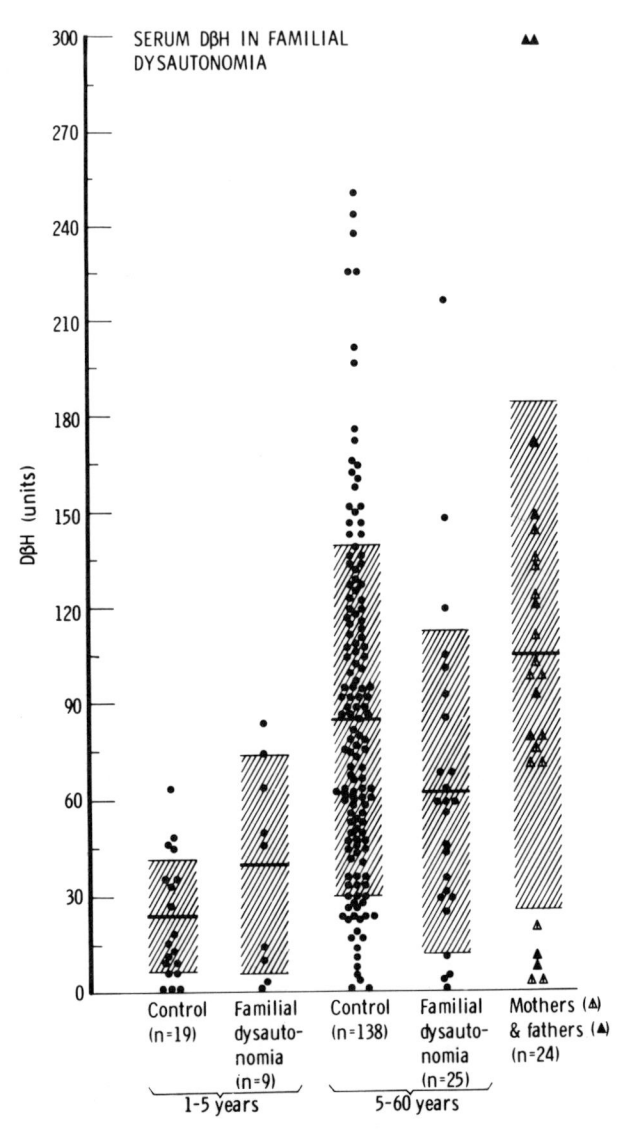

FIG. 4. The distribution of DBH activity in patients with familial dysautonomia and in their parents as well as in age-matched controls. The horizontal lines represent the means and the bars represent ±SD. The mean values ±SD are: controls (age group 1 to 5) 24.0 ± 21.06; patients with familial dysautonomia (age group 1 to 5) 34.0 ± 33.12; controls (over 6 years of age) 86.1 ± 54.32; patients with familial dysautonomia (age group over 6 years of age) 62.7 ± 49.61; parents of patients with familial dysautonomia, 105.5 ± 29.68.

enzyme levels while the third has normal levels do not support the previously presented idea that a subgroup of "DBH-negative" patients have a genetic defect involving DBH as the cause of this disease (Weinshilboum and Axelrod, 1971b).

This study was carried out in collaboration with Dr. Felicia Axelrod and Dr. Joseph Dancis from the Department of Pediatrics, New York University Medical Center.

B. Torsion Dystonia

Torsion dystonias (dystonia musculorum deformans) are characterized by irregular sustained involuntary movements not accompanied by dementia or other organ-system disease. Based on genetic and clinical grounds, these disorders were classified as autosomal recessive, autosomal dominant, and acquired form (Eldridge, Ryan, and Brody, 1969). Recently, it was reported that serum DBH activity levels in the autosomal dominant group were higher than those in the autosomal recessive group (Wooten, Eldridge, Axelrod, and Stern, 1973). We have investigated serum DBH activity levels and serum IR-DBH levels in six families in which one member is affected with the autosomal recessive form and in three families in which one or more members are affected with the autosomal dominant form (Fig. 5). The enzyme activity levels are elevated (>50 units) in four of five patients with the autosomal recessive form and in some healthy members of their families. In the autosomal dominant form, some patients and some healthy members of their families have elevated serum DBH levels, while others (Jewish origin; KA, DE) have DBH activity levels within the normal range.

The correlation between enzyme activity and IR-DBH is of the same order of magnitude as in the healthy control population (Ebstein et al., 1973; correlation coefficient = 0.80; $p < 0.01$). This finding indicates that the elevated serum DBH activity levels could not be attributed to diminished inactivation rate of the enzyme in the circulation. A good correlation exists between enzyme activity or IR-DBH levels of the mothers or fathers versus their children (Fig. 6). The intrafamilial correlation in this analyzed population group is similar to the correlation reported in the control population (Weinshilboum et al., 1973). Thus, it seems that the serum DBH levels are primarily influenced by the hereditary factors and not by clinical manifestations of the disease. Although there is a preponderance of sera with high DBH levels in patients as well as in healthy members of their families with both forms of torsion dystonia (the mean values in units ±SEM are: controls 28.7 ± 1.54; patients with the autosomal recessive form and members of their families 52.3 ± 5.60; patients with the autosomal dominant form and members of their families 37.9 ± 5.41), there is no correlation between the elevated DBH levels and the individuals that might carry the mutant gene. Thus, the serum DBH levels seem to be primarily influenced by the inheritance patterns and not by the pathology of the disease.

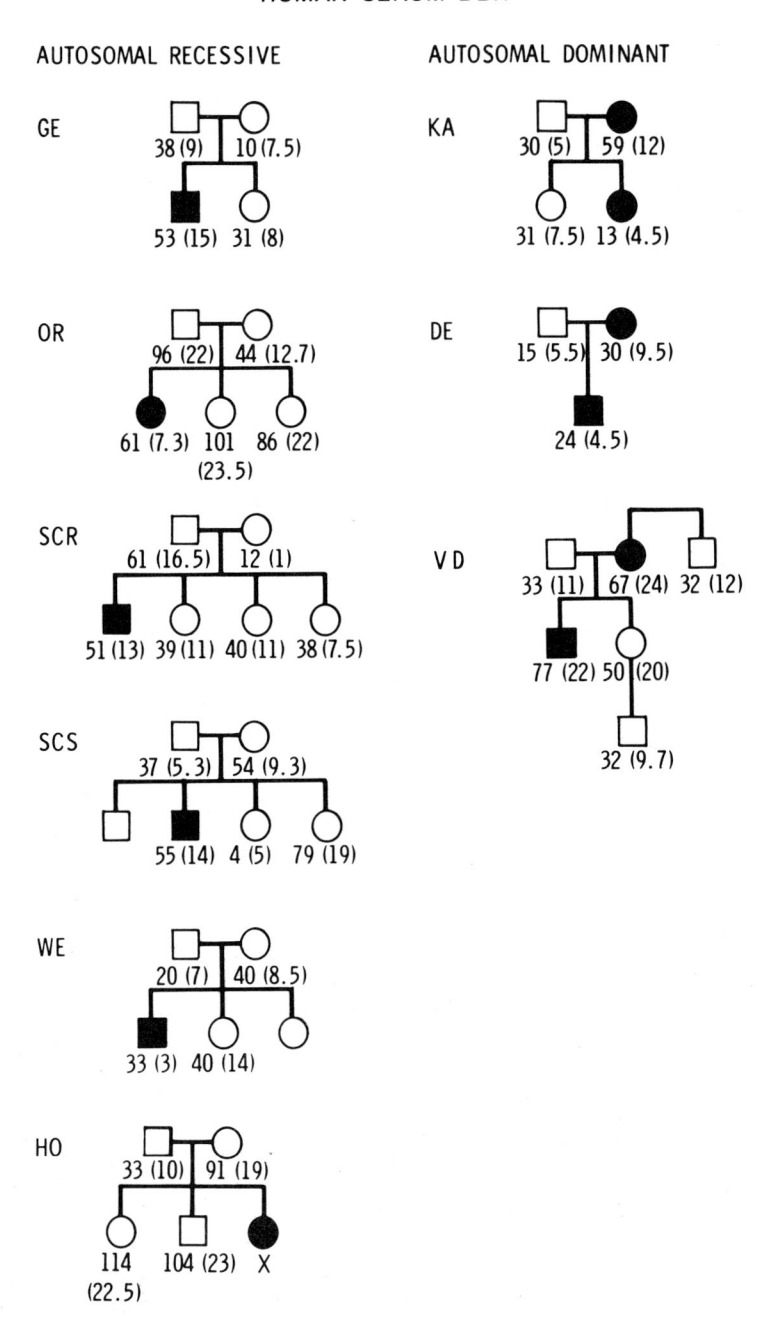

FIG. 5. Pedigrees of families with torsion dystonia. The figures represent enzyme activity (in units) of each individual. The figures in parentheses represent IR-DBH (in micrograms/milliliter) of each individual. The units are expressed as nanomoles of product formed per milliliter per 20 min. The patients are represented as black squares or circles; × represents deceased patient.

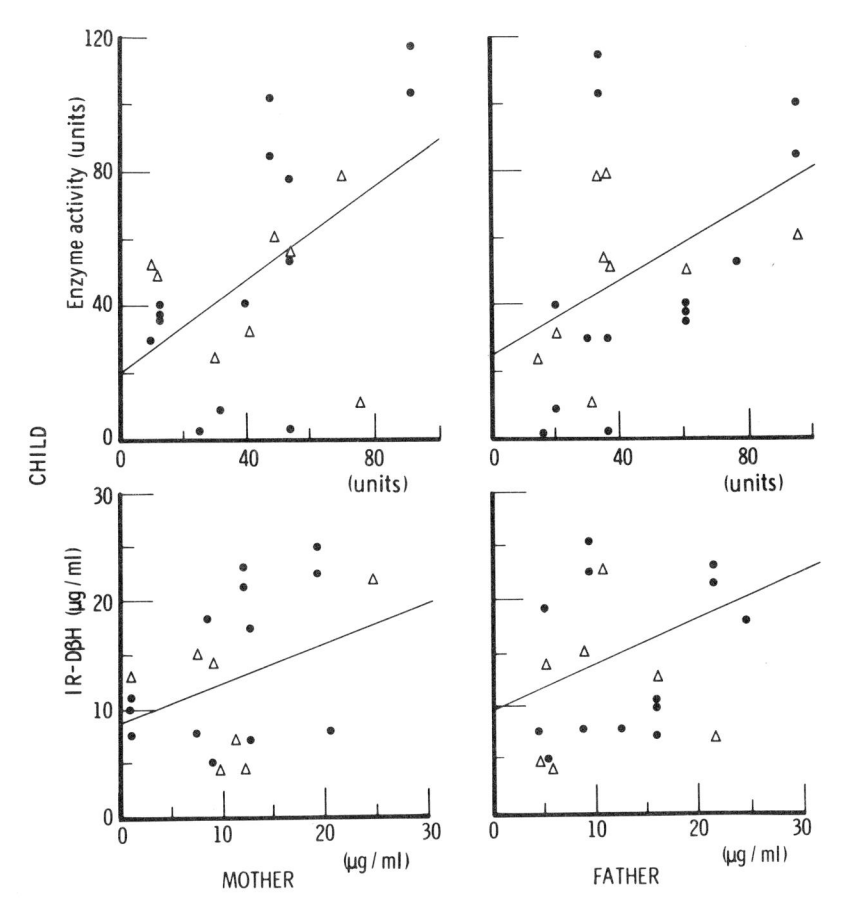

FIG. 6. The relationship of DBH activity and IR-DBH of mothers or fathers vs. their children. Upper left: Enzyme activity (mother vs. child) $r = 0.52$ ($p < 0.05$). Upper right: Enzyme activity (father vs. child) $r = 0.43$ ($p < 0.05$). Lower left: IR-DBH (mother vs. child) $r = 0.44$ ($p < 0.05$). Lower right: IR-DBH (father vs. child) $r = 0.29$; N.S.

The results of our study show that serum DBH levels do not differentiate between patients affected with the autosomal dominant or the autosomal recessive form of the disease. There would appear to be no advantage in classifying patients with one or the other form of the disease according to their serum DBH levels as previously suggested by others (Wooten et al., 1973).

This study was carried out in collaboration with Dr. Mary Coleman of Washington, D.C., Dr. Abraham Lieberman of the Department of Neurology, New York University Medical Center, and Dr. Bernard Pasternack of the Department of Environmental Medicine, New York University Medical Center.

C. Down's Syndrome

It was reported that patients with Down's syndrome have low serum DBH activity (Wetterberg, Gustavson, Backstrom, and Ross, 1972; Coleman, Lodge, Barnett, and Cytryn, 1973). We have now extended these studies to a larger population of patients. The results presented in Fig. 7 show that there is a preponderance of sera with low serum DBH activity among patients with Down's syndrome. The mean serum DBH activity value was significantly lower in patients with Down's syndrome than in either the age-matched normal controls or the age-matched nonmongoloid mentally disturbed children. It is of interest to note that the mean serum DBH activity of the nonmongoloid mentally disturbed children is higher than that of the normal children. However, it is necessary to corroborate this finding in a study of a larger population.

The mean value of serum IR-DBH levels in patients with Down's syndrome is also significantly lower than that of age-matched normal controls. This finding indicates that the diminished serum DBH activity levels in the mongoloid patients cannot be attributed to an increased rate of enzyme inactivation prior to its removal from the circulation. To determine whether the diminished serum DBH levels in patients with Down's syndrome reflect hereditary patterns or are related to trisomy 21, we are investigating the correlation between serum DBH levels of parents and their children. The low serum DBH values in mongoloids may not be attributed to heredity but rather may reflect the chromosomal aberrations characteristic for this disorder. However, if the low serum DBH levels of patients with Down's syndrome are due to familial factors, then the serum enzyme levels of prospective parents might be of prognostic value.

This study was carried out in collaboration with Dr. Mary Coleman of Washington, D.C.

V. DBH AND MENTAL DISORDERS

Sympathetic nervous discharge has been associated with affective states of fear and anxiety in humans. The norepinephrine hypothesis of affective disorders (Schildkraut and Kety, 1967) has been useful for further studies on the etiopathogenesis underlying this group of endogenous illnesses. In our previous studies we found no differences in serum DBH levels between patients with affective states and the age-matched control population (Shopsin, Freedman, Goldstein, and Gershon, 1972). Since heredity factors are important in the etiology of schizophrenia and also influence serum DBH levels, it is of interest to investigate whether there is any correlation between the inheritance patterns of schizophrenics and serum DBH levels. It is possible that variation in serum DBH levels caused by hereditary differences may mask changes caused by the disease.

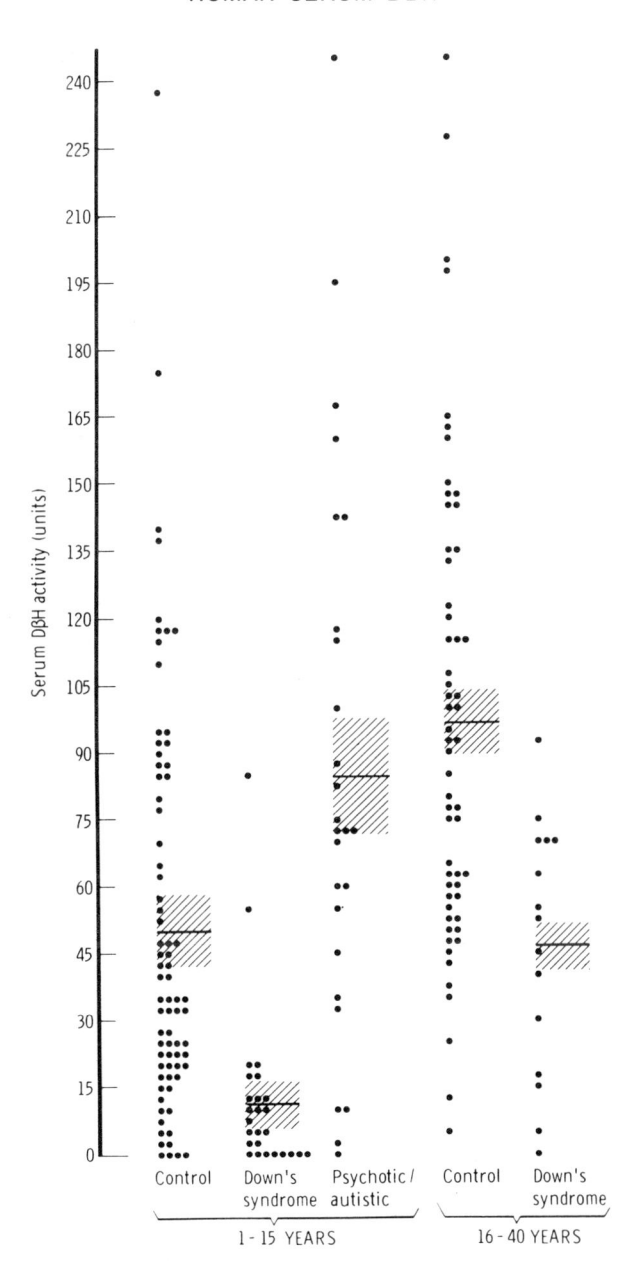

FIG. 7. The distribution of DBH activity in Down's syndrome patients, in normal controls, and in nonmongoloid mentally disturbed patients. The horizontal lines represent the mean and the bars represent ±SEM.

Recently it was reported that DBH is decreased in postmortem brain obtained from schizophrenic patients (Wise and Stein, 1973), and these results were interpreted as consistent with the hypothesis that noradrenergic "reward pathways" are damaged in schizophrenia. Further studies are now required to corroborate or deny this challenging hypothesis.

Inherent in the biochemical study of the human brain is the inability to obtain fresh material such as that used with laboratory animals. The measurement of DBH activity in postmortem brains could yield erroneous results, since enzymatic activity of DBH is labile and could vary with storage time. Furthermore, the enzymatic activity could be influenced by a variety of clinical and postmortem conditions (i.e., drug treatment, interfering disease processes, time interval between death and freezing of the brain). Since radioimmunoassay measures the total DBH protein, which might not be as sensitive to postmortem conditions as the labile enzyme activity, we have modified the radioimmunoassay for measurements of DBH in human postmortem brain and in other tissues.

The results in Table 1 show that DBH activity as well as IR-DBH are present in the norepinephrine-containing areas of the human postmortem brain. DBH was not detectable in the dopamine-containing areas. Thus, DBH could serve as a marker of the noradrenergic system in the CNS. Preliminary results indicate that IR-DBH levels are not altered during prolonged storage periods of the postmortem brain at −20°C. The radioimmunoassay for measuring DBH in postmortem brain could represent a new approach for investigating the role of the noradrenergic system in schizophrenia.

TABLE 1. *Dopamine-β-hydroxylase activity and immunoreactive-dopamine-β-hydroxylase levels in various regions of a human postmortem brain*

Region	DBH activity[a]	IR-DBH[b] (μg/g)
Hypothalamus	1.50	1.15
Mesencephalon	0.70	1.05
Amygdala	N.D.	0.25
Striatum	N.D.	N.D.

The brain was dissected 24 hr after death.

[a] The activity is expressed as nanograms of product formed per gram of wet tissue per hour of incubation time.

[b] The IR-DBH levels are expressed as micrograms per gram of wet tissue. For the radioimmunoassay the enzyme was precipitated in 80% $(NH_4)_2SO_4$. (Details of the procedure will be published elsewhere.)

N.D. = not detectable.

VI. SUMMARY AND CONCLUSIONS

Human serum dopamine-β-hydroxylase (DBH) activity and immunoreactive (IR)-DBH levels were determined in various physiological and pathological states. There is a significant correlation between enzyme activity levels and IR-DBH levels in the normal control population.

The effect of pressor response in humans on circulatory IR-DBH and on DBH activity was investigated. The results indicate that serum DBH levels correlate with other changes in sympathetic function. It is concluded that the use of radioimmunoassay for the measurement of serum DBH as an index of sympathetic activity represents an advance over the measurement of DBH activity.

Serum DBH levels were measured in patients with various disorders and in healthy members of their families. The results of our studies show that the previously reported abnormal levels of serum DBH activity in patients with familial dysautonomia and in patients with torsion dystonia reflect familial influences rather than the pathology of the disease. Serum DBH cannot be used as a diagnostic tool for classification of the two forms of torsion dystonia, namely autosomal recessive and autosomal dominant forms.

The serum DBH activity and serum IR-DBH levels are significantly lower in patients with Down's syndrome than in age-matched controls. The low serum DBH values in patients with Down's syndrome may not be attributed to familial influences but may rather reflect the chromosomal aberrations characteristic for this disorder.

Serum DBH levels failed to differentiate among patients with various psychiatric disorders, and the values were compatible with those obtained for normal control subjects. The correlation between the inheritance patterns of schizophrenics and human serum DBH levels is now under investigation.

The availability of the radioimmunoassay for measuring DBH in human postmortem brain may be useful in monitoring changes in DBH levels in the postmortem brains of schizophrenics.

Based on our own and other studies, we conclude that serum DBH levels can only be interpreted for clinical purposes when (1) an evaluation is made of the influence of familial factors in control of basal serum DBH levels or (2) when individual serum DBH levels are monitored serially in longitudinal studies.

ACKNOWLEDGMENTS

This work was supported by U.S. Public Health Service grant MH-02717 and National Science Foundation grant GB-27603.

REFERENCES

Coleman, M., Lodge, A., Barnett, A., and Cytryn, L. (1973): *Serotonin in Down's Syndrome.* North-Holland Publishing Co., Amsterdam.
Ebstein, R. P., Park, D. H., Freedman, L. S., Levitz, S. M., Ohuchi, T., and Goldstein, M. (1973): A radioimmunoassay of human circulatory dopamine-β-hydroxylase. *Life Sciences,* 13:769–774.
Eldridge, R., Ryan, E., and Brody, J. (1969): Dystonia musculorum deformans: Evidence for two hereditary forms. In: *Progress in Neuro-Genetics: Proceedings of the Second International Congress of Neuro-Genetics and Neuro-Opthamology of the World Federation of Neurology, Montreal, September 17–22, 1967,* Vol. 1 (International Congress Series No. 175), edited by A. Barbeau and J.-R. Brunette. Excerpta Medica, Amsterdam, pp. 772–778.
Freedman, L. S., Ebstein, R. P., Park, D. H., Levitz, S. M., and Goldstein, M. (1973b): The effect of cold pressor test in man on serum immunoreactive dopamine-β-hydroxylase and on dopamine-β-hydroxylase activity. *Research Communications in Chemical Pathology and Pharmacology,* 6:873–879.
Freedman, L. S., Ohuchi, T., Goldstein, M., Axelrod, F., Fish, I., and Dancis, J. (1972): Changes in human serum dopamine-β-hydroxylase with age. *Nature,* 236:310–311.
Freedman, L. S., Roffman, M., and Goldstein, M. (1973a): Changes in human serum dopamine-β-hydroxylase in various physiological and pathological states. In: *Frontiers in Catecholamine Research.* Pergamon Press, Elmsford, N.Y., pp. 1109–1114.
Geffen, L. B., and Livett, B. G. (1971): Synaptic vesicles in sympathetic neurons. *Physiological Reviews,* 51:98–157.
Gitlow, S. E., Bertani, L. M., and Wilk, E. (1970): Excretion of catecholamine metabolites by children with familial dysautonomia. *Pediatrics,* 46:513–522.
Goldstein, M., Freedman, L. S., and Bonnay, M. (1971): An assay for dopamine-β-hydroxylase activity in tissues and in serum. *Experientia* (Basel), 27:632–633.
Greene, M. A., Boltax, A. J., Lustig, G. A., and Rogow, E. (1965): Circulatory dynamics during the cold pressor test. *American Journal of Cardiology,* 16:54–60.
Ohuchi, T., Joh, T. H., Freedman, L. S., and Goldstein, M. (1972): *Fifth International Congress on Pharmacology,* Abstract 1022, p. 171.
Pearson, J., Feingold, M., and Butzilovich, G. (1970): The nervous system in familial dysautonomia. *Pediatrics,* 45:739–745.
Planz, G. and Palm, D. (1973): Acute enhancement of dopamine-β-hydroxylase activity in human plasma after maximum work load. *European Journal of Clinical Pharmacology,* 5:255–258.
Roffman, M., Freedman, L. S., and Goldstein, M. (1973): The effect of acute and chronic swim stress on dopamine-β-hydroxylase activity. *Life Sciences,* 12:369–376.
Ross, S. B., Wetterberg, L., and Myrhed, M. (1973): Genetic control of plasma dopamine-β-hydroxylase. *Life Sciences,* 12:529–532.
Rush, R. A., and Geffen, L. B. (1972): Radioimmunoassay and clearance of circulating dopamine-β-hydroxylase. *Circulation Research,* XXXI:444–452.
Schildkraut, J. J., and Kety, S. S. (1967): Biogenic amines and emotions. *Science,* 156:21–30.
Shopsin, B., Freedman, L. S., Goldstein, M., and Gershon, S. (1972): Serum dopamine-β-hydroxylase (DβH) activity and affective states. *Psychopharmacologia* (Berlin), 27:11–16.
Smith, A. A., Taylor, T., and Wortis, S. B. (1963): Abnormal catechol amine metabolism in familial dysautonomia. *New England Journal of Medicine,* 268:705–707.
Smith, A. D. (1971): Secretion of proteins (chromogranin A and dopamine-β-hydroxylase) from a sympathetic neurone. *Philosophical Transactions of the Royal Society of London* (B), 261:363–370.
Viveros, O. H., Arqueros, L., and Kirshner, N. (1968): Release of catecholamines and dopamine-β-oxidase from the adrenal medulla. *Life Sciences,* 7:609–618.
Weinshilboum, R., and Axelrod, J. (1971a): Serum dopamine-β-hydroxylase: Decrease after chemical sympathectomy. *Science,* 173:931–934.
Weinshilboum, R. M., and Axelrod, J. (1971b): Reduced plasma dopamine-β-hydroxylase activity in familial dysautonomia. *New England Journal of Medicine,* 285:938–942.
Weinshilboum, R. M., Kvetňanský, R., Axelrod, J., and Kopin, I. J. (1971b): Elevation of serum dopamine-β-hydroxylase activity with forced immobilization. *Nature New Biology,* 230:287–288.

Weinshilboum, R. M., Raymond, F. A., Elveback, L. R., and Weidman, W. H. (1973): Serum dopamine-β-hydroxylase activity: Sibling-sibling correlation. *Science*, 181:943–945.

Weinshilboum, R. M., Thoa, N. B., Johnson, D. G., Kopin, I. J., and Axelrod, J. (1971a): Proportional release of norepinephrine and dopamine-β-hydroxylase from sympathetic nerves. *Science*, 174:1349–1351.

Wetterberg, L., Gustavson, K.-H., Backstrom, M., Ross, S. B., and Froden, O., (1972): Low dopamine-β-hydroxylase activity in Down's syndrome. *Clinical Genetics*, 3:152–153.

Wise, C. D., and Stein, L. (1973): Dopamine-β-hydroxylase deficits in the brains of schizophrenic patients. *Science*, 181:344–347.

Wooten, G. F., and Cardon, P. V. (1973): Plasma dopamine-β-hydroxylase activity. *Archives of Neurology*, 28:103–106.

Wooten, G. F., Eldridge, R., Axelrod, J., and Stern, R. S. (1973): Elevated plasma dopamine-β-hydroxylase activity in autosomal dominant torsion dystonia. *New England Journal of Medicine*, 288:284–287.

Neuropsychopharmacology of Monoamines and Their Regulatory Enzymes, edited by E. Usdin. Raven Press, New York © 1974.

Evaluation of Serum Dopamine-β-hydroxylase Activity as an Index of Sympathetic Nervous Activity in Man

Walter Lovenberg, Eleanor A. Bruckwick, R. Wayne Alexander, David Horwitz, and Harry R. Keiser

Experimental Therapeutics Branch, National Heart and Lung Institute, National Institutes of Health, Bethesda, Maryland 20014

I. INTRODUCTION

The final enzyme in the biosynthesis of norepinephrine is dopamine-β-hydroxylase (DBH). This enzyme is localized in the synaptic vesicles of sympathetic nerves and is released together with the neurotransmitter by an exocytotic process upon appropriate stimulation. Although most of the released norepinephrine is either taken up by the nerve ending or metabolized, the fate of the enzyme is not clearly established. It is apparent from the work of Weinshilboum and Axelrod (1971a), Goldstein, Freedman, and Bonnay (1971), and Rush and Geffen (1972) that at least a portion of the enzyme appears in the general circulation. In most animal species the level of DBH activity in serum is relatively low; however, in man this enzyme is present in much higher levels. Assuming that the enzyme in human serum has about the same specific activity as the enzyme purified from the bovine adrenal gland, it is possible to calculate that normal adults can have up to 10 μg/ml of active enzyme protein in serum. The presence of large amounts of easily measured enzyme in serum and the fact that this enzyme is introduced into serum as a result of sympathetic discharge invited study of many physiologic processes thought to involve the sympathetic system.

II. LEVELS OF DBH ACTIVITY IN HUMAN SERUM

A survey of serum or plasma levels of DBH activity in a large number of subjects by several laboratories (Weinshilboum and Axelrod, 1971a; Freedman, Ohuchi, Goldstein, Axelrod, Fish, and Dancis, 1972; Nagatsu and Udenfriend, 1972; Wetterberg, Aberg, Ross, and Froden, 1972; Horwitz, Alexander, Lovenberg, and Keiser, 1973; Weinshilboum, Raymond, Elveback, and Weidman, 1973) showed that a very large variation in enzyme level existed in apparently normal individuals. In our laboratory serum DBH activity in normal subjects extended over a range of 100-fold (5 to 500 units/ml). The frequency distribution of enzyme activity in 141 subjects is shown in Fig. 1. Although this group contains a higher proportion of blacks and hypertensives than the general population, the distribution of levels is similar to that seen in a random population reported by Weinshilboum et al. (1973). The apparent bimodal distribution reported by Schanberg, Stone, Kirschner, Gunnells, and Robinson (1973) was not apparent in this study.

Since there was such a wide variation of serum DBH activity among individuals from a random sample, it was of interest to determine whether individuals also showed significant variations at different sampling times. We found that individuals had remarkably constant levels when measured over time periods up to 6 months (Horwitz et al., 1973). This would suggest that the level of serum DBH may be a genetic characteristic of the individual. In a limited examination of this point, Horwitz et al. (1973) found that a high proportion of siblings of subjects with low enzyme levels also had low levels of serum DBH. In a much more sophisticated study of this point, Weinshilboum et al. (1973) found a highly statistically significant correlation of the enzyme levels among siblings.

The extremely wide range in serum enzyme levels makes it difficult to conceive that this enzyme is an accurate reflection of normal sympathetic nerve activity. A study of the cardiovascular responses that involve the sympathetic nervous system in individuals having either high or low serum DBH activity clearly indicated that there was no difference in the subjects' sympathetically mediated physiologic responses (Horwitz et al., 1973). Although it is believed that the conditions for enzyme activity occur only intracellularly, it was of interest to examine subjects with widely varying serum enzyme activity for *in vivo* beta hydroxylation. Using the test system

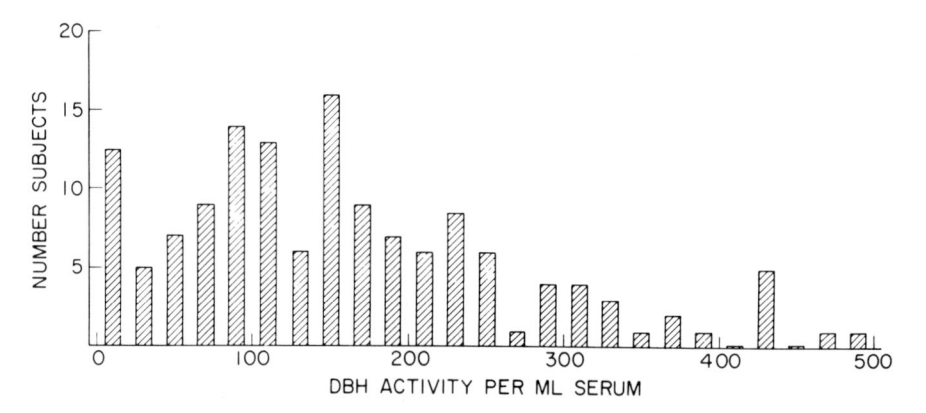

FIG. 1. Frequency distribution of serum dopamine-β-hydroxylase activity in human subjects. Data are from Horwitz et al. (1973).

devised by Sjoerdsma and von Studnitz (1963), in which the conversion of p-hydroxyamphetamine to p-hydroxynorephedrine is measured, it was found that individuals with high serum DBH activity were similar to those with low activity in their conversion ratios. Therefore, it is not likely that the circulating enzyme is active *in vivo*.

III. EFFECTS OF MANIPULATION OF SYMPATHETIC NERVE ACTIVITY ON SERUM DBH ACTIVITY

While resting levels of serum DBH do not appear to reflect normal sympathetic nerve activity, changes from control levels may be a reasonable index of increases or decreases in the basal rates of sympathetic nerve discharge. Three experimental approaches have been attempted to examine this concept. First, the cold pressor test, which elicits an increase in both heart rate and blood pressure, was not found to change serum DBH activity in two subjects examined in our laboratory. Wooten and Cardon (1973), however, observed variable but significant increases in plasma DBH activity following the cold pressor tests in six subjects. Second, bicycle exercise sufficient to increase heart rate to 150 beats/min was found to give

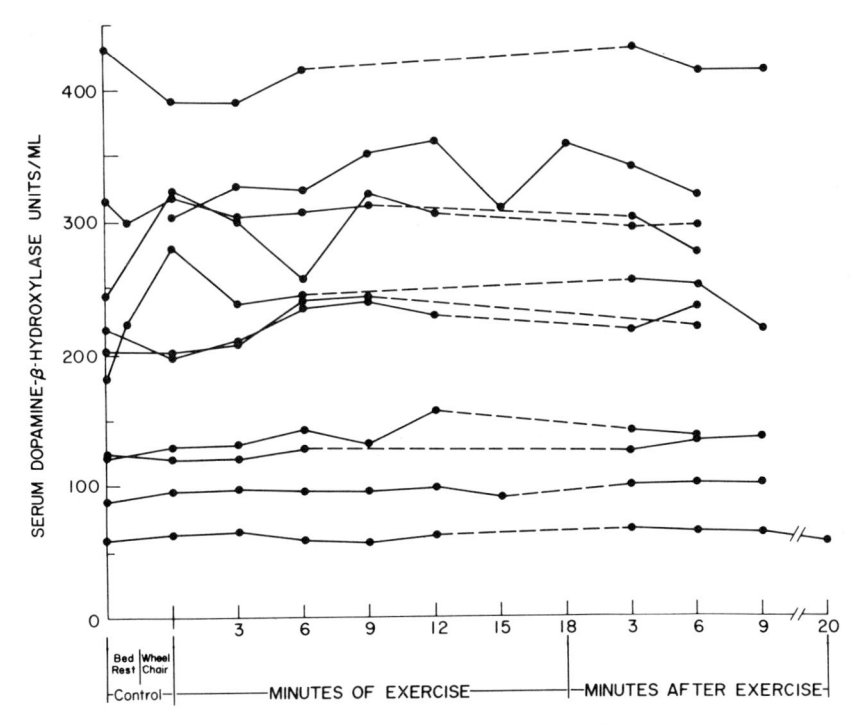

FIG. 2. Effect of bicycle exercise on serum dopamine-β-hydroxylase activity. Data are from Horwitz et al. (1973).

equivocal results. In some subjects a small but significant increase in enzyme activity occurred; however, in others no elevation of serum DBH activity was apparent (Fig. 2). Conflicting results have been reported by two other laboratories. Wooten and Cardon (1973) found qualitatively similar results, whereas Wetterberg et al. (1972) found absolutely no change with vigorous exercise. The third approach was to attempt to reduce sympathetic activity by the infusion of large amounts of normal saline; in this study Alexander, Gill, Yamabe, Lovenberg, and Keiser (1973) found a significant decrease in enzyme activity.

Based on the above experiments, we must conclude that while some changes in circulating DBH activity can be measured following some maneuvers that change sympathetic nerve activity, the enzyme cannot be used as a sensitive or accurate reflection of such *in vivo* activity.

IV. POTENTIAL RELATIONSHIP OF SERUM DBH ACTIVITY TO HYPERTENSION

A question of utmost importance that was recognized and investigated by several groups is the possible relationship between serum DBH levels and essential hypertension. In a study involving 160 subjects, our laboratory (Horwitz et al., 1973) reported no apparent relationship between serum enzyme activity and systolic or diastolic blood pressure. The results of our study are summarized in Table 1. When all normotensives (diastolic ≤ 90 mm Hg) were compared to hypertensives (diastolic ≥ 99 mm Hg), there was no significant difference in serum DBH activity. However, when the subjects were divided on the basis of sex and race, mean DBH levels were significantly lower in black than in white subjects ($p < 0.01$) and in men than in women ($p < 0.05$). It should be noted that there was extensive overlapping among all the subgroups and that DBH levels could not be used to identify any of these groups. Three other laboratories have reported somewhat different findings. Louis, Doyle, and Anavekar (1973) and Geffen, Rush, Louis, and Doyle (1973a) found a positive correlation between blood

TABLE 1. *Serum dopamine-β-hydroxylase activity in normotensive and hypertensive humans*

	Mean serum DBH (range) units/ml	
	Normotensive	Hypertensive
Men		
White	190 (6–474)	224 (76–420)
Black	149 (10–399)	115 (3–305)
Women		
White	210 (5–476)	232 (53–422)
Black	204 (23–581)	159 (54–410)
Total	187 (5–581)	167 (3–422)

pressure and both plasma catecholamines and serum DBH activity. Schanberg et al. (1973) appeared to obtain a bimodal frequency distribution of serum DBH activity in their total population and reported that persons with high levels of DBH are the ones likely to exhibit labile hypertension. Wetterberg et al. (1972) found a statistically significant elevation in serum DBH activity in a small group of hypertensives. There appear to be significant differences which are not readily explained in the findings of these four laboratories.

The major differences may reside in the patient populations from which the hypertensive groups were selected and the manner of subgrouping. For example, in the study of Horwitz et al. (1973), the white hypertensives had a higher (statistically insignificant) mean serum DBH activity than the white normotensive group, whereas the situation was reversed with the black patients. Schanberg et al. (1973) divided the hypertensives into several subgroups, and it is the group described as "labile hypertensives" that has the highest DBH activity. It should be noted that presumably Geffen et al. (1973a), Schanberg et al. (1973), and Wetterberg et al. (1972) were using primarily white subjects in their hypertensive groups. It is entirely possible that determination of serum DBH activity may be useful in categorizing hypertensive patients. However, the studies to date suggest that serum DBH activity alone would not be sufficient for a diagnostically meaningful categorization.

Another possible explanation for differing results might be that each of the groups used slightly different assay techniques. A comparison of the two major types of assay (Fig. 3) indicates that this is not a likely explanation. A potentially more useful approach is the measurement of total DBH protein by a radioimmune type of assay. This was first used by Rush and Geffen (1972) and recently by Ebstein, Park, Freedman, Levitz, Ohuchi, and Goldstein (1973). The first procedure used antibodies directed against sheep antigen, and measured what appeared to be very low levels of enzyme protein (100 to 400 ng/ml) when compared to the enzyme activity (up to 10 μg/ml) (Rush and Geffen, 1972). On the other hand, Ebstein et al. (1973), using antibodies directed against antigen obtained from human pheochromocytoma, found protein levels some two to four times higher than those detected by enzyme assay and nearly 50 times greater than those found by Rush and Geffen (1972). The work of Ebstein et al. (1973) showed a direct relationship between total protein and measured enzyme activity. However, it would appear that the "specific activity" of DBH protein in subjects with low serum DBH enzyme activity was lower than in those with high DBH activity. This would suggest that the enzyme may be more rapidly degraded in individuals with low serum DBH levels. It would appear that the interpretation of the results from any study utilizing either the enzyme or immunoassay would be comparable and that analytical differences would not be a satisfactory explanation for differences in experimental results.

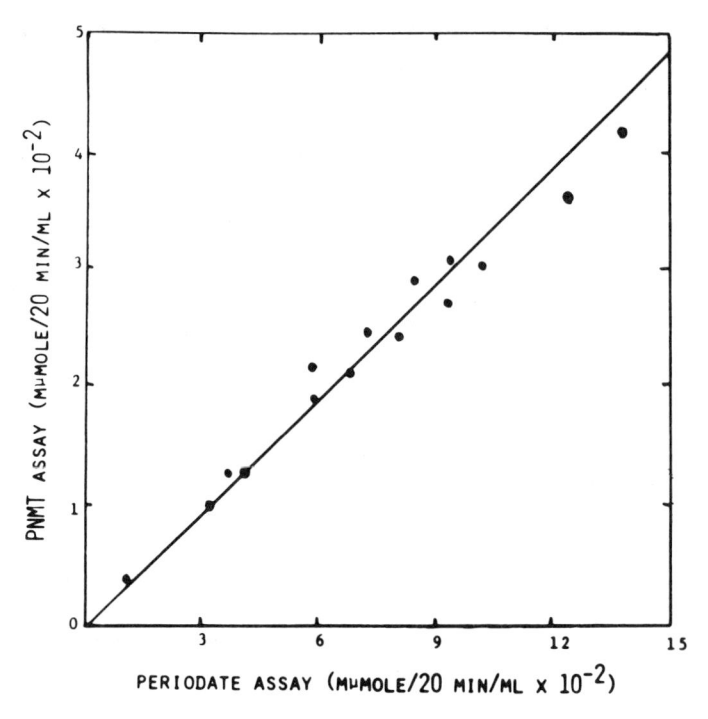

FIG. 3. Comparison of the PNMT and periodate assays for serum dopamine-β-hydroxylase activity. Assays were done on a random selection of samples using the methods exactly as described by Horwitz et al. (1973) and Nagatsu and Udenfriend (1972).

V. SERUM DBH ACTIVITY IN OTHER DISEASE STATES

The wide range in normal values for serum DBH activity has severely limited the use of this enzyme as a diagnostic marker for disease states. Although Wise and Stein (1973) have proposed an attractive hypothesis for the etiology of schizophrenia based on the deficiency of DBH in certain brain regions, studies to date have not revealed any apparent differences in serum DBH activity in these patients (Wetterberg et al., 1972; Dunner, Cohn, Weinshilboum, and Wyatt, 1973). Studies of Wetterberg et al. (1972) also show no statistical differences of serum DBH in various forms of manic-depressive illness, although increased plasma catecholamine concentrations have been reported in patients with depression (Portnoy, Engelman, and Wyatt, 1969). The enzyme has been reported to be consistently low in patients with either familial dysautonomia or Down's syndrome, although it is well within the overall normal range. Conversely, serum DBH activity appears to be slightly elevated in some patients with autosomal dominant torsion dystonia, Huntington's chorea, neuroblastoma, and pheochromocytoma (Weinshilboum and Axelrod, 1971b; Goldstein, Freedman, Bohuon, and Guerinot, 1972; Horwitz et al., 1973; Wetterberg, Gustavson, Bäckström, Ross, and Froden, 1973; Wooten, Eldridge, Axelrod, and Stern, 1973). In six pheochromocytoma patients examined in our laboratory, all had serum enzyme levels that were from the upper range of normal to about two times this upper limit. Two of these patients underwent surgery for removal of the tumor during the course of our study and had significantly lower serum DBH activity following surgery (500 units/ml to 50 units/ml and 600 units/ml to 250 units/ml). In the latter patient the kinetics of the decrease were examined and the half-life of the plasma activity found to be about 8 hr. Two other studies of serum DBH activity in pheochromocytoma have been reported. Bohuon, Guerinot, Tcherdakoff, and Bonnay (1973) found results comparable to those reported above. Geffen et al. (1973b), however, report that serum levels of the enzyme are not supranormal in this disease, although there is a small drop in enzyme activity following removal of the tumor. Since patients with this disease have disproportionately high levels of plasma catecholamines, they suggested that catecholamines and DBH may not be released from the tumor by an exocytotic process. This would support the concept of Winkler and Smith (1972) that these tumors have a defect in the control of synthesis and that norepinephrine continues to be synthesized after storage sites are filled. The neuroamine then simply diffuses from the cells.

VI. CONCLUDING REMARKS

Serum DBH activity may be a potentially useful parameter in the study of physiological or pathological processes related to the sympathetic nervous system. However, it is clear from the studies to date that considerable caution must be exercised in attempting to correlate serum DBH activity in man with particular functions or dysfunctions of the sympathetic nervous system. The wide range of normal values appear to be regulated primarily by genetic factors thus far not recognized as being associated with any disease state. The variable and somewhat equivocal results obtained in tests designed to alter sympathetic nerve activity also place limits on the use of this enzyme marker as an index of sympathetic nerve function. Certainly a great deal more must be learned about the half-life and mechanisms of degradation before this enzyme can be used as a definitive marker for sympathetic nerve activity.

REFERENCES

Alexander, R. W., Gill, J. R., Jr., Yamabe, H., Lovenberg, W., and Keiser, H. R. (1973): Effects of dietary sodium and of acute saline infusion on the interrelationship between dopamine excretion and adrenergic activity in man. *Journal of Clinical Investigations (submitted for publication).*
Bohuon, C., Guerinot, F., Tcherdakoff, P. H., and Bonnay, M. (1973): Dopamine-β-hy-

droxylase activity in five cases of pheochromocytoma. *Biochemical Society Transactions,* 1:152–153.

Dunner, D. L., Cohn, C. K., Weinshilboum, R. M., and Wyatt, R. J. (1973): The activity of dopamine-beta-hydroxylase and methionine-activating enzyme in blood of schizophrenic patients. *Biological Psychiatry,* 6:215–220.

Ebstein, R. P., Park, D. H., Freedman, L. S., Levitz, S. M., Ohuchi, T., and Goldstein, M. (1973): A radioimmunoassay of human circulatory dopamine-β-hydroxylase. *Life Sciences,* 13:769–774.

Freedman, L. S., Ohuchi, T., Goldstein, M., Axelrod, F., Fish, I., and Dancis, J. (1972): Changes in human serum dopamine-β-hydroxylase activity with age. *Nature,* 236:310–311.

Geffen, L. B., Rush, R. A., Louis, W. J., and Doyle, A. E. (1973a): Plasma dopamine-β-hydroxylase and noradrenaline amounts in essential hypertension. *Clinical Science,* 44:617–620.

Geffen, L. B., Rush, R. A., Louis, W. J., and Doyle, A. E. (1973b): Plasma catecholamines and dopamine-β-hydroxylase amounts in phaeochromocytoma. *Clinical Science,* 44:421–424.

Goldstein, M., Freedman, L. S., Bohuon, A. C., and Guerinot, F. (1972): Serum dopamine-β-hydroxylase activity in neuroblastoma. *New England Journal of Medicine,* 286:1123–1125.

Goldstein, M., Freedman, L. S., and Bonnay, M. (1971): Assay for dopamine-β-hydroxylase in tissues and serum. *Experientia,* 27:632–633.

Horwitz, D., Alexander, R. W., Lovenberg, W., and Keiser, H. R. (1973): Human serum dopamine-β-hydroxylase. Relationship to hypertension and sympathetic activity. *Circulation Research,* 32:594–599.

Louis, W. J., Doyle, A. E., and Anavekar, S. (1973): Plasma norepinephrine levels in essential hypertension. *New England Journal of Medicine,* 288:599–601.

Nagatsu, T., and Udenfriend, S. (1972): Photometric assay of dopamine-β-hydroxylase in human blood. *Clinical Chemistry,* 18:980–983.

Portnoy, B., Engelman, K., and Wyatt, R. (1969): Plasma catecholamines in hypertensive and psychiatric disorders. *Clinical Research,* 17:258.

Rush, R. A., and Geffen, L. B. (1972): Radioimmunoassay and clearance of circulating dopamine-β-hydroxylase. *Circulation Research,* 31:444–452.

Schanberg, S., Stone, R., Kirschner, N., Gunnells, J. C., and Robinson, R. R. (1973): Plasma DBH: An aid in the diagnosis and study of hypertension. *Pharmacologist,* 15:211.

Sjoerdsma, A., and von Studnitz, E. (1963): Dopamine-β-oxidase activity in man, using hydroxyamphetamine as substrate. *British Journal of Pharmacology,* 20:278–284.

Weinshilboum, R., and Axelrod, J. (1971a): Serum dopamine-β-hydroxylase activity. *Circulation Research,* 28:307–315.

Weinshilboum, R. M., and Axelrod, J. (1971b): Reduced plasma dopamine-β-hydroxylase activity in familiar dysautonomia. *New England Journal of Medicine,* 285:938–942.

Weinshilboum, R. M., Raymond, F. A., Elveback, L. R., and Weidman, W. H. (1973): Serum dopamine-β-hydroxylase activity: Sibling-sibling correlation. *Science,* 181:943–945.

Wetterberg, L., Aberg, H., Ross, S. B., and Froden, O. (1972): Plasma dopamine-β-hydroxylase activity in hypertension and various neuropsychiatric disorders. *Scandinavian Journal of Clinical and Laboratory Investigations,* 30:283–289.

Wetterberg, L., Gustavson, K. H., Bäckström, M., Ross, S. B., and Froden, O. (1973): Low dopamine-β-hydroxylase in Down's syndrome. *Clinical Genetics,* 3:152–153.

Winkler, H., and Smith, A. D. (1972): Phaeochromocytoma and other catecholamine producing tumours. In: *Handbook of Experimental Pharmacology.* Springer-Verlag, Berlin, 33:900–933.

Wise, C. D., and Stein, L. (1973): Dopamine-beta-hydroxylase deficits in the brains of schizophrenic patients. *Science,* 181:344–347.

Wooten, G. F., and Cardon, P. V. (1973): Plasma dopamine-β-hydroxylase activity. *Archives of Neurology,* 28:103–106.

Wooten, G. F., Eldridge, R., Axelrod, J., and Stern, R. S. (1973): Elevated plasma dopamine-β-hydroxylase activity in autosomal dominant torsion dystonia. *New England Journal of Medicine,* 288:284–287.

Neuropsychopharmacology of Monoamines and
Their Regulatory Enzymes, edited by E. Usdin.
Raven Press, New York © 1974.

Serum Dopamine-β-hydroxylase: A Possible Aid in the Evaluation of Hypertension

Saul M. Schanberg, Richard A. Stone, Norman Kirshner, J. Caulie Gunnells, and Roscoe R. Robinson

Duke University Medical Center, Durham, North Carolina 27706

Since it was first suggested that the soluble portion of dopamine-β-hydroxylase (DBH; EC 1.14.2.1), which is located primarily in the synaptic vesicles of postganglionic sympathetic neurons, might be released along with the release of norepinephrine, it was evident that a measure of plasma DBH activity might possibly serve as an index of the activity of the sympathetic nervous system. Thus far, however, the wide range of values reported to occur in normal subjects has limited its application to clinical studies (Nagatsu and Udenfriend, 1972; Freedman, Chuchi, Goldstein, Axelrod, Fish, and Dancis, 1972).

The present study was undertaken to define the range of plasma DBH activity in a group of apparently healthy subjects, and to establish its relationship to the quantitative excretion of urinary catecholamines and the day-to-day lability of blood pressure. Plasma DBH activity was measured in 82 apparently healthy subjects (ages 22 to 35 years) after the method of Nagatsu and Udenfriend (1972) (Table 1). The plasma activity of DBH ranged widely from 2 to 100 International Units, but a non-normal pattern of distribution was evident (chi-square test for goodness of fit, $p < 0.001$). Sixty-two subjects (76%) had values below 35 units, with a mean of 18 ± 1 unit, while 13 of the remaining 20 subjects (16%) had values above 60 units,

TABLE 1. *Distribution of plasma dopamine-β-hydroxylase activities[a] in a group of control subjects*

Range	Mean	±	SEM	N	Percent total N
2–100 (total)	31		3	82	100
2–35	18		1	62	76
36–59	48		2	7	9
60–100[b]	80		5	13	16

[a] International Units (μmole/min/liter plasma at 37°C).
Fifty-three of the 62 values in this group were below 25 ("low DBH group").
[b] "High DBH group."

with a mean of 80. Actually, 53 of the 62 subjects mentioned in the first group had values which fell below 25 units. Since the observed pattern of distribution was consistent with the possibility that more than one population might be included within this group of apparently healthy subjects, we randomly chose six subjects from the "low" DBH group, that is, those with DBH activity less than 25 units, and six subjects from the "high" group, that is, those with plasma DBH activity greater than 60 units, and carried out some further studies. Blood pressures were obtained for 7 consecutive days. Blood samples were obtained on days 2 and 5, and a 24-hr urine collection was obtained on day 2. Urinary concentrations of norepinephrine and epinephrine were determined fluorometrically (von Euler and Lishajko, 1961) and are expressed as micrograms per gram of creatinine. The results are shown in Table 2.

It can be seen that the DBH activity was constant from day to day in each individual and group, but in those subjects whose DBH activity was high the values for urinary catecholamines were at least twice those of the individuals exhibiting low DBH activity. Also, in subjects with high DBH activity the systemic arterial blood pressure varied widely from day to day,

TABLE 2. *Relation between plasma dopamine-β-hydroxylase activity,[a] urinary catecholamines, and blood pressure*

Subjects	DBH day 2	DBH day 5	μg CA g creat. 24 hr	ΔSystolic BP[d] range (mm Hg) / ΔDiastolic BP[d] range (mm Hg)	No. BP readings > 130/85
2–25 range[b]					
MT	10	9	19	20/20	0
WS	18	23	34	10/15	0
BJ	12	12	40	50/10[c]	1[c]
RH	18	14	28	28/20	0
GM	12	11	33	20/10	0
Mean ± SEM	14 ± 1	14 ± 2	33 ± 3	26 ± 5/15 ± 2	0.2 ± 0.2
60–100 range[b]					
KH	99	86	84	55/30	4
RN	57	70	55	40/35	4
BB	89	79	98	40/45	5
LB	79	82	74	60/45	8
LK	66	72	67	35/20	5
LF	61	63	52	45/20	6
Mean ± SEM	75 ± 6	75 ± 3	72 ± 6	46 ± 3/33 ± 4	5.3 ± 0.6
	$p < 0.001$	$p < 0.001$	$p < 0.001$	$p < 0.01/p < 0.005$	$p < 0.001$

[a] International Units (μmole/min/liter of plasma at 37°C).
[b] See Table 1.
[c] First reading was high systolic.
[d] ΔSystolic BP = Systolic$_{high}$ − Systolic$_{low}$ (for individual patient on particular day); similar for ΔDiastolic BP.

and on the average each of these individuals exhibited a blood pressure greater than 130/85 mm Hg on five occasions. In contrast, the systemic blood pressure exhibited much less lability from day to day in subjects with low DBH activity: only one observation was over 130/85, and that occurred when the subject had her blood pressure taken for the first time.

The correlation between plasma DBH activity and urinary catecholamines was excellent, as can be seen in Fig. 1. The correlation coefficient was $r = 0.93$. When only norepinephrine is plotted, the correlation is the same—the curve shifts slightly to the left but still has the same straight-line relationship. This degree of correlation between 24-hr urinary catecholamine excretion and plasma DBH activity supports the notion that each of these parameters is reflecting sympathetic nervous system activity and not just individual subject differences in their catabolism of these substances.

It is well recognized clinically that labile fluctuating arterial blood pressures occur in apparently healthy subjects as well as in patients with known fixed hypertension (Windesheim, Roth, and Hines, 1955; Doyle and Smirk, 1955; Esler and Goulston, 1973). The observed correlation between the

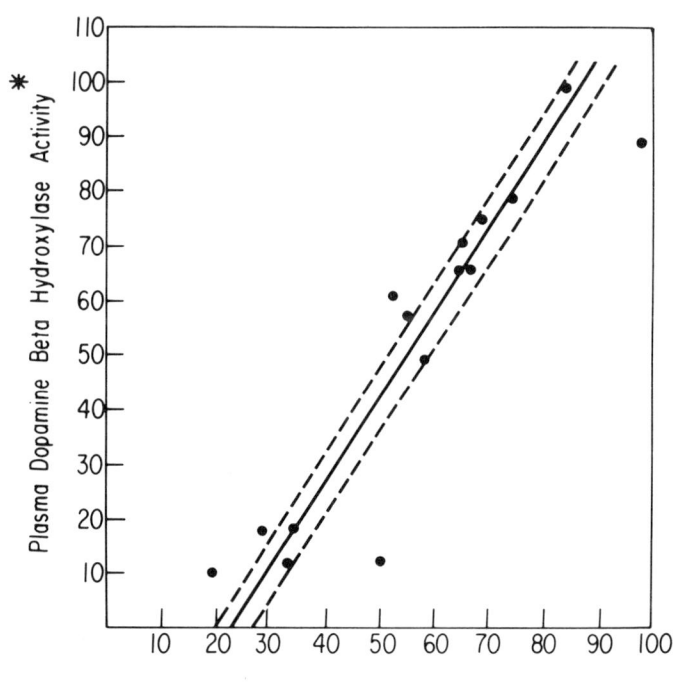

FIG. 1. Relationship of plasma dopamine-β-hydroxylase activity and 24-hr urinary catecholamines. The solid line was determined by linear regression analysis. The broken line represents the standard deviation of the intercept. The linear correlation coefficient is 0.93. *International units (μmole/min/liter of plasma at 37°C).

presence of high plasma DBH activity and greater lability of arterial blood pressure often exceeding the value of 130/85 supported the idea that measurements of plasma DBH activity might be used for an evaluation of patients with hypertension. As a preliminary test of this hypothesis, plasma DBH activity was determined in 50 consecutive patients with hypertension who had been referred to the Duke University Medical Center for evaluation. Results are shown in Table 3.

It is apparent that the DBH activity was highest in those patients who were thought to exhibit fixed or labile essential hypertension, whereas it was much lower in those patients with secondary forms of hypertension. It is of interest that DBH activity was low even in those patients with severely elevated plasma renin activity due to renal vascular disease. Also, it should be noted that in long-standing essential hypertension complicated secondarily by bilateral renal disease (called mixed essential-renal), the DBH values are still markedly increased but are significantly lower than the DBH values in those patients with the single diagnosis of essential hypertension.

Recently, several groups utilizing a double isotope derivative assay for plasma catecholamines have reported that the plasma concentration of catecholamines may be elevated in patients with labile or fixed essential hypertension (Pickering, 1972; de Quattro and Chan, 1972; Louis, Doyle, and Anavekar, 1973). Such an observation suggests that increased activity of the sympathetic nervous system may play a role in the genesis of certain forms of hypertension. The data reported here are consistent with this hypothesis. Furthermore, the finding of higher values for DBH activity in apparently healthy subjects with labile blood pressure of insufficient degree to warrant the clinical diagnosis of labile hypertension and the observation of similar values of DBH activity in patients with definitive labile and fixed hypertension is consistent with the thesis of Eich et al. (1966) that labile

TABLE 3. *Plasma dopamine-β-hydroxylase activity[a] in patients with hypertension*

	Diagnosis age (years)	N	Duration (range, years)	Plasma renin activity	DBH
Labile	25 ± 1	8	<1 → 3	153 ± 21	70 ± 2[b]
Essential	33 ± 3	8	2 → 12	156 ± 17	59 ± 3[c]
Renovascular	48 ± 6	10	<1 → 10	949 ± 424[b]	15 ± 3
Renal parenchymal	48 ± 3	12	1 → 13	262 ± 46	13 ± 2
Endocrine	53 ± 7	3	<1 → 12	38 ± 14[b]	15 ± 9
Mixed essential-renal	51 ± 2	9	10 → 30	219 ± 40	42 ± 2[b]

[a] International units (μmole/min/liter of plasma at 37°C).
[b] $p < 0.001$, different from the other groups.
[c] $p < 0.05$, different from the other groups.

hypertension may represent an early phase in the development of fixed essential hypertension.

In summary, these data appear to indicate that the range of plasma DBH activities is small in young adults with normal stable blood pressure and that measurements of plasma DBH activities could prove useful in the diagnostic evaluation of patients with various types of hypertension as well as provide the means to help further our understanding of the physiological mechanisms involved in these diseases.

ACKNOWLEDGMENTS

This work was supported by National Institute of Mental Health grant MH-13688. Dr. Saul M. Schanberg is the recipient of Research Scientist Award K5-6489.

We wish to express our appreciation to Ms. Agnes Crist and Ms. Edith Harris for their technical assistance.

REFERENCES

DeQuattro, V., and Chan, S. (1972): Raised plasma-catecholamines in some patients with primary hypertension. *Lancet*, 1:806–809.

Doyle, A. E., and Smirk, F. H. (1955): The neurogenic components in hypertension. *Circulation*, 12:543–552.

Eich, R. H., Cuddy, R. P., Smulyan, H., and Lyons, R. H. (1966): Hemodynamics in labile hypertension. Follow-up study. *Circulation*, 34:299–307.

Esler, M. D., and Goulston, K. J. (1973): Levels of anxiety in colonic disorders. *New England Journal of Medicine*, 288:16–20.

Freedman, L. S., Chuchi, T., Goldstein, M., Axelrod, F., Fish, I., and Dancis, J. (1972): Changes in human serum dopamine-β-hydroxylase activity with age. *Nature*, 236:310–311.

Louis, W. J., Doyle, A. E., and Anavekar, S. (1973): Plasma norepinephrine levels in essential hypertension. *New England Journal of Medicine*, 288:599–601.

Nagatsu, T., and Udenfriend, S. (1972): Photometric assay of dopamine-β-hydroxylase activity in human blood. *Clinical Chemistry*, 18:980–982.

Pickering, G. (1972): Hypertension. Definitions, natural histories and consequences. *American Journal of Medicine*, 52:570–583.

Von Euler, V. S., and Lishajko, F. (1961): Improved techniques for the fluorimetric estimation of catecholamines. *Acta Physiologica Scandinavica*, 51:348–355.

Windesheim, J. H., Roth, G. M., and Hines, E. A., Jr. (1955): Direct arterial study of blood pressure response to cold of normotensive subjects and patients with essential hypertension before and during treatment with various antihypertensive drugs. *Circulation*, 11:878–888.

Short-Term Control of Tyrosine Hydroxylase

A. Carlsson, W. Kehr, and M. Lindqvist

Department of Pharmacology, University of Göteborg, Göteborg, Sweden

The ability of the sympatho-adrenomedullary system to maintain relatively constant catecholamine stores in spite of marked variations in nerve-impulse flow has long been a matter of interest. The main reason for this appears to be that catecholamine synthesis can be rapidly and efficiently adjusted to need, mainly by regulation of the first, rate-limiting step, that is, the conversion of tyrosine to DOPA by means of tyrosine hydroxylase.

We will limit this presentation to the short-term control of tyrosine hydroxylase. It is generally agreed that this control occurs without any change in the total number of enzyme molecules, which is in contrast to the long-term adjustment which involves enzyme induction (for review see Weiner, 1970).

I. END-PRODUCT INHIBITION?

From studies of brain catecholamine levels and their O-methylated basic metabolites after treatment with monoamine oxidase inhibitors, we concluded (Carlsson, Lindqvist, and Magnusson, 1960): "The biosynthesis of catecholamines (and 5-HT) seems to slow down as soon as the amines accumulate beyond the maximum capacity of the storage mechanism." Several years later, after the isolation of tyrosine hydroxylase, Nagatsu, Levitt, and Udenfriend (1964) observed that this enzyme was inhibited by norepinephrine and other catechols, and they suggested that tyrosine hydroxylase might be controlled by end-product inhibition. However, Udenfriend and his colleagues have repeatedly pointed out that the concentrations required for inhibition are relatively high (10^{-5} M or higher).

Costa and Neff (1965) confirmed the inhibitory action of a monoamine oxidase inhibitor on catecholamine synthesis and proposed end-product inhibition of tyrosine hydroxylase as the mechanism involved. Moreover, they suggested that by means of this mechanism the body can maintain a steady state level of transmitter despite changes in sympathetic tone. It should be pointed out, however, that this last-mentioned suggestion involves an extrapolation, because their observations, like our earlier findings, were limited to conditions with increased catecholamine levels after monoamine oxidase inhibition.

The conversion of tyrosine to DOPA can now be directly studied *in vivo* by measuring the accumulation of DOPA after the administration of an inhibitor of the aromatic amino acid decarboxylase (3-hydroxybenzylhydrazine = NSD 1015). The procedure has been described in detail (Carlsson et al., 1972). Figure 1 shows that the conversion of tyrosine to DOPA is inhibited by pargyline. The effect is slow in onset, supporting the view that it is indirect and mediated by the accumulating catecholamines. The question of whether a direct end-product inhibition of tyrosine hydroxylase is involved cannot be answered on the basis of the available data. It should be recalled that a similar, though less marked inhibitory action of monoamine oxidase inhibitors on tryptophan hydroxylase, apparently mediated via accumulating 5-hydroxytryptamine (5-HT), has been observed *in vivo*, despite the fact that no end-product inhibition of this enzyme can be demonstrated *in vitro* (see Carlsson and Lindqvist, 1973). This suggests that simple end-product inhibition is not the only factor involved in the short-term regulation of this group of closely related, pterin-dependent oxygenases.

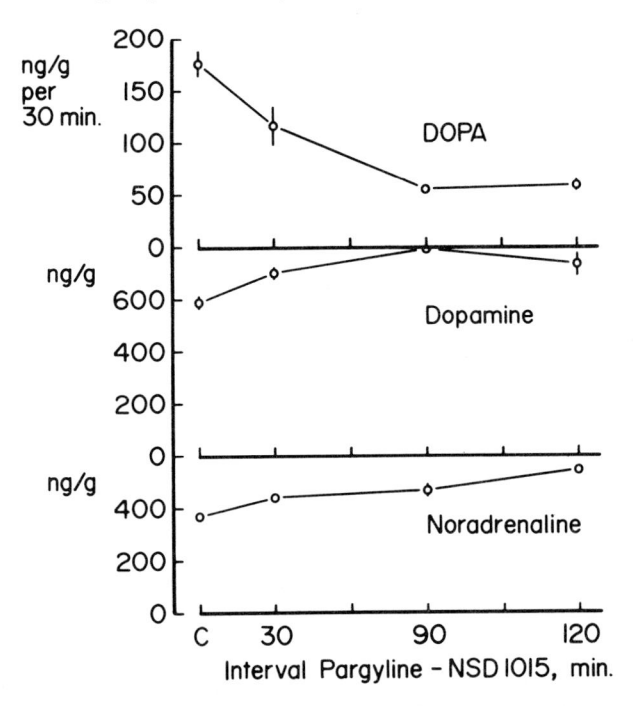

FIG. 1. Effect of MAO inhibitor pargyline on the rate of conversion of tyrosine to DOPA in rat brain.

Pargyline (75 mg/kg, i.p.) was given at various intervals before the inhibitor of the aromatic amino acid decarboxylase, NSD 1015 (100 mg/kg, i.p.). The animals were killed 30 min after the NSD 1015 injection. The levels of DOPA, dopamine, and norepinephrine (noradrenaline) were measured.

Note. The monoamine oxidase inhibitor causes a reduction in DOPA formation. The effect is slow in onset, apparently coinciding with the accumulation of catecholamines.
(Unpublished data of this laboratory.)

Perhaps the receptor-mediated feedback mechanisms to be discussed below play at least a contributory role.

If end-product inhibition played a dominating role in the regulation of tyrosine hydroxylase activity, one would invariably find an increase in catecholamine levels to be accompanied by a decrease in synthesis. However, two experiments will be briefly described in which increased catecholamine levels are actually accompanied by enhanced catecholamine synthesis.

II. SIMULTANEOUS INCREASE IN THE LEVEL AND IN THE SYNTHESIS OF STRIATAL DOPAMINE AFTER NIGROSTRIATAL AXOTOMY

After cutting the nigrostriatal dopamine-carrying nerve fibers, there is a rapid initial increase in the striatal dopamine level (Andén et al., 1972). This is accompanied, not by a decrease, but by an increase in the rate of conversion of tyrosine to DOPA, as measured by the NSD 1015 method (Fig. 2). This increase is initially very pronounced, but later drops to a slightly higher than normal conversion rate (Fig. 3). The last-mentioned decrease may be due to end-product inhibition, in view of the considerable rise in the striatal dopamine level that occurs during the first hour after axotomy. However, the initial rise in DOPA formation evidently operates counter to any

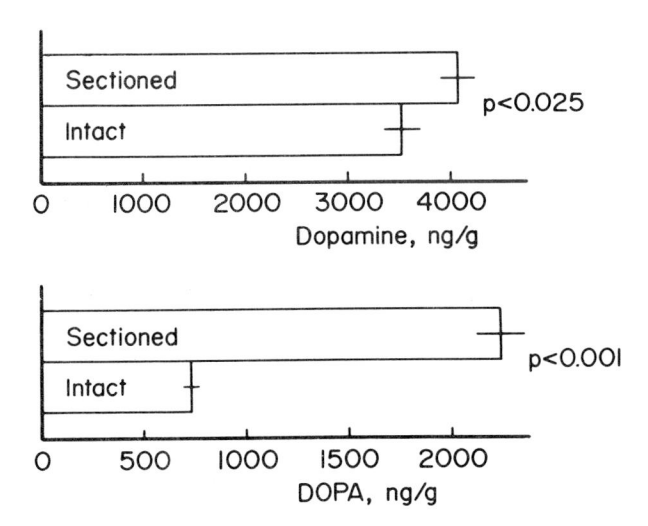

FIG. 2. Effect of cutting nigrostriatal axons (cerebral hemisection) on the synthesis and turnover of dopamine in rat striatum.
  Hemisection and i.p. injection of NSD 1015 were performed simultaneously, and the rats were killed 30 min later. Striatal levels of DOPA and dopamine were measured.
  Note. Dopamine levels were significantly elevated on sectioned side, indicating decreased turnover. A marked increase in the rate of conversion of tyrosine to DOPA took place simultaneously.
  (Unpublished data of this laboratory.)

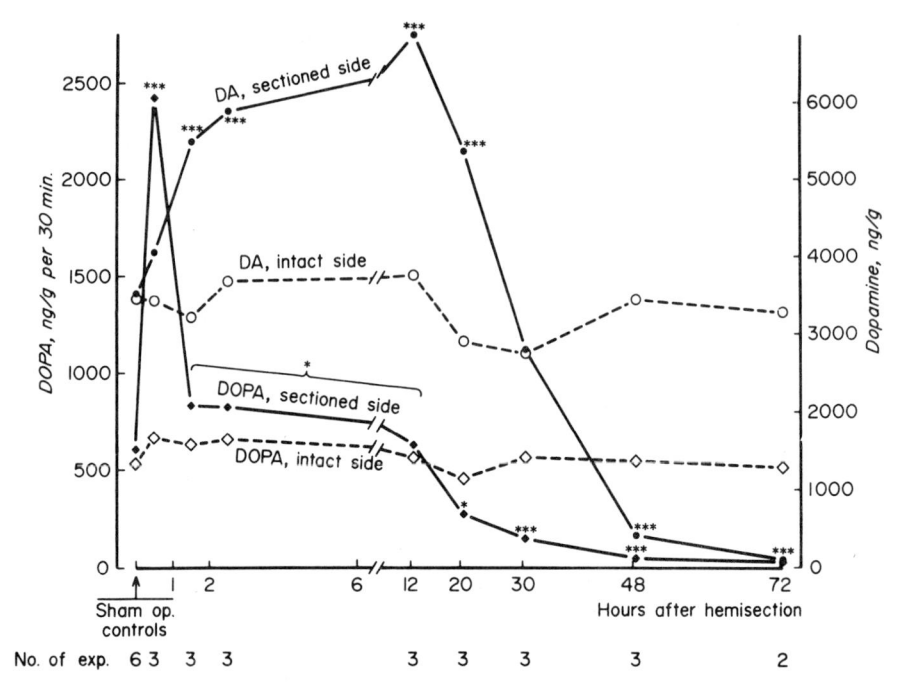

FIG. 3. Effect of cutting nigrostriatal axons (cerebral hemisection) on the rate of conversion of tyrosine to DOPA and on dopamine levels in rat striatum. *(Unpublished data of this laboratory.)*

possible end-product inhibition. We believe that the mechanism underlying this initial response is a receptor-mediated feedback control of dopamine neurons. It has been shown that blocking dopamine receptors enhances the synthesis and turnover of dopamine, whereas stimulation of dopamine receptors has the opposite effect (Carlsson and Lindqvist, 1963; Andén et al., 1967; for review see Carlsson, 1971). This receptor-mediated feedback mechanism is apparently capable of controlling striatal tyrosine hydroxylase activity even after axotomy (Fig. 4). Therefore, it seems logical to conclude that the initial rise in tyrosine hydroxylase activity after axotomy is due to a decrease in dopamine receptor activity, resulting from the interruption of dopamine release by the nerve impulse flow.

III. DISSOCIATION BETWEEN THE RELEASE AND THE ENHANCED SYNTHESIS OF ADRENOMEDULLARY CATECHOLAMINES, INDUCED BY NEUROGENIC STIMULATION

In the adrenal gland, dopamine constitutes about 1% of the total catecholamine content. The half-life of dopamine in this tissue, measured by following its monoexponential disappearance after tyrosine hydroxylase inhibition,

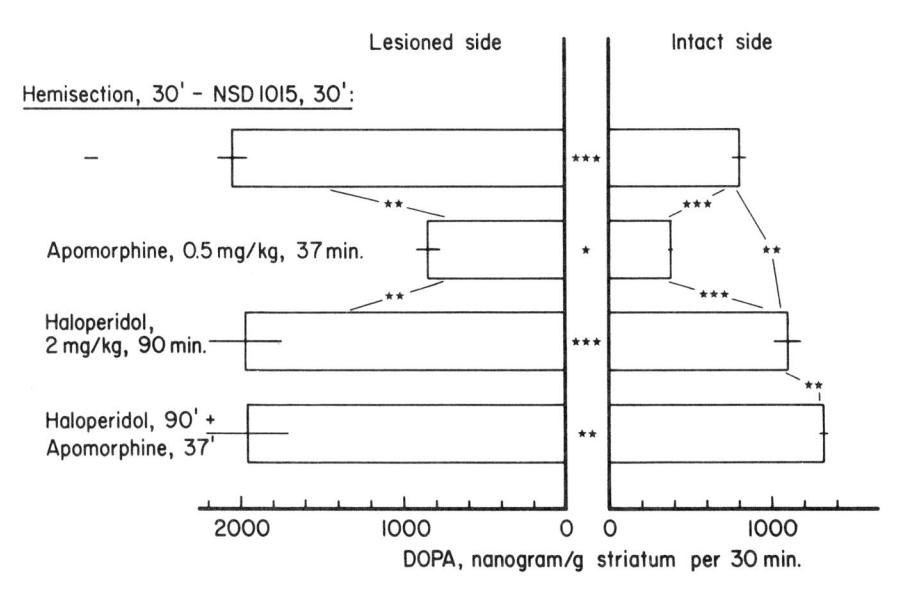

FIG. 4. Effect of a dopamine receptor agonist (apomorphine), of an antagonist (haloperidol), and of both drugs in combination on the rate of conversion of tyrosine to DOPA in rat striatum.
Haloperidol, 2 mg/kg, and apomorphine, 0.5 mg/kg, were given i.p. 60 and 7 min, respectively, before a cerebral hemisection. NSD 1015, 100 mg/kg, was given at the time of hemisection. The animals were killed 30 min later, and striatal DOPA levels were measured separately on the lesioned and on the intact side.
(Unpublished data of this laboratory.)

is 70 to 80 min and is thus roughly 100 times shorter than that of epinephrine (about 1 week). By means of enzyme inhibitors it has been shown that the main metabolic pathway of adrenal dopamine is the conversion to norepinephrine by dopamine-β-hydroxylase. Thus dopamine serves primarily as a precursor to norepinephrine and epinephrine in the adrenal medulla (see Carlsson, Snider, Almgren, and Lindqvist, 1973).

Stimulation of the adrenal medulla by insulin or by physostigmine (potentiated by the α-adrenergic blocking agent phentolamine) caused release of epinephrine and norepinephrine into the bloodstream, accompanied by a decrease in adrenomedullary β-hydroxylated catecholamines. In agreement with earlier observations, the release was neurogenic and mediated via supraspinal centers and adrenomedullary nicotinic receptors: The response was blocked by a high spinal transection as well as by the ganglionic blocking agent chlorisondamine (alone or combined with atropine).

However, both insulin and physostigmine (again potentiated by phentolamine) caused a two- to threefold *increase* in adrenal dopamine, accompanied by an enhanced turnover rate of this precursor. These observations were taken as evidence for an activation of the rate-limiting synthetic enzyme tyrosine hydroxylase. This effect was neurogenic, too, because it was completely blocked by a high spinal transection. On the contrary, it was *not* blocked by chlorisondamine (either alone or combined with atropine).

This dissociation between neurogenic catecholamine release and activation of tyrosine hydroxylase suggests that the two phenomena are not closely linked to each other, as might be expected if end-product inhibition were the predominating mechanism regulating catecholamine synthesis. It would rather seem as if different adrenomedullary receptors were involved in mediating the two responses.

In the animals pretreated with chlorisondamine, the adrenal levels of β-hydroxylated catecholamines tended to be elevated, and this became statistically significant in the animals pretreated with chlorisondamine plus atropine. Nevertheless, the dopamine-elevating action of neurogenic stimulation persisted (Fig. 5). Catecholamine synthesis thus appeared to be

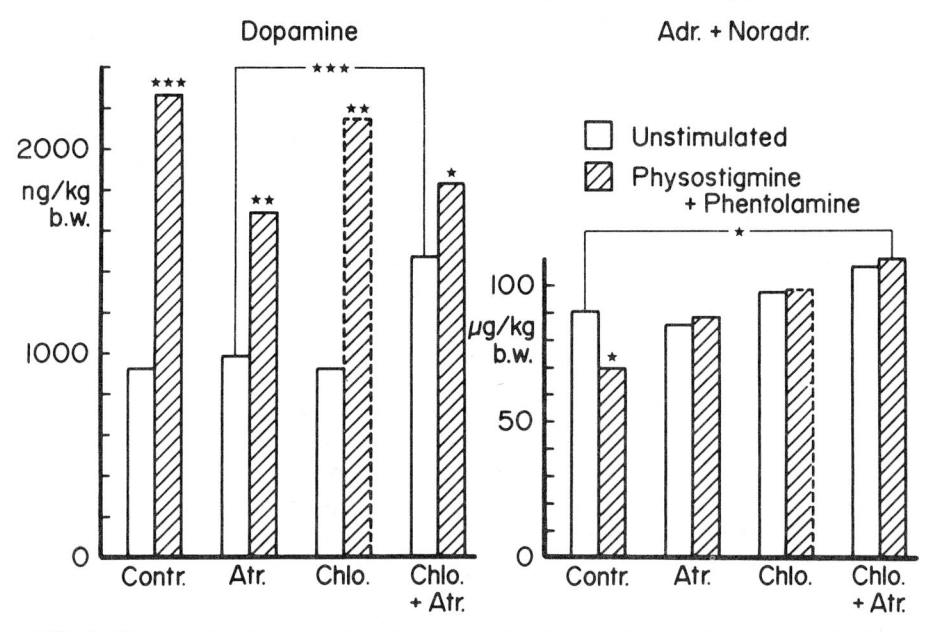

FIG. 5. Dissociation between the dopamine-elevating and the epinephrine-releasing action of neurogenic stimulation (induced by physostigmine) of the rat adrenal medulla, observed after pretreatment with chlorisondamine (Chlo.) and atropine (Atr.).
Physostigmine, 1 mg/kg, and phentolamine, 10 mg/kg, were given i.p. 30 min before death. Chlorisondamine, 5 mg/kg, and atropine, 40 mg/kg, were given i.p. as indicated, 3 hr and 40 min, respectively, before death. Shown are the adrenomedullary levels of dopamine and of epinephrine + norepinephrine (Adr. + Noradr.).

Note. The decrease in epinephrine + norepinephrine induced by physostigmine was prevented by chlorisondamine and atropine, but the increase in dopamine was still evident. A significant increase in epinephrine + norepinephrine as compared to control levels was observed after treatment with chlorisondamine + atropine followed by physostigmine, and was accompanied by an elevated dopamine level.

(Carlsson and Lindqvist, *to be published.*)
* $p < 0.05$; ** $p < 0.01$; *** $p < 0.001$.

enhanced in spite of elevated total catecholamine levels. This strongly argues against end-product inhibition as being a major mechanism controlling adrenal tyrosine hydroxylase activity.

IV. COMMENT

The two experiments briefly described above have certain features in common. In both cases two experimental groups were compared, one in which catecholamine release by nerve impulses apparently took place, and another one in which such release had been interrupted, either by cutting the nigrostriatal axons or by blocking adrenomedullary nicotinic receptors. In both cases this interruption of release caused an increase in total catecholamine levels. In neither case did this increase result in a decreased catecholamine synthesis. In the striatum, the axotomy actually caused an increased conversion of tyrosine to DOPA. In the adrenal medulla the increased catecholamine synthesis induced by neurogenic stimulation remained elevated even when release was prevented by means of ganglionic blockade.

These observations indicate that the short-term regulation of tyrosine hydroxylase is mediated by some factor other than direct end-product inhibition. It seems logical to suggest that the enzyme can be controlled by some modulator. The availability of this modulator is presumably influenced by variations in the activity of post- or presynaptic receptors, as observed in the striatal dopamine system, or by a special type of non-nicotinic receptors responding to specific neurogenic stimuli, as found in the adrenomedullary cells. The nature of this modulator and the mechanism controlling its availability remain to be elucidated.

REFERENCES

Andén, N.-E., Bedard, P., Fuxe, K., and Ungerstedt, U. (1972): Early and selective increase in brain dopamine levels after axotomy. *Experientia*, 28:300–301.

Andén, N.-E., Rubenson, A., Fuxe, K., and Hökfelt, T. (1967): Evidence for dopamine receptor stimulation by apomorphine. *Journal of Pharmacy and Pharmacology*, 19:627.

Carlsson, A. (1971): Basic concepts underlying recent developments in the field of Parkinson's disease. In: *Recent Advances in Parkinson's Disease*, edited by F. H. McDowell and C. H. Markham. F. A. Davis Company, Philadelphia, pp. 1–31.

Carlsson, A., Davis, J. N., Kehr, W., Lindqvist, M., and Atack, C. V. (1972): Simultaneous measurement of tyrosine and tryptophan hydroxylase activities in brain *in vivo* using an inhibitor of the aromatic amino acid decarboxylase. *Naunyn-Schmiedeberg's Archives of Pharmacology*, 275:153–168.

Carlsson, A., and Lindqvist, M. (1963): Effect of chlorpromazine or haloperidol on formation of 3-methoxytyramine and normetanephrine in mouse brain. *Acta Pharmacologica et Toxicologica*, 20:140–144.

Carlsson, A., and Lindqvist, M. (1973): *In-vivo* measurements of tryptophan and tyrosine hydroxylase activities in mouse brain. *Journal of Neural Transmission*, 34:79–91.

Carlsson, A., Lindqvist, M., and Magnusson, T. (1960): On the biochemistry and possible functions of dopamine and noradrenaline in brain. In: *Ciba Symposium on Adrenergic Mechanisms*, edited by J. R. Vane, G. E. W. Wolstenholme, and M. O'Connor, J. & A. Churchill, London, pp. 432–439.

Carlsson, A., Snider, S. R., Almgren, O., and Lindqvist, M. (1974): The neurogenic short-term control of catecholamine synthesis and release in the sympathoadrenal system, as reflected in the levels of endogenous dopamine and β-hydroxylated catecholamines. In: *Frontiers in Catecholamine Research,* edited by E. Usdin and S. Snyder. Pergamon Press, Elmsford, New York, 1973, pp. 551–556.

Costa, E., and Neff, N. H. (1965): Isotopic and non-isotopic measurements of the rate of catecholamine biosynthesis. In: *Biochemistry and Pharmacology of the Basal Ganglia,* edited by E. Costa, L. J. Coté, and M. D. Yahr. Raven Press, New York, pp. 141–155.

Nagatsu, T., Levitt, M., and Udenfriend, S. (1964): Tyrosine hydroxylase: The initial step in norepinephrine biosynthesis. *Journal of Biological Chemistry,* 239:2910–2917.

Weiner, N. (1970): Regulation of norepinephrine biosynthesis. *Annual Review of Pharmacology,* 10:273–290.

Neuropsychopharmacology of Monoamines and Their Regulatory Enzymes, edited by E. Usdin. Raven Press, New York © 1974.

A Critical Assessment of Methods for the Determination of Monoamine Synthesis Turnover Rates In Vivo

Norman Weiner

Department of Pharmacology, University of Colorado School of Medicine, Denver, Colorado 80220

A great deal of information about the synthesis and turnover rates of biogenic amines has been accumulated in the last decade, since Udenfriend and co-workers first established the existence of tyrosine hydroxylase (Nagatsu et al., 1964), indicated that it is the rate-limiting enzyme in the biosynthesis of norepinephrine (Levitt et al., 1965), and demonstrated that the activity of the enzyme could be inhibited by 3,4-dihydroxyphenylalanine (DOPA) and catecholamines (Ikeda et al., 1966; Spector et al., 1967). Both *in vitro* and *in vivo* studies have provided evidence that norepinephrine synthesis and turnover is accelerated by nerve stimulation and inhibited when catechol products are allowed to accumulate in the vicinity of tyrosine hydroxylase (Oliverio and Stjärne, 1965; Alousi and Weiner, 1966; Gordon et al., 1966; Neff and Costa, 1966, 1968; Sedvall and Kopin, 1967; Weiner and Selvaratnam, 1968; Kopin et al., 1969; Weiner et al., 1972).

There is an intense interest in the role of biogenic amine-containing neurons in the genesis and modulation of many peripheral and central nervous system processes, both physiological and pathological. Among these are the relationships of biogenic amines to hypertension, atherosclerosis, stress, hormone release, temperature regulation, motivation, affect, depression, hallucinations, schizophrenia, and narcotic dependence. Since most of these phenomena can be expressed and examined directly only in intact organisms, techniques which allow us to study the metabolism of biogenic amines and the activity of the amine-containing neurons *in vivo* would be expected to contribute greatly to our knowledge of the role of these neural systems in many functions. It should be emphasized, however, that altered neurotransmitter metabolism associated with alterations in a specific aspect of the physiology of the organism does not prove a causal relationship between the two phenomena. The altered amine metabolism in discrete neural systems may be a specific or nonspecific response to the induced behavioral change rather than a mediator of it (Weiner, 1974).

The first estimates of *in vivo* turnover of norepinephrine in specific tissues were conducted by Udenfriend and Zaltzman-Nirenberg (1963) and Burack and Draskóczy (1964), who followed the rates of synthesis and turnover of norepinephrine from ³H-dihydroxyphenylalanine (DOPA) in different tissues. Montenari et al. (1963) studied the turnover of norepinephrine in heart *in vivo* by following the decline in specific activity of ³H-norepinephrine after injection of a tracer dose. Brodie et al. (1966) subsequently introduced a method of measuring norepinephrine turnover *in vivo* by following the rate of decline of endogenous norepinephrine in the absence of synthesis after injection of α-methyl-para-tyrosine. These, and more recently introduced methods, have been used extensively by many investigators desirous of learning how biogenic amine turnover is modified by various stresses (Costa and Neff, 1966; Costa, 1970).

The various methods of estimating biogenic amine turnover *in vivo* which are currently in use are listed in Table 1. They may be divided into "steady state" and "nonsteady state" techniques. Each has particular virtues and, unfortunately, significant theoretical defects inherent to them. I shall confine my remarks to the techniques applied to catecholamines, although the general principles and criticisms apply to analogous studies of other biogenic amines.

A. *Steady state methods.* These offer the theoretical advantage over the nonsteady state methods that the analysis does not necessarily involve a perturbation of the system.

1. *³H-Norepinephrine tracer technique.* In this technique, ³H-norepinephrine, in tracer quantities, is injected into animals and the rate of decline of the specific activity is determined. Most, but not all, studies conclude that this rate of decline follows first-order kinetics (Brodie et al., 1966; Costa, 1970). However, the rate of decline may be more complex than this, depending on the amount of tracer employed (Neff et al., 1968) or

TABLE 1. *Methods for estimation of turnover of biogenic amines*

Steady state methods
 1. Prelabel stores with tracer amount of amine and follow decay.
 2. Follow rate of formation of labeled amine after injection or infusion of amino acid precursor.
 3. Using MS-GLC, determine dopamine conversion to norepinephrine, both formed from labeled tyrosine.

Nonsteady state methods
 4. Inhibit synthesis of amine and follow decline in tissue amine levels.
 5. Block degradation of amine and follow initial rate of rise of tissue amine levels.
 6. Inhibit aromatic-L-amino acid decarboxylase and follow rate of accumulation of hydroxylated amino acid.
 7. In CNS, block secretion of acid metabolites from spinal fluid with probenecid and follow rate of accumulation of acid in brain and CSF.

the manner in which the relationship of the ^{3}H-norepinephrine specific activity decay with time is interpreted *(vide infra)*.

The method assumes that the ^{3}H-norepinephrine labels all the compartments of endogenous norepinephrine uniformly. It also assumes that newly synthesized amine equilibrates rapidly with all amine pools and ^{3}H-amine. Thus, as with all steady state techniques, it is assumed that all the amine in the tissue behaves kinetically as though it were fully equilibrated in a single, homogeneous compartment. However, there is considerable evidence to suggest that ^{3}H-norepinephrine is not completely equilibrated with endogenous norepinephrine (Crout et al., 1962; Chidsey and Harrison, 1963; Weiner, 1970; Thierry et al., 1971, 1973; Weiner et al., 1973). Kopin et al. (1958) and Gewirtz and Kopin (1970) have reported studies that indicate that newly synthesized norepinephrine is preferentially released from the isolated spleen during nerve stimulation, and Thoa et al. (1971) have observed similar results with the vas deferens preparation. In contrast, Bove et al. (1973) and Weiner et al. (1973) have found that preferential release of newly synthesized norepinephrine occurs only with very high frequencies of nerve stimulation or in the presence of phenoxybenzamine.

There is also considerable evidence that norepinephrine in the central nervous system is compartmentalized and turns over at different rates in different brain regions (Iversen and Glowinski, 1966). Furthermore, there may be preferential release of newly synthesized amine which turns over much more rapidly than the bulk of the norepinephrine in brain (Thierry et al., 1973).

2. *Rate of formation of labeled amine after injection of the labeled precursor amino acid.* In this method, labeled tyrosine is administered by either pulse injection or intravenous infusion. The rate of formation of labeled biogenic amine is determined in the tissue and the specific activity of the labeled precursor is followed either in the plasma or in the tissue. From these data, the rate of synthesis and, therefore, at steady state, the turnover of the amine can be deduced (Gordon et al., 1966; Sedvall et al., 1968; Neff et al., 1969; Costa, 1970). This technique assumes that the system is not perturbed by the administration of a tracer amount of the amino acid. In view of the large endogenous levels of tyrosine (or tryptophan), this is a reasonable assumption. However, this assumption is not valid when the hydroxylated amino acids (DOPA or 5-hydroxytryptophan) are employed as "tracer" precursors, since the endogenous pool of these amino acids normally is extremely small.

The method also assumes that the specific activity of the precursor at the site of hydroxylation can be determined. This contention has yet to be demonstrated. In view of the unique pathway for metabolism of the amino acid (i.e., amine synthesis) in the neuron, it is highly likely that total tissue amino acid specific activity does not reflect accurately the specific activity of the substance in the amine-containing neuron. Many investigators employ plasma amino acid specific activity for the calculations, on the assumption that the amino acid may be preferentially taken up by the amine-containing neurons and may therefore be in more rapid equilibrium with the plasma amino acid. The use of plasma specific activity would therefore give a minimal value of amine synthesis since, theoretically, the neuron specific activity of the amino acid cannot exceed that in the plasma (Spector et al., 1963; Sedvall et al., 1968). Costa et al. (1972) have presented evidence that indicates that, after a pulse injection of labeled tyrosine, the specific activity of this substance in adrenergic neurons is considerably greater than that in the remainder of the tissue.

A third assumption required of this method is that the newly synthesized amine is not lost from or metabolized in the tissue. This assumption is hardly likely, particularly if newly synthesized amine is more labile and is either preferentially released from the tissue (Kopin et al., 1968; Sedvall et al., 1968) or preferentially metabolized (Thierry et al., 1973; Weiner et al., 1973). To the degree that the loss or metabolism of the newly synthesized transmitter is not considered, the rate of synthesis and turnover will be underestimated by this procedure.

3. *Combined mass fragmentography-gas chromatography analysis of the rate of conversion of labeled dopamine to norepinephrine after pulse injection of labeled tyrosine.* In this elegant procedure, ^{3}H-tyrosine is injected into an animal and the changes in specific activities of dopamine and norepinephrine are followed over a short time period of about 30 min. This method allows one to calculate the rate of efflux of norepinephrine from a tissue and, at steady state, the turnover rate of the amine. Costa et al. (1972) have indicated that the specific activity of tyrosine in the adrenergic neuron must exceed that in the whole tissue, since dopamine specific activity exceeds that of total tissue tyrosine by a considerable amount 10 min after the tyrosine injection, when their measurements were first taken. The authors attempt to calculate the "zero-time" specific activity of tyrosine in the neuronal compartment and its fractional rate constant. However, to make this calculation, it must be assumed that tissue specific activities of tyrosine rise very rapidly to maximal values after pulse injection, that they decline exponentially after peaking, and that maximal values of specific activity are directly proportional to turnover rate of the amino acid. The first two assumptions (particularly the first) can be considered unlikely, even in a relatively simple system such as rat heart (Spector et al., 1963; Sedvall et al., 1968; Weiner and Rabadjija, 1968).

The mass fragmentography procedure also assumes that endogenous dopamine levels are negligible and that dopamine is quantitatively converted into norepinephrine. The method cannot be applied to brain tissue, since dopamine-containing neurons are present in this tissue and their processes probably innervate most brain regions; thus, the first assumption is not valid for this tissue. The presence of dopamine-containing cells in sympathetic ganglia (Schümann, 1956; Björklund et al., 1970) and in other tissues suggests that the method may have limited applicability for peripheral tissues as well. Furthermore, it is not likely that dopamine is quantitatively converted to norepinephrine. Oxidative deamination of dopamine appears to be a major alternative pathway for this amine (Rutledge and Weiner, 1967), and direct O-methylation of the catecholamine may also introduce a significant error with this procedure (Rutledge and Jonason, 1967). Finally, this method, like all methods which involve estimations of norepinephrine synthesis from tyrosine *in vivo*, assumes that the labeled norepinephrine formed will not be lost from, or metabolized in, the tissue. As noted above, there are serious questions about the validity of this assumption.

B. *Nonsteady state methods.* All of the following methods possess the major theoretical disadvantage of perturbing the system in a profound way. Perturbations of a delicately regulated system might be expected to modify the behavior of that system and distort steady state metabolism.

1. *Inhibition of amine synthesis to determine the rate of decline, and therefore turnover, of the endogenous amine.* For calculations of norepinephrine and dopamine turnover, α-methyl-para-tyrosine is used to block synthesis, and the rate of decline of endogenous norepinephrine is followed (Spector et al., 1965; Brodie et al., 1966; Spector, 1966). The method assumes complete inhibition of catecholamine synthesis, a requirement which seems to be adequately satisfied if high doses of the amino acid are given at sufficiently short intervals (Brodie et al., 1966; Spector et al., 1966). The method must presume that blockade of synthesis does not affect the activity of the neuronal systems deprived of their capacity for amine synthesis. In view of the profound pharmacological effects of this α-methyl-amino acid and in view of our knowledge of the servomechanisms that may come into play to modify neural activity when normal functions are compromised (Weiner, 1974), this seems like an assumption of dubious validity. An additional complication might arise because the loss of endogenous norepinephrine without impairment of the catecholamine storage sites might be expected to facilitate the storage of the remaining amine and slow turnover correspondingly. Finally, this procedure obviously eliminates any estimates of the turnover of newly synthesized amine. If newly synthesized amine does not readily equilibrate with the stored amine or if it is preferentially released, either spontaneously (Bove et al., 1973; Weiner et al., 1973) or with nerve stimulation (Kopin et al., 1968; Sedvall et al., 1968), the turnover rate of the amine will be underestimated.

2. *Blockade of amine degradation and estimate of synthesis rate from the rate of accumulation of the amine.* In this procedure, a monoamine oxidase inhibitor is employed and the initial rate of accumulation of the biogenic amine is measured. The method is most appropriate for estimates of 5-hydroxytryptamine accumulation (Neff and Tozer, 1968), since the major

pathway of metabolism of this amine is via deamination. With the catecholamines, O-methylation is a second major pathway of metabolism (Axelrod, 1959), and neglecting the formation of this product may introduce a significant underestimate of catecholamine turnover.

The major problem with this procedure is the possible effects which accumulating amines may have on the turnover of endogenous amine. Since tyrosine hydroxylase activity is critically affected by the level of free intraneuronal norepinephrine (Weiner, 1970; Weiner et al., 1972), the accumulating amine may reduce its own biosynthesis by end-product feedback inhibition (Alousi and Weiner, 1966; Ikeda et al., 1966). Furthermore, tyramine and other phenylethylamines may serve as substrates for dopamine-β-hydroxylase, and these products can be stored in the vesicles in adrenergic neurons (Fischer et al., 1965; Musacchio et al., 1965) and appear to be released at a high rate (Molinoff and Axelrod, 1972). Inhibition of monoamine oxidase would facilitate the accumulation of these false neurotransmitters (Kopin, 1968), which may in turn affect catecholamine storage and turnover.

Dopamine may serve as the precursor of norepinephrine or it may be directly deaminated to dihydroxyphenylacetaldehyde. Block of monoamine oxidase could perturb the system by favoring β-hydroxylation of dopamine to a degree that does not pertain under normal circumstances (Weiner, 1970).

Even amino acid precursor availability may be affected by block of monoamine oxidase. Mandel (1974) has shown that tyrosine aminotransferase is inhibited by norepinephrine. Thus, accumulation of the catecholamine may enhance the availability of precursor by inhibiting an alternative pathway of metabolism of the amino acid. In addition, various stresses, including administration of drugs which influence catecholamine metabolism, can affect the levels of tyrosine aminotransferase in tissues (Black and Axelrod, 1968; Govier et al., 1969).

3. *Inhibit aromatic-L-amino acid decarboxylase and follow rate of accumulation of hydroxylated amino acid.* This method involves injection of large doses of an inhibitor of aromatic-L-amino acid decarboxylase, such as [N-(DL-seryl)-N'-(2,3,4-trihydroxybenzyl)]hydrazine (Ro-4-4602) or 3-hydroxybenzylhydrazine (NSD-1015) (Carlsson et al., 1972). When given in sufficiently high concentrations, these inhibitors appear to block completely the decarboxylase enzyme. The initial accumulation of either 5-hydroxytryptophan or DOPA, which are normally not detectable in tissues, is followed. Since the rate of accumulation of the hydroxylated amino acid is linear with time for a few hours, it is assumed that this provides an accurate estimate of amine synthesis.

Like the preceding method with monoamine oxidase inhibitors, this procedure has the theoretical defect of perturbing the system by allowing a normal substance to accumulate in abnormal amounts. In the case of DOPA, this catechol would be expected to inhibit tyrosine hydroxylase by end-product feedback inhibition. The method also assumes that alternative pathways for metabolism of the accumulating amino acid are not affected. An active DOPA transaminase has been demonstrated in brain (Fonnum and Larsen, 1965). The activity of this enzyme may be affected by hydrazine decarboxylase inhibitors, since hydrazines are known to inhibit a variety of aminotransferases and decarboxylases (Sourkes, 1966). Thus, the elimination of an alternative pathway of metabolism of DOPA may result in an overestimation of catecholamine synthesis with this technique. Conversely, if the aminotransferases are not inhibited, the accumulation of DOPA will favor this alternative pathway of metabolism because of increased substrate availability, and the method may underestimate the rate of catecholamine synthesis. Furthermore, other aspects of precursor metabolism, such as the uptake or transamination of tyrosine or tryptophan, may be influenced by these decarboxylase inhibitors and thus influence biogenic amine synthesis. Pyridoxine antagonists are known to inhibit tyrosine and tryptophan aminotransferases (Braunstein, 1960).

The administration of the decarboxylase inhibitor will also lead to a reduction in any rapidly turning over pools of norepinephrine, dopamine, or 5-hydroxytryptamine which may exist in the tissue. Such pools may be very important from a functional standpoint and may directly or indirectly influence amine synthesis and turnover.

Since neither DOPA nor 5-hydroxytryptophan is known to be stored in neurons, it is possible that the amino acid will leave the tissue in proportion to its accumulation. An underestimate of synthesis would be made in proportion to the degree of loss.

4. *Accumulation of acid or glycol metabolites in cerebrospinal fluid after probenecid.* In this procedure the egress of the major metabolites of catecholamine or indoleamine metabolism from the central nervous system is blocked by an inhibitor of the organic acid secretion mechanism (Neff and Tozer, 1968). The method assumes that the acids or glycols are the major or exclusive metabolites of the biogenic amines in brain. Although approximately correct for 5-hydroxytryptamine, normetanephrine and methoxytyramine may be significant metabolites of norepinephrine and dopamine, respectively, in the brain (Rutledge and Jonason, 1967; Schildkraut et al., 1971). The method also presumes that the egress of the glycols and acids from the brain is entirely by the organic acid secretion mechanism. Further, the dose of probenecid is assumed to be adequate for complete blockade of the secretion process. High doses are required for the full blockade of this process, and such doses are difficult to achieve in man without intolerable side effects. In studies that have been performed, none of the above assumptions has been fully substantiated.

Thus, with all of the commonly employed methods for estimating biogenic amine synthesis and turnover *in vivo,* there are either real or potential shortcomings that have not been excluded. The methods owe their popularity to the ease with which most of the procedures can be performed and because they have been found to be empirically useful. In spite of their inherent problems, the methods all seem to provide an estimate of relative rates of biogenic amine synthesis and turnover when two or more experimental situations are being compared.

Perhaps the two most commonly used procedures for estimating catecholamine turnover rates are those which involve following either the decline in endogenous catecholamine after blockade of tyrosine hydroxylase or the decline of ^{3}H-catecholamine after injection of a tracer dose of the labeled amine. Both of these techniques assume (as do the other techniques) that the catecholamine is present in an open system which behaves kinetically as though it were a single compartment. The major basis for this premise is the observation that the decay of either ^{3}H-amine or the endogenous amine follows first-order kinetics; that is, when the logarithms of the respective values are plotted against time, a straight-line relationship is obtained.

The mathematical model is difficult to accommodate to our knowledge of the complex structure of neurons and the presence of norepinephrine in various types of vesicles in various regions of the neuron. Different specific activities of catecholamine pools have been demonstrated after labeled tyrosine (Sedvall et al., 1968) or after labeled norepinephrine (Weiner et al., 1973) in peripheral tissues, and multiple compartments of norepinephrine in the brain also have been proposed (Iversen and Glowinski, 1966; Schildkraut et al., 1971; Thierry et al., 1971). The mathematical model of a single biogenic amine compartment in brain is most difficult to fathom in view of the existence of multiple neural pathways which contain the same neurotransmitter but which subserve different functions (Weiner, 1974). In addition, cell bodies, axons, and nerve terminals are all present in the brain for each of these neural systems, and biogenic amine turnover would not be expected to be similar in all segments of the neuron. Why, then, does the turnover of amines in the various tissues appear to obey, at least in some studies, simple log-linear kinetics? The compliance with this simple relationship may be only apparent, as indicated in the following analysis.

Studies which indicate that the logarithm of the decline of ^{3}H-amine or endogenous amine with time is linear are, on close inspection, not wholly convincing. In most of these studies, the variability of the results is large. The variation is largely concealed, albeit not deliberately, by the method of presentation. When results are plotted on a logarithmic scale, great differences appear relatively small. Furthermore, in the studies with ^{3}H-amine, early time points deviate (upward) from linearity to the greatest degree, and these are generally ignored (or the early time studies are not performed) because it is claimed that the ^{3}H-amine is not uniformly labeling the endogenous stores at that early time. A minimum period of time is required for the tissues to rid themselves of extraneuronal amine. In most of the studies performed, it is equally easy, and, *a priori,* equally justifiable, to construct a curvilinear relationship (concave upward) as it is to assume a linear relationship.

If one constructs a hypothetical multicompartment system and follows the decay of either labeled or endogenous amine, some interesting relationships can be deduced. If one assumes that there are several independent compartments, each of which obeys first-order kinetics, then the sum of these compartments may also behave kinetically as a single compartment which obeys first-order kinetics (Fig. 1). Even when the several compart-

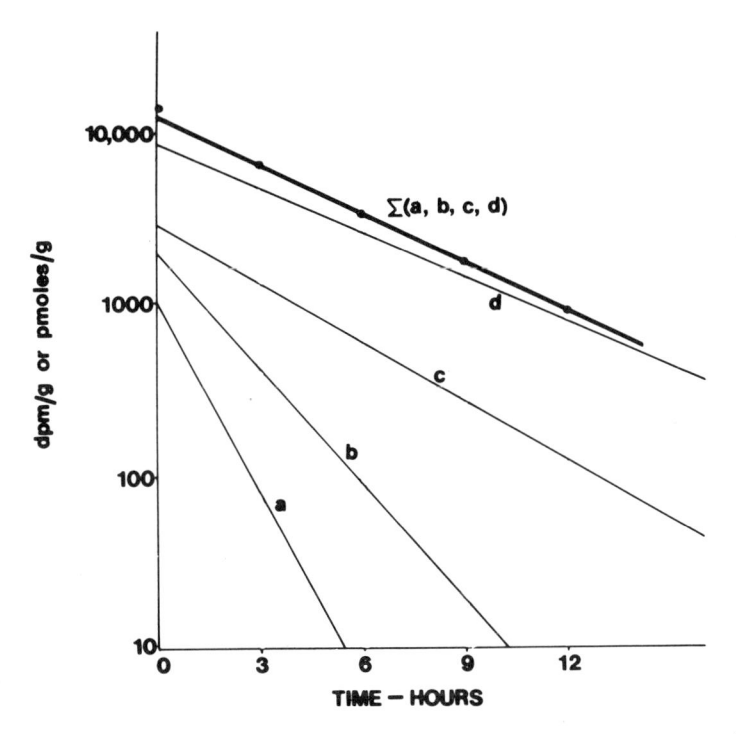

FIG. 1. Analysis of a hypothetical system composed of five amine compartments which are of different size and which turn over at different rates, but all of which obey first-order kinetics. The solid circles and heavy line indicate the algebraic sum of the individual compartments. Note that the composite line deviates upward from a first-order relationship only at the "zero-time" point, a point generally ignored in turnover studies. Ordinate represents, on a logarithmic scale, either dpm/g (in ^{3}H-amine tracer studies) or decay of endogenous amine, in pmoles/g (in studies following inhibition of synthesis).

ments are constituted so that they do not obey first-order kinetics, the composite of these systems appears to follow kinetics more similar to a first-order relationship than do most of the individual hypothetical compartments (Fig. 2). In both instances, the greatest deviations are seen at the earliest time points. These deviations, like those reported by many investigators in

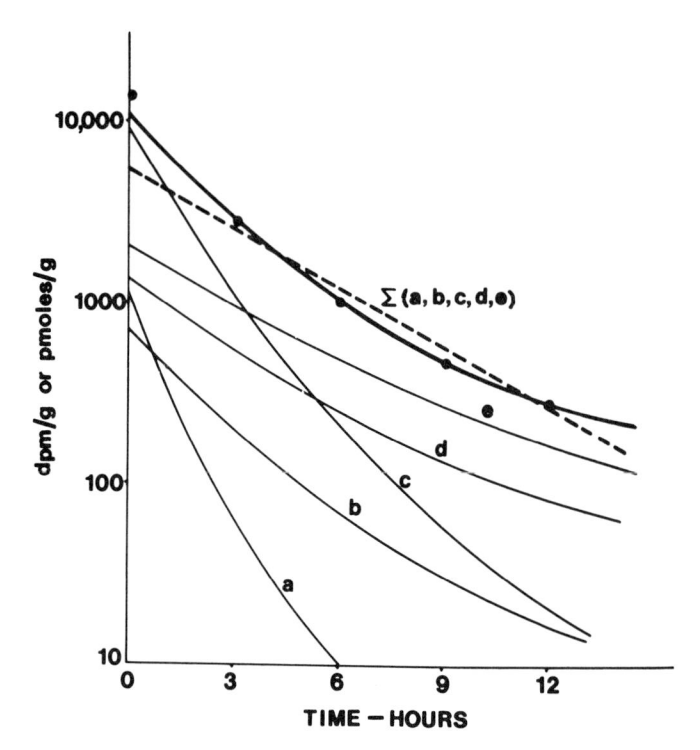

FIG. 2. Analysis of a hypothetical system composed of five amine compartments of different size and which turn over at different rates in a manner suggesting multiple subcompartments in each compartment. The solid circles indicate the algebraic sum of the individual compartments and the heavy, solid line is a visual "best-fit" curve of the composite system. The dashed line is a straight line which would suggest a first-order kinetic relationship, that is, that the system behaves kinetically as a single compartment. Ordinate represents, on a logarithmic scale, either dpm/g (in ^{3}H-amine tracer studies) or decay of endogenous amine, in pmoles/g (in studies following inhibition of synthesis). Note that, except for the (usually ignored) "zero-time" point, the remaining points do not deviate much from the straight line. In real experimental studies, where "biological variation" would be sufficiently great, the standard errors of each point would generally overlap the dashed line; therefore, drawing such a straight line could be quite readily justified (and is regularly done).

experiments of this type with ^{3}H-amines, are upward, and are often ignored in the calculation of synthesis or turnover rates because they are presumed to represent delayed removal from the tissue of ^{3}H-amine present in extraneuronal or nonphysiological neuronal sites.

It would appear from this analysis that the appearance of a first-order kinetic relationship for the turnover of the amine does not allow one to conclude that the system can be represented as a single compartment. One can just as easily (and more reasonably) conclude that the system is indeed multicompartmental and that the turnover of each compartment may or may not obey first-order kinetics. The most obvious instance where the turnover of the composite system will exhibit marked deviations from first-order kinetics would be when the system is composed of a major compartment which turns over very rapidly and a minor component which turns over slowly. In general, however, systems which contain relatively large amounts of amines tend to turn over more slowly than smaller size systems (compare adrenal medulla or vas deferens with brain or heart catecholamine turnover). One must therefore conclude that, if apparent first-order kinetics for the turnover of a monoamine are demonstrable, the results may indeed represent an overall average for the turnover of many compartments in the tissue, each of which turns over at a distinct rate.

In Fig. 3 is indicated an even more disturbing possibility which may be missed in turnover studies of a multicompartment system. Let us assume that, for example, in brain, we are dealing with three neuronal systems each of which contains norepinephrine as its neurotransmitter. Let us arbitrarily assume that each neuron system has cell bodies which, in the aggregate, contain relatively small amounts of norepinephrine which are turning over at high rates (a, b, and c). [Consistent with this supposition is the observation of Costa (1970) that norepinephrine turnover in adrenergic cell bodies (superior cervical ganglion) is several times faster than that in nerve terminal regions (iris and salivary gland).] The three neuronal systems have nerve terminals that, in the aggregate, have larger amounts of norepinephrine and are turning over at slower rates than those in the cell bodies (A, B, and C). Using either the α-methyl-para-tyrosine or the labeled amine technique for determining turnover, one might obtain the hypothetical results depicted in Fig. 3. In the figure, the line $\Sigma(a, b, c)$ is the calculated sum of the individual lines for the "cell bodies," $\Sigma(A, B, C)$ is the calculated sum of the individual lines for the "nerve terminals," and the line $\Sigma(A, B, C, a, b, c)$ represents the total system. Note that, for the summed lines, between zero time and 3 hr, the dashed lines represent the true "calculated" sum of the individual compartments, whereas the solid lines in the early time period represent extrapolations of the straight lines to "zero time," a procedure routinely carried out in order to determine initial tissue amine content or radioactivity. The deviation of these lines, which appears small on a logarithmic scale, will considerably influence the calculated turnover rates because these deviations are, arithmetically, rather large.

Analysis of the six tissue amine compartments is presented in Table 2. Multiplying the initial norepinephrine content or radioactivity (C_0) by the fractional rate constant of decay of either the amine content or radioactivity (K_{amine}) yields a value for the amine synthesis rate. It is apparent from the table that the small compartments (a, b, and c), because of their high turnover rate, contribute over 50% to the overall tissue amine synthesis rate. However, in the composite system, because the larger systems ($A + B + C$) exhibit slower turnover rates, they are the major determinants of the overall turnover rate in the tissue. The neglect of the smaller, rapidly turning over systems is even more serious because extrapolated zero-time radioactivity or amine level values generally are employed in the calculations, rather than the actual zero-time values. Thus, the overall synthesis rate of the system is very similar to that of the larger size components, and the small, rapidly turning over components tend to be neglected. In general, it can be shown

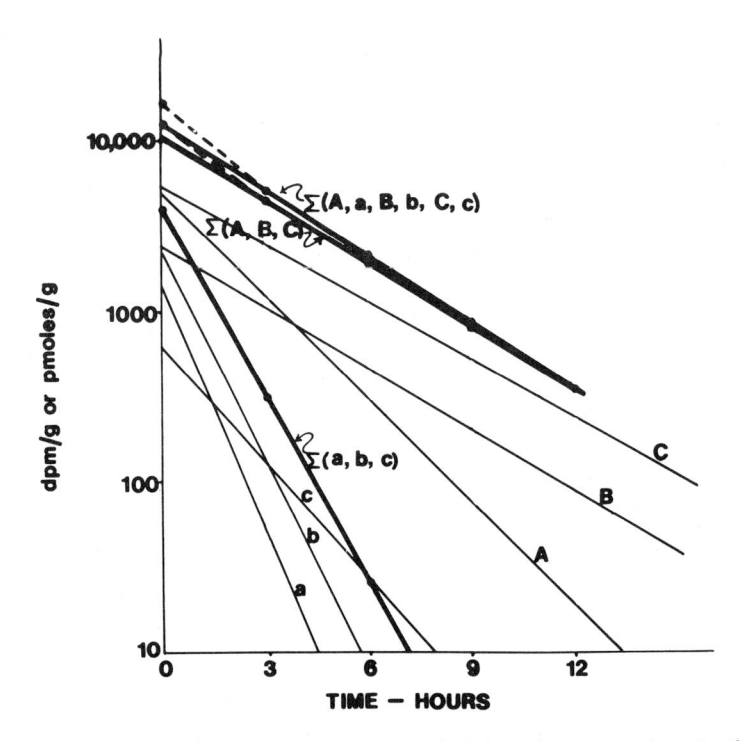

FIG. 3. Analysis of a hypothetical system composed of six amine compartments; three are relatively large in size and turn over rather slowly (A, B, C) and three are smaller amine compartments which turn over quite rapidly (a, b, c). For simplicity, the turnover of each is shown to obey first-order kinetics. The solid line, Σ(a, b, c), is the algebraic sum of the three smaller compartments. The solid line, Σ(A, B, C), is the algebraic sum of the three larger compartments. The solid line, Σ(A, B, C, a, b, c), almost superimposed on Σ(A, B, C), is the algebraic sum of all six compartments. Ordinate represents, on a logarithmic scale, either dpm/g (in ^{3}H-amine tracer studies) or decay of endogenous amine, in pmoles/g (in studies following inhibition of synthesis). The straight lines, Σ(A, B, C) and Σ(A, B, C, a, b, c), are extrapolated back to "zero time," a standard procedure to determine the initial compartment size ("observed C_0") in turnover studies. The dashed lines are lines drawn between the "true C_0" points for these two lines and their respective "3-hr" values (the first "experimental value" in this hypothetical system). Note that, although the deviations appear small on a logarithmic scale, the actual deviations are considerable. For Σ(A, B, C), the "observed C_0" is 10,980; the "true C_0" is 12,320. For the entire system, the "observed C_0" is 12,320; the "true C_0" is 16,740. Multiplying each C_0 value by its respective fractional rate constant of decay (K_{amine}) yields a value for the turnover of each compartment and the composite compartments. For this analysis, refer to Table 2. Note that, when the individual compartments are summed, a value for "true turnover" is obtained (12,110). This value is considerably higher than the "observed turnover" (5,170), largely because smaller compartments with higher turnover rates are ignored when the system is regarded as a single compartment obeying first-order kinetics of decay.

by analyses of this nature that small compartments which undergo rapid turnover tend to be lost in the overall analysis of the tissue, even when they contribute considerably to overall synthesis rates of the tissue. In studies with labeled precursor, similar problems can be encountered.

TABLE 2. *Analysis of hypothetical tissue with six amine compartments, each of which obeys first-order kinetics*

Compartment	$C_0{}^a$	$K_{amine}{}^b$ hr^{-1}	"Observed" synthesis rate^c
A	4.90	0.67	3.28
B	2.40	0.42	1.01
C	5.38	0.37	1.99
a	1.32	1.59	2.10
b	2.14	1.43	3.06
c	0.60	1.11	0.67
A + B + C	10.98	0.42	4.62
a + b + c	4.08	1.16	4.73
A + B + C + a + b + c	12.32	0.42	5.17

^{a} C_0 = size of compartment at "zero time," in pmoles or dpm per g × 10^{-3}, extrapolated from the solid lines depicted in Fig. 3.

^{b} K_{amine} = fractional rate constant of loss of the endogenous or labeled amine, obtained from the slope of the lines from 3 to 12 hr (the "experimental points").

^{c} Synthesis rate, the product of C_0 and K, in pmoles or dpm per g per hr × 10^{-3}.

The sums of the individual synthesis rates (i.e., the "true synthesis" rates for the composite compartments) are (in pmoles or dpm per g per hr × 10^{-3}):

$\Sigma A, B, C = 6.28$
$\Sigma a, b, c = 5.83$
$\Sigma A, B, C, a, b, c, = 12.11$

For additional explanation, refer to legend, Fig. 3, and text.

This analysis emphasizes that one must be extremely cautious in the interpretation of turnover studies. The kinetic analysis probably provides little insight into the number and nature of the amine compartments and the dynamic interactions among these compartments. Furthermore, particularly in brain, marked changes in the functional activity and neurotransmitter turnover of a specific, relatively small neural system could take place without any demonstrable alteration in the overall turnover or synthesis rate of the amine in that tissue.

From the foregoing remarks, the following conclusions may be drawn: (1) It has not been established that studies of biogenic amine turnover or synthesis rates by any of the current methods provide an accurate estimate of *absolute* turnover or synthesis rates. Most of the possible errors would tend to lead to underestimates of the true value, and the errors may be considerable. (2) The first-order kinetic relationships which biogenic amine turnover rates appear to obey do not exclude the possibility that the overall turnover represents a composite of multicompartmentalized amine pools, each with considerably different turnover rates. (3) The turnover studies, when carefully performed, are of value for estimates of *comparative* overall turnover rates in different experimental conditions. (4) Large changes in the neurotransmitter turnover rate or activity of a single neuronal system, especially if it is one of small size, may be completely missed in a turnover study. This may be particularly critical in studying specific physiological processes in the central nervous system, where a small neuronal system may regulate the process and where altered activity in that system may be demonstrable only if that neural pathway is anatomically dissected out for analysis. It would be less critical in studying most drug effects, since the effects are usually demonstrable in all the neurons which utilize that particular neurotransmitter, irrespective of the functions mediated by the neural systems.

ACKNOWLEDGMENTS

The analyses and conclusions summarized in this report are based on research supported by U.S. Public Health Service grants NS 07642, NS 07927, and NS 09199.

REFERENCES

Alousi, A., and Weiner, N. (1966): The regulation of norepinephrine synthesis in sympathetic nerves. Effect of nerve stimulation, cocaine and catecholamine releasing agents. *Proceedings of the National Academy of Sciences* (U.S.), 56:1491–1496.
Axelrod, J. (1959): Metabolism of epinephrine and other sympathomimetic amines. *Physiological Reviews*, 39:751–776.
Björklund, A., Cegrell, L., Falck, B., Ritzén, M., and Rosengren, E. (1970): Dopamine containing cells in sympathetic ganglia. *Acta Physiologica Scandinavica*, 78:334–338.
Black, I. B., and Axelrod, J. (1968): Elevation and depression of hepatic tyrosine transaminase activity by depletion and repletion of norepinephrine. *Proceedings of the National Academy of Sciences* (U.S.), 59:1231–1234.
Bove, F. C., Langer, S. Z., and Weiner, N. (1973): The contributions of mobilization and *de novo* synthesis of norepinephrine (NE) to transmitter release in the isolated, perfused cat spleen. *Federation Proceedings*, 32:739 Abs.
Braunstein, A. E., (1960): Pyridoxal phosphate. In: *The Enzymes*, 2nd ed., Vol. 2., edited by P. D. Boyer, H. Lardy, and K. Myrback. Academic Press, New York, pp. 113–184.
Brodie, B. B., Costa, E., Dlabac, A., Neff, N. H., and Smookler, H. H. (1966): Application of steady state kinetics to the estimation of synthesis rate and turnover time of tissue catecholamines. *Journal of Pharmacology and Experimental Therapeutics*, 154:493–498.
Burack, W. R., and Draskoczy, P. R. (1964): The turnover of endogenously labeled catecholamines in several regions of the sympathetic nervous system. *Journal of Pharmacology and Experimental Therapeutics*, 144:66–75.
Carlsson, A., Kehr, W., Lindqvist, M., Magnusson, T., and Atack, C. V. (1972): Regulation of monoamine metabolism in the central nervous system. *Pharmacological Reviews*, 24:371–384.
Chidsey, C. A., and Harrison, D. C. (1963): Studies on the distribution of exogenous norepinephrine in the sympathetic neurotransmitter store. *Journal of Pharmacology and Experimental Therapeutics*, 140:217–223.
Costa, E. (1970): Simple neuronal models to estimate turnover rate of noradrenergic neuro-

transmitters *in vivo*. In: *Biochemistry of Simple Neuronal Models, Advances in Biochemical Psychopharmacology*, Vol. 2, edited by E. Costa and E. Giacobini. Raven Press, New York, pp. 169–204.

Costa, E., Green, A. R., Koslow, S. H., LeFevre, H. F., Revuelta, A. V., and Wang, C. (1972): Dopamine and norepinephrine in noradrenergic axons: A study *in vivo* of their precursor product relationship by mass fragmentography and radiochemistry. *Pharmacological Reviews*, 24:167–190.

Costa, E., and Neff, N. H. (1966): Isotopic and non-isotopic measurements of the rate of catecholamine biosynthesis. In: *Biochemistry and Pharmacology of the Basal Ganglia*, edited by E. Costa, L. Côté, and M. D. Yahr. Raven Press, New York, pp. 141–155.

Crout, J. R., Muskus, A. J., and Trendelenburg, U. (1962): Effect of tyramine on isolated guinea-pig atria in relation to their noradrenaline stores. *British Journal of Pharmacology*, 18:600–611.

Fischer, J. E., Horst, W. D., and Kopin, I. J. (1965): β-Hydroxylated sympathomimetic amines as false neurochemical transmitters. *British Journal of Pharmacology*, 24:477–482.

Fonnum, F., and Larsen, K. (1965): Purification and properties of dihydroxyphenylalanine transaminase from guinea pig brain. *Journal of Neurochemistry*, 12:589–598.

Gewirtz, G. P., and Kopin, I. J. (1970): Effect of intermittent nerve stimulation on norepinephrine synthesis and mobilization in the perfused cat spleen. *Journal of Pharmacology and Experimental Therapeutics*, 175:514–520.

Gordon, R., Reid, J. V. O., Sjoerdsma, A., and Udenfriend, S. (1966): Increased synthesis of norepinephrine in the cat heart on electrical stimulation of the stellate ganglion. *Molecular Pharmacology*, 2:606–613.

Govier, W. C., Lovenberg, W., and Sjoerdsma, A. (1969): Studies on the role of catecholamines as regulators of tyrosine aminotransferase. *Biochemical Pharmacology*, 18:2661–2666.

Ikeda, M., Fahien, L. A., and Udenfriend, S. (1966): A kinetic study of bovine adrenal tyrosine hydroxylase. *Journal of Biological Chemistry*, 241:4452–4456.

Iversen, L. L., and Glowinski, J. (1966): Regional studies of catecholamines in the rat brain—II. Rate of turnover of catecholamines in various brain regions. *Journal of Neurochemistry*, 13:671–682.

Kopin, I. J. (1968): False adrenergic transmitters. *Annual Review of Pharmacology*, 8:377–393.

Kopin, I. J., Breese, G. R., Krauss, K. R., and Weise, V. K. (1968): Selective release of newly synthesized norepinephrine from the cat spleen during sympathetic nerve stimulation. *Journal of Pharmacology and Experimental Therapeutics*, 161:271–278.

Kopin, I. J., Weise, V. K., and Sedvall, G. C. (1969): Effect of false transmitters on norepinephrine synthesis. *Journal of Pharmacology and Experimental Therapeutics*, 170:246–252.

Levitt, M., Spector, S., Sjoerdsma, A., and Udenfriend, S. (1965): Elucidation of the rate-limiting step in norepinephrine biosynthesis in the perfused guinea-pig heart. *Journal of Pharmacology and Experimental Therapeutics*, 148:1–8.

Mandel, P. (1974): Tyrosine aminotransferase in the rat brain. In: *Ciba Foundation Symposium on Aromatic Amino Acids in the Brain*, edited by R. J. Wurtman, G. E. W. Wolstenholme, and D. FitzSimons. Associated Scientific Publishers, Amsterdam.

Molinoff, P. B., and Axelrod, J. (1972): Distribution and turnover of octopamine in tissues. *Journal of Neurochemistry*, 19:157–163.

Montenari, R., Costa, E., Beaven, M. A., and Brodie, B. B. (1963): Turnover rates of norepinephrine in hearts of intact mice, rats and guinea pigs using tritiated norepinephrine. *Life Sciences*, 2:232–240.

Musacchio, J. M., Kopin, I. J., and Weise, V. K. (1965): Subcellular distribution of some sympathomimetic amines and their β-hydroxylated derivatives in the rat heart. *Journal of Pharmacology and Experimental Therapeutics*, 148:22–28.

Nagatsu, T., Levitt, B. G., and Udenfriend, S. (1964): Tyrosine hydroxylase: The initial step in norepinephrine biosynthesis. *Journal of Biological Chemistry*, 238:2910–2917.

Neff, N. H., and Costa, E. (1966): The influence of monoamine oxidase inhibition on catecholamine synthesis. *Life Sciences*, 5:951–959.

Neff, N. H., and Costa, E. (1968): Application of steady-state kinetics to the study of catecholamine turnover after monoamine oxidase inhibition or reserpine administration. *Journal of Pharmacology and Experimental Therapeutics*, 160:40–47.

Neff, N. H., Ngai, S. H., Wang, C. T., and Costa, E. (1969): Calculation of the rate of cate-

cholamine synthesis from the rate of conversion of tyrosine-^{14}C to catecholamines. Effect of adrenal demedullation on synthesis rates. *Molecular Pharmacology,* 5:90–99.
Neff, N. H., and Tozer, T. N. (1968): *In vivo* measurement of brain serotonin turnover. *Advances in Pharmacology,* 6A:97–108.
Neff, N. H., Tozer, T. N., Hammer, W., Costa, E., and Brodie, B. B. (1968): Application of steady-state kinetics to the uptake and decline of H^3-NE in the rat heart. *Journal of Pharmacology and Experimental Therapeutics,* 160:48–52.
Oliverio, A., and Stjärne, L. (1965): Acceleration of noradrenaline turnover in the mouse heart by cold exposure. *Life Sciences,* 4:2339–2343.
Rutledge, C. O., and Jonason, J. (1967): Metabolic pathways of dopamine and norepinephrine in rabbit brain *in vitro*. *Journal of Pharmacology and Experimental Therapeutics,* 157:493–502.
Rutledge, C. O., and Weiner, N. (1967): The effect of reserpine upon the synthesis of norepinephrine in the isolated rabbit heart. *Journal of Pharmacology and Experimental Therapeutics,* 157:290–302.
Schildkraut, J. J., Draskóczy, P. R., and Sun Lo, P. (1971): Norepinephrine pools in rat brain: Differences in turnover rates and pathways of metabolism. *Science,* 172:587–589.
Schümann, H. J. (1956): Nachweis von Oxytyramin (Dopamin) in sympatischen Nerven und Ganglien. *Naunyn-Schmeideberg's Archives of Experimental Pathology and Pharmacology,* 227:566–573.
Sedvall, G. C., and Kopin, I. J. (1967): Acceleration of norepinephrine synthesis in the rat submaxillary gland *in vivo* during sympathetic nerve stimulation. *Life Sciences,* 6:45–51.
Sedvall, G. C., Weise, V. K., and Kopin, I. J. (1968): The rate of norepinephrine synthesis measured *in vivo* during short intervals; influence of adrenergic nerve impulse activity. *Journal of Pharmacology and Experimental Therapeutics,* 159:274–282.
Sourkes, T. L. (1966): DOPA decarboxylase: Substrates, coenzyme, inhibitors. *Pharmacological Reviews,* 18:53–60.
Spector, S. (1966): Inhibitors of endogenous catecholamine biosynthesis. *Pharmacological Reviews,* 18:599–609.
Spector, S., Gordon, R., Sjoerdsma, A., and Udenfriend, S. (1967): Endproduct inhibition of tyrosine hydroxylase as a possible mechanism for regulation of norepinephrine synthesis. *Molecular Pharmacology,* 3:549–555.
Spector, S., Sjoerdsma, A., and Udenfriend, S. (1965): Blockade of endogenous norepinephrine synthesis by α-methyltyrosine, an inhibitor of tyrosine hydroxylase. *Journal of Pharmacology and Experimental Therapeutics,* 147:86–95.
Spector, S., Sjoerdsma, A., Zaltzman-Nirenberg, P., Levitt, M., and Udenfriend, S. (1963): Norepinephrine synthesis from tyrosine-C^{14} in the isolated perfused guinea pig heart. *Science,* 139:1299–1301.
Thierry, A. M., Blanc, G., and Glowinski, J. (1971): Effect of stress on the disposition of catecholamines localized in various intraneuronal storage forms in the brain stem of the rat. *Journal of Neurochemistry,* 18:449–461.
Thierry, A. M., Blanc, G., and Glowinski, J. (1973): Further evidence for the heterogeneous storage of noradrenaline in central noradrenergic terminals. *Naunyn-Schmiedeberg's Archives of Experimental Pathology and Pharmacology,* 279:255–266.
Thoa, N. B., Johnson, D. G., and Kopin, I. J. (1971): Selective release of newly synthesized norepinephrine in the guinea pig vas deferens during hypogastric nerve stimulation. *European Journal of Pharmacology,* 15:29–35.
Udenfriend, S., and Zaltzman-Nirenberg, P. (1963): Norepinephrine and 3,4-dihydroxyphenylethylamine turnover in guinea pig brain *in vivo*. *Science,* 142:394–396.
Weiner, N. (1970): Regulation of norepinephrine biosynthesis. *Annual Review of Pharmacology,* 10:273–290.
Weiner, N. (1974): Neurotransmitter systems in the central nervous system. In: *Drugs and the Developing Brain,* edited by A. Vernadakis and N. Weiner. Plenum Press, New York.
Weiner, N., Bove, F. C., Bjur, R., Cloutier, G., and Langer, S. Z. (1973): Norepinephrine biosynthesis in relationship to neural activity. In: *New Concepts in Neurotransmitter Regulation,* edited by A. J. Mandell. Plenum Press, New York, pp. 89–113.
Weiner, N., Cloutier, G., Bjur, R., and Pfeffer, R. I. (1972): Modification of norepinephrine synthesis in intact tissue by drugs and during short-term adrenergic nerve stimulation. *Pharmacological Reviews,* 24:203–221.

Weiner, N., and Rabadjija, M. (1968): The effect of nerve stimulation on the synthesis and metabolism of norepinephrine in the isolated guinea-pig hypogastric nerve-vas deferens preparation. *Journal of Pharmacology and Experimental Therapeutics,* 160:61–71.

Weiner, N., and Selvaratnam, I. (1968): The effect of tyramine on the synthesis of norepinephrine. *Journal of Pharmacology and Experimental Therapeutics,* 161:21–33.

Short- and Long-Term Regulation of Tyrosine Hydroxylase

E. Costa, A. Guidotti, and B. Zivkovic

Laboratory of Preclinical Pharmacology, National Institute of Mental Health, Saint Elizabeths Hospital, Washington, D.C. 20032

I. INTRODUCTION

Since neuronal regulation relies on continuous changes of catecholamine turnover rate to obtain a balance between transmitter availability and the changing rates of neuronal activity, a present trend in neuropharmacology attempts to uncover the molecular nature of the mechanisms involved in obtaining this balance. The ultimate goal is that of establishing the basis on which to develop new drugs that can selectively restore a balance where disease processes have reduced efficiency of regulation. When catecholaminergic neurons are involved, our attention is focused on tyrosine hydroxylase (TH), the rate-limiting enzyme in their biosynthesis. The current understanding of its regulation can be summarized by making reference to three modalities of control which promote changes in the rate of TH. These are TH regulation by (a) product inhibition, (b) changes in the kinetic constants of TH, and (c) increase of synthesis rate of TH molecules. Before proceeding further in our discussion, we note that these three modalities of TH regulation differ from each other with regard to the time-constant characteristics of their operation. Feedback control by product inhibition has the shortest time lag, while the induction of TH requires 24 hr or longer depending on the tissue.

II. REGULATION OF TH BY PRODUCT INHIBITION

The norepinephrine (NE) precursors L-3,4-dihydroxyphenylalanine (DOPA) or dopamine (DA) can compete with the pteridine cofactor TH (Udenfriend et al., 1965). This finding has suggested to various investigators that such a competition might have a role in the regulation of the rate-limiting enzyme in catecholaminergic neurons and chromaffin cells. The evidence in support of this possibility is now abundant (Alousi and Weiner, 1966; Costa and Neff, 1966*a, b;* Spector et al., 1967), but the nature of the experimental evidence remains circumstantial. Recently Weiner and collaborators (1973) have reappraised the validity of the control of TH by product inhibition using as a model the vas deferens preparation of the guinea pig. By using nerve stimulation to activate the involvement of product inhibition in the control of TH, they pragmatically isolated an axoplasmic pool of catechols (NE, DA, or DOPA?) where these products have access to TH. It is presumably this pool of catechols which is responsible for the end-product feedback inhibition of the first step in NE synthesis.

The functional role of feedback control by product inhibition is not equally valued by various groups involved in catecholamine research. For instance, Carlsson's group (1972) measures catecholamine turnover rate *in vivo* by injecting an inhibitor of the aromatic amino acid decarboxylase with a rapid onset of action in a dose which virtually inhibits the enzyme completely. The initial accumulation of the substrate (DOPA) is assumed to indicate the rate of formation *in vivo*. The question then arises: Can DOPA accumulation in tissues resulting from decarboxylase inhibition reduce its own rate of formation? There is no doubt that DOPA can inhibit tyrosine hydroxylation *in vitro* (Dairman and Udenfriend, 1971; Tarver et al., 1971). However, Carlsson maintains that the concentrations of DOPA increase linearly up to 30 min. Because of this linearity, he concludes that measurements of DOPA concentrations 30 min after the almost complete inhibition of decarboxylase is an appropriate measurement of the turnover rate of striatal DA or brain NE. This linearity is based on the assumption that the DOPA concentrations in the brain are virtually zero or very close to zero. Doteuchi et al. (1974) have performed experiments in which rats were killed by focusing a microwave radiation on the top of the skull and then striatal DOPA was measured by a fluorometric assay after purification of DOPA by ion exchange and Al_2O_3 chromatography. The microwave generator, operated at 2.5 kW and 25 GHz, causes an inactivation of the tissue enzymes in 1 sec. The results of these experiments are reported in Table 1.

The data of Table 1 clearly show that striatum contains measurable concentrations of DOPA. Thus, if one subtracts the values of DOPA concentrations reported in Table 1 from the values reported by Kehr et al. (1972)

TABLE 1. *Steady state concentrations of tyrosine, DOPA, and DA in rat striatum*

Compound	nmoles/g ± SEM
Tyrosine	76 ± 1.4
DOPA	0.30 ± 0.0062
Dopamine	63 ± 4.2

Each average refers to eight determinations; each determination included the striatum of eight rats killed in 1 sec by microwave radiation.

(1.1 nmoles/g at 30 min after blockade of decarboxylase), we obtain a value for the turnover rate of DA in the rat forebrain of about 2.2 nmoles/g/hr. By an isotopic method Doteuchi et al. (1974) have calculated that the k for DA is 0.34/hr. Assuming that the k of DA is similar in the rats used in the two groups of investigations, from the values of DA concentrations in the forebrain reported by Kehr et al. (1972), one can estimate a turnover rate of DA of about 3.4 nmoles/g/hr. We realize that this comparison only approximates the discrepancy between the isotopic method and the method proposed by Carlsson et al. (1972); we mention this estimation to clarify two important points: (a) The measurement of DOPA accumulation after decarboxylation inhibition does not yield absolute estimation of catecholamine turnover rate; (b) assuming that our comparison of synthesis rate is valid, it appears that DOPA accumulation depresses tyrosine hydroxylation.

The importance of end-product inhibition in the regulation of monoamine oxidase (MAO) catecholamine biosynthesis can be documented in brain and in heart using the MAO inhibitors (Costa and Neff, 1966). In brain, however, because of the rapid synthesis of the monoamines, one can never exclude that the drug affects the synthesis rate by a direct action. In heart, on the other hand, because of the slow synthesis rate of catecholamines, it is possible to show that they accumulate more slowly than in untreated rats even after the MAO inhibitor has disappeared from the body (Ngai et al., 1968).

If tyrosine hydroxylase in nerve terminals is located outside the synaptic vesicles, then only the "free" catecholamines can function in the regulation of the transmitter biosynthesis by end-product inhibition. After MAO inhibition, the extravesicular pool of catecholamines may increase; the significance of such an increase in the regulation of catecholamine turnover is reflected by the reduction of biosynthesis described after MAO inhibitors (Costa and Neff, 1966b). In dopaminergic terminals of striatum, the DA concentration has been calculated to be at least 10^{-2} M (Andén et al., 1966). Since the k_i of striatal tyrosine hydroxylase for DA is smaller than 10^{-4} M (Kuczenski and Mandell, 1972a), if only 1% of the DA were free, a change in DA concentration could regulate the activity of TH. Since DA competes with the pteridine cofactor for binding to TH (Udenfriend et al., 1965), the concentration of this cofactor plays a role in establishing whether any given concentration of DA can inhibit TH.

III. REGULATION OF TH THROUGH CHANGES OF THE AFFINITY CONSTANTS OF ENZYME

A mechanism of feedback regulation of TH different from product inhibition appears to occur in brain dopaminergic nerves (Carlsson and Lindqvist, 1963). Chlorpromazine and haloperidol were found to increase the brain turnover rate of DA and NE metabolites without changing the concentration of these amines. This suggested that the two catecholamines were synthesized at rates faster than normal as a result of the drug injection. A more direct demonstration of the selective stimulation of DA synthesis by chlorpromazine was successively obtained using the initial rate of amine decline after TH inhibition as a method to estimate synthesis rate (Neff and Costa, 1966). It was then suggested that chlorpromazine and other neuroleptics were increasing DA turnover as a result of blockade of postjunctional DA receptors. In other words, the blockade of the receptor elicits transsynaptically a stimulation of dopaminergic cell bodies (for a review, see Costa and Meek, 1974). In support of this feedback control by interneurons of dopaminergic cells are direct electrophysiological measurements showing that haloperidol and chlorpromazine increase the rate of firing of dopaminergic nerves (Bunney and Aghajanian, 1973). Moreover, these authors have also shown that apomorphine, which stimulates DA receptors, slows the firing rate of dopaminergic neurons. Apomorphine also reduces the rate of striatal DA synthesis *in vivo* (Carenzi et al., 1974). Carlsson and his collaborators (1972) have attempted to abolish the rate of firing of dopaminergic axons by axotomy and have then studied tyrosine hydroxylase activity *in vivo* by estimating the rate of DOPA accumulation. They have compared these rates in monolaterally axotomized hemisphere against the contralateral intact after complete inhibition of decarboxylase. The results of these studies and of those of others using a similar technique (Kehr et al., 1972; Stock et al., 1973) indicate that cerebral hemisection increases the DOPA accumulation in striatum after blockade of decarboxylase and also causes a rapid rise of the DA concentration. This activation of DOPA accumulation cannot be due to a release of a direct end-product inhibition, because the axotomy actually raised the DA level. A more reasonable explanation appears to be that the interruption of impulse flow leads to a diminished release of DA into the synaptic cleft and to a depletion of DA at the receptor sites. A model can then be proposed based on a hypothetical system of interneuronal loops postjunctional to DA nerve terminals and on the presence of presynaptic dopaminergic receptors. The interneuronal loop is activated when dopaminergic receptors are blocked, and the activation of this loop results in an increase of dopaminergic neuron firing rate. In contrast, the activation of presynaptic dopaminergic receptors decreases the rate of DA release.

To test this model, we decided to measure the TH activity of striatal homogenates in rats receiving various neuroleptics. The results of these experiments are reported in Table 2. These data show that the various neuroleptics tested all increase tyrosine hydroxylase activity of striatal homogenates. We have studied the dose-response relationship with methiothepin and haloperidol. Both methiothepin and haloperidol increased striatal TH activity when injected intraperitoneally 30 min before the assay in doses of about 1 μmole/kg. The increase of TH activity lasted for about

4 to 6 hr. Both drugs injected in these doses elicited ptosis. We characterized this increase of striatal TH elicited by neuroleptics in three ways: (1) location in subcellular fractions, (2) location in brain parts, and (3) change of TH affinity constants for tyrosine and $DMPH_4$.

TABLE 2. *Tyrosine hydroxylase of rat striata*

Neuroleptic	Dose (μmoles/kg, i.p.)	Tyrosine hydroxylase (nmoles of DOPA/hr/mg of protein)
None (saline)	—	2.6 ± 0.17
Reserpine	16	6.2 ± 0.56[a]
Methiothepin	10	6.6 ± 0.50[a]
Haloperidol	24	6.1 ± 0.44[a]
Pimozide	11	3.6 ± 0.22[a]

Animals were killed 30 min after methiothepin, 1 hr after haloperidol or pimozide, and 2 hr after reserpine. Tyrosine hydroxylase was measured by the method of Waymire et al. (1971), using a 0.32 M sucrose homogenate containing 50 μl 0.4 to 0.5 mg of protein. Standard assay mixture (110 μl) contained 20 mM Tris acetate buffer (pH 6), 90 mM KPO_4 buffer (pH 6.2), 10 mM $NaPO_4$ buffer (pH 7), 0.5 mM $DMPH_4$, 0.2 mM NADH, 0.25 mM pyridoxal-5-phosphate, 7 units of hog kidney L-aromatic amino acid decarboxylase, 1000 units of catalase, 10 μl of sheep liver pteridine reductase (0.125 mg protein).
[a] $p < 0.02$.

A. Localization of the Neuroleptic-Induced Increase of TH in Subcellular Fractions

Tyrosine hydroxylase in striatum appears to occur partly in a soluble form and partly bound to membranes (Nagatsu et al., 1971; Kuczenski and Mandell, 1972a). The enzyme molecules are presumed to have the same amino acid sequence since the fraction that is bound depends on the homogenization conditions (Kuczenski and Mandell, 1972a). According to Kuczenski and Mandell (1972a), the membrane-bound form of tyrosine hydroxylase, when compared to the soluble form, exhibits a smaller apparent k_m for $DMPH_4$ and a lower k_i for DA. They also found that the kinetic parameters of the soluble enzyme became like those of the particulate form if heparin is added (Kuczenski and Mandell, 1972a,b). We have found that the increase of TH activity elicited by neuroleptics is localized only in the soluble fraction; the activity of TH that sediments from a striatal homogenate (0.25 M, sucrose) at 11,000 × 20 min is not increased after pretreatment with neuroleptics. Even when the TH is eluted from the pellet by homogenization in 0.05 M Tris acetate buffer, pH 6, containing 0.2% Triton X 100, the activity of pellet-bound TH is equal in striata of saline- or neuroleptic-injected rats.

B. Location of the TH Increase Elicited by Neuroleptics in Brain Parts

We have measured TH activity in various brain parts of rats receiving reserpine and methiothepin (Table 3) and found that the increase of TH activity is selectively located in the catecholaminergic neurons of the striatum that are mostly dopaminergic. In this experiment, we used neuroleptics at several times the threshold dose to elicit the increase of striatal TH. The selectivity of the site of action of neuroleptics on TH appears to be supported by these experiments.

TABLE 3. *Tyrosine hydroxylase activity in brain parts of rats injected with reserpine and methiothepin*

Treatment	Tyrosine hydroxylase (nmoles of DOPA/mg of protein/hr)		
	Striatum	Brainstem	Hypothalamus
Saline	2.2 ± 0.11	0.67 ± 0.051	0.93 ± 0.060
Methiothepin	6.2 ± 0.56[a]	0.76 ± 0.077	1.1 ± 0.10
Reserpine	6.3 ± 0.73[a]	0.65 ± 0.085	1.2 ± 0.13

Methiothepin (10 μmoles/kg) or reserpine (16 μmoles/kg) were injected intraperitoneally and the animals were killed 30 min after methiothepin or 2 hr after reserpine. Assay performed in 9000 g × 10 min supernatant of striata homogenized with 0.05 M Tris acetate buffer, pH 6, containing 0.2% Triton X 100 according to Waymire et al. (1971).
[a] $p < 0.05$.

C. Change of TH Affinity Constant Elicited by Neuroleptics

Our study indicated that a shift of TH from particle bound to soluble form could not readily account for the increase of TH elicited by neuroleptics. We then tested whether, similarly to heparin, neuroleptics could modify the kinetic properties of the enzyme protein, thereby changing the affinity constants of the enzyme. We tested whether the V_{max} and k_m for tyrosine were changed in striata of rats receiving haloperidol (26 μmoles/kg, i.p.). The results of these experiments are shown in Fig. 1, where the data are reported as a reciprocal plot of substrate concentrations (tyrosine) versus velocity of DOPA formation (nmoles/hr/mg of protein). From this plot the k_m can be calculated to be equal for the TH of saline- and haloperidol-treated rats. The V_{max} of striatal homogenates from rats receiving haloperidol is twice that of rats receiving saline. We then proceeded to the measurement of the affinity constant of $DMPH_4$ in striata of rats receiving haloperidol (26 μmoles/kg, i.p.) or saline. The results of these experiments are shown in

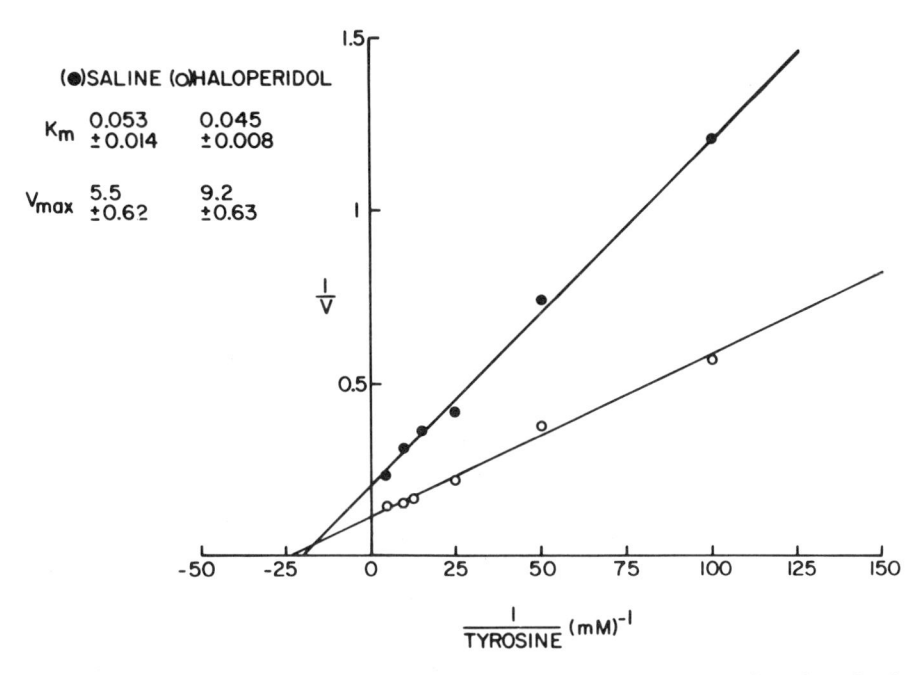

FIG. 1. Reciprocal plot of reaction velocity versus substrate concentrations in striatal homogenates of rats receiving either saline (5 ml/kg, i.p.) or haloperidol (26 μmoles/kg, i.p.). The striata are homogenized as described in Table 3 and the TH activity measured as proposed by Waymire et al. (1971) with the modifications described in Table 2. Computation of enzyme kinetic constants was performed as proposed by Cleland (1963). The V_{max} for tyrosine is reported in nmoles/hr/mg of protein, and calculated for reaction times of 20 min.

Fig. 2. It appears that the k_m for $DMPH_4$ is decreased by fourfold in striatum of rats injected with haloperidol. Our results tend to suggest that neuroleptics which increase the firing rates of dopaminergic neurons change the affinity of the enzyme for $DMPH_4$. We have obtained similar results by using $6MPH_4$ as a cofactor, supporting our contention that TH can be regulated by nerve impulses through a change of its affinity for the pteridin cofactor. This brings into focus the possibility that feedback control by product inhibition has to be viewed as one of the alternative modalities for fast regulation of TH activity. If the natural concentration of the cofactor were to be in the range of its k_m, then by changing the affinity of TH for the cofactor we could also lower the k_i for DA inhibition; therefore, as a result of the action of neuroleptics, the TH has acquired a greater than normal V_{max} through an increased k_m for the pteridine cofactor and can still be controlled by the product of the reaction with an efficiency at least equal to that of the normal enzyme. In other words, the change in affinity for the pteridine cofactor has brought about a greater V_{max} of the enzyme. This in-

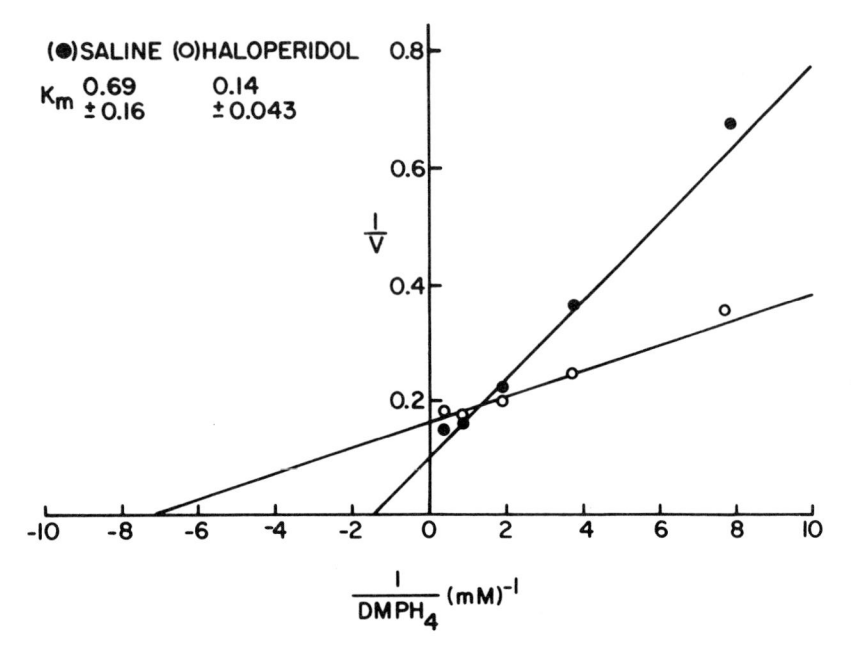

FIG. 2. Reciprocal plot of reaction velocity versus DMPH$_4$ concentrations in striatal homogenates of rats receiving either saline (5 ml/kg, i.p.) or haloperidol (26 μmoles/kg, i.p.). Other specifications as in Fig. 1.

creased V_{max} helps the system cope with the increased rate of neuronal firing, but this change has not released the enzyme from control by product inhibition.

IV. INCREASE OF SYNTHESIS RATE OF TH MOLECULES

Several lines of evidence suggest that cyclic nucleotides can control various metabolic steps involved in the regulation of protein synthesis (Martelo et al., 1970; Johnson and Allfrey, 1972; Atalay et al., 1973; Walton and Gill, 1973). Since neuronal activity regulates intracellular concentrations of 3',5'-adenosine monophosphate (cAMP) Weiss and Costa, 1967; Ferendelli et al., 1970; McAfee et al., 1971, McAfee and Greengard, 1972) or 3',5'-guanosine monophosphate (cGMP) (Ferendelli et al., 1970; George et al., 1970; Lee et al., 1972), it now seems plausible to entertain the view that cAMP may be one of the effectors that promote trans-synaptic induction of enzyme proteins.

An increase of impulse traffic in axons impinging either sympathetic ganglia or adrenal chromaffin cells results in a delayed induction of tyrosine hydroxylase (Meuller et al., 1969; Thoenen et al., 1969; Thoenen, 1972; Guidotti and Costa, 1973; Guidotti et al., 1973a; Hanbauer et al., 1973).

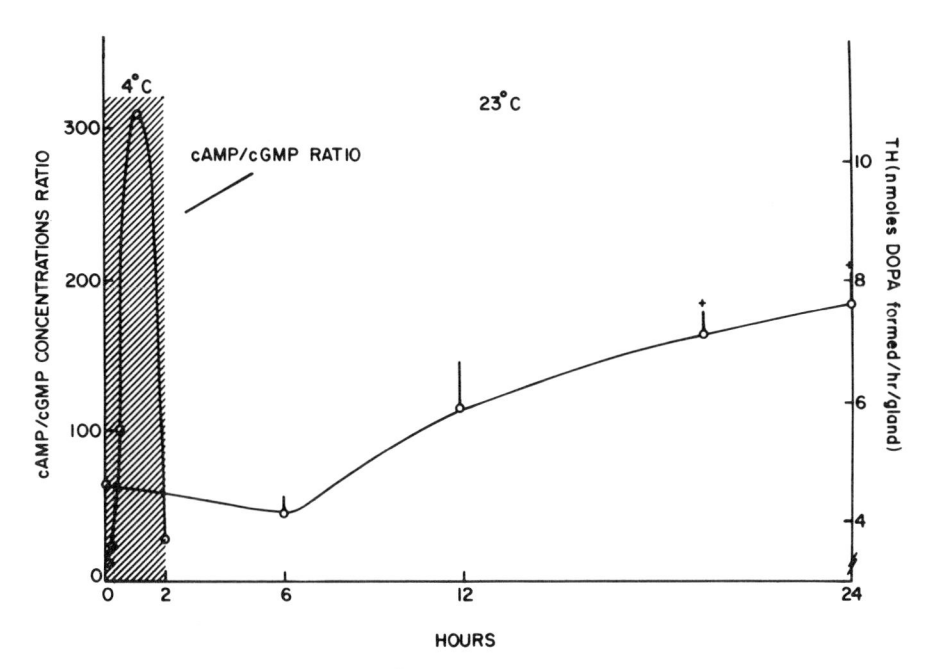

FIG. 3. Rats (200-g Sprague-Dawley, male) are exposed to 4°C for 2 hr. At various times during cold exposure the medullary concentration of cAMP and cGMP is assayed according to the method proposed by Mao and Guidotti (1974). TH activity is assayed according to Waymire et al. (1971) as described in Table 2. Concentrations of cAMP and cGMP in adrenal medulla of rats kept at 23°C were 31 ± 4 and 3.2 ± 0.4 pmoles/mg of protein, respectively.

We have used the trans-synaptically elicited induction of TH in adrenal medulla as a model to study the molecular nature of the events that transform stimulation of receptors located in the membrane of postjunctional cells into an intracellular effector stimulus that enhances the synthesis rate of TH in these cells. The data reported in Fig. 3 have identified three phases in the sequence of intracellular events that bring about the delayed induction of TH activity in adrenal medulla elicited by cold exposure. When rats are exposed to 4°C, the cAMP/cGMP concentration ratio increases by about 30-fold, reaching a peak at about 1 hr and then returning to normal after about 2 hr of exposure to cold. After 2 hr of cold exposure, the animals can be returned to 23°C, but, despite the cessation of the stimulus, after a lag time of about 12 hr, the TH activity is induced. The events that occur during this lag time are independent of neuronal activity, because at the end of the cold exposure the rats can be injected repeatedly with nicotinic receptor blockers without reducing the induction of TH (Guidotti et al., 1973a). In contrast, a single injection of the nicotinic receptor blocker just prior to cold exposure impairs the immediate increase of cAMP/cGMP concentration ratios in medulla and the delayed induction of TH. This increase of TH activity is, in fact, an induction, because by using antibodies to TH one can show that the actual number of enzyme molecules is increased (Joh et al., 1973). We believe that this increase of cAMP/cGMP concentration ratio is mediated trans-synaptically because it can be abolished by surgical decentralization of adrenal medulla (Guidotti et al., 1973a). This denervation also abolishes the induction of TH activity (Guidotti et al., 1973b). Both the immediate increase of cAMP/cGMP concentration ratio and the delayed increase of TH can be elicited in denervated adrenal by injection of carbamylcholine (Guidotti et al., 1973a). The increase of cAMP/cGMP concentration ratio elicited by cold in medulla is due to an increase of cAMP and to a decrease of cGMP concentrations (Guidotti et al., 1973a). Since cold exposure elicits an increase secretion of ACTH from pituitary and ACTH injected into rats can increase the concentration of cAMP in adrenal cortex and medulla (Costa and Guidotti, 1973), we have studied the action of ACTH and carbamylcholine on the concentration of cAMP of medullary and cortical slices. The results of these experiments are reported in Table 4. The data show that carbamylcholine elicits an increment of the cAMP concentrations in medulla but not in cortex slices. Conversely, ACTH increments the cAMP concentrations in cortex but not in medulla. The conditions of the *in vitro* assay are not ideal because the increment *in vivo* of cAMP content in cortex and medulla under conditions of maximal hormonal and neuronal stimulation proceeds at a rate of about 12 pmoles/min/mg of protein, whereas in the data reported in Table 4, the maximal rates recorded when the tissue is maximally activated *in vitro* were about 4 pmoles/min/mg of protein.

To test whether ACTH could be invoked as a mediator of the decrease in medullary cGMP concentration we have measured in adrenal medulla of rats exposed to cold (Guidotti et al., 1973a), we injected 250 mIU/kg, i.v.,

TABLE 4. *Increment of cAMP in slices of rat adrenal medulla or cortex incubated with ACTH or carbamylcholine*

Addition	cAMP increment (pmoles/mg of protein/min)	
	Cortex	Medulla
None	0	0
Carbamylcholine		
1×10^{-6}	0.02	1.6
5×10^{-5} M	0.04	3.0
5×10^{-4} M	0	4.4
ACTH		
500 mIU/ml	2.4	0.6

Slices of paired adrenal medulla or cortex were incubated for 5 min in Krebs Ringer bicarbonate (pH 7.2, 30°C) with and without addition. The concentrations of cAMP after 5 min incubation with only Krebs Ringer bicarbonate were 12 ± 1.2 and 21 ± 1.8 pmoles/mg of protein in cortex and medulla, respectively.

of ACTH and followed the change with time of the concentrations of cGMP in cortex and medulla. The results of these experiments are reported in Fig. 4. These data show that cGMP concentrations in medulla are not affected by ACTH injections. The polypeptide increases the cGMP concentration in adrenal cortex, but this change is short lasting.

In conclusion, our experiments suggest that in adrenal medulla the adenylate cyclase can be regulated by the degree of stimulation of nicotinic receptors. We suggest that the trans-synaptic induction of TH elicited by the activation of these receptors is mediated by an immediate increase of the cAMP/cGMP concentration ratio. The evidence available indicates that if this increase persists for about 2 hr, then the TH induction will occur even if nicotinic receptors are pharmacologically blocked during the intervening 12 to 16 hr. One might speculate that these long-lasting increases of the cAMP/cGMP concentration ratio in chromaffin cells reflect the attempt of the extracellular stimulus to gain control of the function of intracellular organelles. It is conceivable that a maximal activation of adenylate cyclase bound to the cell membrane would render obligatory the ribosomal participation to the response; the rate of protein synthesis thereby becomes influenced by stimuli acting on the cell membrane. The possibility that cAMP can regulate ribosomal function is in keeping with the presence in ribosomes of a cAMP-dependent protein kinase. Adrenocortical ribosomes, for

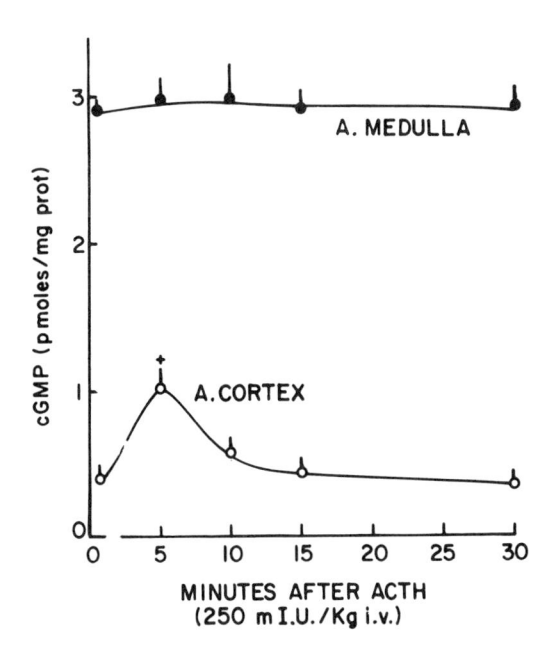

FIG. 4. Rats were injected with ACTH (250 mIU/kg, i.v.) and killed at various times thereafter. cGMP was measured in medulla and cortex according to the method of Mao and Guidotti (1974).

instance, contain proteins which function as substrates for a ribosomal cAMP-dependent protein kinase (Walton and Gill, 1973). When ribosomal proteins are phosphorylated, their binding affinity to the intracellular organelle decreases. Presently, we have a limited understanding of the functional significance of ribosomal phosphorylation and we cannot interpret the change in the composition of ribosomal proteins in terms of the resulting change in the rate of messenger RNA translation. However, these processes and the rate of ribosome formation are presently being investigated in our laboratory as possible mechanisms involved in the trans-synaptic regulation of protein synthesis.

V. CONCLUSIONS

We have reported on the present understanding of the long-term and short-term regulation of TH as it transpires from studies conducted *in vivo* in model systems. The models were the dopaminergic terminals of the rat striatum and the chromaffin cells of the adrenal medulla. The former exemplifies a form of immediate short-term control of TH, the latter a delayed long-term control of synthesis rate of TH molecules.

It was known for a long time that neuroleptics can increase the turnover rate of striatal DA selectively (Carlsson and Lindqvist, 1963; Neff and Costa, 1966). We now demonstrate that this increase is associated with a change of the affinity constants of the soluble TH. The k_m for $DMPH_4$ or for $6MPH_4$ is reduced by four- to fivefold and the V_{max} for DOPA formation is doubled. The affinity constants of the particle-bound TH remain unchanged. The change of the two affinity constants were verified for haloperidol, pimozide, methiothepin, and reserpine. For other neuroleptics, the affinity constant of striatal TH is currently being investigated. In the case of haloperidol and methiothepin, several doses have been studied; the threshold dose for the increase of TH activity in striatum is 1 and 0.5 μmole/kg, i.p., respectively. In rats receiving neuroleptics, the time duration of this response and its dose relation coincide with ptosis. We have studied the TH activity of brainstem and hypothalamus and have found it unaltered. Moreover, Dr. Carenzi in our laboratory has studied the TH activity in striata of rats receiving morphine or amphetamine. Both drugs increased the turnover rate of striatal DA measured *in vivo* but failed to increase TH activity of soluble and particle-bound enzyme. Since morphine and amphetamine, unlike haloperidol, fail to increase the pulse flow rate in dopaminergic axons, we propose that the change of the affinity constant for TH may be the result of the increased neuronal activity. The long-term increase of TH in medulla was used to study the correlation between the induction of TH and the immediate stimulus-coupled increase of cAMP/cGMP concentration ratio. Adrenal medulla is an ideal tissue to study this relationship, because (1) it contains only cholinergic nerves that almost exclusively activate nicotinic receptors; (2) the majority of the cells are chromaffin cells that synthesize TH and contain relatively high concentrations of cAMP; and (3) the tissue can be conveniently denervated. We have established the following: (1) *in vitro* carbamylcholine causes an increment of cAMP concentrations in slices of adrenal medulla but not in slices of adrenal cortex; (2) *in vitro* ACTH causes an increment of cAMP concentrations in adrenal cortex but not in adrenal medulla; (3) cold exposure causes an increase of cAMP and a decrease of cGMP in adrenal medulla that is stimulus-coupled for about 2 hr, after which time the stimulus is no longer capable of increasing the cAMP/cGMP concentration ratio and prolongation of the stimulus fails to elicit a greater induction of TH; (4) pretreatment with nicotinic receptor blockers abolishes the stimulus-coupled increase of cAMP/cGMP concentration ratio and the delayed induction of TH; injections of these drugs after the termination of the cyclic nucleotide response fail to decrease TH induction; (5) denervation abolishes the stimulus-coupled nucleotide response elicited by cold exposure and blocks the delayed induction; (6) injections of carbamylcholine in rats with decentralized medulla elicit a stimulus-coupled cyclic nucleotide response and a delayed long-term increase of TH; (7) the *in vivo* effects of ACTH on cyclic nucleotides can be dissociated from those elicited by cold exposure because cold exposure decreases cGMP concentrations in medulla while ACTH fails to alter these conditions. We suggest that cyclic nucleotides may act on ribosomal function, facilitating messenger translation by activating ribosomal protein kinase.

REFERENCES

Alousi, A., and Weiner, N. (1966): The regulation of norepinephrine synthesis in sympathetic nerves: Effect of nerve stimulation, cocaine and catecholamine releasing agents. *Proceedings of the National Academy of Sciences* (U.S.), 56:1491–1496.

Andén, N. E., Fuxe, K., Hamberger, B., and Hökfelt, T. (1966): A quantitative study on nigro-neostriatal neuron system in the rat. *Acta Physiologica Scandinavica*, 67:306–312.

Atalay, A., Erhan, S., Reishner, S., and Rutman, R. J. (1973): Phosphorylation of the wedge-presumed initiator of DNA replication by cyclic AMP dependent protein kinase. *Physiol. Chem. Phys.* 5:69–74.

Bunney, B. S., and Aghajanian, G. K. (1973): Electrophysiological effects of amphetamine on dopaminergic neurons. In: *Frontiers in Catecholamine Research*, edited by E. Usdin and S. Snyder. Pergamon Press, Elmsford, New York, pp. 957–962.

Carenzi, A., Gerhard, W., Revuelta, A., Guidotti, A., and Costa, E. (1974): *In preparation*.

Carlsson, A., Kehr, W., Lindqvist, M., Magnusson, T., and Atack, C. V. (1972): Regulation of monoamine metabolism in the central nervous system. *Pharmacological Reviews*, 24:371–384.

Carlsson, A., and Lindqvist, M. (1963): Effect of chlorpromazine or haloperidol on the formation of 3-methoxytyramine and normetanephrine in mouse brain. *Acta Pharmacologica et Toxicologica*, 20:140–144.

Cleland, W. W. (1963): Computer programs for processing enzyme kinetic data. *Nature*, 198:463–465.

Costa, E., and Guidotti, A. (1973): The role of 3′,5′-cyclic adenosine monophosphate in the regulation of adrenal medullary function. In: *New Concepts of Neurotransmitter Regulation,* edited by A. J. Mandell, Plenum Press, New York, pp. 135–152.
Costa, E., and Meek, J. L. (1974): Regulation of catecholamines and serotonin in the CNS. *Annual Review of Pharmacology (in press).*
Costa, E., and Neff, N. H. (1966a): The dynamic process for catecholamine storage as a site for drug action. In: *Proceedings of the Fifth International Congress of the Collegium Internationale Neuropsychopharmacologicum.* Excerpta Medica Foundation, Amsterdam, pp. 757–764.
Costa, E., and Neff, N. H. (1966b): Isotopic and nonisotopic measurements of the rate of catecholamine biosynthesis. In: *Biochemistry and Pharmacology of the Basal Ganglia,* edited by E. Costa, L. J. Cote, and M. D. Yahr. Raven Press, New York, p. 141.
Dairman, W., and Udenfriend, S. (1971): Decrease in adrenal tyrosine hydroxylase and increase in norepinephrine synthesis in rats given L-DOPA. *Science,* 171:1022–1024.
Ferendelli, J., Steiner, A. L., McDougal, D. R., and Kipnis, D. M. (1970): The effect of oxotremorine and atropine on cGMP and cAMP levels in mouse cerebral cortex and cerebellum. *Biochemical Biophysiological Research Communications,* 41:1061–1067.
George, W. J., Polson, J. B., O'Toole, A. G., and Goldberg, N. D. (1970): Elevation of guanosine 3′,5′-cyclic phosphate in rat heart after perfusion with acetylcholine. *Proceedings of the National Academy of Sciences* (U.S.), 66:398–403.
Guidotti, A., and Costa, E. (1973): Involvement of adenosine 3′,5′-monophosphate in the activation of tyrosine hydroxylase elicited by drugs. *Science,* 179:902–904.
Guidotti, A., and Costa, E. (1974): A role for nicotinic receptors in the regulation of the adenylate cyclase of adrenal medulla. *Journal of Pharmacology and Experimental Therapeutics (in press).*
Guidotti, A., Mao, C. C., and Costa, E.: (1973a): Trans-synaptic regulation of tyrosine hydroxylase in adrenal medulla: Possible role of cyclic nucleotides. In: *Frontiers of Catecholamine Research,* edited by E. Usdin and S. Synder. Pergamon Press, Elmsford, New York, pp. 231–236.
Guidotti, A., Zivkovic, B., Pfeiffer, R., and Costa, E.: (1973b): Involvement of 3′,5′-cyclic adenosine monophosphate in the increase of tyrosine hydroxylase activity elicited by cold exposure. *Naunyn-Schmiedeberg's Archives of Pharmacology,* 278:195–206.
Hanbauer, I., Kopin, I. J., and Costa, E.: (1973): Mechanisms involved in the trans-synaptic increase of tyrosine hydroxylase and dopamine-β-hydroxylase activity in sympathetic ganglia. *Naunyn-Schmiedeberg's Archives of Pharmacology,* 280:39–48.
Joh, T. H., Geghman, C., and Reis, D. (1973): Immunochemical demonstration of increased accumulation of tyrosine hydroxylase protein in sympathetic ganglia and adrenal medulla elicited by reserpine. *Proceedings of the National Academy of Sciences* (U.S.), 70:2767–2771.
Johnson, E. M., and Allfrey, V. G. (1972): Differential effects of cyclic adenosine 3′,5′-monophosphate on phosphorylation of rat liver nuclear acidic proteins. *Archives of Biochemistry and Biophysics,* 152:786–794.
Kehr, W., Carlsson, A., Lindqvist, M., Magnusson, T., and Atack, C. (1972): Evidence for a receptor mediated feedback control of striatal tyrosine hydroxylase activity. *Journal of Pharmacy and Pharmacology,* 24:744–747.
Kuczenski, R. T., and Mandell, A. J. (1972a): Regulatory properties of soluble and particulate rat brain tyrosine hydroxylase. *Journal of Biological Chemistry,* 247:3114–3122.
Kuczenski, R. T., and Mandell, A. J. (1972b): Allosteric activation of hypothalamic tyrosine hydroxylase by ions and sulphated mucopolysaccharides. *Journal of Neurochemistry,* 19:131–137.
Lee, T. P., Kuo, J. K., and Greengard, P. (1972): Role of muscarinic cholinergic receptors in the regulation of guanosine 3′,5′-cyclic monophosphate content in mammalian brain, heart, muscle, and intestinal smooth muscle. *Proceedings of the National Academy of Sciences* (U.S.), 69:3287–3291.
Mao, C. C., and Guidotti, A. (1974): Simultaneous isolation of adenosine 3′,5′-cyclic monophosphate (cAMP) and guanosine 3′,5′-cyclic monophosphate (cGMP) in small tissue samples. *Analytical Biochemistry,* 59:63–68.
Martelo, O. J., Woo, S. L. C., Reiman, E. M., and Davie, E. W. (1970): Effect of protein kinase on ribonucleic acid polymerase. *Biochemistry,* 9:4807–4813.

McAfee, D. A., and Greengard, P. (1972): Adenosine 3',5'-monophosphate: Electrophysiological evidence for a role in synaptic transmission. *Science,* 178:310–312.

McAfee, D. A., Schorderet, M., and Greengard, P. (1971): Adenosine 3',5'-monophosphate in the nervous tissue: Increase associated with synaptic transmission. *Science,* 171:1156–1158.

Mueller, R. A., Thoenen, H., and Axelrod, J. (1969): Inhibition of trans-synaptically increased tyrosine hydroxylase activity by cycloheximide and actinomycin D. *Molecular Pharmacology,* 5:463–469.

Nagatsu, T., Sudo, Y., and Nagatsu, I. (1971): Tyrosine hydroxylase in bovine caudate nucleus. *Journal of Neurochemistry,* 18:2179–2189.

Neff, N. H., and Costa, E. (1966): Effect of tricyclic antidepressants and chlorpromazine on brain catecholamine synthesis. In: *Proceedings of the First International Symposium on Antidepressant Drugs,* International Congress Series 122. Excerpta Medica, Amsterdam, pp. 28–34.

Ngai, S. H., Neff, N. H., and Costa, E. (1968): Effect of pargyline treatment on the rate of conversion of tyrosine ^{14}C to norepinephrine ^{14}C. *Life Sciences,* 7:847–855.

Spector, S., Gordon, R., Sjoerdsma, A., and Udenfriend, S. (1967): End product inhibition of tyrosine hydroxylase as a possible mechanism for the regulation of norepinephrine synthesis. *Molecular Pharmacology,* 3:549–555.

Stock, G., Magnusson, T., and Andén, N. E. (1973): Increase in brain dopamine after axotomy or treatment with gamma-hydroxybutyric acid due to elimination of the nerve impulse flow. *Naunyn Schmiedeberg's Archives of Pharmacology,* 278:347–361.

Tarver, J., Berkowitz, B., and Spector, S. (1971): Alteration in tyrosine hydroxylase and monoamine oxidase activity in blood vessels. *Nature New Biology,* 231:252–253.

Thoenen, H. (1970): Induction of tyrosine hydroxylase in peripheral and central adrenergic neurons by cold exposure of rats. *Nature,* 228:861–862.

Thoenen, H., Mueller, R. A., and Axelrod, J. (1969): Trans-synaptic induction of adrenal tyrosine hydroxylase. *Journal of Pharmacology and Experimental Therapeutics,* 169:249–254.

Udenfriend, S., Zaltzman-Nirenberg, P., and Nagatsu, T. (1965): Inhibitors of purified beef adrenal tyrosine hydroxylase. *Biochemical Pharmacology,* 14:837–845.

Walton, G. M., and Gill, G. N. (1973): Adenosine 3',5'-monophosphate and protein kinase dependent phosphorylation of ribosomal protein. *Biochemistry,* 12:2604–2611.

Waymire, J. C., Bjur, R., and Weiner, N. (1971): Assay of tyrosine hydroxylase by coupled decarboxylation of DOPA formed from 1-^{14}C-L-tyrosine. *Analytical Biochemistry,* 43:588–600.

Weiner, N., Bove, F. C., Bjur, R., Cloutier, G., and Langer, S. Z. (1973): Norepinephrine biosynthesis in relationship to neural activation. In: *New Concepts in Neurotransmitter Regulation,* edited by A. J. Mandell. Plenum Press, New York, pp. 89–113.

Weiss, B., and Costa, E. (1967): Adenyl cyclase activity in rat pineal gland: Effects of chronic denervation and norepinephrine. *Science,* 156:1750–1752.

Neuropsychopharmacology of Monoamines and Their Regulatory Enzymes, edited by E. Usdin.
Raven Press, New York © 1974.

Regulation of Function of Tryptophan Hydroxylase

Arnold J. Mandell and Suzanne Knapp

Department of Psychiatry, School of Medicine, University of California, San Diego, La Jolla, California 92037

Over the past few years we have been studying the regulation of serotonin biosynthesis in the rat brain. Having shifted our attention from the regulation of serotonergic transmission by substrate supply, release, reuptake, and inactivation by monoamine oxidase, we are focusing on the regulation of the biosynthetic capacity of the serotonergic system. Figure 1 summarizes our programmatic approach. We are looking at (1) receptor sensitivity to infused biogenic amines, (2) intrasynaptosomal changes that can be induced *in vitro* or *in vivo* by drugs, ions, or other treatments, hopefully to reveal the roles of cofactor and substrate supply, as well as the conformation of the enzyme, (3) low- and high-affinity tryptophan uptake, the latter of which we have shown to be drug-sensitive, and (4) and (5) alteration in synthesis and concomitant flow of biosynthetic rate-limiting enzymes from the cell bodies to the nerve endings.

In this chapter we shall discuss the latter pair of phenomena, alterations in apparent amount of enzyme in the regions of the rat brain where serotonergic cell bodies or nerve endings predominate, and the time bases of their responses to drugs. Previously we have demonstrated alterations in tryptophan hydroxylase that take several days or weeks to develop and several days or weeks to wane in differentiated regions of the brain. In addition, we have observed other alterations in biosynthetic capacity that have short latency (hours) and disappear within a day. We refer to these colloquially as "slow-track" and "fast-track" biosynthetic adaptations in response to drugs. Most of our data come from measuring four aspects of serotonergic biosynthetic capacity: (1) the capacity of synaptosomes from various regions of the brain to convert tryptophan to serotonin; (2) the capacity of synaptosomes to take up radioactive tryptophan; (3) the capacity of regions rich in cell bodies to manifest soluble enzyme activity in an assay coupling decarboxylase with hydroxylase, as first described by Ichiyama et al. (1968, 1970); and (4) the manifestation of soluble enzyme activity in lysed synaptosomes. Our specific techniques for these measurements have been detailed elsewhere (Knapp and Mandell, 1972*a*, *b*; 1973*a*, *b*, 1974).

Our original work 3 years ago suggested that chronic administration of

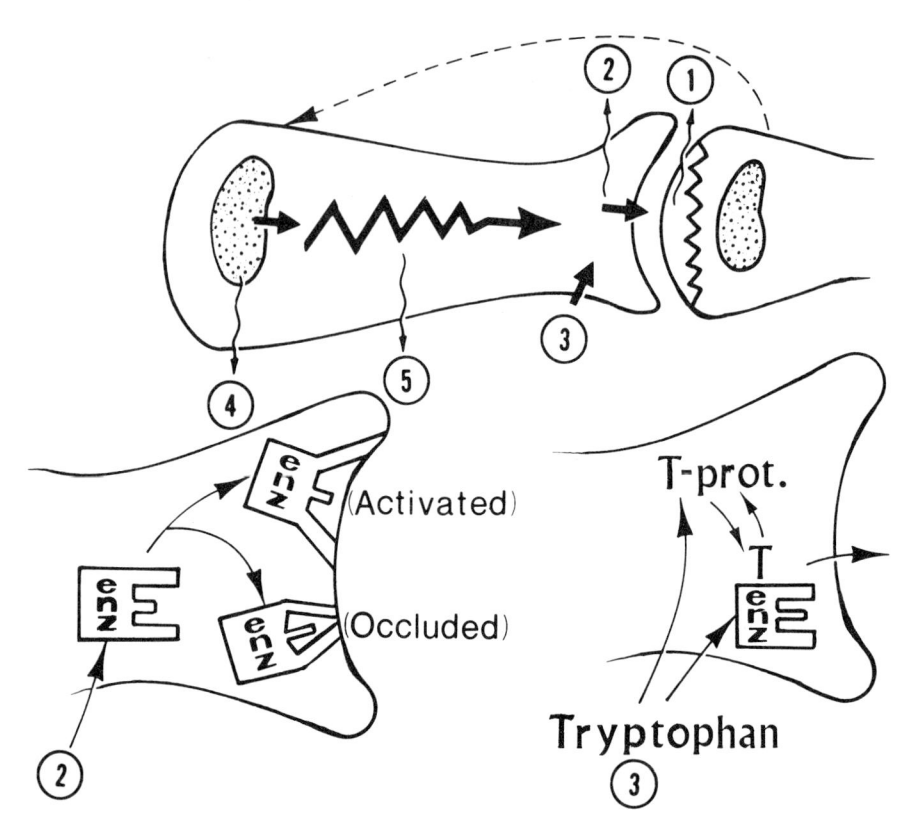

FIG. 1. Macromolecular mechanisms involved in the regulation of neurotransmitter synthesis and efficacy. (1) The receptor area is thought to increase or decrease in sensitivity to infused neurotransmitter. (2) Enzymes in the nerve ending may be inhibited by occlusion or activated by alterations in physical conformation resulting in decreased K_m or increased V_{max}. (3) Uptake is thought to involve a drug-sensitive mechanism, a storage pool, and a direct conversion pathway. (4) Nuclear enzyme synthesis or degradation is probably affected by intra- or interneuronal feedback communication. (5) Axoplasmic flow is thought to affect the latency of enzymatic activity increase or decrease in the nerve ending.

various agents produced alterations initially in the cell body enzyme and eventually in nerve ending conversion of substrate to transmitter. In our first experiments we examined the consequences of a single injection of parachlorophenylalanine (PCPA) in the rat (250 mg/kg, Fig. 2). There was an initial decrease of conversion of tryptophan to serotonin by synaptosomes, which we have attributed to competition between PCPA and tryptophan for uptake into the nerve ending. Thereafter the nerve ending conversion rate returned to normal, but a gradual undialyzable inhibition of soluble enzyme peaked in the midbrain in 2 days and gradually moved out over the next several days. This apparent deficiency in enzymatic

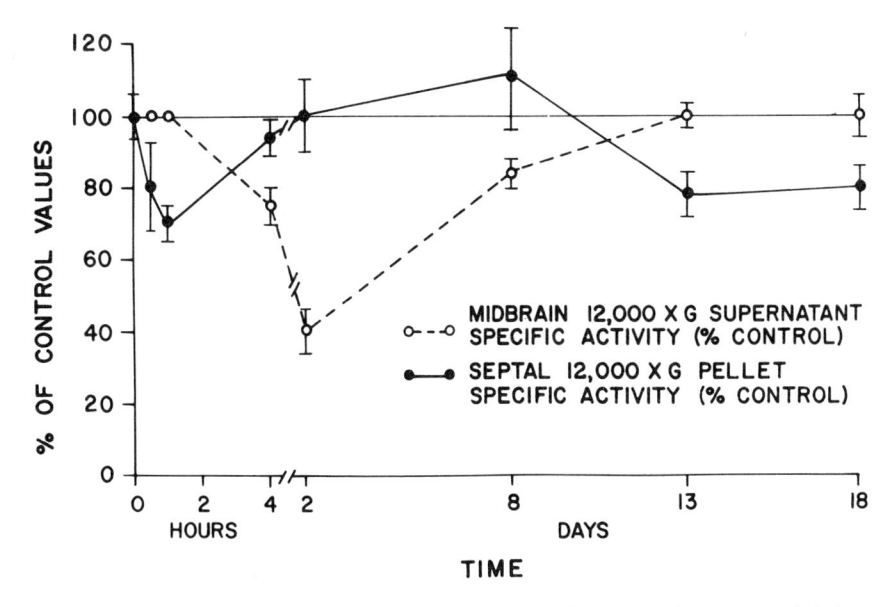

FIG. 2. The activity of midbrain and septal tryptophan hydroxylase after acute administration of PCPA (300 mg/kg) in the rat. Initial reversible decrease in septal enzyme activity was followed by return to control levels and then a delayed persistent decrease. Midbrain activity decreased more slowly and more profoundly, and the return to control levels was delayed. The data are presented as percent control specific activity. Septal control conversion: 75 pmoles/mg of protein/hr; midbrain control enzyme activity: 75 to 85 pmoles/mg of protein/hr.

capacity arrived at the nerve ending sometime between 13 and 18 days. Twenty-one days after injection a slight decrease was still present.

Our studies with morphine also demonstrated long time constants associated with alteration in cell body enzyme activity and later conversion changes at the nerve ending (Fig. 3). Conversion of tryptophan to serotonin by the synaptosomes was inhibited almost immediately after acute administration of morphine, but following chronic administration of the drug there was an increase in conversion in preparations from the same regions. We had shown previously that morphine in a concentration of 500 μM inhibits tryptophan hydroxylase directly. Furthermore, Doris Clouet (Scrafani et al., 1969) had shown that synaptosomes concentrate morphine about 10-fold. We administered a range of doses to achieve an intrasynaptosomal level of about 500 μM in the rats, which theoretically would inhibit the enzyme directly. As we understand the phenomenon, this chronic inhibition of the enzyme at the nerve ending led in 7 to 10 days to a compensatory change in the capacity of the synaptosomes to convert tryptophan to serotonin.

We have reported a similar long-latency change with the chronic treatment of rats with lithium. At 5 mEq/kg per day, resulting in a blood level

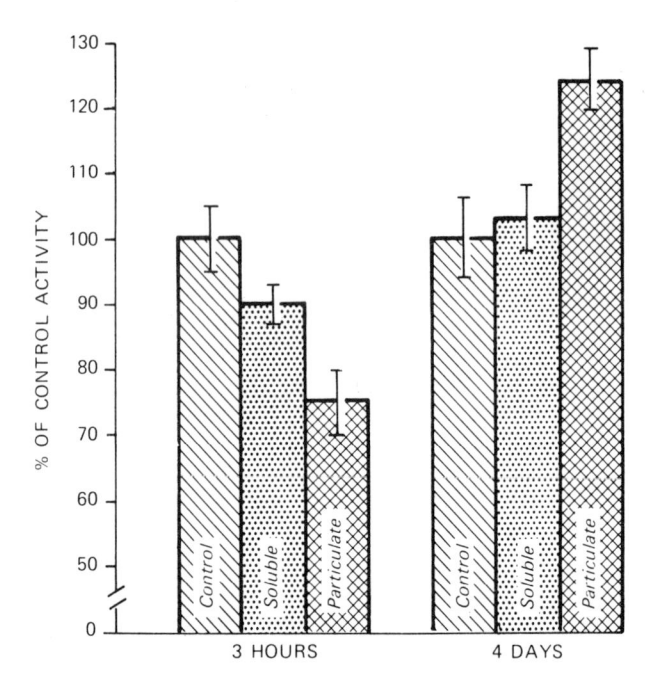

FIG. 3. The effect of short- and long-term morphine administration on regional rat brain tryptophan hydroxylase. For short-term treatment morphine (10 mg/kg) was administered subcutaneously 3 hr before sacrifice. For long-term treatment tablets of 75 mg morphine (obtained from Dr. E. L. Way) were implanted subcutaneously for 5 days. Midbrain control soluble enzyme activity: 100 to 120 pmoles/mg of protein/45 min; control particulate enzyme (septal synaptosomal conversion of tryptophan to serotonin) activity: 120 pmoles/mg of protein/45 min. For conversion activity $p < 0.05$ by the Mann Whitney U test.

of 0.6 mEq in the rats 24 hr after the last drug administration, lithium initially stimulated tryptophan uptake into synaptosomes and consequently enhanced conversion of tryptophan to serotonin by the synaptosomes. Shortly thereafter midbrain soluble enzyme activity decreased, and that decrease took 2 to 3 weeks to reach nerve ending regions such as the striatum or the septum, at which time the synaptosomal conversion rate returned to control levels (Fig. 4). We interpreted that return to control levels as a joint function of the stimulated uptake of substrate and the decrease in the biosynthetic enzyme. Our recent work with soluble enzyme activity in lysed synaptosomes has confirmed our interpretation.

When we began to study acute drug effects on synaptosomal conversion, we found changes with extremely short latency. Within 1 hr of a single high dose of methamphetamine (10 mg/kg in the rat), the conversion of tryptophan to serotonin in synaptosomes decreased markedly. The kinetics of uptake indicate that amphetamine did not alter the uptake of substrate (Fig. 5). If product-feedback inhibition were regulatory, as Glowinski et al. (1972) have suggested, the blockade of reuptake and release of serotonin by amphetamine would be expected to increase conversion, not decrease it. To confirm the reality of a "fast-track" change in synaptosomal conversion rate induced by a drug, we administered 5 mg/kg of reserpine to rats 2 hr before sacrifice and found a marked increase in synaptosomal conversion of

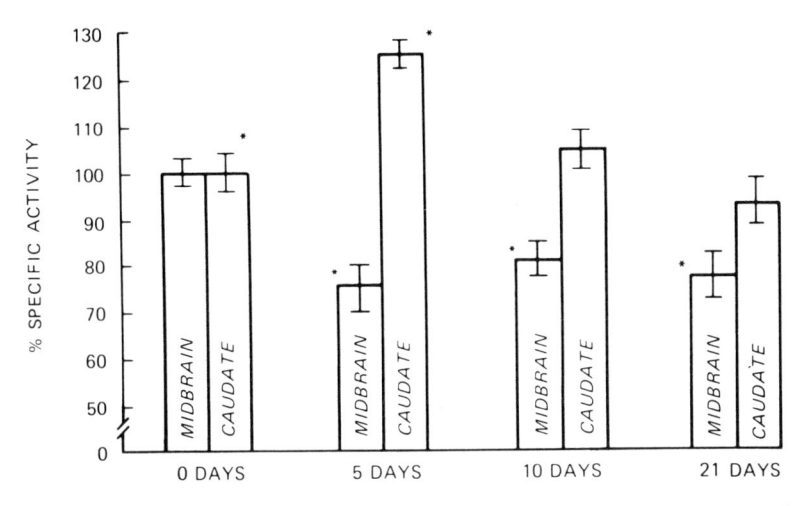

FIG. 4. The effects of short- and long-term lithium chloride treatment on rat midbrain tryptophan hydroxylase and striatal synaptosomal conversion of tryptophan to serotonin as a function of time. The markers on the bars represent the SE, and * indicates that the difference between those values and control values attained statistical significance ($p < 0.005$). Control midbrain activity: 120 pmoles/mg of protein/hr; control striatal (caudate) conversion activity: 150 pmoles/mg of protein/hr.

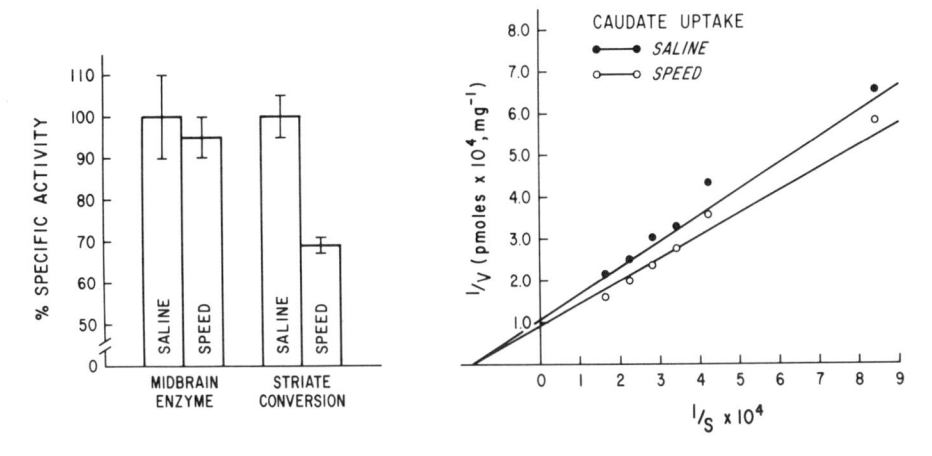

FIG. 5. The effect of methamphetamine (10 mg/kg) on rat brain serotonergic biosynthesis. Methamphetamine ("Speed") pretreatment resulted in a 30 to 35% decrease in striatal conversion of tryptophan to serotonin over that in saline-pretreated controls ($p < 0.002$). Control midbrain enzyme activity: 140 pmoles/mg of protein/45 min; control striate conversion: 125 pmoles/mg of protein/45 min.

tryptophan to serotonin, again without an alteration in the uptake of radioactive substrate (Fig. 6).

So, we had two kinds of alterations in the capacity of synaptosomes to synthesize transmitter that appeared to move in the direction opposite to the acute effect on neurotransmission of whatever drug we administered. One kind required chronic drug administration, altered synthesis of the rate-limiting enzyme, axoplasmic flow of that altered molecule to the nerve ending, and, only then, alterations in the rate of conversion of substrate to transmitter. The other, induced by such drugs as amphetamine and reserpine, was of short latency and left us a bit lost for explanations. We wondered if the physical state of the enzyme could affect its affinity for substrate or cofactor. Ronald Kuczenski, working in our laboratories, has reported such changes in tyrosine hydroxylase, the rate-limiting enzyme in the biosynthesis of catecholamines, in which heparin, a stereospecific mucopolysaccharide, mimicked certain aspects of enzyme activation by means of membrane binding (Kuczenski and Mandell, 1972a, b; Kuczenski, 1973a, b). We have developed an ion bath with which we alter synaptosomal conversion rates *in vitro* with substances such as potassium to check out the possibility that an intact physical state affords manipulable dynamic alterations in enzyme activity that an enzyme in solution cannot undergo.

During the intital phases of our work we were using the conventional artificial cofactor for tryptophan hydroxylase, 2-amino-4-hydroxy-6,7-

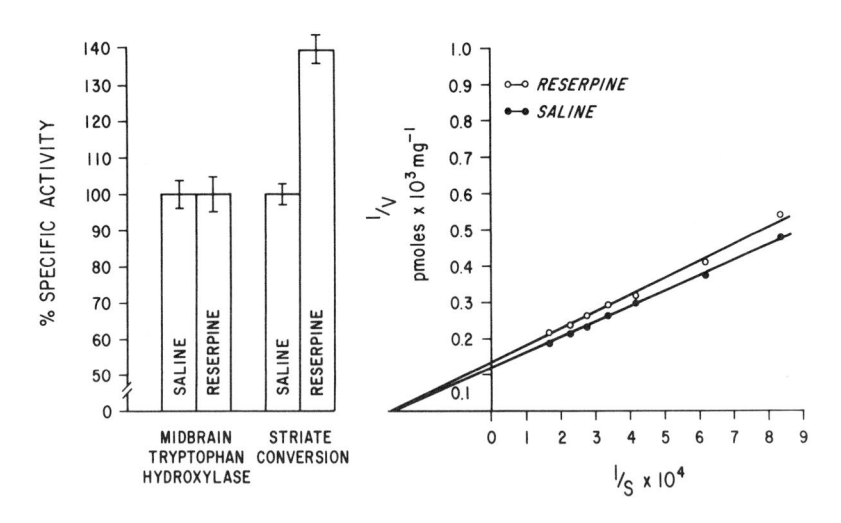

FIG. 6. The effect of reserpine (5.0 mg/kg, 2 hr before sacrifice) on midbrain tryptophan hydroxylase, striatal conversion (left), and striatal tryptophan uptake (right). The markers on the bars represent the SEM. Reserpine pretreatment resulted in an elevation of striatal conversion activity to 140% of control activity ($p < 0.002$). Control midbrain enzyme activity: 125 pmoles/mg of protein/hr; control striatal conversion activity: 125 pmoles/mg of protein/hr.

dimethyl-5,6,7,8-tetrahydropteridine (DMPH$_4$), and although we were able to measure soluble enzyme activity in those regions of the rat brain rich in cell bodies (e.g., the raphe nuclei), we could only get variable low counts in lysed preparations from synaptosomal regions. The recent advent of two more efficient cofactors, tetrahydrobiopterin and 2-amino-4-hydroxy-6-methyl-5,6,7,8-tetrahydropteridine (6MPH$_4$) made it possible to increase the efficiency and relative velocity in our assay of such lysed preparations four- to sixfold. As a result we could measure soluble enzyme from lysed synaptosomes and do more than conjecture about the meaning of alterations in the conversion rate.

We could study soluble synaptosomal enzyme from very small regions of the rat brain (e.g., septal nuclei, hippocampal cortex, sections of the hypothalamus) to check out our speculations about the mechanisms of the short-latency changes. Although we had speculated that the changes were not in amount of enzyme but in conformation, that was not the case. The administration of dextroamphetamine (10 mg/kg) markedly decreased synaptosomal conversion of tryptophan to serotonin (Table 1). Using the

TABLE 1. *Effects of amphetamines[a] on conversion of tryptophan to serotonin in rat striate synaptosomes*

	Percent change from control specific activity $\pm$ SEM[b]	
Control	100.0 $\pm$ 4.0	(50)
D-Amphetamine	−35.0 $\pm$ 4.6	(30)
Methamphetamine	−28.0 $\pm$ 4.0	(18)
p-Chloroamphetamine	−21.0 $\pm$ 5.5	(30)
L-Amphetamine	−17.5 $\pm$ 5.3	(18)
Fenfluramine	+11.0 $\pm$ 2.0	(18)

[a] All drugs were administered subcutaneously 2 hr before sacrifice. The doses of D-amphetamine SO$_4$, methamphetamine HCl, p-chloroamphetamine HCl, and L-amphetamine SO$_4$ were 10 mg/kg. The dose of fenfluramine HCl was 20 mg/kg.
[b] Control specific activity = 125 $\pm$ 6.3 pmoles of $^{14}CO_2$ formed per mg of protein per 45 min with L[1-^{14}C]-tryptophan as substrate (see text). The number in parentheses is the number of determinations performed. The mean and standard error of the mean for each treatment is the average of all experiments.

6-monomethyl cofactor, we were able to lyse the synaptosomes and look at the soluble enzyme activity 1 hr after the drug has been administered (Fig. 7). We saw a decrease in soluble synaptosomal enzyme activity that paralleled the alteration in synaptosomal conversion rate: The ratio of conversion rate to soluble enzyme activity was about 1.4 to 1 in both control and treated samples. Control specific activity for conversion was 160 pmoles/mg of protein/45 min; for soluble enzyme, 120 pmoles/mg of protein/45 min. Thus the short-latency change caused by amphetamine

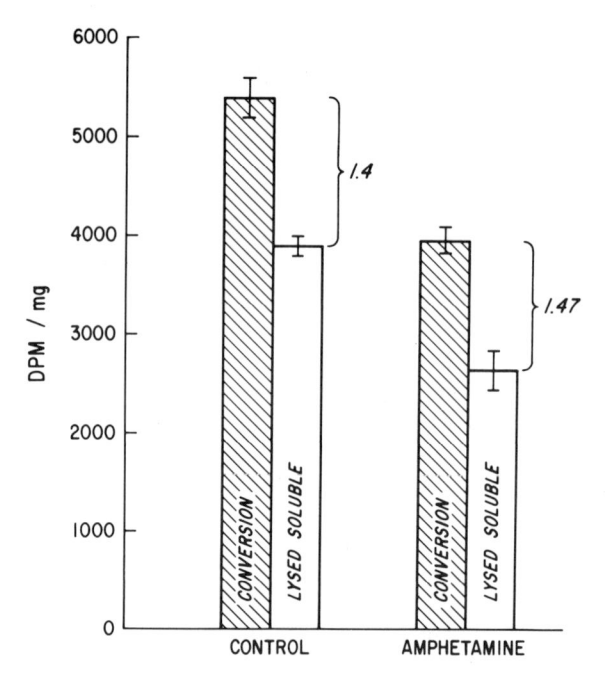

FIG. 7. D-Amphetamine sulfate (10 mg/kg) was administered to rats subcutaneously 2 hr before sacrifice. Brackets indicate the ratio of conversion specific activity to soluble enzyme specific in striatal synaptosomes. In the amphetamine-treated animals both measures of activity were significantly lower than those in the control animals ($p < 0.005$ for conversion; $p < 0.01$ for soluble enzyme). The data are the means of 16 measurements, and are presented as DPM per mg of protein per 45 min.

apparently did not involve a change in the physical state of the enzyme, but an incredibly fast decrease in amount of soluble enzyme in the nerve ending. In two experiments performed since then, the short-latency decrease was reversed in 8 to 24 hr. Reserpine, again in contrast to amphetamine, increases the conversion rate without increasing the soluble enzyme within 2 hr (Fig. 8). If anything there is a slight but insignificant decrease in soluble enzyme activity. We are again left with a mystery: With amphetamine, short-latency changes in conversion seem to involve very fast changes in the amount of soluble enzyme in nerve endings, but the fast alteration of conversion rate by reserpine does not. Amphetamine or reserpine in concentrations up to 100 μM had no effect on the enzyme activity *in vitro*. Decreases in soluble synaptosomal enzyme from striata and septa of rats treated *in vivo* occurred promptly with a wide variety of amphetamines (Table 1), and the decreases returned to normal within 24 hr except in the brains of rats that had received *p*-chloroamphetamine. Those values remained low for as long as we followed them.

In simultaneous comparisons among soluble enzyme activity in the medial

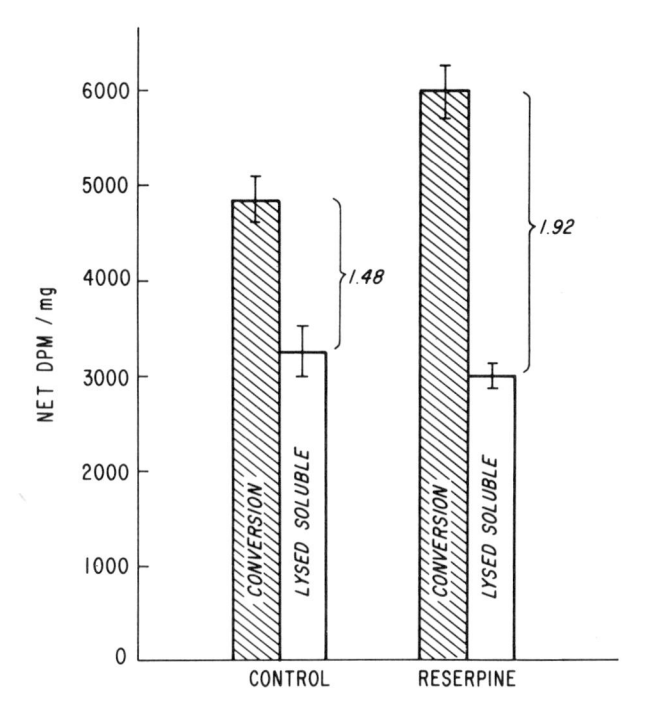

FIG. 8. The effect of reserpine (5 mg/kg, 2 hr before sacrifice) on synaptosomal conversion of tryptophan to serotonin and on soluble synaptosomal tryptophan hydroxylase. Brackets indicate the ratio of conversion specific activity to soluble enzyme specific activity. In the reserpine-treated animals only conversion activity (net DPM per mg protein) was significantly different from control value ($p < 0.005$). These data are the means of 12 measurements.

or lateral midbrain, conversion activity in the striatum, and soluble enzyme activity in the striatum after the administration of amphetamine, we saw intricate early changes (Fig. 9). It is not surprising that the tryptophan hydroxylase level in the medial midbrain did not change because most histofluorescent maps show input to the striatum coming from the lateral raphe nuclei. First, within 15 min there was a decrease in the nerve ending conversion rate without an alteration in the soluble enzyme. At 30 min enzyme activity went down in the cell bodies of the lateral midbrain, and that apparent decrease in enzyme activity moved to the striatum within 1 hr. Perhaps not all of these early changes can be attributed exclusively to alterations in the level of soluble enzyme, especially the decrease in conversion that occurred before the soluble synaptosomal enzyme level decreased. We may be dealing with a complex mechanism that involves changes in the enzyme already available in the nerve ending as well as the alteration in enzyme synthesis and transport.

Using 6MPH$_4$ and the more sensitive assay for soluble synaptosomal

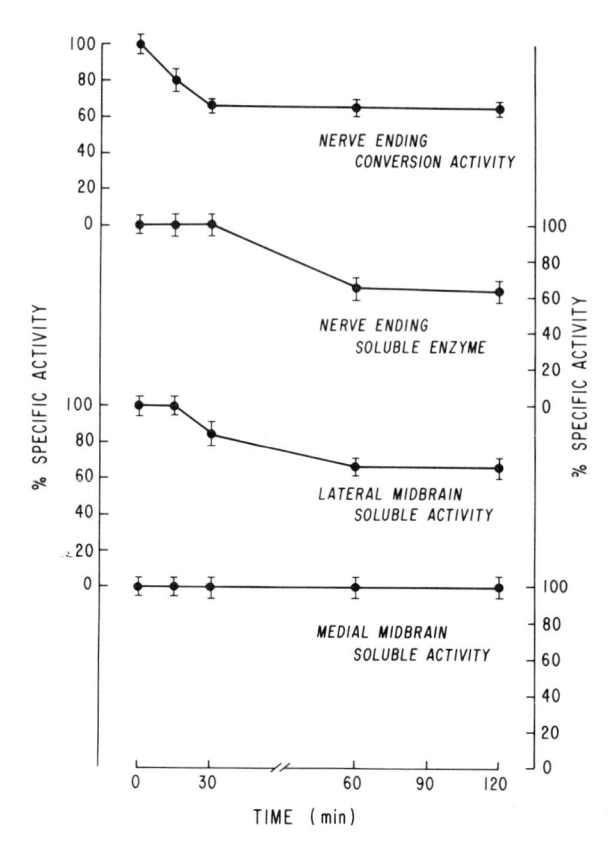

FIG. 9. The effect of D-amphetamine (10 mg/kg) administered to rats at various times before sacrifice. Nerve ending serotonin biosynthesis was measured as striatal conversion of tryptophan to serotonin and soluble enzyme activity in lysed striatal synaptosomes. Cell body enzyme activity measured in two areas of the midbrain: lateral and medial. The data are presented as percent of control specific activity (represented by "0" time). The decrease in conversion activity was significant at 15 min ($p < 0.05$), and at all later times ($p < 0.001$). Nerve ending soluble enzyme activity was decreased at 60 min ($p < 0.02$). Lateral midbrain soluble enzyme activity was decreased at 30 min ($p < 0.05$) and at all later times ($p < 0.02$). Medial midbrain enzyme activity did not change during the experiment. Control specific activities in pmoles per mg of protein per 45 min: lateral midbrain, 200; medial midbrain, 1000; synaptosomal conversion, 157; solubulized synaptosomal, 120. Each value is the mean of at least 12 determinations.

enzyme, we were able to show that even the effects of lithium upon cell body and nerve ending enzyme begin to occur within days rather than weeks. In the brains of rats treated with lithium (5 mEq/kg) daily we found a progressive increase in synaptosomal conversion of tryptophan to serotonin associated with stimulated substrate uptake (Fig. 10). However, by the second day there were already compensatory decreases occurring in both the lateral midbrain and the soluble synaptosomal enzyme activity.

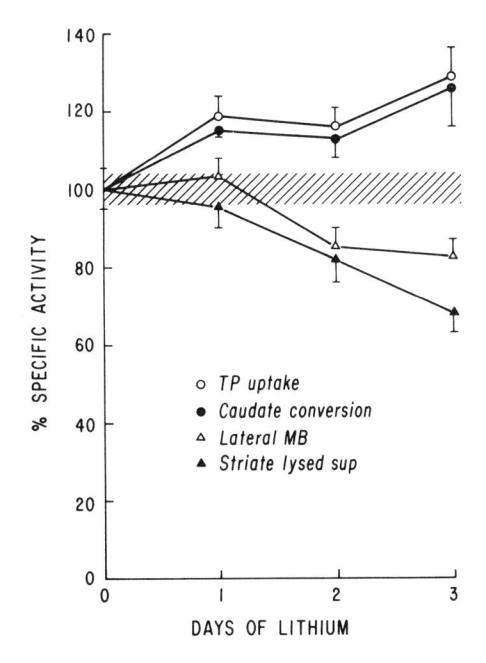

FIG. 10. The effect of lithium treatment on regional serotonin biosynthesis and ^{14}C-tryptophan uptake. The retention of ^{14}C-tryptophan and conversion activity were measured in intact synaptosomes; the lateral midbrain enzyme activity was measured in a 40,000 × g supernate from a 0.001 M Tris homogenate; and striatal nerve ending soluble enzyme was obtained in the 40,000 × g supernate from hypotonically lysed striatal synaptosomes. Control striatal conversion activity: 175 pmoles/mg of protein/45 min; control lateral midbrain enzyme activity: 200 pmoles/mg of protein/45 min; control synaptosomal solubulized enzyme: 110 pmoles/mg of protein/45 min.

Three years ago we felt that there were two kinds of biosynthetic adaptation to drugs that affect serotonergic transmission: those of long latency that involved enzyme synthesis and transport, and those of short latency that involved changes in intrasynaptosomal dynamics among substrate, cofactor, and even the enzyme itself. With further drug studies and more sensitive measurements of synaptosomal enzyme activity, we have found that even very prompt adaptation (within 1 hr) can involve changes in the apparent amount of enzyme activity. Now, with lithium, we are beginning to see changes with intermediate time constants, that is, a few days.

All of this suggests to us that, especially for the alterations in enzyme synthesis and flow, there may be a whole family of curves of latency, magnitude, and duration that have to do with signal thresholds or particular demands for compensatory activity. We believe this is the first time that changes in biosynthetic enzymes with such short half-lives have been reported in the brain. Recently Segal, Kuczenski, and Mandell have shown a similar change in tyrosine hydroxylase from the locus coeruleus within

30 min of the administration of various agents *(in preparation)*. The protein synthetic apparatus involving biosynthetic enzymes and their transport may not be just an adjustment to chronic perturbation but may be involved too in hour-to-hour changes in the function of biogenic amines in the brain.

Much more work, with other agents, at differential doses, over various periods of time, will be necessary for the full elucidation of these kinds of enzymatic adaptation. Then, as is the case for tyrosine hydroxylase, the functional implications in relation to transmitter synthesis in an intact, behaving brain await our attention.

ACKNOWLEDGMENTS

This work was supported by National Institute of Mental Health grants DA-00265-02 and DA-00046-04 and Friends of Psychiatric Research of San Diego, Inc.

REFERENCES

Glowinski, J., Hamon, M., and Hery, F. (1973): Regulation of 5-HT synthesis in central serotonergic neurons. In: *New Concepts in Neurotransmitter Regulation,* edited by A. J. Mandell. Plenum Publishing Corp., New York.

Ichiyama, A., Nakamura, S., Nishizuka, Y., and Hayaishi, O. (1968): Tryptophan hydroxylase in mammalian brain. *Advances in Pharmacology,* 6A:5-17.

Ichiyama, A., Nakamura, S., Nishizuka, Y., and Hayaishi, O. (1970): Enzymic studies on the biosynthesis of serotonin in mammalian brain. *Journal of Biological Chemistry,* 245:1699-1709.

Knapp, S., and Mandell, A. J. (1972a): Parachlorophenylalanine: Its three phase sequence of interactions with the two forms of brain tryptophan hydroxylase. *Life Sciences,* 11:761-771.

Knapp, S., and Mandell, A. J. (1972b): Narcotic drugs: Effects on serotonin biosynthetic systems of the brain. *Science,* 177:1209-1211.

Knapp, S., and Mandell, A. J. (1973a): Some drug effects on the functions of the two physical forms of tryptophan-5-hydroxylase: Influence on hydroxylation and uptake of substrate. In: *Serotonin and Behavior,* edited by J. Barchas and E. Usdin. Academic Press, New York.

Knapp, S., and Mandell, A. J. (1973b): Short- and long-term lithium administration: Effects on the brain's serotonergic systems. *Science,* 180:645-647.

Knapp, S., and Mandell, A. J. (1974): Effects of amphetamines on regional tryptophan hydroxylase activity and synaptosomal conversion of tryptophan to 5-HT in rat brain. *Journal of Pharmacology and Experimental Therapeutics (in press).*

Kuczenski, R. (1973a): Soluble, membrane-bound, and detergent-solubulized rat striatal tyrosine hydroxylase: pH-dependent cofactor binding. *Journal of Biological Chemistry,* 258:5074-5080.

Kuczenski, R. (1973b): Striatal tyrosine hydroxylases with high and low affinity for tyrosine: Implications for the multiple-pool concept of catecholamines. *Life Sciences,* 13:247-255.

Kuczenski, R., and Mandell, A. J. (1972a): Allosteric activation of hypothalamic tyrosine hydroxylase by ions and sulfated mucopolysaccharides. *Journal of Neurochemistry,* 19:131-137.

Kuczenski, R., and Mandell, A. J. (1972b): Regulatory properties of soluble and particulate rat brain tyrosine hydroxylase. *Journal of Biological Chemistry,* 247:3114-3122.

Scrafani, J. T., Williams, N., and Clouet, D. N. (1969): Binding of dihydromorphine to subcellular fractions of rat brain. *Pharmacologist,* 11:256.

Neuropsychopharmacology of Monoamines and
Their Regulatory Enzymes, edited by E. Usdin.
Raven Press, New York © 1974.

Comparative Properties of Soluble and Particulate Catechol-O-methyl Transferases from Rat Red Blood Cells: Preliminary Observations

Mark Roffman, Thomas G. Reigle, Paul J. Orsulak, and Joseph J. Schildkraut

Neuropsychopharmacology Laboratory, Massachusetts Mental Health Center, Department of Psychiatry, Harvard Medical School, Boston, Massachusetts 02115

I. INTRODUCTION

Catechol-O-methyl transferase (COMT) inactivates catecholamines by O-methylation (Axelrod, 1957). Initially COMT was found to be a soluble enzyme (Axelrod and Tomchik, 1958), but recent reports indicate that a membrane-bound (particulate) enzyme, also capable of O-methylating catecholamines, exists in liver (Bohuon and Assicot, 1973), brain (Broch and Fonnum, 1972), and red blood cells (Assicot and Bohuon, 1971). In addition, there is evidence that the soluble and particulate enzymes of red blood cells differ in several properties, including K_m, pH optima, and immunochemical reactivity (Assicot and Bohuon, 1971; Bohuon and Assicot, 1973). Our studies were undertaken to confirm and extend these findings that particulate as well as soluble O-methylating enzymes are present within rat red blood cells.

II. METHODS

Heparinized blood was centrifuged at $1000 \times g$ at 4°C for 10 min. The plasma was removed by aspiration and discarded. The red blood cells were lysed by hypo-osmotic shock, that is, by diluting 10-fold with ice-cold 5×10^{-4} M dithiothreitol (DTT). After standing on ice for 10 min with intermittent shaking, the lysed red cells were centrifuged for 20 min at $2000 \times g$. The supernatant was used as the source of soluble enzyme. The pellet was then washed four times by resuspension in a volume of DTT (5×10^{-4} M) equivalent to that used to lyse the red cells. The resuspended pellet was then shaken vigorously and centrifuged at $2500 \times g$ for 10 min. After discarding the supernatant of the final wash, the pellet was resuspended in 0.002 M phosphate buffer, pH 7.8, containing 0.1% Triton

X-100 and 5×10^{-4} M DTT in a volume equivalent to the original volume of red cells. The resuspended pellet was then shaken vigorously for 3 min on a vortex mixer and centrifuged at $2000 \times g$ for 10 min. The resulting supernatant was used as a source of the particulate enzyme. Soluble and particulate COMT were assayed by modifications of the methods described by McCaman (1965) and Axelrod and Cohn (1970).

III. RESULTS

After observing soluble and particulate O-methyl transferase activity in preliminary experiments, the relative substrate specificities of both enzyme preparations were determined.

Data presented in Table 1 indicate relatively similar substrate specificities for the two enzyme preparations using catechols as substrates. That is,

TABLE 1. *Relative substrate specificities of soluble and particulate catechol-O-methyl transferases*

Substrate	Concentration (M)	COMT activity (cpm × 10⁻⁴/ml rbc/20 min) Soluble	Particulate
3,4-Dihydroxybenzoic acid	1×10^{-3}	102,109	0.52,0.37
	1×10^{-4}	39,42	0.29,0.22
	Average % decrease[a]	62	43
Dopamine	1×10^{-3}	40,47	0.22,0.19
	1×10^{-4}	7.9,9.4	0.08,0.10
	Average % decrease[a]	80	56
3,4-Dihydroxy-norephedrine	1×10^{-3}	14,16	0.07
	1×10^{-4}	2.4,2.7	0.04
	Average % decrease[a]	83	43
Octopamine	1×10^{-3}	0	0

[a] Average % decrease = the mean difference between the COMT activity at 1×10^{-3} M and 1×10^{-4} M divided by the mean activity at 1×10^{-3} M multiplied by 100.

both soluble and particulate enzyme activity was greatest with 3,4-dihydroxybenzoic acid (DBA) and poorest with 3,4-dihydroxynorephedrine. However, particulate enzyme activity was less sensitive to changes in substrate concentration than was soluble enzyme activity (Table 1). Neither enzyme preparation was capable of O-methylating the noncatechol octopamine, indicating that both the soluble and particulate enzymes are specific catechol-O-methyl transferases (Table 1).

The optimal pH for each enzyme preparation was determined by varying the pH of the reaction mixture from 6.7 to 8.7. Soluble enzyme activity was optimal within a pH range of 7.9 to 8.3, whereas particulate enzyme activity was optimal within a range of 7.9 to 8.1. The particulate enzyme retained relatively more activity in the lower pH range than did the soluble enzyme, but the activity of both enzyme preparations declined rapidly as pH increased above the optimum.

The double reciprocal Lineweaver-Burk plots shown in Fig. 1 illustrate the difference in Michaelis constants (K_m). The apparent K_m for the soluble enzyme preparation $(1.7 \pm 0.20 \times 10^{-4}$ M$)$ is significantly higher $(p < 0.005)$ than the apparent K_m of the particulate enzyme preparation $(4.7 \pm 0.70 \times 10^{-5}$ M$)$.

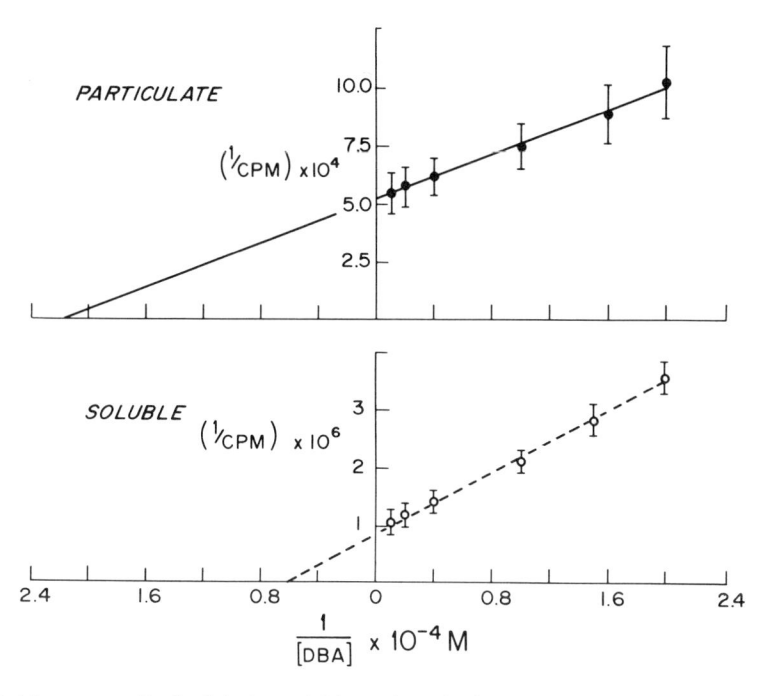

FIG. 1. Lineweaver-Burk plots for soluble and particulate catechol-O-methyl transferase. The concentration of substrate, 3,4-dihydroxybenzoic acid, was varied from 5×10^{-5} to 1×10^{-3} M. Incubations, as described under Methods, were run for 10 min. Velocity is defined as CPM/ml RBC/10 min. Each point represents the mean ±standard error of four determinations.

Because of the differences in substrate kinetics, we determined the apparent K_m of each enzyme preparation for the metal cofactor magnesium. The apparent K_m was 6×10^{-6} M for preparations of both the soluble and particulate enzymes.

In order to determine the relative stabilities of the enzymes to heat, each enzyme preparation was pre-incubated for 5 min at various temperatures above the normal incubation temperature (37°C). Data presented in Fig. 2 indicate that as the temperature was increased, the activity of the

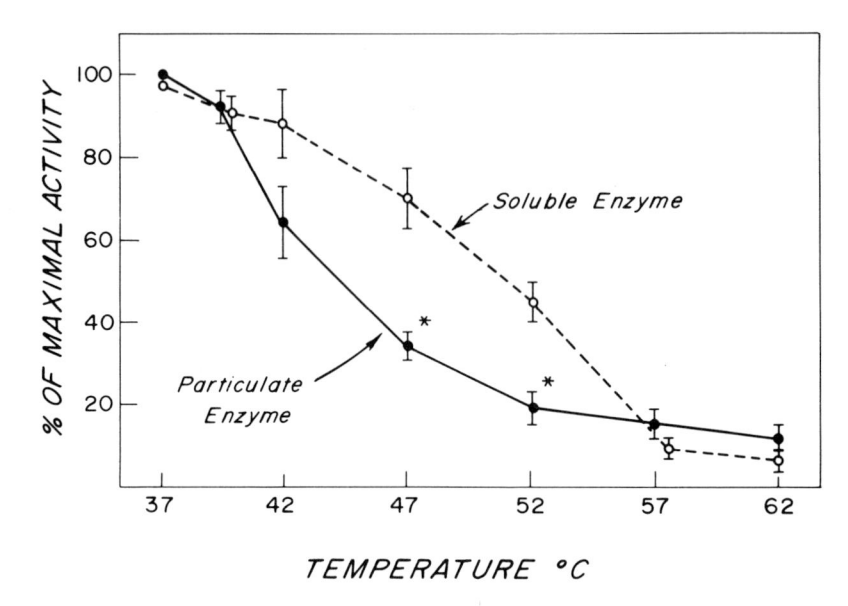

FIG. 2. Percent of maximal activity of soluble and particulate catechol-O-methyl transferases after pre-incubation at various temperatures. Enzymes were pre-incubated for 5 min at each temperature before incubation for 20 min at 37°C using 3,4-dihydroxybenzoic acid as substrate as described under Methods. The activity at 37°C was taken as maximal (100%). Each point represents the mean ± standard error of four determinations. * = $p < 0.05$ for difference between particulate and soluble enzyme preparations.

particulate enzyme preparation declined more rapidly than that of the soluble enzyme preparation.

IV. DISCUSSION

The results of these studies indicate that rat red blood cells contain at least two forms of COMT which differ in substrate kinetics, pH characteristics, and thermal stability but which have similar pH optima, substrate specificities, and magnesium kinetics. The existence in rat red blood cells of a soluble and membrane-bound COMT differing in several physical and chemical properties confirms the earlier work of Assicot and Bohuon (1971).

It may be argued that the observed differences in the physical and chemical properties of the two enzyme preparations were due to an artifact of the binding of the soluble enzyme to the membranes. However, the fact that the particulate enzyme preparation, released partially or completely from the membrane by detergent, still exhibited properties which were different from those of the soluble enzyme preparation does not favor this possibility.

In order to control for nonspecific effects of the detergent in our kinetic studies, the soluble enzyme was diluted to approximately equal activity with the same Triton-containing buffer as the particulate enzyme preparation. Thus, while the possibility still remains that the Triton affected the physical or chemical properties of the enzyme, it seems an unlikely possibility.

The results of the thermal stability study indicate that the particulate enzyme, once solubilized, was highly sensitive to temperature changes. Thus, it may be argued that the differences in substrate kinetics between the two enzyme preparations were the result of an instability of the particulate enzyme. However, the fact that product formation by both enzyme preparations was linear for 80 min at 37°C does not favor this possibility.

There are indications that COMT is closely associated with the adrenergic receptor (Axelrod, 1966). If true, it might be expected that a membrane-bound enzyme would more closely resemble this receptor than would a soluble enzyme. In support of this hypothesis, a recent report indicates that propranolol, a beta-adrenergic blocking agent, partially inhibits particulate enzyme activity at concentrations which only slightly affect soluble enzyme activity (Bohuon and Assicot, 1973). In addition, a synaptosomal enzyme has been demonstrated in the central nervous system (Broch and Fonnum, 1972), and further studies are required to determine whether the physical and chemical properties of the particulate enzyme of the red blood cell will resemble those of the synaptosomal enzyme.

Several investigators have examined soluble red blood cell COMT activity in various psychiatric disorders (Cohn et al., 1970; Matthysse and Baldessarini, 1972; Shopsin et al., 1973). One study has noted differences in the K_m (for DBA) of the soluble COMT in patients with paranoid schizophrenia (Shopsin et al., 1973), while another study observed alterations in soluble red blood cell COMT activity in females with primary affective disorders (Cohn et al., 1970). Studies are now in progress to determine if a membrane-bound enzyme can be identified in human red blood cells, since measures of the activity or kinetic properties of this enzyme may be of further interest in biochemical studies of patients with various psychiatric and addictive disorders.

V. SUMMARY

In studies using the rat red blood cell, we have observed soluble and membrane-bound (particulate) forms of catechol-O-methyl transferase (COMT) having similar substrate specificities, affinities for magnesium, and nearly similar pH optima, but differing in substrate (3,4-dihydroxybenzoic acid) kinetics and thermal stabilities. Of particular note is the difference in K_m between the two enzyme preparations (4.7 ± 0.70 × 10^{-5} M for particulate and 1.7 ± 0.20 × 10^{-4} M for soluble), indicating that the affinity for substrate of the particulate enzyme preparation is greater than that of the soluble enzyme preparation. These studies, which confirm the work of Assicot and

Bohuon (1971), indicate that two forms of COMT, differing in several physical and chemical properties, exist within the rat red blood cell. We are now examining human red blood cells in an effort to identify a comparable membrane-bound COMT.

ACKNOWLEDGMENTS

This work was supported by U.S. Public Health Service grants MH 15413 and DA 00257.

REFERENCES

Assicot, M., and Bohuon, C. (1971): *Biochimie,* 53:871–874.
Axelrod, J. (1957): *Science,* 126:400–401.
Axelrod, J. (1966): *Pharmacological Reviews,* 18:95–113.
Axelrod, J., and Cohn, K. (1970): *Journal of Pharmacology and Experimental Therapeutics,* 176:650–654.
Axelrod, J., and Tomchik, R. (1958): *Journal of Biological Chemistry,* 233:702–705.
Bohuon, C., and Assicot, M. (1973): In: *Frontiers in Catecholamine Research,* edited by E. Usdin and S. Snyder. Pergamon Press, Elmsford, N.Y., pp. 107–112.
Broch, O. J. Jr., and Fonnum, F. (1972): *Journal of Neurochemistry,* 19:2049–2055.
Cohn, K., Dunner, D. L., and Axelrod, J. (1970): *Science,* 170:1323–1324.
McCaman, R. E. (1965): *Life Sciences,* 4:2353–2359.
Matthysse, S., and Baldessarini, R. J. (1972): *American Journal of Psychiatry,* 128:1310–1312.
Shopsin, B., Wilk, S., Gershon, S., Roffman, M., and Goldstein, M. (1973): In: *Frontiers in Catecholamine Research,* edited by E. Usdin and S. Snyder. Pergamon Press, Elmsford, N.Y. pp. 1173–1179.

Neuropsychopharmacology of Monoamines and
Their Regulatory Enzymes, edited by E. Usdin.
Raven Press, New York © 1974.

Genetic Aspects of Monoamine Mechanisms

Jack D. Barchas, Roland D. Ciaranello,* Jerome A. Dominic,**
Takeo Deguchi,[†] Elaine Orenberg, Jean Renson,
and Seymour Kessler

Department of Psychiatry, Stanford University School of Medicine, Stanford, California 94305

Variation based on genetic factors of mechanisms involving biogenic amines could be important both in terms of behavioral states and effects of pharmacological agents. The possibility of genetic variation in biogenic amine mechanism is suggested by several lines of evidence. For example, in animals, strain and subline differences have been reported in the amounts of biogenic amines in brain regions of mice (Maas, 1962, 1963; Sudak and Maas, 1964a; Schlesinger, Boggan, and Freedman, 1965; Karczmar and Scudder, 1967) and rats (Sudak and Maas, 1964b; Miller, Cox, and Maickel, 1968), and in the utilization and uptake of cardiac norepinephrine in mice (Page, Kessler, and Vesell, 1970). In humans, a variety of forms of pheochromocytoma have been shown to be associated with familial factors (Rimoin and Schimke, 1971). The known genetic variation in adrenocortical function (Badr and Spickett, 1965; Stempfel and Tomkins, 1966; Hamburg and Kessler, 1967) and in adrenocortical and adrenomedullary structure (Chai and Dickie, 1966; Shire, 1970) further suggest that the search for genetic variation in the enzymes involved in biogenic biosynthesis would be fruitful.

The use of inbred mouse strains is particularly advantageous in studies of genetic variation and provides a powerful tool for subsequent behavioral and genetic analysis. Each inbred strain represents a distinct genotypic constellation; differences found between strains, when environmental conditions are held constant, suggest that genetic variation may be present. By appropriate matings, the nature of this variation can be elucidated.

The first aspect of our studies involved the synthesis of catecholamines. Two enzymes were of particular interest to us, tyrosine hydroxylase (TH) and phenylethanolamine N-methyl transferase (PNMT). Tyrosine hydroxylase is the first step in the formation of catechols; from the work of the Uden-

---

Current addresses: * Laboratory of Clinical Science, National Institute of Mental Health, Bethesda, Maryland 20014; ** University of Kentucky Medical School, Lexington, Kentucky 40506; † Tokyo Metropolitan Institute for Neuroscience, 2-6 Mushashidai, Fuchu-City, Tokyo, Japan.

friend group and of others, tyrosine hydroxylase is generally considered to be the rate-limiting step in catecholamine formation, and thus assumes an important role in regulating the levels and activity of catecholamines in various tissues (Nagatsu, Levitt, and Udenfriend, 1964; Spector, Gordon, Sjoerdsma, and Udenfriend, 1967; Sedvall, Weise, and Kopin, 1968).

The final synthetic enzyme in the pathway, PNMT, converts norepinephrine to epinephrine. Enzyme activity is high only in the adrenal medulla (Kirshner and Goodall, 1957; Axelrod, 1962), although low levels of epinephrine formation have been found in mammalian brain (Barchas, Ciaranello, and Steinman, 1969; Ciaranello, Barchas, Byers, Stemmle, and Barchas, 1969; Deguchi and Barchas, 1971). The conversion of norepinephrine to epinephrine by PNMT has been shown to be altered by a number of factors, including adrenal steroids (Coupland, 1953; Wurtman and Axelrod, 1966; Molinoff and Axelrod, 1971). Studies have demonstrated that PNMT also is under partial neuronal control. When the splanchnic nerves to the adrenal are stimulated by administration of the drug 6-hydroxydopamine, PNMT activity rises slowly (Mueller, Thoenen, and Axelrod, 1969). This rise can be blocked by denervation, suggesting that the response is neuronally triggered. Hypophysectomy has no effect on the enzyme response to 6-hydroxydopamine, indicating that the pituitary-adrenocortical axis is not involved in this aspect of PNMT regulation (Thoenen, Mueller, and Axelrod, 1970). Similarly, these same investigators demonstrated that denervation has no effect on the response of the enzyme to dexamethasone in hypophysectomized animals. These results are interpreted to mean that the neuronal and hormonal controls on the enzymes are mediated via different routes.

The first set of studies involved the comparison of the levels of tyrosine hydroxylase and PNMT in strains of mice (Kessler, Ciaranello, Shire, and Barchas, 1971; Ciaranello, Barchas, Kessler, and Barchas, 1972). As shown in Table 1, there are marked differences in the levels of the enzymes in the brains and adrenal in the different strains. We tested a number of strains, and the relative rank order of the enzymes in the strains was identical for both adrenal and brain tyrosine hydroxylase, with the notable exception of C57Bl/Ka, which showed the lowest adrenal tyrosine hydroxylase in combination with a relatively high activity of brain tyrosine hydroxylase. This finding raises the possibility that structurally different forms of this enzyme and/or differences in the rates of turnover exist in the two tissues.

Recently, Ciaranello and Axelrod (Ciaranello and Axelrod, 1973; Ciaranello, 1973) have found that the differences in the activity of PNMT between different strains which they were investigating were due to differences in the rate of degradation of the PNMT.

In other recent studies (Barchas, Erdelyi, and Kessler, 1974), we have investigated strain differences in levels of the enzyme dopamine-β-hydroxylase. The enzyme converts dopamine to norepinephrine. Dopamine-β-

TABLE 1. Activities of enzymes involved in synthesis of catecholamines

	n	Brain Tyrosine hydroxylase[a]	Brain Tyrosine hydroxylase[b]	Adrenal Phenylethanolamine N-methyltransferase[c]
Strain				
BALB/cJ	8	14.6 ± 1.14	5.28 ± 0.81	0.198 ± 0.015
CBA/J	8	9.4 ± 0.66	1.99 ± 0.14	0.124 ± 0.007
C57Bl/Ka	8	11.6 ± 0.99	1.40 ± 0.09	0.085 ± 0.008
F_1 mice				
BALB/cJ × CBA/J	16	12.5 ± 0.66	2.24 ± 0.10	0.119 ± 0.008
CBA/J × C57Bl/Ka	16	11.7 ± 0.51	1.82 ± 0.16	0.143 ± 0.010
C57Bl/Ka × BALB/cJ	16	10.4 ± 0.70	2.17 ± 0.10	0.118 ± 0.006

[a] nmoles of [^{14}C] DOPA formed per hour per gram of brain tissue.
[b] nmoles of [^{14}C] DOPA formed per hour per pair of adrenals.
[c] nmoles of N-[^{14}C] methylphenylethanolamine formed per hour per pair of adrenals.
Activities expressed as the mean ± SE.

hydroxylase activity in the cortex and adrenals of the five studied strains varied. Significant differences were found in the cortex dopamine-β-hydroxylase activity between BRT and BALB, BRT and SJL, and NZB and BALB mouse strains. The strain CSW was significantly different from the other strains, with higher enzyme activity in the cortex, and lower activity in the adrenal.

In the few studies concerned with strain differences in levels of catecholamines (e.g., strain differences in levels of brain catecholamines), the demonstration of a genetic basis of the differences was not attempted. Therefore, we crossed several of the strains (Kessler, Ciaranello, Shire, and Barchas, 1972) and measured enzyme activity in the resulting progenies. From this work (Table 1), a number of interesting points are apparent. Genetic factors play an important role in the determination of the differences between the strains. All the reciprocal crosses involving a given pair of strains yielded similar values; therefore pre- and postnatal biological and cultural maternal effects and also X-linkage could be eliminated as important determinants of the observed differences. In Table 2, the reciprocal hybrid

TABLE 2. Comparison of factors which increase PNMT activity in different mouse strains

	Cold exposure	Glucocorticoid	Neuronal	Direct ACTH
DBA/2J	+	+	+	0
C57Bl/Ka	+	0	0	+
CBA/J	+	+	0	0

0 = no increase.
+ = increase.

combinations were pooled because they were not significantly different. In the adrenal gland, PNMT and tyrosine hydroxylase activities were correlated over all the genotypic groups studied, suggesting that in these strains, the two enzymes may be controlled by the same genetic factor(s), possibly due to close linkage on the same chromosome or to coordinate gene regulation. These possibilities might be further tested by studying whether the correlation between the enzyme activities is maintained or breaks down in segregating generations (F_2 and backcrosses) (Shire, 1969).

The pattern of inheritance of brain tyrosine hydroxylase differs from that found for adrenal tyrosine hydroxylase; the gene(s) of C57Bl/Ka are dominant to those of the other two strains. The cross between BALB/cJ and CBA/J produced hybrids with brain tyrosine hydroxylase activity intermediate between the parental strains.

The work presented demonstrates clear genetically determined differences between strains in the ability to synthesize catecholamines. While static enzyme levels provide limited information, the physiological regulation of enzymatic activity provides a more dynamic means of investigation of control processes. In this regard, we have been particularly interested in the role of PNMT in converting norepinephrine to epinephrine as a model for investigation.

PNMT activity in the intact organism can be changed by prolonged stress in the rat (Vernikos-Danellis, Ciaranello, and Barchas, 1968; Ciaranello, Barchas, and Vernikos-Danellis, 1969; Milkovic, Deguchi, Winget, Barchas, Levine, and Ciaranello, 1974) and by short-term stress in the mouse (Ciaranello, Dornbusch, and Barchas, 1972a). Because of the lack of rapid changes in PNMT in intact rats, genetic studies that have attempted to focus on control of the enzyme or stress-induced fluctuations in enzyme activity have not been fruitful. Because of our demonstration that PNMT levels varied widely in the adrenals of several inbred mouse strains, it was of interest to study these strains further to determine the characteristics of the changes in activity of PNMT in response to stress and to detect strain differences in the physiological regulation of the enzyme.

The nature of the physiological regulatory mechanisms of PNMT activity was compared in inbred mouse strains (Ciaranello, Dornbusch, and Barchas, 1972b). The strains utilized, DBA/2J, C57Bl/Ka, and CBA/J, showed marked and unexpected differences. All three strains responded to cold stress with an elevation of adrenal PNMT activity, and in all strains the pituitary was involved in the regulation of enzyme activity. However, the mechanism of regulation, in terms of neuronal control of the enzyme, differed among the strains.

In the DBA/2J strain, cold exposure, glucocorticoid administration, and phenoxybenzamine administration were all effective in increasing enzyme activity. The results suggest that, in this strain, PNMT activity is under both glucocorticoid and neuronal control.

In the C57Bl/Ka strain, only cold exposure and adrenocorticotropic hormone (ACTH) were effective in the induction of PNMT. Exogenous glucocorticoid administration had no effect on enzyme activity. In this strain, ACTH appears to exert a direct regulatory effect on the enzyme without the mediation of the adrenal glucocorticoids. No evidence for neuronal control of PNMT activity was found in this strain.

In the CBA/J strain, cold exposure increased the level of the enzyme. The enzyme responded to ACTH, but the response was mediated by glucocorticoids. There is no evidence of neuronal control of the enzyme.

The half-life of the enzyme was estimated to be 1 hr in the DBA strain, 3 hr in the C57Bl/Ka strain, and 7 hr in the CBA/J strain. The rate of increase of enzyme activity following induction is 10 times greater in the DBA/2J strain than in either the C57Bl/Ka or CBA/J strains.

Taken together, the differences in regulation among the strains (Table 2) suggest powerful differences in regulation mechanisms, which could be important in response to stress or drugs.

A comparison series of studies to these dealing with catecholamines are currently in progress for the indoleamine pathway. In that pathway, tryptophan hydroxylase appears to have an important regulatory role (Kaufman, 1963; Deguchi and Barchas, 1972a, b, 1973; Sinha and Barchas, 1973; Renson, 1973), although the mechanisms of regulation are not clearly established.

We examined differences in the activity of tryptophan hydroxylase and other parameters of serotonin metabolism among several strains of mice (Table 3). Comparing C57Bl/10J and BALB/cJ, it can be concluded that the latter strain has a higher enzyme activity in the brainstem but a lower value in the cortex. Considering the pair DBA/2J and SJL/J, the latter strain proved to have greater enzyme activity in both brain areas.

The emergence of these enzymatic differences raised questions concern-

TABLE 3

Strain	Cortex Tryptophan hydroxylase (nmoles/g/hr)	Brainstem Tryptophan hydroxylase (nmoles/g/hr)	Tryptophan (μg/g)	5-HT (μg/g)	5-HIAA (μg/g)
C57Bl/10J	1.04 ± 0.05 (11)	3.36 ± 0.15 (12)	2.41 ± 0.17	0.91 ± 0.02	0.55 ± 0.04
BALB/cJ	0.64 ± 0.03[a] (19)	4.10 ± 0.24[b] (10)	2.36 ± 0.09	0.91 ± 0.03	0.72 ± 0.04[a]
DBA/2J	0.75 ± 0.03 (17)	3.31 ± 0.17 (11)	2.45 ± 0.13	0.81 ± 0.05	0.45 ± 0.03
SJL/J	1.22 ± 0.06[a] (12)	4.00 ± 0.27[c] (6)	2.72 ± 0.13	1.03 ± 0.04[a]	0.80 ± 0.04[a]

The number of mice is indicated in parentheses.
[a] $p < 0.01$.
[b] $p < 0.02$.
[c] $p < 0.05$.

ing what relationship tryptophan hydroxylase activity may have to other parameters used to assess the functional state of serotonergic neurons in the brain.

When the steady state concentrations of tryptophan, serotonin, and 5-hydroxyindoleacetic acid (5-HIAA) were analyzed in the brainstem, no consistent pattern was apparent. C57BL/10J and BALB/cJ showed similar concentrations of tryptophan and serotonin, but BALB/cJ had a higher level of 5-HIAA. In the other pair, the SJL/J strain had greater concentrations of both serotonin and 5-HIAA. Relating these data to tryptophan hydroxylase activity, it is evident that the member of each pair demonstrating higher enzyme activity in the brainstem (BALB/cJ and SJL/J) also had higher steady state levels of 5-HIAA.

Although certain significant differences were found, no relationship between these data and enzyme activity can be demonstrated. For example, in C57BL/10J and SJL/J, the members of each pair of mouse strains exhibiting the higher enzyme activity did not show consistent differences in the concentrations of tryptophan, serotonin, or 5-HIAA.

In order to obtain a more dynamic measure of serotonergic activity in the brain, we are conducting studies of the formation of ^{3}H-serotonin from ^{3}H-tryptophan. From the results to date, it is evident that the higher tryptophan hydroxylase activity in BALB/cJ and SJL/J was not associated with a change in the formation of ^{3}H-serotonin. In the cortex, in C57BL/10J and SJL/J, the members of their respective pairs with the greater enzyme activity also had significantly great formation of ^{3}H-serotonin. SJL/J demonstrated a greater synthesis of ^{3}H-serotonin despite a lower amount of ^{3}H-tryptophan. The dynamic aspects of these differences in relation to stress and drug treatment remains to be investigated.

Another aspect of biogenic amine mechanisms that can be studied in terms of genetic processes involves cyclic adenosine 3',5'-monophosphate (cyclic AMP). Cyclic AMP has been found to function as a second messenger for various hormones and neurotransmitters, including norepinephrine (Sutherland, Robison, and Butcher, 1968). Considerable evidence suggests that the nucleotide may participate in some molecular events underlying synaptic transmission within the mammalian central nervous system. Cyclic AMP together with adenyl cyclase and phosphodiesterase, the enzymes responsible for its synthesis and degradation, occur in the brain, and more specifically in synaptosomal membranes. The formation of cyclic AMP has been shown to be enhanced by several putative central neurotransmitters, the biogenic amines norepinephrine, dopamine, and serotonin, which stimulate adenyl cyclase activity (Burkard, 1972). We have found evidence of the existence of differences in brain cyclic AMP content among inbred strains of mice which would suggest that the nucleotide should perhaps be considered as one of the parameters in the modulation of neuroregulatory mechanisms which are subject to genetic variation.

In our studies, cyclic AMP was determined by the protein-binding assay procedure of Gilman (1970) on male mice of four inbred strains (BALB/J, C57Bl/6J, CBA/J and A/J, and CAF$_1$/J, the F$_1$ generation of BALB/J female × A/J male) purchased from Jackson Laboratory, Bar Harbor, Maine. Differences in strain characteristics in open-field behavior of activity, defecation, and aggressiveness, as well as preference of alcohol and avoidance conditioning have been reported. For example, the strains selected for evaluation in this study represent extremes of aggressive behavior in open-field testing according to the Jackson strain rankings (Southwick and Clark, 1966). BALB/J and C57Bl/J are highly aggressive and CBA/J and A/J low aggressors. In addition, BALB/J and C57Bl/6J are known to show a preference for alcohol.

Mice were housed six to a cage on a 12-hr light and dark cycle and killed at 8 to 9 weeks of age by microwave irradiation as a means of enzyme inactivation of brain adenyl cyclase and phosphodiesterase (Schmidt, Schmidt, and Robison, 1971; Schmidt, Hopkins, Schmidt, and Robison, 1972). Significant differences ($p < 0.01$) in cyclic AMP concentration of whole-brain homogenates were found among the four strains. The C57Bl/J and BALB/J strains with 2.66 ± 0.20 and 2.40 ± 0.18 pmoles cyclic AMP/mg wet weight were not different from one another, but interestingly, had significantly higher levels ($p < 0.01$) of the nucleotide than did CBA/J and A/J strains with 1.25 ± 0.09 and 1.80 ± 0.11, respectively. Analyses were conducted on animals which are hybrids of a high and low cyclic AMP level strain, CAF$_1$ (BALB/J female × A/J male) to determine the mode of inheritance of cyclic AMP levels. The cyclic AMP concentration of the hybrid was found to be similar to the low cyclic AMP level parent (Orenberg, Renson, Kessler, and Barchas, 1974).

The biogenic amines have been related to a number of forms of behavior both normal and abnormal (see reviews of Barchas, Ciaranello, Stolk, Brodie, and Hamburg, 1972; Hamburg, Hamburg, and Barchas, 1974; Barchas, Ciaranello, Kessler, and Hamburg, 1974). The data presented here suggest that many of the basic mechanisms involving brain amines may vary from strain to strain and have powerful genetic controls. Such information could have important ramifications for theories involving amines in behavior and also for individual responses to stress in varying forms and to drugs.

ACKNOWLEDGMENTS

We should like to thank Dr. Usdin for his encouragement of this paper. We appreciate the long-term support and involvement in these and related studies of Dr. David Hamburg. We wish to thank Mrs. Elizabeth Erdelyi and Mrs. Pam Angwin for their research assistance, and Florence Parma and Rosemary Schmele for their secretarial assistance. This work was supported by U.S. Public Health Service grants MH 23861 and AA 00498.

REFERENCES

Axelrod, J. (1962): The enzymatic N-methylation of serotonin and other amines. *Journal of Pharmacology and Experimental Therapeutics,* 138:28-33.
Badr, F. M., and Spickett, S. G. (1965): Genetic variation in the biosynthesis of corticosteroids in *Mus musculus. Nature,* 205:1088-1090.
Barchas, J. D., Ciaranello, R. D., Kessler, S., and Hamburg, D. A. (1974): Genetic aspects of catecholamine synthesis. In: *Genetics and Psychopathology,* edited by R. Fieve, H. Brill, and D. Rosenthal. Johns Hopkins University Press, Baltimore *(in press).*
Barchas, J. D., Ciaranello, R. D., and Steinman, A. M. (1969): Epinephrine formation and metabolism in mammalian brain. *Biological Psychiatry,* 1:31-48.
Barchas, J. D., Ciaranello, R. D., Stolk, J. M., Brodie, H. K. H., and Hamburg, D. A. (1972): Biogenic amines and behavior. In: *Hormones and Behavior,* edited by S. Levine. Academic Press, New York, pp. 235-329.
Barchas, J. D., Erdelyi, E., and Kessler, S. (1974): Strain differences in dopamine β-hydroxylase activity in mice. *(In preparation.)*
Burkard, W. P. (1972): Catecholamine induced increased of cyclic adenosine 3',5'-monophosphate in rat brain *in vivo. Journal of Neurochemistry,* 19:2615.
Chai, C. K., and Dickie, M. M. (1966): In: *Biology of the Laboratory Mouse,* edited by E. L. Green. McGraw-Hill Book Company, New York, p. 387.
Ciaranello, R. D. (1973): Regulation of phenylethanolamine N-methyltransferase. In: *Frontiers in Catecholamine Research,* edited by E. Usdin and S. Snyder. Pergamon Press, Elmsford, N.Y.
Ciaranello, R. D., and Axelrod, J. (1973): Genetically controlled alterations in the rate of degradation of phenylethanolamine N-methyltransferase. *Journal of Biological Chemistry,* 248:5616-5623.
Ciaranello, R. D., Barchas, R. E., Byers, G. S., Stemmle, D. W., and Barchas, J. D. (1969): Enzymatic synthesis of adrenaline in mammalian brain. *Nature,* 221:368-369.
Ciaranello, R. D., Barchas, R., Kessler, S., and Barchas, J. D. (1972): Catecholamines: Strain differences in biosynthetic enzyme activity in mice. *Life Sciences,* I, 11:565-572.
Ciaranello, R. D., Barchas, J. D., and Vernikos-Danellis, J. (1969): Compensatory hypertrophy and phenylethanolamine N-methyl transferase (PNMT) activity in the rat adrenal. *Life Sciences,* 8:401-407.
Ciaranello, R. D., Dornbusch, J. N., and Barchas, J. D. (1972*a*): Rapid increase of phenylethanolamine-N-methyltransferase by environmental stress in an inbred mouse strain. *Science,* 175:789-790.
Ciaranello, R. D., Dornbusch, J. N., and Barchas, J. D. (1972*b*): Regulation of adrenal phenylethanolamine N-methyltransferase activity in three inbred mouse strains. *Molecular Pharmacology,* 8:511-520.
Coupland, R. E. (1953): On the morphology and adrenaline-noradrenaline content of chromaffin tissue. *Journal of Endocrinology,* 9:194-203.
Deguchi, T., and Barchas, J. D. (1971): Inhibition of transmethylations of biogenic amines by S-adenosylhomocysteine. *Journal of Biological Chemistry,* 246:3175-3181.
Deguchi, T., and Barchas, J. D. (1972*a*): Effect of *p*-chlorophenylalanine on tryptophan hydroxylase in rat pineal. *Nature New Biology,* 235:92-93.
Deguchi, T., and Barchas, J. D. (1972*b*): Effect of *p*-chlorophenylalanine on hydroxylation of tryptophan in pineal and brain of rats. *Molecular Pharmacology,* 8:770-779.
Deguchi, T., and Barchas, J. (1973): Comparative studies on the effect of *para*chlorophenylalanine on hydroxylation of tryptophan in pineal and brain of rat. In: *Serotonin and Behavior,* edited by J. Barchas and E. Usdin. Academic Press, New York, pp. 33-47.
Deguchi, T., Sinha, A. K., and Barchas, J. D. (1973): Serotonin biosynthesis in rat raphe nuclei: Effect of *p*-chlorophenylalanine. *Journal of Neurochemistry,* 20:1329-1336.
Dominic, J. A., Deguchi, T., Kessler, S., and Barchas, J. D. (1974): Differences in serotonin formation and metabolism in inbred strains of mice. *(In preparation.)*
Gilman, A. G. (1970): A protein binding assay for adenosine 3',5-cyclic monophosphate. *Proceedings of the National Academy of Sciences* (U.S.), 67:305.
Hamburg, D. A., Hamburg, B., and Barchas, J. D. (1974): Anger and depression: Current psychobiological approaches. In: *Parameters of Emotion,* edited by L. Levi. Raven Press, New York *(in press).*

Hamburg, D., and Kessler, S. (1967): A behavioral-endocrine-genetic approach to stress problems. In: *Memoirs of the Society for Endocrinology No. 15: Endocrine Genetics*, edited by S. Picket. Cambridge University Press, London, pp. 249-270.

Karczmar, A. G., and Scudder, C. L. (1967): Behavioral responses to drugs and brain catecholamine levels in mice of different strains and genera. *Federation Proceedings*, 26:1186-1191.

Kaufman, S. (1963): The structure of the phenylalanine-hydroxylation cofactor. *Proceedings of the National Academy of Sciences* (U.S.), 50:1085.

Kessler, S., Ciaranello, R. D., Shire, J. G. M., and Barchas, J. D. (1971): Genetic variation in catecholamine synthesizing enzyme activities. *Genetics*, 68:S33 (Suppl.).

Kessler, S., Ciaranello, R. D., Shire, J. G. M., and Barchas, J. D. (1972): Genetic variation in catecholamine synthesizing enzyme activities. *Proceedings of the National Academy of Sciences* (U.S.), 69:2448-2450.

Kirshner, N., and Goodall, M. (1970): Formation of adrenaline from noradrenaline. *Biochimica Biophysica Acta*, 24:658-659.

Maas, J. W. (1962): Neurochemical differences between two strains of mice. *Science*, 137:621-622.

Maas, J. W. (1963): Neurochemical differences between two strains of mice. *Nature*, 197:255-257.

Milkovic, K., Deguchi, T., Winget, C., Barchas, J., Levine, S., and Ciaranello, R. (1974): The effect of maternal manipulation on the phenylethanolamine N-methyltransferase activity and corticosterone content of the fetal adrenal gland. *American Journal of Physiology (in press)*.

Miller, F. P., Cox, R. H., Jr., and Maickel, R. P. (1968): Intrastrain differences in serotonin and norepinephrine in discrete areas of rat brain. *Science*, 162:463-464.

Molinoff, P. B., and Axelrod, J. (1971): Biochemistry of catecholamines. *Annual Review of Biochemistry*, 40:465-500.

Mueller, R. A., Thoenen, H., and Axelrod, J. (1969): Adrenal tyrosine hydroxylase: Compensatory increase in activity after chemical sympathectomy. *Science*, 158:468-469.

Nagatsu, T., Levitt, M., and Udenfriend, S. (1964): Tyrosine hydroxylase: The initial step in norepinephrine biosynthesis. *Journal of Biological Chemistry*, 239:2910-2917.

Page, J. G., Kessler, R. M., and Vesell, E. S. (1970): Strain differences in uptake, pool size and turnover rate of norepinephrine in hearts of mice. *Biochemical Pharmacology*, 19:1381-1386.

Renson, J. (1973): Assays and properties of brain L-tryptophan-5-hydroxylase. In: *Serotonin and Behavior*, edited by J. Barchas and E. Usdin. Academic Press, New York, pp. 19-32.

Rimoin, D., and Schimke, R. (1971): *Genetic Disorders of the Endocrine Glands*. C. V. Mosby Company, St. Louis, pp. 251-257.

Schlesinger, K., Boggan, W. O., and Freedman, D. X. (1965): Genetics of audiogenic seizures: I. Relation to brain serotonin and norepinephrine in mice. *Life Sciences*, 4:2345-2351.

Schmidt, M. J., Hopkins, J. T., Schmidt, D. E., and Robison, G. A. (1972): Cyclic AMP in brain areas: Effects of amphetamine and norepinephrine assessed through the use of microwave radiation as a means of tissue fixation. *Brain Research*, 42:465.

Schmidt, M. J., Schmidt, D. E., and Robison, G. A. (1971): Cyclic adenosine monophosphate in brain areas: Microwave irradiation as a means of tissue fixation. *Science*, 173:1142.

Sedvall, G. C., Weise, V. K., and Kopin, I. J. (1968): The rate of norepinephrine synthesis measured *in vivo* during short intervals; influence of adrenergic nerve impulse activity. *Journal of Pharmacology and Experimental Therapeutics*, 159:274-282.

Shire, J. G. M. (1970): Genetic variation in adrenal structure: Quantitative measurements on the cortex and medulla in hybrid mice. *Journal of Endocrinology*, 48:419-431.

Southwick, C. H., and Clark, L. H. (1966): Aggressive behavior and exploratory activity in fourteen mouse strains. *American Zoologist*, 6:559 (Abs.).

Spector, S., Gordon, R., Sjoerdsma, A., and Udenfriend, S. (1967): End-product inhibition of tyrosine hydroxylase as a possible mechanism for regulation of norepinephrine synthesis. *Molecular Pharmacology*, 3:549-555.

Stempfel, R. S., and Tomkins, G. M. (1966): Congenital virilizing adrenocortical hyperplasia (the adrenogenital syndrome). In: *The Metabolic Basis of Inherited Disease*, edited by J. B. Stanbury, J. B. Wyngaarden, and D. S. Fredrickson. McGraw-Hill Book Company, New York, pp. 635-664.

Sudak, H. S., and Maas, J. W. (1964a): Central nervous system serotonin and norepinephrine localization in emotional and nonemotional strains in mice. *Nature*, 203:1254-1256.

Sudak, H. S., and Maas, J. W. (1964b): Behavioral-neurochemical correlation in reactive and nonreactive strains of rats. *Science,* 146:418–420.

Sutherland, E. W., Robinson, G. A., and Butcher, R. W. (1968): Some aspects of the biological role of adenosine 3',5'-monophosphate (cyclic AMP). *Circulation,* 37:279.

Thoenen, H., Mueller, R. A., and Axelrod, J. (1970): Neuronally dependent induction of adrenal phenylethanolamine-N-methyltransferase by 6-hydroxydopamine. *Biochemical Pharmacology,* 19:669–674.

Vernikos-Denellis, J., Ciaranello, R., and Barchas, J. (1968): Adrenal epinephrine and phenylethanolamine N-methyl transferase (PNMT) activity in the rat bearing a transplantable pituitary tumor. *Endocrinology,* 83:1357–1358.

Wurtman, R. J., and Axelrod, J. (1966): Control of enzymatic synthesis of adrenaline in the adrenal medulla by adrenal cortical steroids. *Journal of Biological Chemistry,* 241:2301–2305.

Open Discussion: Monoaminergic Enzymes Other Than MAO

Reporter: Earl Usdin

Molinoff. In reply to a question of Dr. Lovenberg on the specificity of PNMT for the D or L form of β-hydroxylated amines, Dr. Molinoff replied that L-norepinephrine is the best substrate for PNMT. Its K_m (10 μM) is $2\frac{1}{2}$ times less than that of the D stereoisomer.

Goldstein. Dr. Brodie postulated that because of its high molecular weight, DBH might have to enter the circulation via the lymph before getting into the plasma. Dr. Goldstein replied that serum DBH has about the same molecular weight as adrenal DBH, but the bulk of DBH is inactive and may already be broken down. Dr. Weiner added that work in his laboratory on rate of egress of DBH from isolated spleen by nerve stimulation tends to indicate that the rate at which the enzyme exits from the spleen is dependent on the pressor response. Blocking the pressor response with phenoxybenzamine or a smooth-muscle relaxant increases considerably the rate of release. Pretreatment of the spleen with cocaine enhances the pressor response and delays release. This is not seen with NE, whose rate of egress appears to be independent of the pressor response.

Dr. Breese mentioned the observation that patients given spinal anesthesia do not have changes in DBH activity. Dr. Lieberman described some patients in whom there was a correlation between serum DBH levels and coma. These patients were comatose from a variety of reasons: cerebral trauma, hypoxia, sepsis. All of the patients had suffered an episode of coma, defined as a period of absence of cerebral responsiveness with absence of spontaneous respiration requiring artificial respiration of 15 min or more. DBH levels determined drawn within 24 hr after the cerebral insult. However, during general anesthesia, while the patients were also "cerebrally unresponsive," there was no correlation between DBH levels and depth of anesthesia.

Lovenberg and Schanberg (combined discussion). Dr. Meltzer wondered whether it was necessary to do a radioimmunoassay as a measure of total DBH protein in human serum since activity correlates with DBH protein. Dr. Goldstein replied that it probably was not necessary, but further studies were needed before this could be definitively ascertained.

In reply to Dr. Goldstein's remark that he failed to see the advantage to the method used by Dr. Schanberg, Dr. Schanberg replied that on theoretical grounds it was better to look at the initial product than a degradative product. He felt that the ratio between DBH values obtained by chemical assay and radioimmunoassay might be of value in studying the biodegradation of DBH. Dr. Goldstein pointed out that although there is a high correlation between the levels of immunoreactive DBH and DBH activity in the sera of the normal population, it is conceivable that in some cases this correlation may not exist. Therefore the simultaneous determination of DBH by the radioimmunoassay and by the enzymatic assay might be of value. In some individuals the inactivation of DBH in the circulation may occur at a faster rate than in others. DBH is a copper enzyme, and the cupric copper of the enzyme is reduced by ascorbate or other reducing agents. The reduced form of the enzyme is less stable than the oxidized form. Thus, dietary factors such as ingestion of large amounts of ascorbate will cause a more rapid inactivation of the enzyme, and the inactive form of the enzyme can be measured only by the radioimmunoassay. Dr. Goldstein disagreed, since he felt that degradation may be nonenzymatic as well as enzymatic, and therefore determining DBH activity alone may not be sufficient at the time of release. Dr. Schanberg stated that if disappearance of inactive DBH protein varies significantly in different people, levels of total DBH protein might lack correlation with release phenomena. Dr. Brodie stated that if active DBH had a biological half-life of several hours, the determination of activity would be an appropriate measure for this kind of study. Dr. Schanberg stated that his interest was in the release of DBH, and that therefore the level of plasma DBH activity would seem to be the most appropriate. Dr. Schanberg concluded this interchange by pointing out that DBH activity measurements had been able to differentiate between essential and other forms of hypertension in the patients studied, but as yet no one had been successful in the use of radioimmunoassay for differential diagnosis.

In reply to a question on what blood DBH values he classified as normal, Dr. Schanberg replied probably somewhere between 8 and 25. He then stated that of the many subjects he has studied, only three had apparently normal blood pressures and DBH values over 60, and two of these showed sympathetic hyperreactivity as tested by forearm blood-flow methods. After stating that his data were expressed in terms of standard error of the mean, Dr. Schanberg admitted that even he was surprised at the low error figures in his data. However, he stated that not only were the determinations run blind by his technicians, but he also had encouraged collaborators to send in duplicate samples identified with different numbers and the results always checked.

Dr. Costa inquired about the concordance of results obtained by radioimmunoassay and chemical assay. Dr. Goldstein pointed out that the former required a specific antibody and that none was available for the rat. In human postmortem brain he had found a good correlation between enzyme activity and immunoassay.

Dr. Meltzer pointed out a seeming discrepancy between the results of

DISCUSSION

Drs. Lovenberg and Schanberg: Dr. Lovenberg found that blacks tended to have lower DBH levels, while Dr. Schanberg reported that labile hypertensives have high DBH levels and it has been reported that blacks have a higher degree of labile hypertension than the average population. Dr. Schanberg said that a number of conditions could possibly explain this apparent discrepancy. Essential hypertension is a "wastepaper basket" diagnosis; it is therefore quite possible that the diagnosis of essential hypertension in the black population actually covers diseases of different etiologies. Also, since it appears that plasma DBH concentration is to a large extent dependent upon genetic factors, it is possible that the values of plasma DBH in the black population could show the same relative individual differences, but at the same time have the total population curve set at a lower level.

In reply to a question by Dr. Goldstein on the significance of his data *vis-à-vis* number of samples, Dr. Lovenberg said that he had studied six patients with pheochromocytoma. Whereas a normal population showed serum DBH values ranging from 50 to 500 units, the six patients had values of 450 to 1000 units, generally higher than the members of the normal population. Furthermore, one patient went from 600 units before surgical removal of the tumor to 250 units several days after the operation, and another went from 500 units to 50 units.

Dr. Brodie stated that, since the enzyme had such a long half-life and so large a pool, he could see no theoretical reason for measuring the inactive form as well. Dr. Goldstein pointed out that the enzyme might be altered under stressful conditions. For example, half-life is changed during the cold pressor test.

In reply to a question of Dr. Neff as to whether or not there was DBH activity in the spinal fluid, Dr. Goldstein said that the very small amounts of DBH activity he had measured in CSF were probably the result of contamination by blood in the samples.

Carlsson and Weiner (combined discussion). In reply to Dr. Moore's question as to whether newly synthesized NE is preferentially released from the cat spleen in the absence of nerve stimulation, Dr. Weiner stated that he had originally made such an interpretation, but now thought it was also possible that the material which was spontaneously released was coming from deeper pools; that is, there may be spontaneous release from preterminal regions or regions further up the axons. Dr. Moore also wondered if there were vesicles lying on the membrane. Dr. Weiner indicated that with high-frequency stimulation, AMPT does not seem to inhibit NE synthesis as completely as with low-frequency stimulation; TH in the intact tissue appears to become less sensitive to inhibitors.

Next, Dr. Moore asked about the experiments in which cocaine was given to block uptake, and Dr. Weiner replied that the reuptake process does not conserve NE. On nerve stimulation, NE seems to be released intact, but is subsequently taken up and deaminated in the nerve terminal. No great degree of storage of this released NE is noted. The cocaine-sensitive reuptake process seems to function as a system for eliminating NE from the synaptic gap, thereby reducing the receptor response, but it does not appear to function as an amine conservation mechanism. Dr. Weiner pointed out that in spleens labeled with ^{3}H-NE, the total ^{3}H coming out of the spleen after nerve stimulation in the presence or absence of cocaine was about the same, but that in the presence of cocaine the label collected was primarily ^{3}H-NE and, in the absence of cocaine, more left the spleen as deaminated metabolite.

Dr. Costa questioned Dr. Carlsson about interpretation of some of his data, concluding that comparisons of half-lives does not document a change in the rate of synthesis when the steady state of dopamine fluctuates. Dr. Carlsson replied that he had made the kinetic calculations alluded to by Dr. Costa and found considerable differences, particularly in comparing spinal transected animals with stimulated ones. However, Dr. Carlsson emphasized that such kinetic calculations must assume one homogeneous pool. He concluded that what he wanted to demonstrate was that stimulation resulted in an increase in dopamine and that this increase does not cause a slowing of turnover as expected according to the postulate of control by product inhibition. Actually, when dopamine is increased there is an increase in turnover. Dr. Costa next questioned Dr. Carlsson on his data regarding 5-HT synthesis. Dr. Costa's published data indicate that there is no control of 5-HT synthesis by product inhibition, but Dr. Carlsson's data indicate such inhibition; he wondered whether Dr. Carlsson had measured tryptophan, since it did not seem valid to estimate turnover from radioactive tryptophan without reporting data on tryptophan specific activity. Dr. Carlsson replied that he had measured tryptophan and that it had not changed during this period.

Dr. Brodie pointed out that when stimulation at a high frequency is initiated during measurements of turnover taken at supposed steady state condition, such conditions no longer apply. Under such stimulation, a number of granules stick to the membrane, with the result that they preferentially release NE, and newly synthesized NE will preferentially go into these particular granules. Thus, if an organ is labeled and then stimulated at high frequency, first, product will come out with a high specific activity and this will be followed by a lower specific activity product. This was shown about 15 years ago by Chidsey, who labeled dog heart with radioactive NE. Thus, it should be remembered that when high-frequency stimulation is started, a new compartment is created and steady state conditions are no longer operative.

Dr. Weiner agreed with Dr. Brodie and added that the point he wished to make was that the specific activity of NE coming out under certain circumstances was different from that which remains in the spleen. Thus, there must be different compartments of NE which are not in equilibrium. Dr.

Brodie countered that even though each granule might be considered a compartment anatomically, to a kineticist, when all the compartments are in equilibrium, he can treat them as one compartment; this is not valid during high-frequency stimulation. Dr. Brodie concluded that both Drs. Weiner and Costa were getting correct answers, but they were asking different questions: Dr. Weiner is studying the steady state while stimulating, Dr. Costa is studying the steady state under relatively nonchanging stimulation conditions.

Costa. In reply to a question as to whether he had tried to stimulate guanylcyclase, Dr. Costa said that preliminary experiments have been done and that he feels the ratio of guanyl to adenyl nucleotides may be the signal which controls protein synthesis. In reply to a question as to whether he had tested the prostaglandins, he said that he had tried the E series but not yet the F series.

A series of questions on temperature effects led Dr. Costa to state that his results explain the failure of Thoenen to see a correlation between the increase of cyclic AMP and induction of tyrosine hydroxylase. As long as there is hypothermia, the cyclic AMP in medulla cannot increase. If one waits for about half an hour, until the body temperature reaches about 30°C, then the cyclic AMP goes up. Thus, swimming stress is not an exception to the association between increase of cyclic AMP and interaction of tyrosine hydroxylase in adrenal medulla. Dr. Costa emphasized that he used exactly the same conditions as those used by Dr. Thoenen. In reply to a question on the results of swimming stress on TH activity keeping the temperature of the water higher, Dr. Costa replied that a much lower temperature drop was observed at 25°C and the increase of cyclic AMP was not delayed and when cyclic AMP increased the long-term induction of TH was observed. His data demonstrate that the exhaustion of the animals is the result of hypothermia. Since the volume/surface ratio is a factor in the hypothermia of swimming stress, it is possible to explain why Thoenen could see the lack of apparent association between induction of TH and the increase of cyclic AMP after swimming stress only when he used 100-g rats; he anticipates that Thoenen would have obtained an even more dramatic dissociation if he had used mice.

Dr. Rech expressed interest in the fact that the change in level of tyrosine hydroxylase did not seem any greater with a short-term cold stress than for a long term (up to 24 hr) and wondered if a 5-day stress would give a greater change or whether Dr. Costa would anticipate a leveling off. Dr. Costa said that he had not done the experiment, but that Dr. Hanbauer had run her animals up to 2 days of cold exposure. She found that adrenal medulla TH levels at 48 hr were about double those found after 20 hr. Longer exposure did not seem to produce any additional increase. This seems to agree with the idea that the regulatory phenomenon occurs at a certain time.

Dr. Mandel inquired whether the animals were shaved; Dr. Costa replied that they were not, but that the furs were wetted. He said that with wet fur it was possible to observe the changes in cyclic-AMP after 3 or 4 min, but the group variability is greater without such wetting.

In reply to Dr. Mandel's question as to whether the decrease in cyclic AMP is due to the induction of diesterase, Dr. Costa mentioned that his data show that cyclic AMP levels and adenylcyclase activity in medulla are down 10 days after denervation, although no changes in cyclic AMP levels and a 30% decrease in enzyme activity were found after 5 days. Dr. Costa is currently studying the relationship of denervation to adenylcyclase and guanylcyclase activity. He feels that this may be critical, since in animals which have been denervated for a long time, the injection of carbamylcholine seems to have greatly reduced the ability to induce tyrosine hydroxylase and to increase the levels of cyclic AMP in adrenal medulla. This indicates that the nerves have the capability of regulating enzyme level.

Dr. Weiner inquired of Dr. Hanbauer if her results implied that there was a muscarinic receptor involved in the tyrosine hydroxylase induction in sympathetic ganglia. She replied that the experiments were inconclusive; she suggested that the injection of β-blockers in the ganglion made a difference, but that interpretation was difficult.

In reply to Dr. Barchas' question on the manner in which the animals were killed, Dr. Costa said that a modified microwave oven was used which killed the rats in 1 to 2 sec.

Mandell and Roffman (combined discussion). In reply to a question on the inhibitors used by him, Dr. Roffman replied that he had not studied these but that he was interested in such studies not only because of their possible value in differentiating the two enzymes of the RBC, but also because they would help discern any similarities or differences between the membrane-bound enzymes of the brain and the RBC. If these enzymes are shown to have similar physical and chemical properties, then measurements of the properties of the membrane-bound RBC COMT, in clinical studies, may provide an indication of properties of the membrane-bound enzyme in the brain.

Neuropsychopharmacology of Monoamines and
Their Regulatory Enzymes, edited by E. Usdin.
Raven Press, New York © 1974.

Dopamine and Psychotic States: Preliminary Remarks

B. Angrist and S. Gershon

Neuropsychopharmacology Research Unit, Department of Psychiatry, New York University Medical Center, New York, New York 10016

An increasing body of evidence supports the possibility that some psychotic states may be related to central dopaminergic activity. To review this evidence in detail would be redundant, and it would be more profitable to present new data. Accordingly, we shall allude to these lines of evidence only superficially, in the interest of remaining brief.

A variety of drugs, which might be loosely classified as dopamine agonists but whose pharmacological actions are in some ways different, have been shown to be capable of inducing psychotic states which bear a striking clinical similarity to some forms of naturally occurring psychosis (Connell, 1958; Ellinwood, 1967; Angrist and Gershon, 1970). Such drugs are L-DOPA, the central nervous system stimulants (amphetamine, phenmetrazine, methylphenidate, cocaine), and possibly pemoline and apomorphine (Takas, 1965; Rylander, 1969; Spensley and Rockwell, 1972; Strian, Michler, and Benkert, 1972; Sathananthan, Angrist, and Gershon, 1973).

When administered to animals, these drugs induce stereotyped behavior (Randrup and Munkvad, 1967; Willner, Samach, Angrist, Wallach, and Gershon, 1970; Wallach and Gershon, 1972). The extensive body of data demonstrating that this behavior is dopaminergically mediated has been reviewed by Randrup and Munkvad (1967, 1970). This behavior has been proposed as an animal model for certain psychotic states (Randrup and Munkvad, 1967, 1970; Ellinwood, 1971).

Furthermore, the fact that neuroleptic drugs, which block central dopamine receptors (Nybäck and Sedval, 1969; Andén, Butcher, Corrode, Fuxe, and Ungerstedt, 1970), not only block drug-induced stereotypy (Randrup and Munkvad, 1970; Willner et al., 1970) but are also efficacious in human stimulant psychoses (Angrist, Lee, and Gershon, 1974) and in naturally occurring psychotic states (Klein and Davis, 1969) suggests this relationship between central dopaminergic hyperactivity and some forms of psychosis. This is so much the case that antagonism of animal stereotypy has become a standard preclinical index of potential neuroleptic activity (Janssen, Neimegeers, and Schellekens, 1965; van Rossen, 1966).

This evidence suggests an important role for dopaminergic mechanisms, and does so in a rather global way. For a more precise definition of what these "mechanisms" are, some inconsistencies in the effects of the drugs cited must be explained. Some of the agonistic drugs, whose clinical effects are quite similar, differ when examined in some controlled experimental systems. Similarly, some neuroleptic agents have effects that differ in some respects. We present these problematic differences not as invalidating the hypothesized role of dopamine, but as qualifying factors that must await clear resolution in order for the hypothesis to be presented more precisely.

As has been noted, clinically similar psychotic states have been reported after ingestion of amphetamine, phenmetrazine, methylphenidate, cocaine, L-DOPA, and possibly pemoline and apomorphine. In addition, methylphenidate and L-DOPA have been shown to cause exacerbations of symptomatology in schizophrenics in doses which are nonpsychotogenic in normals (Janowsky, el-Yousef, Davis, and Sekerke, 1973; Angrist, Sathananthan, and Gershon, 1973).

However, when an attempt was made to assess the mode of action of these drugs by evaluating the effects of pretreatment with reserpine and α-methyl-para-tyrosine, important differences between them emerged. In these studies (Wallach and Gershon, 1972; Wallach, Rotrosen, and Gershon, 1973), stereotyped behavior was induced in cats by the administration of D-amphetamine, cocaine, L-DOPA, methylphenidate, pemoline, and phenmetrazine. This behavior was scored on a scale of 0 to 2, in increments of 0.5. Thus, the maximal stereotypy score was 2. A score of 1 represented an animal who had clearly recognizable stereotyped behavior, but in whom this was discontinuous. A score of 0.5 indicated barely perceptible stereotyped behavior, whereas 1.5 was used to describe a state between discontinuous and maximal stereotypy. The effects of pretreatment with saline, α-methyl-para-tyrosine, and reserpine are indicated by the stereotypy scores obtained (Table 1). The numbers in the parentheses indicate how many cats are in each group.

TABLE 1. *Antagonism of stimulant-induced stereotyped behavior after pretreatment with α-methyl-para-tyrosine (α-MPT) and reserpine*

Drug	Dose (mg/kg)	Control	α-MPT (100 mg/kg × 2)	Reserpine (1 mg/kg)
D-Amphetamine	2	2.0 (19)	0[a] (6)	0[a] (6)
Cocaine	4	2.0 (14)	1.0[a] (12)	0[a] (6)
L-DOPA	75	2.0 (6)	1.7 (6)	2.0 (5)
Methylphenidate	4	2.0 (6)	1.8 (6)	0[a] (6)
Pemoline Mg(OH)$_2$	100	2.0 (5)	0.3[a] (6)	0[a] (6)
Phenmetrazine	10	1.8 (18)	0.3[a] (6)	0.2[a] (6)

Figure in parentheses indicates number of cats.
[a] $p < 0.001$ from control group.
From Wallach and Gershon (1972) and Wallach, Rotrosen, and Gershon (1973).

As can be seen, stereotyped behavior induced by amphetamine, phenmetrazine, and pemoline was antagonized by both α-methyl-para-tyrosine and reserpine. α-Methyl-para-tyrosine, however, did not antagonize stereotypy induced by methylphenidate or by L-DOPA, and only partially antagonized this effect of cocaine. Pretreatment with reserpine blocked stereotypy induced by amphetamine, cocaine, methylphenidate, pemoline, and phenmetrazine, but not that induced by L-DOPA. These findings suggest that, even when one excludes the dopamine precursor L-DOPA, the clinically similar, indirect-acting stimulant drugs may affect synaptic events by different mechanisms or different ratios of mechanisms such as release and blockade of reuptake.

In subsequent studies (Wallach, Rotrosen, and Gershon, 1973), an attempt was made to overcome the reserpine-induced blockade of stereotyped behavior by increasing the challenging dose of phenmetrazine and D-amphetamine. In these studies, it was found that a challenging dose of 25 mg/kg of phenmetrazine could effectively overcome the blockade induced by 1 mg/kg of reserpine. When cats were pretreated with 2 mg/kg of reserpine 20 hr before administration of D-amphetamine, a challenge dose of 10 mg/kg elicited a period of stereotyped behavior which lasted approximately 2 hr. (In unpretreated cats, 2 mg/kg of D-amphetamine induced stereotyped behavior lasting for about 6 hr.) This partial efficacy of reserpine in antagonizing stimulant-induced stereotyped behavior in the cat is at variance with prior studies in rodents, in which it was found that pretreatment with α-methyl-para-tyrosine, but not with reserpine, was capable of antagonizing amphetamine-induced stereotypy (Quinton and Holliwell, 1963; Randrup, Munkvad, and Udsen, 1963; Randrup and Munkvad, 1966; Weissman, Koe, and Tenin, 1966).

These findings, however, do correlate closely with observations made on our unit in humans. In these studies (Angrist and Gershon, 1970), four amphetamine abusers were administered amphetamine in large doses (595 to 955 mg per cumulative dose) and their behavioral responses documented. After a drug-free period of at least 10 days, the same subjects received reserpine, 14.5 to 18 mg per cumulative dose, over 3 to 5 days before amphetamine was readministered in comparable doses. In all four subjects, pretreatment with reserpine led to an amelioration of the pathological behavioral effects induced by amphetamine previously. Nonetheless, all subjects still showed signs of central nervous system stimulation and had insomnia throughout the period of amphetamine administration, in spite of the relative absence of the overt pathological behaviors that were seen with amphetamine alone. In retrospect, these observations appear to correlate rather closely with the partial, reversible blockade of amphetamine stereotypy induced by reserpine in cats.

According to the global schema that has been presented, the dopamine receptor stimulant (Ernst, 1967) apomorphine might be expected to be psychotogenic. In fact, Strian et al. (1972) have reported one case of psychosis associated with administration of apomorphine to parkinsonian patients. In this study, patients with Parkinson's disease were treated with a combination of L-DOPA and a decarboxylase inhibitor. Apomorphine was then supra-added to this regimen in large oral doses to assess its effect on residual tremor. Under these conditions, one such patient developed a psychosis that began when apomorphine was added to his regimen and that resolved after apomorphine was discontinued. Thus far, the potent emetic effects of apomorphine have made it difficult for other investigators to test its psychotogenic potential formally. Differences with respect to emetic effects constitute another perplexing difference among dopaminergic agonists.

One is tempted to speculate that these differences might relate more to specific synaptic mechanisms than to overall receptor activity, since emesis is virtually inevitable (in man) with administration of the receptor stimulant apomorphine, is seen rather frequently after the dopamine precursor L-DOPA, and is rather insignificant in the case of the central nervous system stimulants in which indirect presynaptic mechanisms are presumed to be of greatest importance. Why this should be so, however, is problematic.

A comparison of two studies from our laboratories (Rotrosen, Wallach, Angrist, and Gershon, 1972; Klingenstein, Wallach, and Gershon, 1973) underscores the differences between amphetamine stereotyped behavior (ASB) and the stereotypy and emesis induced by apomorphine. In these studies, these effects were induced in the dog by administering amphetamine, 2 mg/kg, and apomorphine, 0.1 to 0.2 mg/kg, and were antagonized by pretreatment with pimozide and thioridazine at varying doses and times before the challenge dose of agonist. The results are summarized in Table 2.

These results can be summarized as follows:
1. Pimozide was effective as an antagonist at much lower doses.
2. Thioridazine, 6 mg/kg, was found to be ineffective as an antagonist of

TABLE 2. *Antagonism of amphetamine stereotypy and apomorphine stereotypy and emesis by thioridazine and pimozide*

	Amphetamine, 2 mg/kg, i.v.	Apomorphine, 0.1 or 0.2 mg/kg, i.v.
Pimozide	0.2 mg/kg blocks (4/4) at 2 hr 0.4 mg/kg blocks (5/6) at 20 hr	0.025 mg/kg blocks: Emesis (4/6) > stereotypy (1/6) at 1 hr Emesis (6/6) and stereotypy (6/6) at 3 hr Stereotypy (3/6) not emesis (0/6) at 20 hr
Thioridazine	6.0 mg/kg blocks (1/8) at $\frac{1}{5}$ hr 6.0 mg/kg blocks (6/6) at 20 hr	4.0 mg/kg blocks: Stereotypy (6/8) and emesis (6/8) at 1 hr Stereotypy (8/8) and emesis (7/8) at 3 hr Stereotypy (7/8) and emesis (6/8) at 20 hr

Figures in parenthesis indicate: (number of dogs protected/number of dogs challenged).
From Rotrosen, Wallach, Angrist, and Gershon (1972) and Klingenstein, Wallach, and Gershon (1973).

ASB at 1.5 hr, but was serendipitously found to be surprisingly effective at 20 hr. In slightly lower doses (4 mg/kg), thioridazine blocked apomorphine effects at 1, 3, and 20 hr.

3. Pimozide blocked apomorphine effects at doses of 0.025 mg/kg, with maximum effects at 3 hr. By contrast, 10 times this dose of pimozide was required to block ASB at 2 hr, and 20 times this dose was required at 20 hr.

Interpretation of these complex findings is difficult. We would tentatively venture the following. The prolonged efficacy of thioridazine may suggest that this drug acts via an active metabolite and may explain its clinical efficacy. While thioridazine has a high ratio of antinoradrenergic/antidopaminergic effects, and pimozide a low ratio of these effects (Janssen, Neimegeers, Schellekens, and Lenaerts, 1967; Janssen, Neimegeers, Schellekens, Dresse, Lenaerts, Pinchard, Schaper, Van Nueten, and Verbruggen, 1968), the neuroleptic efficacy of thioridazine is probably somewhat greater than that of pimozide in acute psychotic states (Janssen, 1973). This efficacy appears to correlate better with long-term blockade of ASB than with that of apomorphine effects. Since norepinephrine probably contributes to amphetamine stereotypy, the derivative possibilities are as follows:

1. Amphetamine stereotyped behavior may be a better model for acute psychotic states than the emesis and stereotypy induced by apomorphine.

2. Norepinephrine may indeed contribute to psychotogenesis.

Finally, the 10:1 ratio of doses of amphetamine and apomorphine for inducing stereotypy, and the similar ratios required for its blockade by pimozide, may suggest some stoichiometric relationships.

Similar problematic relationships exist among the neuroleptic drugs, which have been shown to be effective in amphetamine psychosis (Angrist, Lee, and Gershon, 1974) and which are of well-established efficacy in the treatment of schizophrenia (Klein and Davis, 1969). The demonstration that these drugs block central dopaminergic receptors constitutes important evidence for the involvement of dopamine in the pathogenesis of psychotic states. However, if blockade of dopamine receptors were the sole mechanism of neuroleptic action, then one would expect an inevitable, linear, and arithmetic relationship between extrapyramidal side effects (EPS) and efficacy. In general, the potency of neuroleptic drugs does parallel their potential for producing EPS (Snyder, 1973). However, some neuroleptics appear to combine clinical efficacy with little tendency to produce EPS; some antiemetic phenothiazines produce EPS and yet are not considered to have neuroleptic activity.

The compound that dissociates neuroleptic efficacy and EPS most dramatically, according to several reports (Berzewski, Helmchen, Hippius, and Kanowski, 1967; Gross and Langner, 1969; De Maio, 1972), is clozapine. In an early clinical efficacy study on our unit, we administered the drug, by design, to patients who had developed EPS on other neuroleptics.

No EPS were observed, even in this predisposed population, and some neuroleptic activity was documented. This dissociation has stimulated interest in the mechanism of action of clozapine; several studies have attempted to assess this.

Bartholini, Haefely, Jalfre, Keller, and Pletschner (1972) have suggested that clozapine causes a surmountable blockade of dopamine receptors, while classical neuroleptics block these receptors in an insurmountable way. This concept is consistent with an unquantified observation that was made, during our experience with the drug, of something akin to a tolerance phenomenon. Frequently, patients appeared to improve initially at a given dose only to regress in their status and again to improve when the dose of clozapine was again increased. Sedvall and Nybäck (1974) have shown that clozapine causes increased accumulation of dopamine from ^{14}C-labeled tyrosine. Thus, if clozapine causes a surmountable block, this increased dopamine accumulation would explain this possible "tolerance" phenomenon. Sedvall and Nybäck have noted that clozapine does not increase the conversion of dopamine to homovanillic acid (HVA) and therefore has overall effects on turnover that differ from those of classical neuroleptic drugs.

Andén and Stock (1973) have examined the effects of clozapine, using an approach which combines neurochemical and neuroanatomical methods. These studies have shown that clozapine causes a greater increase in HVA formation in dopaminergic areas of the limbic system than in the striatum. Haloperidol, on the other hand, caused HVA to accumulate equally in both areas. These investigators also noted that the combination of haloperidol plus an anticholinergic drug produced a pattern which was similar to that induced by clozapine alone. This finding led Andén to postulate that these findings might indicate that "determinations of HVA in the corpus striatum and in the limbic system of rabbits may be of value in predicting the ability of neuroleptic drugs to induce in man extrapyramidal and antipsychotic actions, respectively."

Ellinwood, Sudilowsky, and Grabowy (1973) have also used electrophysiologic measurements to assess the effects of dopaminergic agonists in discrete brain areas. Their findings suggest that changes in extrastriatal dopaminergic systems might play an important role in amphetamine psychotogenesis. Moreover, Thierry, Stinus, Blanc, and Glowinski (1973) have recently presented evidence for the existence of dopaminergic neurons in the cortex and point out that their presence "will certainly have to be taken into account to explain some of the pharmacological effects of psychotropic drugs known to act on central dopaminergic neurons."

Thus, differences in effects of dopamine agonist drugs and of neuroleptics might relate to:

  1. differing ratios of effect on synaptic activity in dopaminergic systems in discrete brain areas, perhaps relating to the individual drugs' distribution in brain;

2. different mechanisms of action on synaptic events; and
3. balances between dopaminergic activity and that of other systems, particularly norepinephrine and acetylcholine.

This presentation has intentionally stressed problematic inconsistencies in the effect of drugs affecting dopaminergic systems. We have not done so to diminish in any way the potential importance of dopaminergic mechanisms in psychotic states. Indeed, we are convinced that the relationships among dopamine, the stimulant psychoses, stereotyped behavior and its antagonism, neuroleptic potency, and EPS constitute an overwhelmingly suggestive argument that the relationship of dopamine to some psychotic states is a very real phenomenon. Rather, the purpose of this presentation has been to stress that, when we say that "dopaminergic hyperactivity correlates with psychosis," we are using highly telegraphic language that leaves the specific mechanisms involved rather imprecisely defined. Ratios of activity of dopaminergic systems in various neuroanatomic sites, relationships among dopaminergic activity and that of other neurotransmitters, and specific mechanisms by which synaptic activity is either enhanced or diminished, all might prove to be critical in determining clinical outcome. It is hoped that resolution of these problematic issues can lead to a more precise understanding of the mechanisms involved in this provocative hypothesis.

REFERENCES

Andén, N. E., Butcher, S. G., Corrodi, H., Fuxe, K., and Ungerstedt, U. (1970): Receptor activity and turnover of dopamine and noradrenaline after neuroleptics. *European Journal of Pharmacology*, 11:303–314.

Andén, N. E., and Stock, G. (1973): Effect of clozapine on the turnover of dopamine in the corpus striatum and in the limbic system. *Journal of Pharmacy and Pharmacology*, 25:346–348.

Angrist, B. M., and Gershon, S. (1970): The phenomenology of experimentally induced amphetamine psychosis. Preliminary observations. *Biological Psychiatry*, 2:95–107.

Angrist, B., Lee, H. K., and Gershon, S. (1974): The antagonism of amphetamine-induced symptomatology by a neuroleptic. *American Journal of Psychiatry* (in press).

Angrist, B. M., Sathananthan, G., and Gershon, S. (1973): Behavioral effects of L-DOPA in schizophrenic patients. *Psychopharmacologia*, 31:1–12.

Bartholini, G., Haefely, W., Jalfre, M., Keller, H. H., and Pletscher, A. (1972): Effects of clozapine on cerebral catecholaminergic systems. *British Journal of Pharmacology*, 46:736–740.

Berzewski, von, H., Helmchen, H., Hippius, H., Hoffmann, H., and Kanowski, S. (1969): Das klinische Wirkungsspektrum eines neuen Dibenzodiazepin Derivates. *Arzneimittel-Forschung*, 19:495–496.

Connell, P. H. (1958): *Amphetamine Psychosis*, Maudsley Monographs No. 5. Oxford University Press, London and New York.

De Maio, D. (1972): Clozapine, a novel major tranquilizer. *Arzneimittel-Forschung*, 22:919–923.

Ellinwood, E. H. (1967): Amphetamine psychosis, I. Description of the individuals and process. *Journal of Nervous and Mental Diseases*, 144:273–283.

Ellinwood, E. H. (1971): Effect of chronic methamphetamine intoxication in rhesus monkeys. *Biological Psychiatry*, 3:25–32.

Ellinwood, E. H., Sudilowsky, A., and Grabowy, R. (1973): Olfactory forebrain seizures induced by methamphetamine and disulfiram. *Biological Psychiatry,* 7:89–99.

Ernst, A. M. (1967): Mode of action of apomorphine and dexamphetamine on gnawing compulsion in rats. *Psychopharmacologia,* 10:316–323.

Gross, von, H., and Langner, E. (1969): Klinische qualifikation eines Neuroleticums aus der Dibenzodiazepin-Reihe 8-chlor-11(4'-methyl)-piperazino-5-dibenzo[b,e][1,4] diazepin W108/HF 1854. *Arzneimittel-Forschung,* 19:496–498.

Janowsky, D. S., El-Yousef, K., Davis, J. M., and Sekerke, H. J. (1973): Provocation of schizophrenic symptoms by intravenous administration of methylphenidate. *Archives of General Psychiatry,* 28:185–191.

Janssen, P. A. J. (1973): Haloperidol Conference, New York City, November 10, 1973.

Janssen, P. A. J., Niemegeers, J. E., and Schellekens, K. H. L. (1965): Is it possible to predict the clinical effects of neuroleptic drugs (major tranquilizers) from animal data? Part I. "Neuroleptic activity spectra" for rats. *Arzneimittel-Forschung,* 15:104–117.

Janssen, P. A. J., Niemegeers, C. J. E., Schellekens, K. H. L., and Lenaertes, F. M. (1967): Is it possible to predict the clinical effects of neuroleptic drugs (major tranquilizers) from animal data? Part IV. An improved experimental design for measuring the inhibitory effects of neuroleptic drugs on amphetamine- or apomorphine-induced "chewing" and "agitation" in rats. *Arzneimittel-Forschung,* 17:841–854.

Janssen, P. A. J., Niemegeers, C. J. E., Schellekens, K. H. C., Dresse, A., Lenaerts, F. M., Pinchard, A., Schaper, W. K. A., Van Nueten, J. M., and Verbruggen, F. J. (1968): Pimozide, a chemically novel, highly potent and orally long-lasting neuroleptic drug. *Arzneimittel-Forschung,* 18:261–279.

Klein, D. F., and Davis, J. M. (1969): *Diagnosis and Drug Treatment of Psychiatric Disorders.* Williams and Wilkins Company, Baltimore.

Klingenstein, R. J., Wallach, M. B., and Gershon, S. (1973): A comparison of pimozide and thioridazine as antagonists of amphetamine-induced stereotyped behavior in dogs. *Archives Internationales de Pharmacodynamie et de Thérapie,* 203:67–71.

Nybäck, H. G., and Sedvall, G. (1969): Regional accumulation of catecholamines formed from tyrosine-^{14}C in rat brain: Effect of chlorpromazine. *European Journal of Pharmacology,* 4:245–252.

Quinton, R. M., and Holliwell, G. (1963): Effects of methyldopa and DOPA on the amphetamine excitatory response in reserpinized rats. *Nature,* 200:178–179.

Randrup, A., and Munkvad, I. (1966): The role of catecholamines in the amphetamine excitation response. *Nature,* 211:540.

Randrup, A., and Munkvad, I. (1967): Stereotyped activities produced by amphetamine in several animal species and man. *Psychopharmacologia,* 11:300–310.

Randrup, A., and Munkvad, I. (1970): Biochemical, anatomical and psychological investigation of stereotyped behaviour induced by amphetamines. In: *Amphetamines and Related Compounds,* edited by E. Costa and S. Garattini. Raven Press, New York, pp. 695–713.

Randrup, A., Munkvad, I., and Udsen, P. (1963): Adrenergic mechanisms and amphetamine-induced stereotyped behavior. *Acta Pharmacologica Toxicologica,* 20:145–157.

Rotrosen, J., Wallach, M. B., Angrist, B., and Gershon, S. (1972): Antagonism of apomorphine-induced stereotypy and emesis by thioridazine, haloperidol, and pimozide. *Psychopharmacologia,* 26:185–194.

Rylander, G. (1969): Clinical and medico criminological aspects of addiction to central stimulating drugs. In: *Abuse of Central Stimulants,* edited by F. Sjoqvist and M. Tattie. Raven Press, New York, pp. 251–273.

Sathananthan, B., Angrist, B., and Gershon, S. (1973): Response threshold to L-DOPA in psychiatric patients. *Biological Psychiatry,* 7:139–146.

Sedvall, G., and Nybäck, H. G. (1974): Effect of clozapine and some other antipsychotic agents on synthesis and turnover of dopamine formed from ^{14}C-tyrosine in mouse brain. *(In press.)*

Snyder, S. H. (1973): Amphetamine psychosis: A "model" schizophrenia mediated by catecholamines. *American Journal of Psychiatry,* 130:61–67.

Spensley, J., and Rockwell, D. A. (1972): Psychosis during methylphenidate abuse. *New England Journal of Medicine,* 286:880–881.

Strian, F., Micheler, E., and Benkert, O. (1972): Tremor inhibition in Parkinson syndrome after apomorphine administration under L-DOPA and decarboxylase inhibition basic therapy. *Pharmakopsychiatry,* 5:198–205.

Takas, L. (1965): Psychoses following the abuse of anorexogenic agents. *Psychiatrie, Neurologie und Medizinische Psychologie,* 17:183–185.

Thierry, A. M., Stinus, L., Blanc, G., and Glowinski, J. (1973): Some evidence for the existence of dopaminergic neurons in the rat cortex. *Brain Research,* 50:230–234.

Van Rossum, J. M. (1966): The significance of dopamine receptor blockade for the mechanism of action of neuroleptic drugs. *Archives Internationales de Pharmacodynamie et de Thérapie,* 160:492–494.

Wallach, M. B., and Gershon, S. (1972): The induction and antagonism of central nervous system stimulant-induced stereotyped behavior in the cat. *European Journal of Pharmacology,* 18:22–26.

Wallach, M. B., Rotrosen, J., and Gershon, S. (1973): A neuropsychopharmacological study of phenmetrazine in several animal species. *Neuropharmacology,* 12:541–548.

Weissman, A., Koe, B. K., and Tenen, S. S. (1966): Aintamphetamine effects following inhibition of tyrosine hydroxylase. *Journal of Pharmacology and Experimental Therapeutics,* 151:339–352.

Wilner, J. G., Samach, M., Angrist, B. M., Wallach, M. B., and Gershon, S. (1970): Drug-induced stereotyped behaviour and its antagonism in dogs. *Communications in Behavioral Biology,* 5:135–149.

Neuropsychopharmacology of Monoamines and Their Regulatory Enzymes, edited by E. Usdin. Raven Press, New York © 1974.

Effect of Chronic Treatment with Central Stimulants on Brain Monoamines and Some Behavioral and Physiological Functions in Rats, Guinea Pigs, and Rabbits

Tommy Lewander

Psychiatric Research Center, University of Uppsala, Ulleråker Hospital, Uppsala, Sweden

I. INTRODUCTION

The appearance of the amphetamine psychosis in humans treated with or abusing central stimulants and the various aspects of amphetamine dependence in a sizable population of addicts are challenges for research efforts on the acute and, particularly, chronic biological effects of this class of pharmacological agents (see Lewander, 1972).

Biochemical and pharmacological evidence support the present, almost generally accepted view that amphetamine exerts most of its pharmacological effects through preferential release of newly synthesized catecholamines (CA) from dopaminergic and noradrenergic nerve terminals (see reviews by van Rossum, 1970; Sulser and Sanders-Bush, 1971; Lewander, 1972; Scheel-Krüger, 1972; Weiner, 1972).

The actions of amphetamine on brain serotonin (5-hydroxytryptamine, 5-HT) neurons are less studied, but biochemical *in vitro* data suggest that amphetamine causes an increased release of serotonin (see Azzaro and Rutledge, 1973). Behaviorally, the inhibition of the lordosis response in the female rat seems to be related to an effect of amphetamine on serotonin transmission (Meyerson, 1968).

The prolonged depletion of central and peripheral norepinephrine (NE) levels and tolerance to the hyperthermic effect of DL- and D-amphetamine in the rat has been related to the formation and accumulation of *p*-hydroxynorephedrine in NE neurons as a false transmitter in rats (Brodie, Cho, and Gessa, 1970; Costa and Groppetti, 1970; Lewander, 1970, 1971*a, b*).

However, tolerance does not develop to all effects of D- or DL-amphetamine in rats, and the false transmitter mechanism does not explain the tolerance to the anorexigenic effect of amphetamine (Lewander, 1971*b*). Further, brain and heart CA levels are changed after L-amphetamine in rats and after DL-amphetamine in guinea pigs, in which cases amphetamine is not converted to *p*-hydroxynorephedrine (Lewander, 1971*a, c*). Thus, there are several reasons for continuing research on the effects of chronic amphetamine treatment on brain monoamines and other biochemical, physiological, and behavioral functions in different species.

This chapter contains a short review of previous and new, preliminary data on the effects of acute and chronic amphetamine treatment on brain concentrations of CA, homovanillic acid, tryptophan, serotonin, and its acid metabolites in the rat, guinea pig, and rabbit. Some aspects of tolerance to the hyperthermic effect of amphetamine, such as rate of development, duration, and carry-over of tolerance (cf. Lewander, Moliis, and Brus, 1973), have been related to brain CA levels and brain tryptophan and serotonin levels, respectively. Phenmetrazine, a different central stimulant and anorexigenic drug, has been studied with respect to its effects on brain monoamines in the rat. Finally, a cross-tolerance experiment between L-amphetamine, phenmetrazine, and D-amphetamine is preliminarily reported.

II. METHODS

Brain NE and dopamine (DA) were determined according to Chang (1964) and Carlsson and Waldeck (1958), respectively, in eluates from alumina columns used for isolation of the CA from 0.4 perchloric acid extracts of brain or heart tissue. Brain normetanephrine (NM) and methoxytyramine (MTA) were isolated and assayed fluorimetrically (Häggendal, 1962; Carlsson and Waldeck, 1964) as described by Lewander (1971c). Homovanillic acid (HVA) was determined in caudate nuclei according to Korf, Roos, and Werdinius (1971). Serotonin and 5-hydroxyindoleacetic acid (5-HIAA) were extracted, isolated, and determined as described by Jönsson and Lewander (1970). Brain tryptophan was separated from the same tissue extracts as 5-HT and 5-HIAA on Dowex 50 ion exchange columns (0.5 × 50 mm, H^+-form). Aliquots of the 1 M ammoniumacetate buffer (pH 4.8) eluate were taken for fluorimetric determination of tryptophan according to Denckla and Dewey (1967). Amphetamine in brain tissue was determined as previously described (Lewander, 1971a) with a gas chromatographic method (Änggård, Gunne, and Niklasson, 1970). In some experiments the brains were dissected into the brainstem, including the medulla oblongata, pons, mesencephalon, hypothalamus, and thalamus, and the telencephalon, including the striatum, and so on. Body temperature was measured with an electric thermometer (Ellab, Copenhagen). The thermocouple was inserted into the colon 4 cm from the anal orifice. Food intake was recorded for individual rats in ordinary macrolon cages equipped with wire net baskets built into their covers. Ordinary rat food pellets (Anticimex nr 214, Sollentuna, Sweden) were supplied and weighed to the nearest 0.01 g.

III. RESULTS AND DISCUSSION

A. Brain Catecholamines in Rats, Guinea Pigs, and Rabbits after Acute and Chronic Amphetamine Treatment

It has previously been shown (see Lewander, 1968, 1971a, b) that the rat brain and heart NE levels are decreased to 70% of control levels after a single injection of 16 to 20 mg/kg of D-, L-, or DL-amphetamine. The prolonged depletion of NE after D- or DL-amphetamine (as opposed to L-amphetamine) seems to be due to the incorporation of p-hydroxynorephedrine (a metabolite of amphetamine in this species) into NE storage vesicles (Groppetti and Costa, 1969; Brodie, Cho, and Gessa, 1970; Lewander, 1970, 1971a). Chronic treatment with high doses of DL-amphetamine (16 to 32 mg/kg × 2) for 1 week or more causes a further depletion of brain and heart NE to 40 to 50% of control levels, which might be at least partly explained by further accumulation of p-hydroxynorephedrine (Lewander, 1968, 1971a, d, e).

The rat brain DA level increased by 15 to 25% after a single injection of amphetamine, while control levels or a decrease in DA to 50% of the controls can be observed after chronic treatment, depending on the dose and the duration of treatment (Lewander, 1968, 1971e). It has been suggested that the DA depletion after chronic amphetamine treatment might be due to a reduced storage capacity of DA, since DA synthesis (accumulation of ^{14}C-DA from ^{14}C-tyrosine given intravenously) did not seem to be reduced (Lewander, 1971e).

In vitro studies on isolated NE and DA storage vesicles from pig brain have shown that amphetamine competitively inhibits the ATP-magnesium-dependent uptake into the vesicles of the respective CA (Philippu and Beyer, 1973). This mechanism might at least partly explain the NE- and DA-depleting effect of chronically administered high doses of amphetamine. Studies on guinea pigs and rabbits were initiated in order to obtain further knowledge of the effects on brain NE and DA levels caused by chronic amphetamine treatment. The choice of these species was due to the fact that p-hydroxylation of amphetamine does not occur (Dring, Smith, and Williams, 1970) and, thus, the influence of p-hydroxynorephedrine on NE neurons should be avoided. In fact, p-hydroxynorephedrine could not be detected in brain tissue of guinea pigs after amphetamine; brain NE was, however, depleted after single or repeated amphetamine injections (Lewander, 1971c, Fig. 1B). Preliminary results from similar experiments in rabbits are presented in Fig. 1A, and a summary of the results with guinea pigs from Lewander (1971c) are presented in Fig. 1B for comparison. Norepinephrine was decreased by acute and chronic amphetamine treatment both in the brainstem and the telencephalon in the rabbit as in the guinea pig. Telencephalic (except the caudate nuclei) DA was not changed significantly,

224 ON TOLERANCE TO CENTRAL STIMULANTS

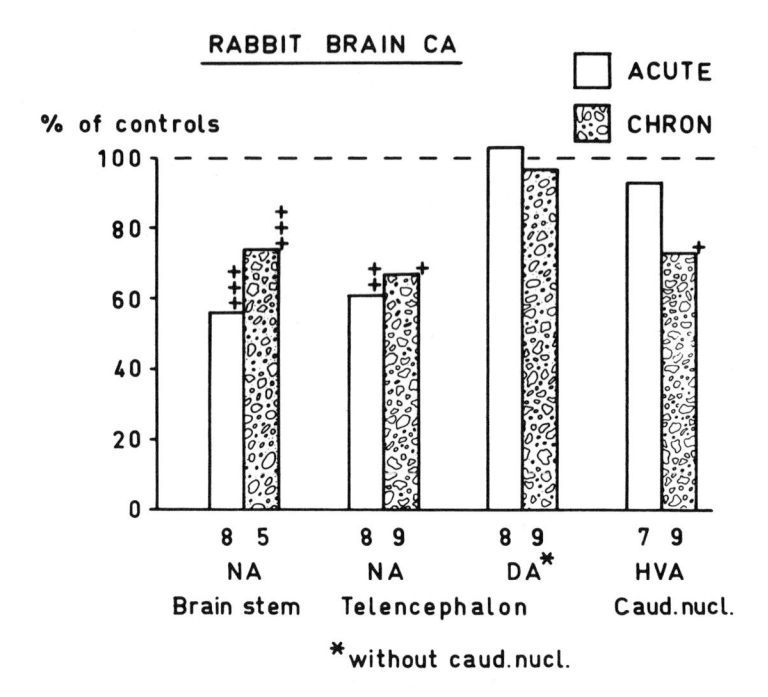

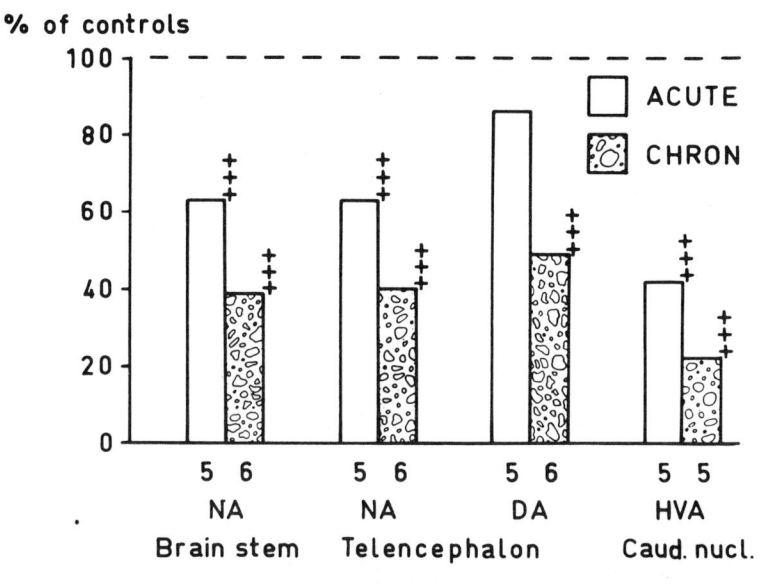

RABBIT BRAIN CA

FIG. 1. Effects of acute and chronic treatment with DL-amphetamine sulfate on brain catecholamine (CA) levels in various parts of the brains of rabbits and guinea pigs. Albino rabbits of both sexes, 1.5 to 2.5 kg body weight, were injected with a single dose of 20

however, which is in contrast to the findings in the guinea pig experiment. HVA in the caudate nuclei of guinea pigs was decreased (Fig. 1B) and maximally so, 40% of the control level, at 4 hr after a single injection of amphetamine. Chronic amphetamine treatment caused a further decrease in HVA. In the rabbit experiment only chronic amphetamine treatment caused a slight decrease in the HVA level (Fig. 1A).

There can be no definite explanation as yet for the decrease in brain NE in guinea pigs and rabbits, in which p-hydroxynorephedrine is not formed. However, a decrease in rat brain NE has been observed previously after L-amphetamine (Lewander, 1971a), which is not converted to p-hydroxynorephedrine, and after D- or DL-amphetamine in rats pretreated with desmethylimipramine (Lewander, 1971a) or iprindole (Freeman and Sulser, 1972), which block p-hydroxylation of amphetamine in rats and thereby the possibility of p-hydroxynorephedrine formation. The discrepancy between the effects of chronic amphetamine treatment on brain DA in guinea pigs and rabbits should be further analyzed, taking into consideration the actual brain concentrations of amphetamine.

The brain HVA levels, used as an indicator of release of brain DA, seem to be differently affected by amphetamine in different species. In rats, mice, and cats, brain HVA is increased after a single injection of amphetamine (Laverty and Sharman, 1965; Jori and Bernardi, 1969), and tolerance to this effect has been observed in rats but not in mice after chronic amphetamine treatment (Jori and Bernardi, 1972). In the rabbit, no effect on the HVA level was observed after a single dose of amphetamine, and in the guinea pig there was a decrease in the HVA level (Lewander, 1971c; Fig. 1B) as well as in the dog (Laverty and Sharman, 1965). Whether a decrease in brain HVA after a single dose of amphetamine in some species and after chronic amphetamine treatment in other species is due to monoamine oxidase inhibition, inhibition of DA synthesis (cf. Harris and Baldessarini, 1973), or other mechanisms cannot be concluded at present.

---

mg/kg of amphetamine subcutaneously (ACUTE) or chronically (CHRON) with 10 mg/kg twice daily for 40 days + a last dose of 20 mg/kg. The animals were killed 3 hr after the (last) injection. Values of norepinephrine (NA), dopamine (DA), and homovanillic acid (HVA) are presented as percentages of control values (mean ± SEM): NA in the brainstem = 0.34 ± 0.02 µg/g, NA in the telencephalon = 0.18 ± 0.02 µg/g, DA in the telencephalon = 0.26 ± 0.02 µg/g, and HVA in the caudate nucleus = 4.69 ± 0.31 µg/g. Figures below bars represent number of observations. +, ++, +++ indicate statistically significant differences at $p < 0.05$, $p < 0.01$, and $p < 0.001$ as compared with control values (Student's t test). (Data from Jönsson and Lewander, 1973, *submitted for publication*.)

Mottled guinea pigs, 275 g body weight, were injected with a single injection of 20 mg/kg amphetamine (ACUTE) or chronically (CHRON) with 20 mg/kg twice daily for 7 days (HVA) or 18 days. The animals were killed 4 hr after the (last) injection. Control values (means ± SEM) of NA in the brainstem = 0.17 ± 0.001 µg/g, NA in the telencephalon = 0.44 ± 0.03 µg/g, DA in the telencephalon (including the caudate nuclei) = 1.01 ± 0.05 µg/g, and HVA in the caudate nucleus = 1.62 ± 0.26 µg/g. Symbols and abbreviations are the same as in Fig. 1A.

B. Effect of Acute and Chronic Amphetamine Treatment on Brain 5-HT and 5-Hydroxyindoleacetic Acid (5-HIAA) in Rats, Guinea Pigs, and Rabbits

Acute and chronic amphetamine treatment has been reported to cause no change or an increase in the brain concentrations of 5-HT in rats (Garattini and Valzelli, 1965; McLean and McCartney, 1961). Rat brain 5-HIAA has been shown to increase after amphetamine (Reid, 1970; Tagliamonte, Tagliamonte, Perez-Cruet, Stern, and Gessa, 1971), and also 5-HT uptake and release mechanisms seem to be affected by high concentrations of the drug *in vitro* (see Azzaro and Rutledge, 1973). An increase in brain 5-HIAA levels in rats has been ascribed to increased brain tryptophan concentrations (Tagliamonte et al., 1971), which might cause an increased turnover of brain 5-HT. However, by use of other techniques the turnover of brain 5-HT has been reported to be decreased by a single injection of amphetamine (Schubert, Fyrö, Nybäck, and Sedvall, 1970; Görlitz and Frey, 1972) or increased after chronic but not acute amphetamine treatment (Diaz and Huttonen, 1972; Sparber and Tilson, 1972). The increase in brain tryptophan concentrations has not so far been adequately explained (Schubert and Sedvall, 1972).

Since data in the literature on the effects of amphetamine on brain 5-HT mechanisms are conflicting, and there is no study on the effects of chronic amphetamine treatment on brain tryptophan concentrations, the following experiment was undertaken. Twenty-five male Sprague-Dawley rats were treated with D-amphetamine, 10 mg/kg, i.p., twice daily for 12 days. An additional lot of 25 animals were given saline according to the same schedule. At 16 hr after the last regular injection, one group of five rats of the chronically amphetamine-treated rats received saline, and the other four groups received 10 mg/kg, i.p., of amphetamine and were killed at 1, 2, 4, and 8 hr afterwards. The chronically saline-treated rats were divided into five equal groups and injected with saline or amphetamine and killed as described for the chronically amphetamine-treated rats. The brain concentrations of tryptophan, 5-HT, and 5-HIAA were determined; the results are shown in Fig. 2. Brain tryptophan increased after a single amphetamine injection, reaching a maximal level at 1 to 2 hr after the injection as previously reported (see above). The last amphetamine injection in the chronically amphetamine-treated rats caused an increase in brain tryptophan after 1 hr, but the brain tryptophan level decreased more rapidly toward the control level in these rats, indicating tolerance to the effect of amphetamine on brain tryptophan. Brain 5-HT was uneffected by a single or the last injection of amphetamine in the chronically treated rats. However, the steady state level of brain 5-HT after chronic amphetamine treatment was reduced to approximately 75% of the control level. Brain 5-HIAA was reduced to 80% of the control level at 1 hr and increased to 130% at 4 hr after a single injection of amphetamine. This biphasic reaction of brain 5-HIAA was not evident after the last amphetamine injection in the chronically amphetamine-treated rats, and, as for brain 5-HT, the 5-HIAA steady state level was reduced after chronic amphetamine treatment.

An initial decrease in brain 5-HIAA in spite of an increase in brain

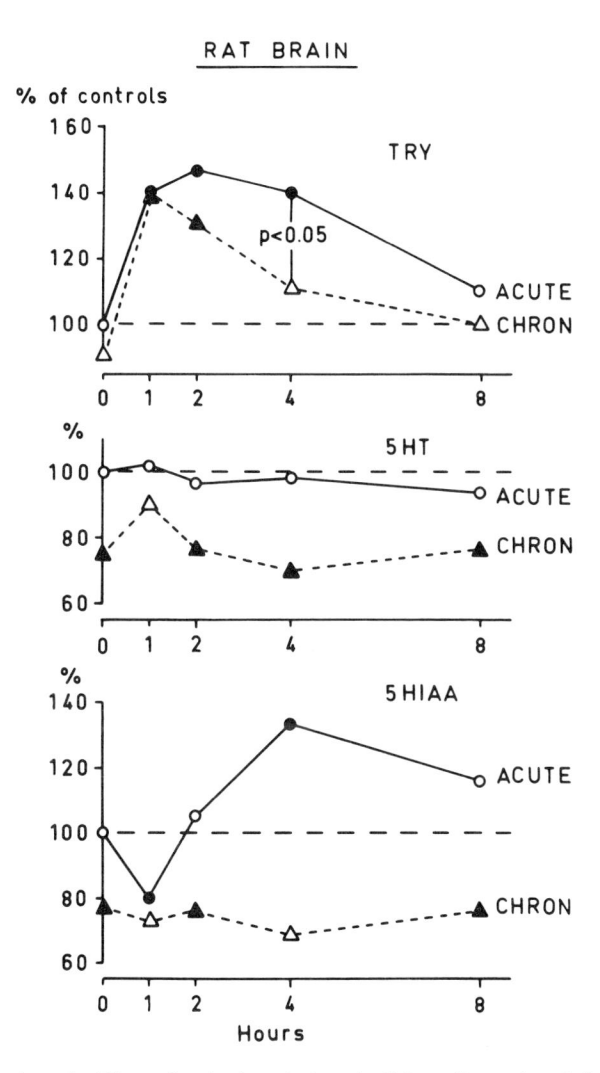

FIG. 2. Effects of acute (10 mg/kg, i.p.) and chronic (10 mg/kg, twice daily for 12 days) treatment with D-amphetamine sulfate on whole-brain levels of tryptophan (TRY), 5-hydroxytryptamine (5HT), and 5-hydroxyindoleacetic acid (5HIAA) in male Sprague-Dawley rats, 275 to 325 g body weight, at various intervals after the (last) amphetamine injection. Filled symbols represent statistically significant differences ($p < 0.05$, Student's t test) from control values (means ± SEM): TRY = 4.79 ± 0.22 µg/g, 5HT = 0.53 ± 0.03 µg/g, and 5HIAA = 0.44 ± 0.02 µg/g. Each value represents the mean of five observations.

RAT BRAIN

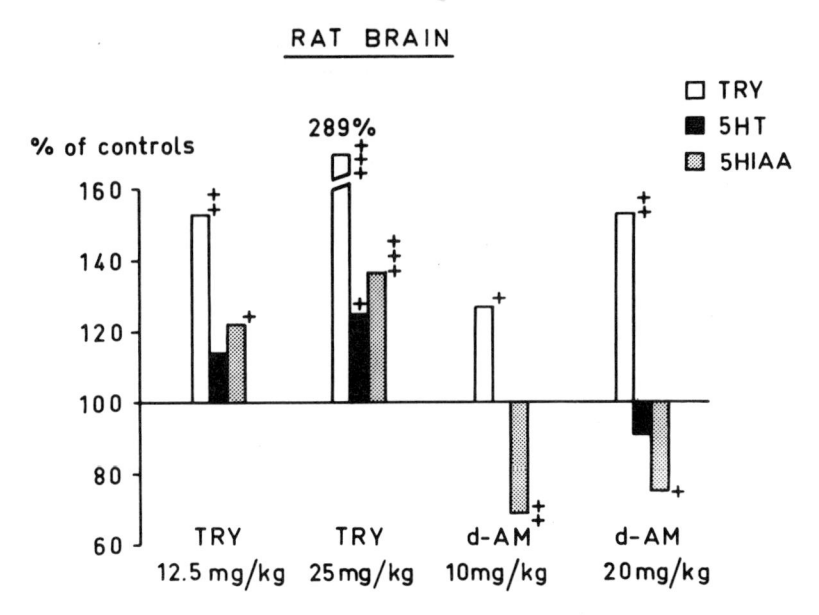

FIG. 3. Brain levels of tryptophan (TRY), 5-hydroxytryptamine (5HT), and 5-hydroxyindoleacetic acid (5HIAA) in female Sprague-Dawley rats, 250 g body weight, at 1 hr after a single i.p. injection of tryptophan or D-amphetamine sulfate in the indicated doses. Each bar represents the mean of six observations. +, ++, or +++ indicates statistically significant differences from control values (means ± SEM): TRY = 8.72 ± 0.44 µg/g, 5HT = 0.65 ± 0.02 µg/g, and 5HIAA = 0.47 ± 0.03 µg/g at $p < 0.05$, $p < 0.01$, and $p < 0.001$ levels, respectively (Student's t test).

tryptophan has not been reported previously. In another experiment (Fig. 3), this result was confirmed. However, an increase in brain tryptophan of the same magnitude as after amphetamine accomplished by tryptophan loading was followed by an increase in brain 5-HIAA in agreement with previous findings (see Fernström and Wurtman, 1971).

In conclusion, these two experiments show that amphetamine has a biphasic action on brain 5-HT neurons. In spite of an induced increase in brain tryptophan, there is an initial and transient decrease in the 5-HIAA level after amphetamine, which might be explained by inhibition of uptake of tryptophan into the 5-HT neurons, inhibition of 5-HT or 5-HIAA formation, or increased 5-HIAA transport from brain tissue caused by amphetamine. At later time points such effects of amphetamine seem to disappear, and brain 5-HIAA increases as after tryptophan loading. In rats chronically treated with amphetamine for 12 days, when tolerance has developed to the hyperthermic and anorexigenic effects of the drug, the effects of amphetamine on brain tryptophan and 5-HIAA seem to have diminished or vanished.

The brainstem concentration of 5-HT in guinea pigs (Table 1) after acute or chronic amphetamine treatment was found to be unaffected. In agree-

TABLE 1. Effect of DL-amphetamine on 5-HT and 5-HIAA in the brainstem of guinea pigs

Treatment		Hours after injection	5-HT % (mean ± SEM)	5-HIAA % (mean ± SEM)
Saline	(13)	—	100 ± 14	100 ± 5
Amphetamine, 20 mg/kg, i.p.	(4)	2	104 ± 14	67 ± 10[a]
	(4)	4	73 ± 5	71 ± 5[a]
	(4)	8	91 ± 5	100 ± 14
	(8)	16	150 ± 27	110 ± 14
	(4)	24	91 ± 10	81 ± 5
Amphetamine, 20 mg/kg × 2 × XVIII	(3)	4	105 ± 14	110 ± 19

5-HT = 0.22 ± 0.03 µg/g; 5-HIAA = 0.21 + 0.01 µg/g.
Male mottled guinea pigs, body weight approximately 275 g, were injected i.p. with DL-amphetamine sulfate in doses and time schedules according to the table. Values of 5-HT and 5-HIAA are given as percentages of the control levels shown below the body of the table. Roman numerals indicate days of chronic treatment. Figures within parentheses represent number of observations.
[a] Statistically significant difference ($p < 0.05$, Student's t test) from the controls.

ment with the observations in rats, however, there was a small decrease in 5-HIAA at 2 to 4 hr after a single amphetamine injection. In the light of the rat experiments (above), these preliminary data must be complemented with determinations of 5-HT and 5-HIAA in diencephalic and telencephalic brain regions, where most of the 5-HT neuron terminals are located. No data on the effects of amphetamine on brain tryptophan in species other than mice and rats are so far available.

A further detailed analysis of the influence of amphetamine on 5-HT neurons in relation to its acute behavioral effects and to amphetamine tolerance seems to be warranted.

C. Development, Disappearance, and Carry-over of Tolerance to the Hyperthermic Effects of Amphetamine in Rats in Relation to Central and Peripheral CA Levels and Brain Tryptophan and Serotonin Concentrations

Tolerance to the hyperthermic effect of amphetamine in rats has been linked with the formation and accumulation into central and peripheral NE neurons of p-hydroxynorephedrine as a false transmitter (for references, see above). In a previous report (Lewander, Moliis, and Brus, 1973) further aspects of tolerance to this effect of amphetamine, viz. rate of development, duration, and carry-over of tolerance into a subsequent second tolerance development phase, were described. In order to evaluate further the false transmitter hypothesis for amphetamine tolerance, brain and heart NE levels were determined at intervals during the development and dis-

TOLERANCE TO AMPH. HYPERTHERMIA
RELATION TO BRAIN □ AND HEART ▨ NA LEVELS

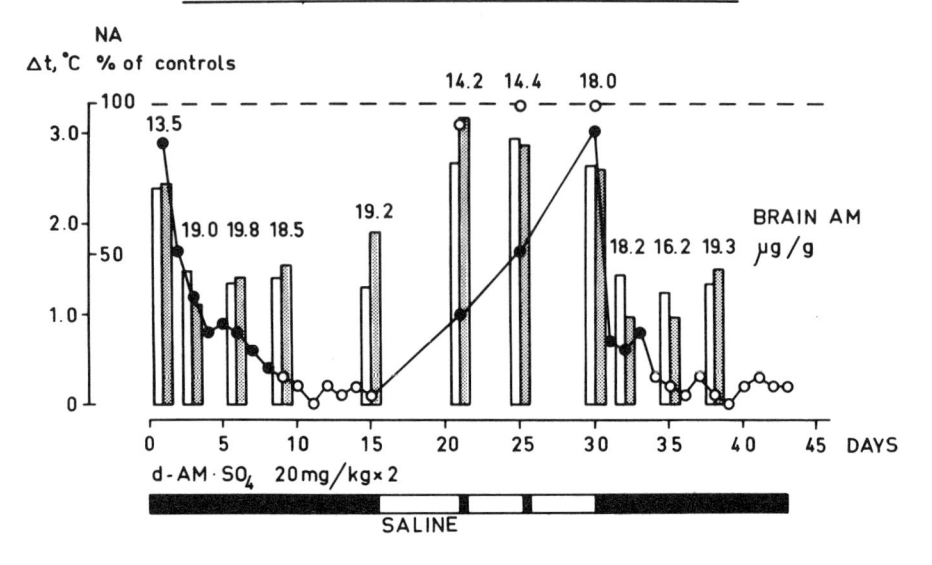

TOLERANCE TO AMPH. HYPERTHERMIA
RELATION TO BRAIN TRY ▨ LEVELS

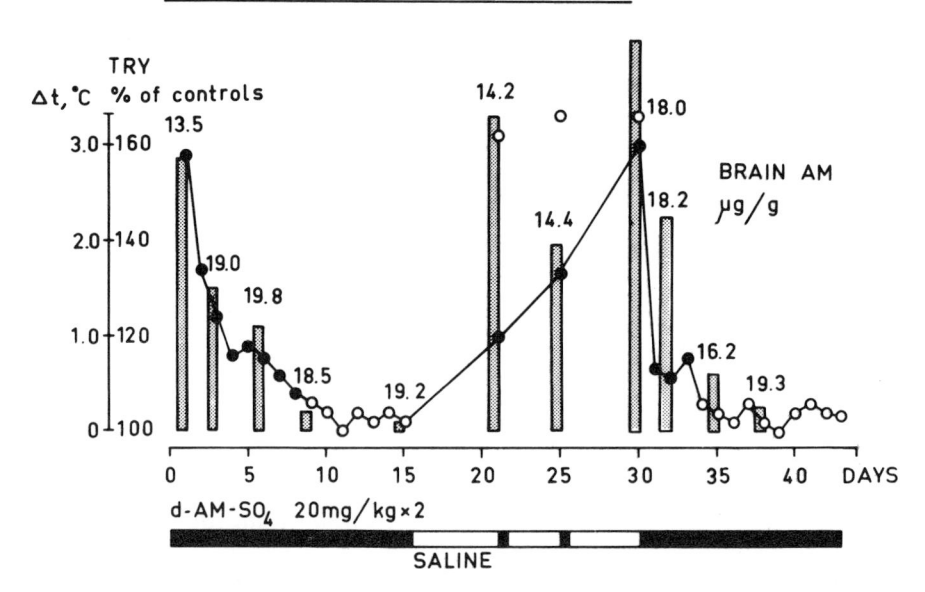

FIG. 4. Relation between the development, disappearance, and carry-over of tolerance to the hyperthermic effect of D-amphetamine and brain and heart norepinephrine (NA) concentrations, brain tryptophan (TRY), and brain amphetamine (AM) concentrations. Male Sprague-Dawley rats, 275 g body weight, were treated twice daily with 20 mg/kg, i.p., of

appearance of tolerance. According to the hypothesis, there should be a close relation between the degree of tolerance and the tissue NE levels during the development and disappearance of tolerance. As shown in Fig. 4A, tolerance to the hyperthermic effect of amphetamine developed to a maximal degree during 9 days. Tissue NE levels decreased to approximately 40% of controls after 3 days and remained there throughout the amphetamine treatment period. A very similar pattern was observed during the second treatment period, when the tolerance to the hyperthermic effect developed more rapidly. Thus, tolerance continued to develop without concomitant changes in central or peripheral NE levels. This finding does not necessarily exclude a relationship between tolerance and the presence of p-hydroxynorephedrine, since the remaining NE stores might not be accessible to amphetamine and its metabolite, or the previously accumulated p-hydroxynorephedrine molecules might be repeatedly released and replaced by newly formed ones as suggested by previous biochemical evidence (Lewander, 1971d).

The part of the figure showing disappearance of tolerance appears to be indicative for a refutation of the hypothesis that p-hydroxynorephedrine should be the only explanation of tolerance to the hyperthermic effect of amphetamine. It is found that in spite of residual tolerance to amphetamine on days 21 and 25 (according to Fig. 4A), the brain and heart NE concentrations have increased to the same level as after the first amphetamine injection, indicating that the false transmitter has disappeared before the initial body temperature response to amphetamine has been reestablished. Further studies for a final evaluation of the false transmitter hypothesis for amphetamine tolerance are in progress.

The figures above the bars in Fig. 4A show that the brain amphetamine concentrations accumulate to a steady state level during chronic amphetamine administration and are not changing in such a direction as to explain the development or disappearance of amphetamine tolerance. The brain DA levels did not change significantly during the experiment.

The brain concentrations of tryptophan and 5-HT were measured in the same animals and are presented in Fig. 4B. The increase in brain tryptophan after the first dose of amphetamine was successively reduced after repeated

---

D-amphetamine sulfate or saline. The increase in body temperature at 1 hr after each morning injection of amphetamine is given as the temperature difference ($\Delta t°C$) between amphetamine-treated and control rats (n = 15 animals in each group except on days 21 and 25 when n = 5). Filled circles indicate statistically significant differences between treated and control rats. Open unconnected circles on days 21, 25, and 30 indicate the temperature response in three groups of five saline-treated rats given a single amphetamine injection. Columns representing brain and heart NA levels (top) and brain TRY levels (bottom) are given as percentages of the control levels (means ± SEM, n = 4 to 6): brain NA = 0.42 ± 0.02 µg/g, heart NA = 0.64 ± 0.03 µg/g, TRY = 3.33 ± 0.22 µg/g. Figures above bars refer to brain amphetamine concentrations (means of four to six observations).

injections with the drug. The vanishing tryptophan response to amphetamine nicely paralleled the weakening temperature response during the development of tolerance. However, the tryptophan concentrations had already increased from the initial value by 6 days after amphetamine withdrawal, in contrast to the reduced temperature response. During the second chronic phase the brain tryptophan response to amphetamine did not follow the more rapid development of tolerance to the hyperthermic effect. There was not a complete parallelism between the amphetamine-induced changes in brain tryptophan and body temperature, and thus these two phenomena did not appear to be causally related. Brain 5-HT was not significantly changed during this experiment. The use of a dose of 20 mg/kg of D-amphetamine might be the reason for the development of a more clear-cut tolerance to the amphetamine-induced increase in brain tryptophan, as was apparent in this experiment compared with that in Fig. 2, in which 10 mg/kg was given.

D. Effects of Acute and Chronic Phenmetrazine Administration on Brain Catecholamines and 5-Hydroxyindoles in the Rat

Phenmetrazine is a methylmorfolin derivative of phenylethylamine. Similar to amphetamine, this compound has central stimulant and anorexigenic effects and induces increased motor activity and stereotype behavior in rats, which effects are blocked by alpha-methyl-*p*-tyrosine (Weissman, Koe, and Tenen, 1966; Scheel, Krüger, 1971). Furthermore, phenmetrazine, like amphetamine, causes an increase in the urinary excretion of catecholamines (Gunne and Lewander, 1968). Baird and Lewis (1964) and Baird (1968) reported that single injections of phenmetrazine caused an increase in brain NE and DA levels in rats. According to the present data (Table 2), 40 mg/kg of the drug caused no consistent changes in brain CA levels. Higher doses of the drug (80 to 120 mg/kg, i.p., 1 or 4 hr after injection) also had no effect on brain CA levels (data not shown). Prolonged chronic treatment of rats with phenmetrazine was followed by a small statistically significant increase in brain NE and brain DA. Thus, phenmetrazine and amphetamine seem to differ markedly from each other with regard to their effects on brain CA levels after acute and chronic administration. The finding of increased levels of NM and MTA, the methoxylated metabolites of NE and DA, respectively, after injection of phenmetrazine in rats pretreated with a monoamineoxidase inhibitor (Fig. 5) supports, however, the hypothesis that phenmetrazine, like amphetamine (see Introduction), acts through release of brain CA. This result agrees with similar observations by Scheel-Krüger (1971).

Brain 5-HT and 5-HIAA did not change after acute or chronic phenmetrazine treatment (Table 2). This is in contrast to the finding of increased 5-HIAA levels after phenmetrazine reported by Tagliamonte et al. (1971).

TABLE 2. *Effect of phenmetrazine on brain catecholamines and 5-hydroxyindoles in rats*

Treatment	Hours after injection	NE % (mean ± SEM)	DA % (mean ± SEM)	5-HT % (mean ± SEM)	5-HIAA % (mean ± SEM)
Saline	—	100 ± 3 (6)	100 ± 3 (6)	100 ± 5 (10)	100 ± 11 (10)
Phenmetrazine,	1	105 ± 2 (5)	113 ± 5 (5)	100 ± 7 (5)	74 ± 7 (10)
40 mg/kg. i.p.	3	105 ± 2 (7)	114 ± 5 (7)	95 ± 10 (5)	104 ± 11 (10)
	6	107 ± 2 (3)	121 (2)	88 ± 7 (5)	111 ± 15 (10)
	9	110 ± 2 (3)	127 ± 3 (3)[a]	—	—
Saline	—	—	—	100 ± 5 (10)	100 ± 13 (10)
Phenmetrazine chron I	3	—	—	114 ± 12 (4)	94 ± 18 (4)
40 mg/kg × 2 × VII					
Saline	—	100 ± 1 (6)	100 ± 4 (4)	—	—
Phenmetrazine chron II	3	107 ± 1 (4)[b]	117 ± 5 (4)[a]	—	—
20 mg/kg × 1 × XIV +					
40 mg/kg × 1 × XIV					
Phenmetrazine chron III	3	116 ± 4 (4)[b]	—	—	—
20 mg/kg × 1 × XIV +					
40 mg/kg × 2 × XIV					

NE = 0.37 ± 0.01 µg/g; DA = 0.91 ± 0.03 µg/g; 5-HT = 0.42 ± 0.02 µg/g; 5-HIAA = 0.27 ± 0.03 µg/g. Male Sprague-Dawley rats, body weight 250 to 350 g, were injected with phenmetrazine HCl in a single injection or chronically (chron) in doses and time schedules as indicated in the table. Values are given as percentages of the control levels shown below the body of the table. Statistically significant differences from controls (Student's *t* test) are indicated by [a]($p < 0.05$) or [b]($p < 0.001$). Figures within parentheses represent number of observations.

RAT BRAIN

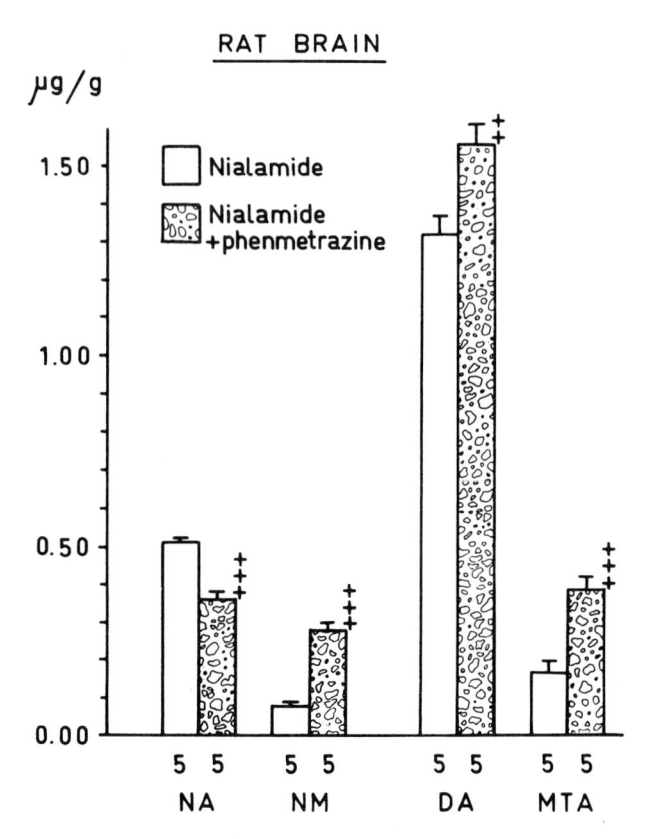

FIG. 5. Effect of phenmetrazine · HCl (40 mg/kg, i.p., 1 hr before sacrifice) on whole-brain concentrations of norepinephrine (NA), normetanephrine (NM), dopamine (DA), and methoxytyramine (MTA) in male Sprague-Dawley rats, 300 g body weight, pretreated with nialamide, 100 mg/kg, i.p., 16 hr before sacrifice. Columns represent mean values ± SEM (vertical bars) of five observations (figures below columns). ++ and +++ indicate statistically significant differences from control values at $p < 0.01$ and $p < 0.001$, respectively (Student's t test).

Brain tryptophan was not measured in the present experiments, but, as after amphetamine, brain tryptophan levels are increased after phenmetrazine in rats (Tagliamonte et al., 1971).

E. Development of Tolerance to the Anorexigenic Effect of D- and L-Amphetamine and Phenmetrazine and the Demonstration of Cross-Tolerance Between These Central Stimulants

Tolerance to the anorexigenic effect of D- or DL-amphetamine develops (see Lewander, 1971b), but does not seem to be dependent on the accumulation of p-hydroxynorephedrine as a false transmitter (Lewander, 1971b). This conclusion is corroborated by the data in Fig. 6, which show

FOOD INTAKE IN RATS g/7 hrs

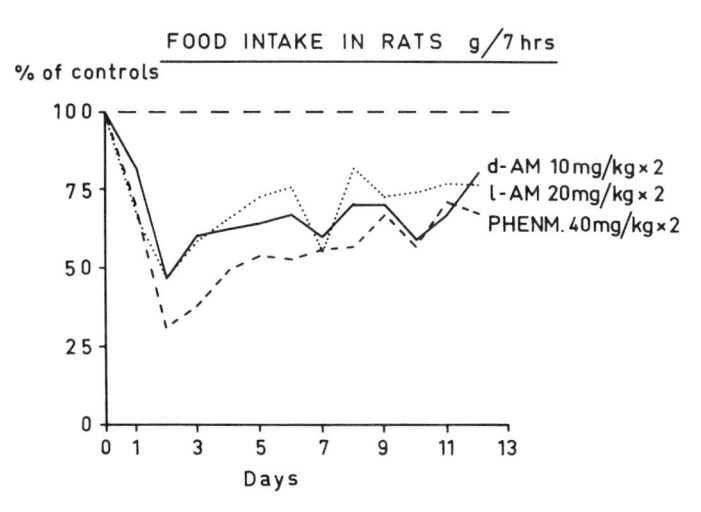

FIG. 6. Tolerance to the anorexigenic effects of D-amphetamine, L-amphetamine, and phenmetrazine. Daily food intake (g/7 hr, from 9 A.M. to 4 P.M.) was measured in male Sprague-Dawley rats, 300 g body weight, treated twice daily with D-amphetamine sulfate (D-AM), 10 mg/kg, i.p., L-amphetamine sulfate (L-AM), 20 mg/kg, i.p., or phenmetrazine · HCl (PHENM), 40 mg/kg, i.p. Before the proper experiment the rats were conditioned for 8 to 10 days to have their food intake at the indicated hours. Values represent means of groups of five rats and are expressed as percentages of the food intake of chronically saline-treated controls run simultaneously.

that tolerance develops to the anorexigenic effect of D-amphetamine but also to L-amphetamine, which is not converted to p-hydroxynorephedrine in rats but still decreases brain NE levels (Lewander, 1971a). Tolerance development was also shown for phenmetrazine, a drug that does not even change brain CA concentrations (see Section D, above). Whether the anorexigenic effects of these drugs are mediated via CA neurons or not has not been definitely proved, but one report supports this contention (Weissman, Koe, and Tenen, 1966). The results shown in Fig. 7, demonstrating cross-tolerance to the anorexigenic effect between L-amphetamine and D-amphetamine as well as between phenmetrazine and D-amphetamine, strongly indicate that the three drugs have a similar mechanism of action in this respect. The experiment also demonstrates that tolerance to the anorexigenic effect of D-amphetamine actually means a shortening of the duration of its anorexigenic action in spite of the fact that brain concentrations of amphetamine are unchanged during chronic treatment (see Fig. 4A), and the disappearance rate of amphetamine seems to be unaffected by chronic treatment in rats (Lewander, 1971e). Tolerance to the hyperthermic effect of D-amphetamine does not seem to be due to a shorter duration of action, but rather to a reduction of the maximal hyperthermic response (Fig. 4A). As pointed out above, tolerance does not seem to develop to the effects of amphetamine on motor behaviors in rats (Lewander, 1971b).

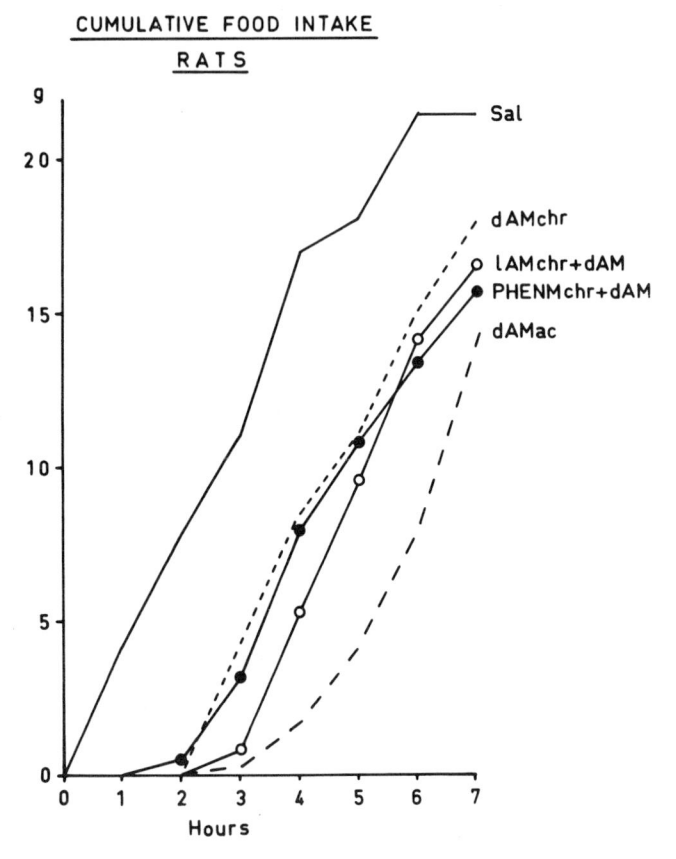

FIG. 7. Cross-tolerance to the anorexigenic effect between L- and D-amphetamine and phenmetrazine and D-amphetamine, respectively. Food intake was measured daily at hourly intervals between 9 A.M. and 4 P.M. in five groups of five animals; the cumulative plots of the mean values are shown. D-AM$_{ac}$ represents the pattern of food intake after the third D-amphetamine injection (corresponding to day 2 in Fig. 6), when maximal anorexia is present, and D-AM$_{chr}$ the pattern after the last (25th) injection of D-amphetamine sulfate, 10 mg/kg, twice daily for 12½ days. The L-AM$_{chr}$ rats received 20 mg/kg, i.p., of L-amphetamine twice daily for 12 days and were tested on day 13 with D-amphetamine, 10 mg/kg. The PHENM$_{chr}$ rats were treated with phenmetrazine 40 mg/kg, i.p., twice daily for 12 days and were tested on day 13 with 10 mg/kg of D-amphetamine.

ACKNOWLEDGMENTS

The author is indebted to Miss Sonja Ahlén, Mr. Sven Ullman, and Mr. Christer Cederberg for skillful technical assistance and to Dr. Lars-Erik Jönsson for permission to include a preliminary report of rabbit data. Phenmetrazine was generously supplied by Boehringer Sohn, Ingelheim, Germany. The investigations have been supported by the Swedish Medical Research Council (project No. B73-04X-1017-09, B74-04P-4288-01A),

Magnus Bergvalls Stiftelse, and the Tricentennial Fund of the Bank of Sweden (project No. 150).

REFERENCES

Änggård, E., Gunne, L.-M., and Niklasson, F. (1970): Gas chromatographic determination of amphetamine in blood, tissue and urine. *Scandinavian Journal of Clinical and Laboratory Investigations*, 26:137–144.
Azzaro, A. J., and Rutledge, C. O. (1973): Selectivity of release of norepinephrine, dopamine and 5-hydroxytryptamine by amphetamine in various regions of rat brain. *Biochemical Pharmacology*, 22:2801–2813.
Baird, J. R. C. (1968): The effects of (+)-amphetamine and (±)-phenmetrazine on the noradrenaline and dopamine levels in the hypothalamus and corpus striatum of the rat. *Journal of Pharmacy and Pharmacology*, 20:234–235.
Baird, J. R. C., and Lewis, J. J. (1964): The effects of cocaine, amphetamine and some amphetamine-like compounds on the in vivo levels of noradrenaline and dopamine in the rat brain. *Biochemical Pharmacology*, 13:1475–1482.
Brodie, B. B., Cho, A. K., and Gessa, G. L. (1970): Possible role of p-hydroxynorephedrine in the depletion of norepinephrine induced by D-amphetamine and in tolerance to this drug. In: *International Symposium on Amphetamines and Related Compounds*, edited by E. Costa and S. Garattini. Raven Press, New York.
Carlsson, A., and Waldeck, B. (1958): A fluorimetric method for the determination of dopamine (3-hydroxytyramine). *Acta Physiologica Scandinavica*, 44:293–298.
Carlsson, A., and Waldeck, B. (1964): A method for the fluorimetric determination of 3-methoxytyramine in tissues and the occurrence of this amine in brain. *Scandinavian Journal of Clinical and Laboratory Investigations*, 16:133–138.
Chang, C. C. (1964): A sensitive method for spectrophotofluorimetric assay of catecholamines. *International Journal of Neuropharmacology*, 3:643–649.
Costa, E., and Groppetti, A. (1970): Biosynthesis and storage of catecholamines in tissues of rats injected with various doses of D-amphetamine. In: *International Symposium on Amphetamines and Related Compounds*, edited by E. Costa and S. Garattini. Raven Press, New York.
Denckla, W. D., and Dewey, H. K. (1967): The determination of tryptophan in plasma, liver and urine. *Journal of Laboratory and Clinical Medicine*, 69:160–166.
Díaz, J.-L., and Huttonen, M. O. (1972): Altered metabolism of serotonin in the brain of the rat after chronic ingestion of D-amphetamine. *Psychopharmacologia* (Berlin), 23:365–372.
Dring, L. G., Smith, R. L., and Williams, R. T. (1970): The metabolic fate of amphetamine in man and other species. *Biochemical Journal*, 116:425–435.
Fernström, J. D., and Wurtman, R. J. (1971): Brain serotonin content: Physiological dependence on plasma tryptophan levels. *Science*, 173:149–152.
Freeman, J. J., and Sulser, F. (1972): Iprindole-amphetamine interactions in the rat: The role of aromatic hydroxylation of amphetamine in its mode of action. *Journal of Pharmacology and Experimental Therapeutics*, 183:307–315.
Garattini, S., and Valzelli, L. (1965): *Serotonin*. Elsevier Publishing Company, Amsterdam, London, New York, p. 279.
Görlitz, B.-D., and Frey, H.-H. (1972): Central monoamines and antinociceptive drug action. *European Journal of Pharmacology*, 20:171–180.
Groppetti, A., and Costa, E. (1969): Tissue concentration of p-hydroxynorephedrine in rats injected with D-amphetamine: Effect of pretreatment with desipramine. *Life Sciences*, 8:653–665.
Gunne, L.-M., and Lewander, T. (1968): Brain catecholamines during chronic amphetamine intoxication. In: *The Addictive States*. A.R.N.M.D., Vol. 46:106–112, Williams and Wilkins Company, Baltimore.
Häggendal, J. (1962): Fluorimetric determination of 3-O-methylated derivatives of adrenaline and noradrenaline in tissues and body fluids. *Acta Physiologica Scandinavica*, 56:258–266.
Harris, J. E., and Baldessarini, R. J. (1973): Amphetamine-induced inhibition of tyrosine hydroxylation in homogenates of rat corpus striatum. *Journal of Pharmacy and Pharmacology*, 25:755–757.

Jönsson, J., and Lewander, T. (1970): A method for the simultaneous determination of 5-hydroxyindoleacetic acid (5-HIAA) and 5-hydroxytryptamine (5-HT) in brain tissue and cerebrospinal fluid. *Acta Physiologica Scandinavica,* 78:43–51.

Jori, A., and Bernardi, D. (1969): Effect of amphetamine and amphetamine-like drugs on homovanillic acid concentration in the brain. *Journal of Pharmacy and Pharmacology,* 21:694–697.

Jori, A., and Bernardi, D. (1972): Further studies on the increase of striatal homovanillic acid induced by amphetamine and fenfluramine. *European Journal of Pharmacology,* 19:276–280.

Korf, J., Roos, B.-E., and Werdinius, B. (1970): Fluorimetric determination of homovanillic acid in tissues using anion exchange separation and mixed solvent elution. *Acta Chemica Scandinavica,* 25:333–335.

Laverty, R., and Sharman, D. F. (1965): Modification by drugs of the metabolism of 3,4-dihydroxyphenylethylamine, noradrenaline and 5-hydroxytryptamine in the brain. *British Journal of Pharmacology,* 24:759–772.

Lewander, T. (1968): Urinary excretion and tissue levels of catecholamines during chronic amphetamine intoxication. *Psychopharmacologia* (Berlin), 13:394–407.

Lewander, T. (1970): Catecholamine turn-over studies in chronic amphetamine intoxication. In: *International Symposium on Amphetamines and Related Compounds,* edited by E. Costa and S. Garattini. Raven Press, New York.

Lewander, T. (1971a): On the presence of *p*-hydroxynorephedrine in the rat brain and heart in relation to changes in catecholamine levels after administration of amphetamine. *Acta Pharmacologica et Toxicologica,* 29:33–48.

Lewander, T. (1971b): A mechanism for the development of tolerance to amphetamine in rats. *Psychopharmacologia* (Berlin), 21:17–31.

Lewander, T. (1971c): Effects of acute and chronic amphetamine intoxication on brain catecholamines in the guinea pig. *Acta Pharmacologica et Toxicologica,* 29:209–225.

Lewander, T. (1971d): Displacement of brain and heart noradrenaline by *p*-hydroxynorephedrine after administration of *p*-hydroxyamphetamine. *Acta Pharmacologica et Toxicologica,* 29:20–32.

Lewander, T. (1971e): Effects of chronic amphetamine intoxication on the accumulation in the rat brain of labelled catecholamines synthesized from circulating tyrosine-^{14}C and DOPA-^{3}H. *Naunyn Schmiedeberg's Archives of Pharmacology,* 271:211–233.

Lewander, T. (1972): Experimental and clinical studies on amphetamine dependence. In: *Biochemical and Pharmacological Aspects of Dependence and Reports on Marihuana Research,* edited by H. M. van Praag. De Erven F. Bohn N.V. Haarlem.

Lewander, T., Moliis, G., and Brus, I. (1973): On amphetamine tolerance and abstinence in rats. In: *Frontiers in Catecholamine Research,* edited by E. Usdin and S. Snyder. Pergamon Press, Elmsford, N.Y.

McLean, J. R., and McCartney, M. (1961): Effect of D-amphetamine on rat brain noradrenaline and serotonin. *Proceedings of the Society for Experimental Biology and Medicine,* 107:77–79.

Philippu, A., and Beyer, J. (1973): Dopamine and noradrenaline transport into subcellular vesicles of the striatum. *Naunyn Schmiedeberg's Archives of Pharmacology,* 278:387–402.

Reid, W. D. (1970): Turn-over rate of brain 5-hydroxytryptamine increased by D-amphetamine. *British Journal of Pharmacology,* 40:483–491.

van Rossum, J. M. (1970): Mode of action of psychomotor stimulant drugs. *International Review of Neurobiology,* 12:307–383.

Scheel-Krüger, J. (1971): Comparative studies of various amphetamine analogues demonstrating different interactions with the metabolism of catecholamines in the brain. *European Journal of Pharmacology,* 14:47–59.

Scheel-Krüger, J. (1972): Some aspects of the mechanism of action of various stimulant amphetamine analogues. *Psychiatria, Neurologia, Neurochirurgia,* 75:179–192.

Schubert, J., Fyrö, B., Nybäck, H., and Sedvall, G. (1970): Effects of cocaine and amphetamine on the metabolism of tryptophan and 5-hydroxytryptamine in mouse brain *in vivo. Journal of Pharmacy and Pharmacology,* 22:860–862.

Schubert, J., and Sedvall, G. (1972): Effect of amphetamines on tryptophan concentrations in mice and rats. *Journal of Pharmacy and Pharmacology,* 24:53–62.

Sparber, S. B., and Tilson, H. A. (1972): The releasability of central norepinephrine and serotonin by peripherally administered D-amphetamine before and after tolerance. *Life Sciences,* 11:1059–1067.

Sulser, F., and Sanders-Bush, E. (1971): Effect of drugs on amines in the CNS. *Annual Review of Pharmacology,* 11:209–230.

Tagliamonte, A., Tagliamonte, P., Perez-Cruet, G., Stern, S., and Gessa, G. L. (1971): Effect of psychotropic drugs on tryptophan concentration in the rat brain. *Journal of Pharmacology and Experimental Therapeutics,* 177:475–480.

Weiner, N. (1972): Pharmacology of central nervous system stimulants. In: *Drug Abuse. Proceedings of the International Conference,* edited by C. J. D. Zarafonetis. Lea and Febiger, Philadelphia.

Weissman, A., Koe, B. K., and Tenen, S. S. (1966): Amphetamine effects following inhibition of tyrosine hydroxylase. *Journal of Pharmacology and Experimental Therapeutics,* 151:339–352.

Neuropsychopharmacology of Monoamines and Their Regulatory Enzymes, edited by E. Usdin. Raven Press, New York © 1974.

Drug-Induced Stereotyped Behavior: Similarities and Differences

Marshall B. Wallach

Department of Pharmacology, Syntex Research, Palo Alto, California 94304

I. INTRODUCTION

A. Stereotyped Behavior

In the last decade a considerable body of literature has been developed concerning what is termed "stereotyped behavior." The first description of what is currently termed "stereotyped behavior" was made in 1939 by Hauschild in the German literature. Today, stereotyped behavior refers to a group of behaviors that can be elicited in most species by agents such as amphetamine or apomorphine. These agents produce a characteristic, repetitive, and apparently purposeless activity. This activity occurs in the absence of other normal behaviors. The characteristics of stereotypy vary from species to species and tend toward greater similarity from animal to animal in lower species. In cats, dogs, monkeys, and man the characteristics of stereotyped behavior are similar; however, the patterns vary from animal to animal. Each has an individual behavior which reappears from day to day following the same or different stereotypy-inducing drugs.

B. Stimulant-Induced Psychoses

The interest in stereotyped behavior has developed due to two different, yet related, observations. The first observation dates back to 1938, when Young and Scoville reported the occurrence of a paranoid or paranoid-hallucinatory syndrome following the ingestion of large doses of amphetamine. In the 1950s, several papers reporting on amphetamine- and related stimulant-induced psychoses suggested that these central nervous system stimulants were not inducing a latent paranoid psychosis, but actually were inducing psychoses in individuals with no overt schizophrenic background (Herman and Nagler, 1954; Connell, 1958).

In the late 1960s, several laboratories became interested in the clinical similarities between the amphetamine-induced paranoid psychoses and schizophrenia, paranoid type. This interest in amphetamine psychoses coincided with the beginnings of the drug abuse era. Psychiatric receiving hospitals began to see a significant number of amphetamine-induced psychoses. These patients were extremely paranoid upon arrival but underwent a miraculous recovery within several days. Clinical experiments were carried out by at least three laboratories to determine if amphetamine, when administered to relatively normal subjects, could induce paranoid psychoses (Griffith, Oates, and Cavanaugh, 1968; Angrist and Gershon, 1970; Griffith, Cavanaugh, Held, and Oates, 1972; Bell, 1973).

In the early 1960s, a series of papers began appearing from a laboratory in Denmark, describing what was termed "stereotyped behavior" in various species and its modification by pharmacological manipulations (Randrup, Munkvad, and Udsen, 1963; Randrup and Munkvad, 1965*a, b*). The stereotyped behavior concept was soon observed in amphetamine abusers. Ellinwood (1967, 1969) reported that amphetamine abusers often exhibited repetitive and useless behaviors, for example, dismantling objects, analyzing and sorting their parts, or performing menial tasks such as continuous polishing of a floor even when it was already clean and shining. These repetitive behaviors began to occur before the psychoses appeared. A report by Rylander (1969) suggested that phenmetrazine abusers also exhibited abnormal behaviors. The association of these stereotyped or repetitive behaviors in man with amphetamine-like stimulant abuse, and in other species under experimental conditions, suggested marked similarities between the states. Many laboratories observed these similarities and began utilizing CNS stimulant-induced stereotyped behavior as a model for both amphetamine-induced psychoses and for schizophrenia, paranoid type.

C. Differentiation of Stimulant and Perception-Distorting Psychotomimetic Drugs

It is immediately apparent that agents which induce psychoses in man cannot all be similar. We have differentiated the CNS stimulants from other psychotomimetic agents based on a combination of clinical, electroencephalographic, and behavioral criteria. Throughout this chapter, the term CNS stimulants will refer to amphetamine-like agents which induce stereotypy and paranoid psychoses. Such stimulants as caffeine are not included in this definition of CNS stimulants, although, strictly considered, caffeine is a CNS stimulant.

Several of the CNS stimulants, which have previously been demonstrated to elicit stereotyped behavior in rodents, have been examined utilizing electroencephalographic techniques in cats. The results of these studies have indicated that D-amphetamine, L-DOPA, cocaine, methamphetamine, and phenmetrazine all elicit a desynchronized EEG (Fig. 1). The duration of the desynchrony in the conscious cat was dose-dependent. Associated with the desynchronized EEG was an elevation of the reticular formation multiple

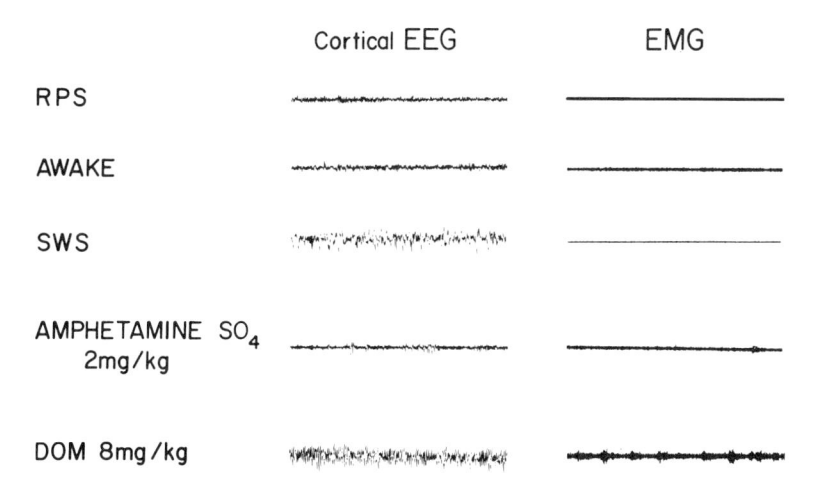

FIG. 1. Cortical EEG and cervical neck muscle EMG during various states. EEG records for rhombencephalic phase of sleep (RPS), awake, and following amphetamine are considered to be desynchronized. Note the differences in the wave forms for slow-wave sleep (SWS) and following 2,5-dimethoxy-4-methylamphetamine (DOM).

unit activity, a reflection of CNS stimulation. All of these agents also disrupted sleep, with a return to normal generally occurring between 12 and 24 hr. After these drugs all sleep, both slow-wave and rhombencephalic, was completely abolished for up to 24 hr (Wallach, Winters, Mandell, and Spooner, 1969a, b; Winters, Mori, Wallach, Marcus, and Spooner, 1969; Wallach and Gershon, 1971; Wallach, Rotrosen, and Gershon, 1973).

The EEG characteristics of these CNS stimulants are different than the EEG patterns obtained following various perception-distorting psychotomimetic agents. Following agents such as 2,5-dimethoxy-4-methylamphetamine (DOM), the EEG is hypersynchronous and the behavioral patterns are markedly different than those observed following various CNS stimulants. Thus a clear dichotomy can be obtained between agents such as the CNS stimulants and the perception-distorting psychotomimetics, for example, DOM, LSD, or mescaline (Winters et al., 1969; Wallach et al., 1969a, b, 1973; Wallach and Gershon, 1971; Wallach, Friedman, and Gershon, 1972).

D. Aims

This chapter will attempt to evaluate critically the experimental stereotypy data collected over the last decade and to correlate it to CNS stimulant-induced psychoses in man and to schizophrenia, paranoid type. Particular attention will be devoted to differences observed among the various stimulants, their mechanisms of action, and species responses.

II. INDUCTION OF STEREOTYPY

A. Description of Stereotypy

Randrup and Munkvad (1967) described stereotypy in many laboratory animals. The rat, following high doses of agents eliciting stereotyped behavior, first became hyperactive with a sniffing behavior. This was followed by licking and biting of the cage wires or the shavings on the cage floor. Occasional bursts of retropulsive activity were observed. All normal behaviors, such as grooming, eating, and sleeping, ceased. If the cage was devoid of both shavings and wires, the rat would gnaw on its own feet, flank, or that of a cage mate (Wallach, *unpublished*). As the effects of the stereotypy-inducing drug declined, normal behaviors reappeared.

Mice generally developed similar behavioral patterns following stereotypy-inducing doses of stimulants; however, the mouse is peculiar in its response to various drugs and various routes of administration. This will be discussed below. The guinea pig, following stereotypy-inducing agents, generally gnawed on the food dish or wires of the cage. Its behavior was disrupted by the presence of an observer (Randrup and Munkvad, 1967).

Cats developed one of two patterns of stereotyped behavior. One pattern consisted of sniffing a confined area of the cage. The other behavior pattern seen in cats was a continuous looking back and forth over its shoulders in an apparent "fearful" or "paranoid-like" manner. This was done to the exclusion of all normal behaviors such as eating, drinking, grooming, and sleeping (Randrup and Munkvad, 1967; Ellinwood, 1969; Wallach and Gershon, 1971).

Following doses of stereotypy-eliciting drugs, dogs developed their own characteristic pattern. One dog stood absolutely still with its eyes shifting back and forth. Another dog lunged back and forth out of a corner of the room, apparently jumping at, or pouncing on, some imaginary prey. A third dog circled continuously in very tight circles. A fourth dog circled in very large circles, moving in a fixed pattern around the periphery of the room. These behaviors were characteristic for individual dogs and could be induced by any one of a series of stereotypy-inducing agents (Willner, Samach, Angrist, Wallach, and Gershon, 1970).

Monkeys, following CNS stimulants, spent prolonged periods of time in characteristic behavior patterns. One monkey sat and rocked continuously, while another spent prolonged periods of time picking at various parts of its body or maneuvering its hand or paw in a specific pattern. These behaviors were very similar to those which have been observed in some humans following large doses of amphetamine or cocaine. The behaviors might be related to the so-called cocaine tics. Thus, it appears that stereotyped behaviors have been identified in a large number of species (Randrup and Munkvad, 1967; Ellinwood, 1971).

B. Drugs Inducing Stereotypy Following Systemic Administration

Many agents have been identified which will induce stereotyped behavior in one or more species. Fog (1969) has identified many agents, including amphetamine, methamphetamine, phenmetrazine, pemoline, lysergic acid diethylamide (LSD), methylphenidate, cocaine, diethylpropion, pipradrol, fencamfamine, apomorphine, and two research drugs, KSW3019 and WIN-25978. He further reported that caffeine and nikethamide did not elicit stereotyped behavior. In another publication, Fog (1970) indicated that chronic morphine treatment will elicit stereotyped behavior in rats. L-DOPA in very high doses also induced stereotyped behavior in rats, cats, and dogs (Randrup and Munkvad, 1966a, b; Wallach and Gershon, 1971; Willner et al., 1970). More recently, ET495 [7-(2''-pyrimidyl)-4-piperonyl-piperazine] also has been demonstrated to elicit stereotypy in rats (Costall and Naylor, 1972a). Table 1 is a list of stereotypy-inducing drugs and the species examined.

TABLE 1. *Drug-induced stereotyped behavior in various species*

Rats	
Amphetamine	Hauschild, 1939; Randrup et al., 1963
Apomorphine	Ernst and Smelik, 1966[c]; Ernst, 1969; Fog, 1969
p-Chloroamphetamine	Scheel-Kruger, 1972
Chlorphencyclane	Randrup and Munkvad, 1964
Cocaine	Fog, 1969; Scheel-Kruger, 1972
Diethylpropion	Fog, 1969
DOPA	Randrup and Munkvad, 1966b
(−) Ephedrine	Langwinski, 1970
Fencamfamine	Fog, 1969
5-Hydroxytryptophan	Randrup and Munkvad, 1966b
KSW 3019	Fog, 1969
LSD[a]	Fog, 1969
Methamphetamine	Fog, 1969; Scheel-Kruger, 1971
Methylphenidate	Fog, 1969; Scheel-Kruger, 1971
α-Methyltryptamine	Randrup and Munkvad, 1964
Morphine[a]	Fog, 1970; Ayhan and Randrup, 1972
NCA[b]	Aceto et al., 1967
Pemoline	Fog, 1969
Phenmetrazine	Fog, 1969; Scheel-Kruger, 1971; Wallach et al., 1973
β-Phenylethylamine	Randrup and Munkvad, 1966b
Pipradrol	Fog, 1969; Scheel-Kruger, 1971
Piribedil (ET-495)	Costall and Naylor, 1972a
Tryptamine	Randrup and Munkvad, 1966b
Win 25978	Fog, 1969
Mice	
Amphetamine	Randrup and Scheel-Kruger, 1966
Apomorphine	Fekete et al., 1970, Scheel-Kruger, 1970
Methylphenidate	Pedersen and Christensen, 1972
Pipradrol	Christensen and Pedersen, 1972

TABLE 1 *(Continued)*

Guinea pigs	
Amphetamine	Randrup and Munkvad, 1967
Methylphenidate	Scrimal and Dhawan, 1970
Cats	
Amphetamine	Randrup and Munkvad, 1967; Wallach and Gershon, 1971
Apomorphine	Wallach, *Unpublished;* Cools, 1971[c]
Cocaine	Wallach and Gershon, 1971
DOPA	Wallach and Gershon, 1971
Dopamine[c]	Cools and Van Rossum, 1970; Cools, 1971
Methamphetamine	Ellinwood, 1971*b*
3-Methoxytyramine[c]	Cools, 1971
Methylphenidate	Wallach and Gershon, 1972
Pemoline Mg(OH)$_2$	Wallach and Gershon, 1972
Phenmetrazine	Wallach et al., 1973
Dogs	
Amphetamine	Randrup and Munkvad, 1967; Willner et al., 1970
Apomorphine	Rotrosen et al., 1972; Nymark, 1972
Cocaine	Willner et al., 1970
DOPA	Willner et al., 1970
Methylphenidate	Wallach, *Unpublished*
Phenmetrazine	Wallach et al., 1973
Monkeys	
Amphetamine	Randrup and Munkvad, 1967; Kjellberg and Randrup, 1972
Methamphetamine	Ellinwood, 1971*a*
Man	
Amphetamine	Randrup and Munkvad, 1967
Cocaine	Redlich and Freedman, 1966
Methamphetamine	Ellinwood, 1969
Phenmetrazine	Rylander, 1969

[a] Pretreatment necessary.
[b] 7-Benzyl-1-ethyl-1,4 dihydro-4-oxo-1,8 naphthyridine-3 carboxylic acid.
[c] Intracerebral administration.

C. Drugs Inducing Stereotypy Following Direct Administration into the CNS

The induction of stereotyped behavior has also been examined following localized central administration of pharmacological agents. Dopamine itself does not cross the blood-brain barrier, and therefore would not be expected to be active following peripheral administration. Stereotyped behavior occurred after administration of dopamine into the caudate nucleus of cats (Cools and Van Rossum, 1970; Cools, 1971). Another report on the effects of intrastriatal administration of dopamine described rotational behavior (Ungerstedt, Butcher, Butcher, Andén, and Fuxe, 1969). Administration of DOPA or apomorphine into the striatum of rats elicited gnawing behavior; however, administration into the substantia nigra was ineffective (Ernst and Smelik, 1966). Cools (1971) demonstrated that amphetamine, apomorphine, and 3-methoxytyramine could elicit stereotyped behavior when injected into the caudate nucleus of cats. Costall, Naylor, and Olley (1972a) have extended the work of Ernst and Smelik in rats. Amphetamine introduced into the caudate-putamen or into the globus pallidus resulted in stereotyped behavior. When administered into the substantia nigra, only a very mild stereotypy occurred, similar to that resulting from saline administration into the same locus. Amphetamine administration into the cerebral cortex, hippocampus, or thalamus did not result in stereotyped behavior (Costall et al., 1972a).

D. Species Differences

Three types of species-dependent differences have been observed with the various CNS stimulants. One difference relates to the relative sensitivity of various species to a group of agents. For example, a rat will develop clear stereotyped behavior following less than 10 mg/kg of D-amphetamine, and a dog requires approximately 2 mg/kg (Randrup et al., 1963; Willner et al., 1970). With phenmetrazine, in excess of 75 mg/kg are required to induce this behavior in rats, yet only 1.25 mg/kg are necessary in the dog (Wallach et al., 1973). A similar phenomenon was observed with cocaine in the rat and the dog. Fog (1969) reported that 600 mg/kg of cocaine elicited stereotyped behavior in the rat, while Willner et al. (1970) found that only 1.5 mg/kg was necessary in the dog. Differences of this nature may reflect metabolic and/or permeability differences of the various compounds in different species. It is, however, somewhat disturbing to see differences of this magnitude.

Another type of interspecies difference relates to the response of the mouse to various CNS stimulants. In the mouse, amphetamine or apomorphine will elicit stereotyped behavior only when administered intravenously. Intraperitoneal or subcutaneous administration results in lethality without clear, reproducible stereotypy. On the other hand, methylphenidate in our hands elicited stereotypy following intraperitoneal or subcutaneous administration; however, intravenous administration is lethal (Pedersen and Christensen, 1972; Wallach, *unpublished*). Phenmetrazine does not elicit stereotyped behavior in the mouse following intraperitoneal dosing. This route specificity was not observed in rats or other species. These differences in the ability of CNS stimulants to elicit stereotypy in the mouse as compared to the rat may reflect differences in the metabolism of either the CNS stimulant itself or the central mechanisms involved in the induction of the stereotyped behavior. A third species difference has been noted. Stereotypy has been observed following 5-hydroxytryptophan in the rat, but no such observations were made in cats and dogs (Randrup and Munkvad, 1966b; Willner et al., 1970; Wallach, *unpublished*). This stereotypy may be the result of biogenic amine displacement or false transmitters if, indeed, it is related to the CNS stimulant-induced stereotypy.

III. MECHANISMS OF STEREOTYPY

A. CNS Stimulant Stereotypies

There are several means of explaining stereotypy, and a great deal of work has been done by various laboratories to define the mechanisms of action of various stimulants. One finding pertinent to the mechanism of action of the stimulant-induced stereotypies is their reliance upon endogenous biogenic amines.

Some of the earliest studies utilized pretreatments such as reserpine or alpha-methyl-p-tyrosine, while others utilized agents which altered receptor sensitivity or antagonized cholinergic and serotoninergic systems. In addition, attempts have been made to eradicate stereotyped behavior by specific lesions at various anatomical sites. Still other investigators attempted to induce stereotyped behavior by the anatomical implantation of various neurohumoral substances in specific brain regions.

1. *Biogenic Amine Depletion.* Shortly after the appearance of initial reports describing stereotyped behavior, reserpine was administered to rats in attempts to antagonize amphetamine-induced stereotyped behavior. Doses of reserpine up to 22.5 mg/kg failed to antagonize the amphetamine-induced behavior (Randrup et al., 1963; Randrup and Jonas, 1967; Scheel-Kruger, 1971), although Herman (1967) was able to detect some reduction in the duration of amphetamine-induced stereotypy. Our own laboratory has demonstrated that reserpine pretreatment antagonized amphetamine-induced stereotypy in cats (Wallach and Gershon, 1972).

There are other discrepancies involving the antagonism of stereotyped behavior induced by various CNS stimulants. Scheel-Kruger (1971) failed to demonstrate reserpine antagonism of phenmetrazine stereotypy in rats; however, in cats this antagonism occurred (Wallach and Gershon, 1972). Reserpine has been demonstrated to antagonize pipradrol-induced stereotypy in both mice and rats, and methylphenidate stereotypy in both rats and cats (Scheel-Kruger, 1971; Christensen and Pedersen, 1972; Wallach and Gershon, 1972). The stereotyped behavior induced by cocaine or pemoline magnesium hydroxide has also been antagonized in reserpinized cats (Wallach and Gershon, 1972).

Not only did reserpine fail to antagonize apomorphine-induced stereotyped behavior, but reserpine has been reported to potentiate the stereotyped behavior induced by apomorphine in rats (Rotrosen, Wallach, Angrist, and Gershon, 1972; Costall and Naylor, 1973). In mice, Fekete, Kurti, and Pribusz (1970) have indicated that reserpine reduced the number of holes gnawed in a paper-lined cage following apomorphine administration; however, they did not discuss the size of the holes produced. This hole-gnawing behavior may be related to the stereotyped behavior elicited in other species by apomorphine. Augmentation of apomorphine-induced stereotypy by reserpine has been observed in rats (Rotrosen et al., 1972; Costall and Naylor, 1973). Rotrosen et al., (1972) demonstrated that this phenomenon is not related either to endogenous catecholamines or to serotonin, because alpha-methyl-*p*-tyrosine, parachlorophenylalanine, or methysergide pretreatments failed to alter apomorphine stereotypy by alpha-methyl-*p*-tyrosine. However, Costall and Naylor (1973), using a longer pretreatment time, observed an augmentation of apomorphine stereotypy by alpha-methyl-*p*-tyrosine. The longer pretreatment time with alpha-methyl-*p*-tyrosine may lead to receptor supersensitivity. This suggests that serotonin and other catecholamines are most likely not involved in the potentiation of the apomorphine stereotypy by reserpine.

2. *Catecholamine Synthesis Inhibition.* Since reserpine's effects on amphetamine-induced stereotyped behavior are unclear, it was of interest to examine the effects of alpha-methyl-*p*-tyrosine pretreatment on stereotypy induced by various CNS stimulants. Weissman, Koe, and Tenen (1965) demonstrated that alpha-methyl-*p*-tyrosine was capable of antagonizing the amphetamine-induced gnawing syndrome in rodents. Similar results have also been obtained by others (Randrup and Munkvad, 1966*a;* Fog, 1967; D'Encarnacao, D'Encarnacao and Tapp, 1969; Ernst, 1969; Scheel-Krüger, 1971). Wallach and Gershon (1972) demonstrated that alpha-methyl-*p*-tyrosine did not antagonize methylphenidate or L-DOPA-induced stereotypy. This agrees with the failure of Scheel-Krüger (1971) to antagonize stereotyped behavior induced by pipradrol and methylphenidate by alpha-methyl-*p*-tyrosine pretreatment. However, Christensen and Pedersen (1972) demonstrated that alpha-methyl-*p*-tyrosine antagonized pipradrol-induced gnawing in mice.

The phenomenon of alpha-methyl-*p*-tyrosine antagonism of amphetamine-stereotyped behavior has been examined further. Either L-DOPA or dopamine, administered into the corpus striatum, reinstitute amphetamine-induced stereotyped behavior in animals previously treated with alpha-methyl-*p*-tyrosine (Randrup and Munkvad, 1966*a;* Fog, 1967; D'Encarnacao et al., 1969). Rats pretreated with alpha-methyl-*p*-tyrosine and dihydroxyphenylserine, when given amphetamine, do not display stereotypy. This suggests that dihydroxyphenylserine was unable to enter the CNS, or that upon entry it was metabolized to norepinephrine and was unavailable for release by amphetamine, or that amphetamine does not rely upon norepinephrine to elicit stereotypy (D'Encarnacao et al., 1969).

3. *Adrenergic Receptor Blockade.* Alpha-adrenergic receptor blocking agents were unable to antagonize amphetamine-induced stereotyped behavior in the rat. Phentolamine, phenoxybenzamine, and dihydroergotamine were ineffective. Similar results were obtained with two beta-adrenergic blocking agents, pronethalol and dichloroisoproterenol; however, propranolol, another beta antagonist, at high doses was observed to antagonize amphetamine-induced stereotyped behavior (Randrup et al., 1963; Herman,

1967; Del Rio and Fuentes, 1969). Very few studies utilizing receptor antagonists have attempted to disrupt stereotyped behavior induced by agents other than amphetamine. Phenoxybenzamine did not antagonize methylphenidate-induced gnawing in mice (Christensen and Pedersen, 1972) or apomorphine stereotypy and emesis in dogs (Nymark, 1972).

4. *Dopaminergic Receptor Blockade.* The dopaminergic receptor blocking agents have been widely examined as antagonists of amphetamine-induced stereotyped behavior. Certain neuroleptics are antagonists of amphetamine-induced behavior. Their activity often parallels their clinical efficacy. Among the agents reported to antagonize amphetamine-induced stereotypy in the rat are chlorpromazine, perphenazine, thioperazine, thioridazine, promazine, haloperidol, dehydrobenzperidol, triperidol, and pimozide (Randrup et al., 1963; Del Rio and Fuentes, 1969; Schiørring and Randrup, 1971). A similar list of antagonists has been developed for the methylphenidate-induced gnawing in the mouse (Pedersen and Christensen, 1972). Apomorphine-induced stereotyped behavior in the dog can be antagonized by many neuroleptics, including fluphenazine, flupenthixol, pimozide, haloperidol, perphenazine, clopenthixol, chlorprothixene, and thioridazine (Nymark, 1972; Rotrosen et al., 1972).

To recapitulate, many neuroleptics are capable of antagonizing stereotypy. Reserpine and alpha-methyl-*p*-tyrosine, agents that reduce the available catecholamine stores, also reduced stereotypy. Inhibitors of dopamine-beta-hydroxylase are ineffective as are both alpha and beta adrengeric receptor antagonists. These findings, when taken in conjunction with the available data, suggest that replacement of dopamine, after synthesis inhibition has occurred, restores the ability of amphetamine and pipradrol to cause stereotypy. Therefore, it would appear that dopamine is deeply involved in the induction of stereotypy.

5. *Induction of Stereotypy Following Lesions.* The effect of specific neuroanatomical lesions upon stereotyped behavior induced by some CNS stimulants has also been examined. Removal of the corpus striatum from rats prevented induction of stereotypy following either amphetamine or a monoamine oxidase inhibitor and DOPA (Fog, Randrup, and Pakkenburg, 1970). Bilateral lesion of the substantia nigra did not antagonize amphetamine stereotypy, but apomorphine stereotypy was antagonized (Costall et al., 1972*b*). Bilateral medial forebrain bundle lesions at the diencephalic level of the subthalamus did not antagonize apomorphine stereotypy and possibly enhanced amphetamine gnawing (Boissier, Etevenon, Piarroux, and Simon, 1971). Lesioning of the nucleus amygdala lateralis, a noradrenergic area, was capable of preventing apomorphine- or piribedil (ET495)-induced gnawing and biting behavior. However, sniffing and forelimb movements characteristically produced by these agents were not antagonized (Costall and Naylor, 1972*a*). These data suggest that the stereotyped behavior is a dopaminergic phenomenon requiring some noradrenergic input.

The primary area appears to be the corpus striatum. Recently, however, McKenzie (1972) has demonstrated that a lesion of the tuberculum olfactorium was capable of preventing stereotyped behavior following subcutaneous apomorphine. The lesioning of the corpus striatum did not reduce apomorphine stereotypy unless the tuberculum olfactorium was simultaneously destroyed.

6. *Directly and Indirectly Acting Dopaminergic Stimulants.* In order to examine further the role of dopamine release or dopamine receptor stimulation by CNS stimulants, we have examined several of these compounds in rats with unilateral 6-hydroxydopamine lesions (Ungerstedt, 1971a, b). This technique involves the use of 6-hydroxydopamine stereotaxically implanted into the nigrostriatal pathway. Unilateral lesions of these dopamine-containing neurons causes differential turning in a rotometer upon administration of agents stimulating the dopaminergic system. Agents blocking dopamine receptors, such a neuroleptics, do not elicit turning. Agents stimulating dopamine receptors directly, that is, without releasing endogenous dopamine stores, elicit a preferential turning contralateral to the lesion. Agents that release endogenous dopamine stores induce ipsilateral rotation. In addition to the data collected by Ungerstedt (1971a, b), which indicated that apomorphine induced contralateral rotation while amphetamine resulted in ipsilateral rotation, we have examined several other CNS stimulants. Methylphenidate, cocaine, pemoline magnesium, and phenmetrazine caused marked ipsilateral rotation (Fig. 2). Caffeine, on the other hand, caused a small amount of turning, also to the ipsilateral side.

In our hands, L-DOPA elicited contralateral rotation. Low doses of RO4-4602 [N(DL-seryl)-N'-(2,3,4-trihydroxybenzyl) hydrazine], an aromatic L-amino acid decarboxylase inhibitor, potentiated the contralateral DOPA-induced rotation, and high doses of RO4-4602 antagonized the DOPA effects. This agrees with similar results obtained by Ungerstedt (1971b). The effects of the RO4-4602 have been explained by Bartholini and Pletscher (1969), who demonstrated that the central decarboxylase is inhibited only by massive doses of RO4-4602. One can infer, therefore, that although the dopaminergic neurons have been destroyed, the L-DOPA is converted to dopamine, which can then act on the supersensitive receptors. This decarboxylation must occur centrally, and infers that the contralateral turning is a result of supersensitive dopaminergic receptor stimulation.

7. *Cholinergic Influences.* The role that noncatecholaminergic neurohumors play in the induction of stereotyped behavior has also been examined. Unfortunately, the results of some of these studies are not completely in agreement. Amphetamine-induced stereotyped behavior can be antagonized by cholinergic agents such as physostigmine, oxotremorine, and arecoline (Arnfred and Randrup, 1968; Costall et al., 1972a); however, Del Rio and Fuentes (1969) failed to observe this antagonism with pilocarpine or

252 STEREOTYPY: SIMILARITIES AND DIFFERENCES

CONTRALATERAL TURNS	TREATMENT mg/kg	IPSILATERAL TURNS
1000 100 10		10 100 1000
	d-AMPHETAMINE	
	0.3 (6)	▯
	1.0 (6)	▭
	3.0 (5)	▭
	10.0 (6)	▭
	APOMORPHINE	
▭ (6)	0.3	
▭ (12)	1.0	
	COCAINE	
	10.0 (5)	▯
	30.0 (6)	▭
	L-DOPA	
▯ (6)	50.0	
▭ (3)	100.0	
▭ (3)	300.0	
▭ (2)	500.0	
	METHYLPHENIDATE	
	3.0 (6)	▭
	10.0 (6)	▭
	PEMOLINE	
	50.0 (3)	▭≈
	100.0 (2)	▭≈
	PHENMETRAZINE	
	10.0 (6)	▭
	30.0 (6)	▭≈

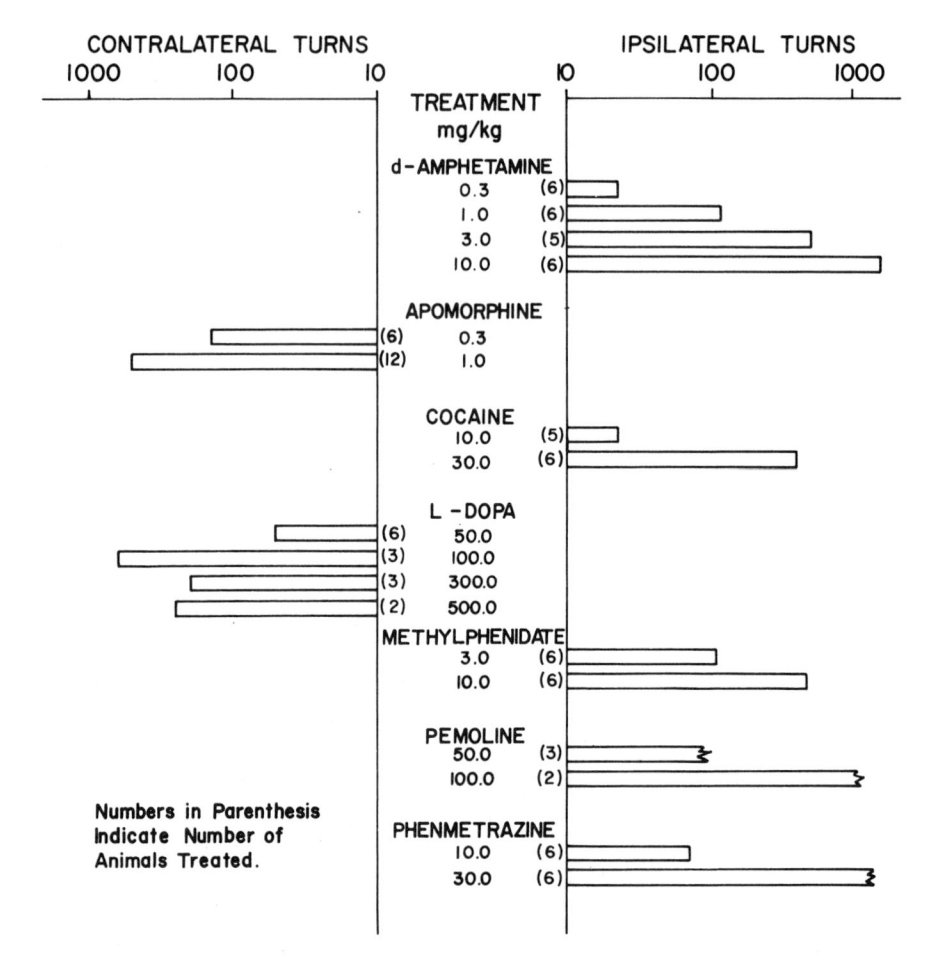

Numbers in Parenthesis Indicate Number of Animals Treated.

FIG. 2. Mean total rotations of rats with unilateral 6-hydroxydopamine lesions of nigrostriatal pathway following various drugs. Rotations were recorded for a maximum of 3½ hr. Broken bars indicate continuing effects beyond the arbitrary termination time.

oxotremorine pretreatments. Anticholinergic agents, such as benzhexol, orphenadrine, benztropine, scopolamine, benzactyzine, hyoscyamine, and atropine, have been reported to potentiate amphetamine stereotypy (Arnfred and Randrup, 1968; Naylor and Costall, 1971; Costall et al., 1972a). Apomorphine stereotypy also has been potentiated by anticholinergic agents and antagonized by cholinergic agents in mice (Scheel-Kruger, 1970). In guinea pigs and dogs, no mention is made of potentiation of apomorphine by atropine or scopolamine (Frommel, Ledebur, and Seydoux, 1965; Nymark, 1972). A similar situation exists for atropine and methylphenidate in mice (Pedersen and Christensen, 1972).

Costall and Naylor (1972) have extensively studied the effects of locally applied anticholinergic agents on peripherally administered amphetamine-induced stereotyped behavior. They suggested that the potentiation or antagonism of stereotyped behavior by anticholinergic agents is related to the individual nature of the agent. Orphenadrine, when placed into the striatum or pallidum, immediately potentiated amphetamine stereotypy. However, atropine and phenglutarimide, when given locally, first decreased and then increased amphetamine stereotypy.

The complex relationships between the cholinergic and dopaminergic nervous systems do not appear to be totally capable of balancing each other. Under normal circumstances, these areas are in balance; however, severe imbalances cannot be corrected by the other system. A clinical reflection of this is the weak antiparkinsonian activities of anticholinergic agents.

8. *Serotoninergic Influences.* Another major biogenic amine that may influence stereotyped behavior is serotonin. Some serotonin-related compounds have been reported to induce stereotyped behavior: alpha-methyl-tryptamine (Randrup and Munkvad, 1964), LSD-25 in extremely high dosage (Fog, 1969), tryptamine with a monoamine oxidase inhibitor, and 5-hydroxytryptophan with pretreatments with both a monoamine oxidase inhibitor and a serotonin receptor antagonist (Randrup and Munkvad, 1966*b*). Unfortunately, these scattered observations of serotonin-related stereotyped behavior have been neither fully examined nor fully documented. Agents such as brom-LSD (BOL) have been reported to antagonize amphetamine-induced stereotyped behavior; however, other serotonin antagonists, such as methysergide and cyproheptadine, do not antagonize stereotyped behavior elicited by amphetamine or alpha-methyltryptamine (Randrup and Munkvad, 1964). This suggests that the alpha-methyltryptamine-induced stereotyped behavior might be related to an amphetamine-like action and that the antagonism by BOL may be nonspecific.

Amphetamine, when administered to young chicks, induced a behavioral syndrome characterized by wing dropping, postural changes, twittering, and aggressive behavior. Pretreatment with parachlorophenylalanine, a serotonin antagonist, antagonized this behavior (Schrold and Squires, 1971). This suggests that the amphetamine-induced response in the chick might be serotoninergic. Similar studies have not been reported in other species.

9. *Histaminic Influences.* Recently, histamine has been suggested as possibly being implicated in stereotyped behavior. Antihistamines have been observed to potentiate stereotyped behavior induced in rats by low doses of amphetamine (Naylor and Costall, 1971). The nature of this potentiation has not been examined in any detail. One possible explanation is that many antihistamines also share anticholinergic activity. The anticholinergic agents have been demonstrated to potentiate amphetamine-induced stereotyped behavior. It is also possible that this potentiation may be related to an inhibition of the metabolism of the amphetamine. The tricyclic antidepressant desmethylimipramine, which possesses anticholiner-

gic properties, also has been demonstrated to increased brain amphetamine levels, most likely by inhibition of the metabolism of amphetamine (Consolo, Dolfini, Garattini, and Falzelli, 1967; Dolfini, Tansella, Falzelli, and Garattini, 1969; Dingell and Bass, 1969). Thus, anticholinergic activity or metabolic competition also may explain antihistaminic augmentation of amphetamine stereotypy. Studies with different stereotypy-inducing agents could solve this question.

10. *Cyclic Adenosine Monophosphate.* The benzodiazepines—chlordiazepoxide, diazepam, nitrazepam, oxazepam, desmethyldiazepam, and methyloxazepam—have been demonstrated to potentiate methamphetamine-induced stereotyped behavior in the rat (Babbini, Montanaro, Strocchi, and Gaiardi, 1971). This potentiation may be related to an inhibition of methamphetamine metabolism. It is also possible to speculate that another mechanism may be interacting here. If a dopaminergic mechanism is involved in the induction of stereotyped behavior, and there have been recent suggestions that the dopaminergic receptor may be related to cyclic adenosine monophosphate (cyclic AMP) (Kebabian and Greengard, 1971; McAfee, Schorderet, and Greengard, 1971; McAfee and Greengard, 1972), then it might be expected that agents altering the metabolism of cyclic AMP would either potentiate or inhibit amphetamine-induced stereotyped behavior. Indeed, Beer, Chasin, Clody, Vogel, and Horovitz (1972) have demonstrated that the benzodiazepines could inhibit phosphodiesterase, an enzyme which inactivates cyclic AMP. Thus, with amphetamine probably releasing endogenous dopamine, which may act via cyclic AMP, the presence of the benzodiazepines inhibiting phosphodiesterase could conceivably result in the augmentation observed. Further work is necessary to prove this speculation convincingly.

E. Non-CNS Stimulant Stereotypy-Inducing Agents

The agents discussed which elicit stereotyped behavior in most cases would be expected either to release endogenous dopamine or to stimulate dopaminergic receptors directly; however, some pharmacological agents can elicit stereotyped behavior and do not appear to belong to either of these two groups of agents. The induction of stereotyped behavior by certain tryptamine derivatives, such as alpha-methyltryptamine, tryptamine, and 5-hydroxytryptophan, as well as observations of stereotypy following LSD and chronic morphine administration, leave unanswered questions (Randrup and Munkvad, 1964, 1966*b*; Fog, 1969, 1970).

Chronic administration of morphine to rats results in stereotyped behavior. Reports suggest that this behavior can be antagonized by pretreatment of the animal with alpha-methyl-*p*-tyrosine, FLA-63, a dopamine-β-hydroxylase inhibitor, or phenoxybenzamine, an alpha adrenergic receptor antagonist. This would imply that the morphine-induced effects are catecholaminergic and, furthermore, probably related to alpha noradrenergic receptors. Spiramide, a dopaminergic antagonist, was only weakly antagonistic. Following FLA-63 antagonism in morphine-tolerant rats, intraventricular norepinephrine administration results in the return of stereotypy. This suggests that morphine-induced stereotyped behavior is mediated noradrenergically (Ayhan and Randrup, 1972). If rats were pretreated with morphine and subsequently administered a stereotypy-inducing dose of amphetamine, catalepsy and not stereotyped behavior occurred. In chronically morphinized rats, amphetamine also failed to elicit stereotyped behavior. In addition, the stereotyped behavior elicited by chronic morphine administration was antagonized by narcotic antagonists (Fog, 1970).

IV. DISCUSSION

This chapter set out to evaluate critically the experimental data and to attempt to correlate it to CNS stimulant-induced psychoses in man and to schizophrenia, paranoid type. Having reviewed the available data on the induction and antagonism of stereotyped behavior, one predominant characteristic is evident: many of the agents have a dopaminergic link. This link may be modified by noradrenergic input, or influenced by anticholinergic agents and benzodiazepines. Almost all of the agents that induce stereotypy can be antagonized by neuroleptics. Currently, most neuroleptics are believed to have a common mode of action, that is, by dopamine receptor antagonism. There are only a few agents which, based on our current knowledge, cannot be fitted into this hypothesis. The most important exception is morphine. The tryptaminergic agents may be influencing dopamine metabolism as false transmitters, however, this is purely speculative.

The parallelism between agents that elicit stereotypy in animals and in man associated with stimulant-induced psychoses is quite interesting. The psychoses, which sometimes are mistakenly identified as schizophrenia, paranoid type, may involve a dopaminergic overstimulation. At present it appears that the stimulant-induced stereotypy is perhaps the best, simple animal model for a human psychiatric disorder. The occurrence of stereotypies associated with or without the psychoses in man suggest the possibility of a separation of two phenomena. Perhaps it would be worthwhile to attempt this discrimination.

Within the data accumulated by many laboratories on stimulant-induced stereotypy, few discrepancies have been observed. Many of these discrepancies can be attributed to different parameters used for the observation of the stereotypy. The ability of only some agents to be antagonized by alpha-methyltyrosine may also be related to the mechanism of action of the behavior-inducing drug; that is, certain stereotypy-inducing agents may release from one catecholamine pool, while other agents release from another pool (Scheel-Kruger, 1971; Wallach and Gershon, 1972). Studies utilizing the unilateral 6-hydroxydopamine lesion of the nigrostriatal pathway infer that most stereotypy-inducing agents are dependent upon the endogenous dopamine. Apomorphine and DOPA elicit contralateral rotation suggestive of direct dopaminergic stimulation following peripheral administration.

Another difference is the relative potency of various agents in different species. The close similarity between the dose of amphetamine that induces stereotypy in rats, cats, and dogs and the psychoses in man (all between 1 and 10 mg/kg) compared to the marked discrepancy between rats and dogs with phenmetrazine (75 versus 1.2 mg/kg, respectively) or cocaine (600 versus 1.0 mg/kg, respectively) raises some questions. Although these differences are possibly related to uptake and/or metabolism of the stimulant, the large magnitude of difference is curious.

The apparent relationship of dopaminergic mechanisms, CNS stimulant-induced stereotypy in animals, drug-induced psychoses, and schizophrenia, paranoid type, in man, compels further investigation. The possibility that this is the final mechanism of action is probably quite low, inasmuch as the knowledge of what a receptor is and how a molecule, by interacting with a receptor, elicits an effect which encompasses a whole animal is yet to be determined. Although there are some differences among the various agents, the overall general pattern appears to be quite consistent. It is likely that the small differences between compounds are of much less importance than the general pattern. Perhaps at the next meeting where these problems are discussed, the interrelationships of the dopaminergic receptor with such substances as cyclic AMP might be a topic for discussion.

ACKNOWLEDGMENTS

The author would like to thank Mr. Mark Dawber, Mr. Richard McGuire, Mrs. Charlotte Rogers, Miss V. Joy Simmons, and Miss Robin Thayer for their skillful technical assistance. Grateful appreciation is also acknowledged to Dr. A. P. Roszkowski for his review and comments on this manuscript. Some earlier parts of the author's research contributions were accomplished in the laboratories of Dr. Samuel Gershon at New York University Medical Center, Department of Psychiatry, under a grant from the National Institute of Mental Health, MH-12383.

REFERENCES

Aceto, M. D., Harris, L. S., Lesher, G. Y., Pearl, J., and Brown, T. G., Jr. (1967): Pharmacologic studies with 7-benzyl-1-ethyl-1,4-dihydro-4-oxo-1,8 naphthyridine-3-carboxylic acid. *Journal of Pharmacology and Experimental Therapeutics,* 158:286–293.

Angrist, B. M., and Gershon, S. (1970): The phenomenology of experimentally induced amphetamine psychosis. *Biological Psychiatry,* 2:95–107.

Arnfred, T., and Randrup, A. (1968): Cholinergic mechanism in brain inhibiting amphetamine induced stereotyped behaviour. *Acta Pharmacologica et Toxicologica* (Kbh.), 26:384–394.

Ayhan, I. H., and Randrup, A. (1972): Role of brain noradrenaline in morphine-induced stereotyped behaviour. *Psychopharmacologia*, 27:203-212.
Babbini, M., Montanaro, N., Strocchi, P., and Gaiardi, M. (1971): Enhancement of amphetamine-induced stereotyped behavior by benzodiazepines. *European Journal of Pharmacology*, 13:330-340.
Bartholini, G., and Pletscher, A. (1969): Effect of various decarboxylase inhibitors on the cerebral metabolism of dihydroxyphenylalanine. *Journal of Pharmacy and Pharmacology*, 21:323-324.
Beer, B., Chasin, M., Clody, D. E., Vogel, J. R., and Horovitz, Z. P. (1972): Cyclic adenosine monophosphate phosphodiesterase in brain: Effect on anxiety. *Science*, 176:428-430.
Bell, D. S. (1973): The experimental reproduction of amphetamine psychosis. *Archives of General Psychiatry*, 29:35-40.
Boissier, J. R., Etevenon, P., Piarroux, M. C., and Simon, P. (1971): Effects of apomorphine and amphetamine in rats with a permanent catalepsy induced by diencephalic lesion. *Research Communications in Chemistry, Pathology, Pharmacology*, 2:829-836.
Christensen, A. V., and Pedersen, V. (1972): The effect of pipradrol in mice and its antagonism by neuroleptic drugs. *Acta Pharmacologica et Toxicologica* (Suppl.) (Kbh.). 31:8.
Connell, P. H. (1958): *Amphetamine Psychosis*. The Institute of Psychiatry, London.
Consolo, S., Dolfini, E., Garattini, S., and Valzelli, L., (1967): Desipramine and amphetamine metabolism. *Journal of Pharmacy and Pharmacology*, 19:253-256.
Cools, A. R. (1971): The function of dopamine and its antagonism in the caudate nucleus of cats in relation to the stereotyped behaviour. *Archives Internationales de Pharmacodynamie et de Thérapie*, 194:259-269.
Cools, A. R., and Van Rossum, J. M. (1970): Caudal dopamine and stereotype behaviour in cats. *Archives Internationales de Pharmacodynamie et de Thérapie*, 187:163-173.
Costall, B., and Naylor, R. J. (1972a): Possible involvement of a noradrenergic area of the amygdala with stereotyped behaviour. *Life Sciences*, 11:1135-1146.
Costall, B., and Naylor, R. J. (1972b): Modification of amphetamine effects by intracerebrally administered anticholinergic agents. *Life Sciences*, 11:239-253.
Costall, B., and Naylor, R. J. (1973): On the mode of action of apomorphine. *European Journal of Pharmacology*, 21:350-361.
Costall, B., Naylor, R. J., and Olley, J. E. (1972a): Stereotypic and anticataleptic activities of amphetamine after intracerebral injections. *European Journal of Pharmacology*, 18:83-94.
Costall, B., Naylor, R. J., and Olley, J. E. (1972b): The substantia nigra and stereotyped behaviour. *European Journal of Pharmacology*, 18:95-106.
Del Rio, J., and Fuentes, J. A. (1969): Further studies on the antagonism of stereotyped behaviour induced by amphetamine. *European Journal of Pharmacology*, 8:73-78.
D'Encarnacao, P. S., D'Encarnacao, P., and Tapp, J. T. (1969): Potentiation of amphetamine induced psychomotor activity by diethyldithiocarbamate. *Archives Internationales de Pharmacodynamie et de Thérapie*, 182:186-189.
Dingell, J. V., and Bass, A. D. (1969): Inhibition of the hepatic metabolism of amphetamine by desipramine. *Biochemical Pharmacology*, 18:1535-1538.
Dolfini, E., Tansella, M., Valzelli, L., and Garattini, S. (1969): Further studies on the interaction between desipramine and amphetamine. *European Journal of Pharmacology*, 5:185-190.
Ellinwood, E. H., Jr. (1967): Amphetamine psychosis: I. Description of the individuals and process. *Journal of Nervous and Mental Diseases*, 144:273-283.
Ellinwood, E. H., Jr. (1969): Amphetamine Psychosis: A multidimensional process. *Seminars in Psychiatry*, 1:208-226.
Ellinwood, E. H., Jr. (1971a): Effect of chronic methamphetamine intoxication in rhesus monkeys. *Biological Psychiatry*, 3:25-32.
Ellinwood, E. H., Jr. (1971b): "Accidental conditioning" with chronic methamphetamine intoxication: Implications for a theory of drug habituation. *Psychopharmacologia*, 21:131-138.
Ernest, A. M. (1969): The role of biogenic amines in the extrapyramidal system. *Acta Physiologica et Pharmacologica Neerlandica*, 15:141-154.
Ernest, A. M., and Smelik, P. G. (1966): Site of action of dopamine and apomorphine on compulsive gnawing behaviour in rats. *Experientia*, 22: 837-838.
Fekete, M., Kurti, A. M., and Pribusz, I. (1970): On the dopaminergic nature of the gnawing

compulsion induced by apomorphine in mice. *Journal of Pharmacy and Pharmacology,* 22:377–379.

Fog, R. L. (1967): Role of the corpus striatum in typical behavioral effects in rats produced by both amphetamine and neuroleptic drugs. *Acta Pharmacologica et Toxicologica* (Suppl.) (Kbh.), 25:59.

Fog, R. (1969): Stereotyped and non-stereotyped behaviour in rats induced by various stimulant drugs. *Psychopharmacologia,* 14:299–304.

Fog, R. (1970): Behavioural effects in rats of morphine and amphetamine and of a combination of the two drugs. *Psychopharmacologia,* 16:305–312.

Fog, R., Randrup, A., and Pakkenberg, H. (1970): Lesions in corpus striatum and cortex of rat brains and the effect on pharmacologically induced stereotyped, aggressive and cataleptic behaviour. *Psychopharmacologia,* 18:346–356.

Frommel, E., Ledebur, I. V., and Seydoux, J. (1965): Is apomorphine's vomiting action in man and dog equivalent to its' chewing effect in guinea pigs? *Archives Internationales de Pharmacodynamie et de Thérapie,* 154:227–230.

Griffith, J. D., Cavanaugh, J. H., and Oates, J. A. (1970): Psychosis induced by the administration of D-amphetamine to human volunteers. In: *Psychotomimetic Drugs,* edited by D. H. Efron. Raven Press, New York.

Griffith, J. D., Cavanaugh, J., Held, J., and Oates, J. A. (1972): Dextroamphetamine: Evaluation of psychotomimetic properties in man. *Archives of General Psychiatry,* 26:97–100.

Hauschild, F. (1939): Zur Pharmakologie des 1-phenyl-2 methylamino propans (Pervitin). *Naunyn-Schmiedeberg's Archives für Pharmakologie,* 191:465–481.

Herman, Z. S. (1967): Influence of some psychotropic and adrenergic blocking agents upon amphetamine stereotyped behaviour in white rats. *Psychopharmacologia,* 11:136–142.

Herman, M., and Nagler, S. H. (1954): Psychoses due to amphetamine. *Journal of Nervous and Mental Diseases,* 120:268–272.

Kebabian, J. W., and Greengard, P. (1971): Dopamine-sensitive adenyl cyclase: Possible role in synaptic transmission. *Science,* 174:1346–1349.

Kjellberg, B., and Randrup, A. (1972): Stereotypy with selective stimulation of certain items of behaviour observed in amphetamine-treated monkeys (cercopithecus). *Pharmako-Psychiatrie Neuro-Psychopharmakologie,* 5:1–12.

Langwinski, R. (1970): Stereotyped behaviour induced by (−) ephedrine in rats. *Journal of Pharmacy and Pharmacology,* 22:874.

Mayer, O., and Eybl, V. (1971): The effect of diethyldithiocarbamate on amphetamine-induced behaviour in rats. *Journal of Pharmacy and Pharmacology,* 23:894–896.

McAfee, D. A., and Greengard, P. (1972): Adenosine 3′,5′-monophosphate: Electrophysiological evidence for a role in synaptic transmission. *Science,* 178:310–312.

McAfee, D. A., Schorderet, M., and Greengard, P. (1971): Adenosine 3′,5′-monophosphate in nervous tissue: Increase associated with synaptic transmission. *Science,* 171:1156–1158.

McKenzie, G. M. (1972): Role of the tuberculum olfactorium in stereotyped behaviour induced by apomorphine in the rat. *Psychopharmacologia,* 23:212–219.

Naylor, R. J., and Costall, B. (1971): The relationship between the inhibition of dopamine uptake and the enhancement of amphetamine stereotype. *Life Sciences,* 10:909–915.

Nymark, M. (1972): Influence of different drugs on apomorphine-induced stereotypy in the dog. *Acta Pharmacologica et Toxicologica* (Suppl.) (Kbh.), 31:9.

Pedersen, V., and Christensen, A. V. (1972): Antagonism of methylphenidate-induced stereotyped gnawing in mice. *Acta Pharmacologica et Toxicologica* (Kbh.), 31:488–496.

Randrup, A., and Jonas, W. (1967): Brain dopamine and the amphetamine-reserpine interaction. *Journal of Pharmacy and Pharmacology,* 19:483–484.

Randrup, A., and Munkvad, I. (1964): On the relation of tryptaminic and serotonergic mechanisms to amphetamine induced abnormal behaviour. *Acta Pharmacologica et Toxicologica* (Kbh.), 21:272–282.

Randrup, A., and Munkvad, I. (1965a): Special antagonism of amphetamine-induced abnormal behaviour: Inhibition of stereotyped activity with increase of some normal activities. *Psychopharmacologia,* 7:416–422.

Randrup, A., and Munkvad, I. (1965b): Pharmacological and biochemical investigations of amphetamine-induced abnormal behaviour. In: *Neuro-psychopharmacology,* Vol. 4, edited by D. Bente and P. B. Bradley. Elsevier Publishing Co., Amsterdam.

Randrup, A., and Munkvad, I. (1966a): Role of catecholamines in the amphetamine excitatory response. *Nature*, 211:540.
Randrup, A., and Munkvad, I. (1966b): DOPA and other naturally occurring substances as causes of stereotypy and rage in rats. *Acta Psychiatrica Scandinavica* (Suppl.), 42:193–199.
Randrup, A., and Munkvad, I. (1967): Stereotyped activities produced by amphetamine in several animal species and man. *Psychopharmacologia*, 11:300–310.
Randrup, A., Munkvad, I., and Udsen, P. (1963): Adrenergic mechanisms and amphetamine induced abnormal behaviour. *Acta Pharmacologica et Toxicologica* (Kbh.), 20:145–157.
Randrup, A., and Scheel-Kruger, J. (1966): Diethyldithiocarbamate and amphetamine stereotyped behaviour. *Journal of Pharmacy and Pharmacology*, 18:752.
Redlich, F. C., and Freedman, D. X. (1966): *The Theory and Practice of Psychiatry*. Basic Books, New York.
Rotrosen, J., Wallach, M. B., Angrist, B., and Gershon, S. (1972): Antagonism of apomorphine-induced stereotypy and emesis in dogs by thioridazine, haloperidol and pimozide. *Psychopharmacologia*, 26:185–194.
Rylander, G. (1969): Clinical and medico-criminological aspects of addiction to central stimulating drugs. In: *Abuse of Central Stimulants*, edited by F. Sjoqvist and M. Tottie. Almqvist and Wiksell, Stockholm.
Scheel-Kruger, J. (1970): Central effects of anticholinergic drugs measured by the apomorphine gnawing test in mice. *Acta Pharmacologica et Toxicologica* (Kbh.), 28:1–16.
Scheel-Kruger, J. (1971): Comparative studies of various amphetamine analogues demonstrating different interactions with the metabolism of the catecholamines in the brain. *European Journal of Pharmacology*, 14:47–59.
Scheel-Kruger, J. (1972): Behavioural and biochemical comparison of amphetamine derivatives, cocaine, benztropine and tricyclic antidepressant drugs. *European Journal of Pharmacology*, 18:63–73.
Schiørring, E., and Randrup, A. (1971): Social isolation and changes in the formation of groups induced by amphetamine in an open-field test with rats. *Pharmako-Psychiatrie Neuro-Psychopharmakologie*, 4:2–12.
Schrold, J., and Squires, R. F. (1971): Behavioural effects of D-amphetamine in young chicks treated with p-Cl-phenylalanine. *Psychopharmacologia*, 20:85–90.
Srimal, R. C., and Dhawan, B. N. (1970): An analysis of methylphenidate induced gnawing in guinea pigs. *Psychopharmacologia*, 18:99–107.
Ungerstedt, U. (1971a): Striatal dopamine release after amphetamine or nerve degeneration revealed by rotational behavior. *Acta Physiologica Scandinavica*, Supplement 367:49–68.
Ungerstedt, U. (1971b): Postsynaptic supersensitivity after 6-hydroxydopamine induced degeneration of the nigro-striatal dopamine system. *Acta Physiologica Scandinavica*, Supplement 367:69–93.
Ungerstedt, U., Butcher, L. L., Butcher, S. G., Andén, N. E., and Fuxe, K. (1969): Direct chemical stimulation of dopaminergic mechanisms in the neostriatum of the rat. *Brain Research*, 14:461–471.
Wallach, M. B., Friedman, E., and Gershon, S. (1972): DOM (2,5 dimethoxy-4-methylamphetamine): A neuropharmacological examination. *Journal of Pharmacology and Experimental Therapeutics*, 182:145–154.
Wallach, M. B., and Gershon S. (1971): A neuropsychopharmacological comparison of D-amphetamine, L-DOPA and cocaine. *Neuropharmacology*, 10:743–752.
Wallach, M. B., and Gershon, S. (1972): The induction and antagonism of central nervous system stimulant-induced stereotyped behavior in the cat. *European Journal of Pharmacology*, 18:22–26.
Wallach, M. B., Rotrosen, J., and Gershon, S. (1973): A neuropsychopharmacological study of phenmetrazine in several animal species. *Neuropharmacology*, 12:541–548.
Wallach, M. B., Winters, W. D., Mandell, A. J., and Spooner, C. E. (1969a): A correlation of EEG, reticular multiple unit activity, and gross behavior following various antidepressant agents in the cat. IV. *Electroencephalography and Clinical Neurophysiology*, 27:563–573.
Wallach, M. B., Winters, W. D., Mandell, A. J., and Spooner, C. E. (1969b): Effect of antidepressant drugs on wakefulness and sleep in the cat. *Electroencephalography and Clinical Neurophysiology*, 27:574–580.
Weissman, A., Koe, B. K., and Tenen, S. S. (1965): Antiamphetamine effects following inhibi-

tion of tyrosine hydroxylase. *Journal of Pharmacology and Experimental Therapeutics,* 151:339–352.

Willner, J. H., Samach, M., Angrist, B. M., Wallach, M. B., and Gershon, S. (1970): Drug induced stereotyped behavior and its antagonism in dogs. *Communications in Behavioral Biology,* Part A, 5:135–142.

Winters, W. D., Mori, K., Wallach, M. B., Marcus, R. J., and Spooner, C. E. (1969): Reticular multiple unit activity during a progression of states induced by CNS excitants. III. *Electroencephalography and Clinical Neurophysiology,* 27:514–522.

Young, D., and Scoville, W. B. (1938): Paranoid psychosis in narcolepsy and the possible danger of benzedrine treatment. *Medical Clinics of North America,* 22:637–646.

Neuropsychopharmacology of Monoamines and Their Regulatory Enzymes, edited by E. Usdin.
Raven Press, New York © 1974.

Central Dopamine Function in Affective Illness: Evidence from Precursors, Enzyme Inhibitors, and Studies of Central Dopamine Turnover

Frederick K. Goodwin and Robert L. Sack

Section on Psychiatry, Laboratory of Clinical Science, National Institute of Mental Health, Bethesda, Maryland

I. INTRODUCTION

This chapter will review some of the studies carried out at the NIMH Clinical Center that relate to the role of central dopamine (DA) function in affective illness. Interest in the possibility that mood disorders may involve abnormalities in central catecholamine neurotransmitter systems began with the discoveries that mood-altering drugs in man have potent effects on central amine metabolism in animals. Thus, reserpine, a depletor of central catecholamines, was associated with the onset of depression in some individuals, while monoamine oxidase inhibitors, which increase brain levels of catecholamines in animals, were reported to have some antidepressant activity. Further, amphetamine, a stimulant in man, was found to increase catecholamine function at the synapse through several mechanisms. Finally, the tricyclic antidepressants were shown to block the presynaptic reuptake of catecholamines, thereby increasing their availability at the receptor site; conversely, lithium, an effective antimanic agent, was found to alter catecholamines in ways that should result in decreases in functionally active transmitter at the receptor. These clinical-pharmacological correlations led to the norepinephrine (NE) hypothesis of affective disorders (Bunney and Davis, 1965; Schildkraut, 1965), in which it was proposed that depression is associated with a functional deficit of NE at critical synapses in the central nervous system, and conversely, that mania is associated with a functional excess of NE.

Although interest has predominantly focused on the role of NE in affective illness, most of the clinical-pharmacological correlations reviewed above apply as well to DA (Carlsson, Rosengren, Bertler, and Nilsson, 1957; Carlsson, Lindqvist, and Magnusson, 1960) or to other noncatecholamines such as serotonin (Coppen, 1967; Glassman, 1969; Lapin and Oxenkrug, 1969). Consequently we have utilized a number of experimental clinical-pharmacological approaches using drugs which have more specific

```
TYROSINE
   │
   ├──── [AMPT]
   ▼
 DOPA ◄──── [L-DOPA]
   │
   ▼
DOPAMINE
   │
   ├──── [FUSARIC ACID]
   ▼
NOREPINEPHRINE
```

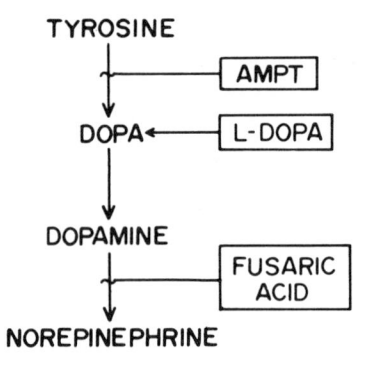

FIG. 1. Schematic illustration of the synthesis of norepinephrine from tyrosine, illustrating the points of pharmacological intervention discussed in the text. (1) AMPT inhibits tyrosine hydroxylase; (2) L-DOPA, an amino acid precusor, increases the synthesis of catecholamines when administered exogenously; (3) fusaric acid inhibits dopamine-β-hydroxylase (DBH).

effects on catecholamine metabolism as well as some differential effects on DA and NE.

Figure 1 illustrates schematically the synthetic pathway for DA and NE, indicating three clinically usable drugs which have relatively specific effects on catecholamine synthesis. Thus, the synthesis of DA [and under some conditions NE (Seiden and Petersen, 1968)] can be increased by the administration of its amino acid precursor, L-DOPA (Iversen, 1967; Butcher and Engel, 1969; Everett and Borcherding, 1970; Bartholini, Constantinidis, Tissot, and Pletscher, 1971). The synthesis of both DA and NE is markedly inhibited by alpha-methyl-para-tyrosine (AMPT), a specific inhibitor of tyrosine hydroxylase, the rate-limiting step in the synthesis of these two catecholamines (Spector, Sjoerdsma, and Udenfriend, 1965). Dopamine-beta-hydroxylase (DBH), the enzyme which converts DA to NE, is localized in noradrenergic neurons (reviewed in Axelrod, 1972); fusaric acid has been found to be a relatively specific and potent inhibitor of DBH (Nagatsu, Hidaka Kuzuya, Takeya, Umezawa, Takeuchi, and Suda, 1970).

In addition to employing drugs that have specific influences on catecholamine synthesis, we have concentrated on developing techniques for the direct measurement of central amine function in man. Of particular relevance to this presentation is the measurement of homovanillic acid (HVA), the major metabolite of DA, in the cerebrospinal fluid (CSF). By inhibiting the transport of HVA out of the CSF by high doses of probenecid, this metabolite accumulates at a rate which has been shown to be proportional to the turnover rate for DA in the brain (Neff, Tozer, and Brodie, 1967; Goodwin, Post, Dunner, Gordon, and Bunney, 1973).

The data to be reviewed focus on clinical-biochemical correlations obtained from manic-depressive patients undergoing treatment with three experimental drugs (L-DOPA, AMPT, and fusaric acid) that affect brain catecholamine synthesis. Results of these pharmacological studies will be integrated with observations on clinical and biological differences in subgroups of patients with affective disorders. We will suggest that differences in the behavioral response to experimental drugs which alter brain catecholamines (particularly brain DA) are determined not only by the biochemical effect of the drug but also by psychobiological differences between subgroups. Further, in relation to the hypothesized role of central amines in schizophrenia as well as in affective illness, our clinical observations lead us to suggest that a disturbance in a given central amine system may not correlate as well with a specific *diagnosis* as it might with specific *symptom* profiles which themselves overlap diagnostic categories.

II. CLINICAL METHODOLOGY

Prior to admission to the NIMH clinical research unit, the diagnosis of primary affective disorder was established by the independent judgment of two psychiatrists and a social worker, utilizing the criteria of Feighner, Robins, Guze, Woodruff, Winokur, and Munoz (1972); the diagnosis was reconfirmed during hospitalization by supplemental information from relatives, previous physicians, and hospital reports, and by psychological testing. For the depressed patients, symptoms ranged from moderately severe to severe, and included psychomotor retardation or agitation, anorexia, weight loss, sleep disturbance, and depressive thought content often of psychotic proportions, including feelings of hopelessness, worthlessness, preoccupation with guilt, and suicidal ideation. The manic patients manifested to varying degrees the clinical features of motor and verbal hyperactivity, with pressure of speech and flight of ideas, provocative intrusiveness in the interpersonal affairs of others, skillful manipulativeness, grandiosity, pseudo-euphoria or mood lability with bursts of depressive affect, poor judgment with lack of insight, inappropriate behavior often with sexual overtones, insomnia, and for some, inappropriate anger, paranoid ideation, and episodes of violent and destructive behavior (Beigel and Murphy, 1971; Goodwin, Murphy, Dunner, and Bunney, 1972; Kotin and Goodwin, 1972).

Patients with histories suggestive of preexisting psychopathology other than affective illness were excluded; we were particularly careful to exclude those with prior histories involving the core symptoms of schizophrenia, especially Schneider's "first rank" symptoms (Schneider, 1959). In relation to the question of possible overlap between mania and schizophrenia, we have previously noted in a longitudinal study of patients with histories of bipolar affective illness a subgroup who experience a severe psychotic stage of the manic episode ("Stage III") dominated by hyperactive bizarre behavior, hallucinations, paranoia, and extreme dysphoria (Carlson and Goodwin, 1973) (see Fig. 2). Although, as we have noted, these patients can be

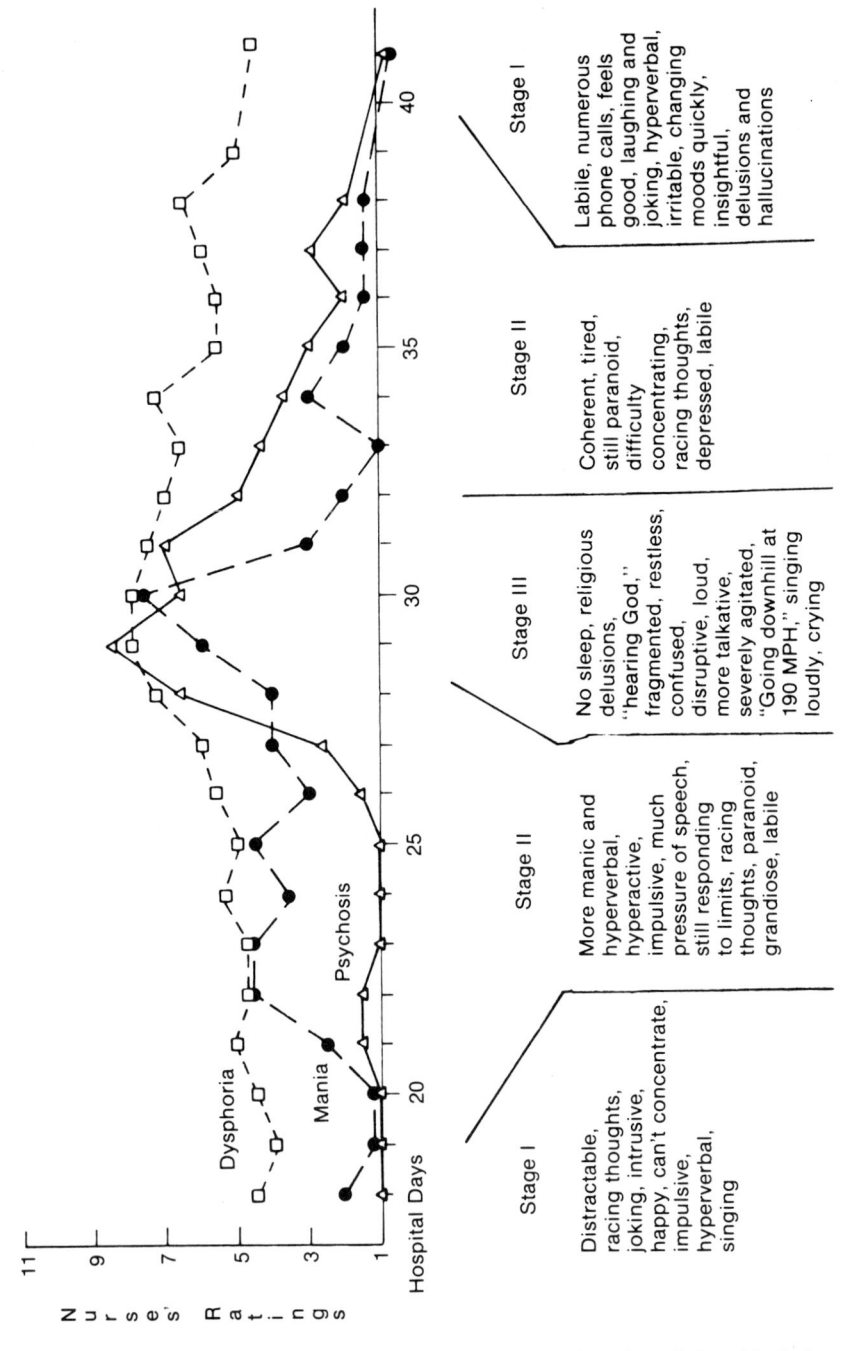

FIG. 2. A complete manic episode in one patient, illustrating the relationship between the daily nurses' global consensus ratings for mania and psychosis. Note the increase in psychosis ratings coincident with the peak of mania (designated Stage III). This pattern was characteristic of about 40% of the patients studied. (From Carlson and Goodwin, 1973.)

difficult to distinguish from acute schizophrenics when viewed cross-sectionally during Stage III, a 2-year follow-up study indicates that they undergo a typical manic-depressive course, both in terms of lack of deterioration and good prophylactic response to lithium (Carlson and Goodwin, 1973).

All patients were rated twice daily by a trained nursing research team using a number of rating instruments, primarily a 15-point multi-item scale (Bunney and Hamburg, 1963). Of particular relevance to this chapter are the global items of depression, mania, and psychosis as outlined in Table 1.

TABLE 1. *Nurses' rating instructions for global consensus items*

Depression
  A. *Expressions of hopelessness and helplessness*
  B. *Expressions of worthlessness and guilt*
  C. *Concern with death or suicide*
  D. *Psychomotor activity*
    1. *Retardation,* as indicated by slowed speech and movement, low monotonous voice, disinterest in surroundings, or
    2. *Agitation*—nervous hyperactivity, hand-wringing, pacing
  E. *Depressed appearance,* including sad expression, crying, drabness in dress and makeup
  F. *For some persons:* (1) physical complaints, (2) loss of appetite, (3) sleep disturbance

Mania
  A. *Mood*—classically "elated but unstable." However, some persons may show a predominance of impatience, irritability and anger
  B. *Pressure of speech*—increased volume or speed, sometimes progressing to flight of ideas with thoughts too rapid for connections to be made
  C. *Increased motor activity*—includes walking, writing, phoning
  D. *Grandiosity or overconfidence*
  E. *Inappropriate language, actions, or dress*
  F. *Intrusive, provocative, or manipulative behavior,* frequently demanding of attention to self

Psychosis
  A. *Loose associations*—thoughts which are not connected, out of context
  B. *Inappropriate affect,* such as when mood and thought do not match; for example, laughing or crying without obvious reason, also, lack of emotional responsiveness and emotional lability
  C. *Paranoid thinking,* ranging from unrealistic suspiciousness to delusions of persecution
  D. *Delusions or hallucinations*
  E. *Bizarre or crazy behavior,* including inappropriate dress

III. RESULTS

A. Studies with L-DOPA

L-DOPA was administered to 26 depressed patients either alone (average dose, 100 mg/kg) or in combination with a peripheral decarboxylase inhibitor (average L-DOPA dose, 13 mg/kg)* (Goodwin, Brodie, Murphy, and Bunney, 1970). The minimum duration of treatment was 3 weeks, with a mean of 4 weeks. Studies of urinary and CSF amine metabolites indicated that this amino acid precursor produced profound effects on catecholamines both peripherally and centrally. Figure 3 illustrates the effect of L-DOPA administration on central dopamine "turnover" as reflected by the accumulation of HVA in the cerebrospinal fluid following probenecid administration. The probenecid-induced accumulation of HVA was more than doubled

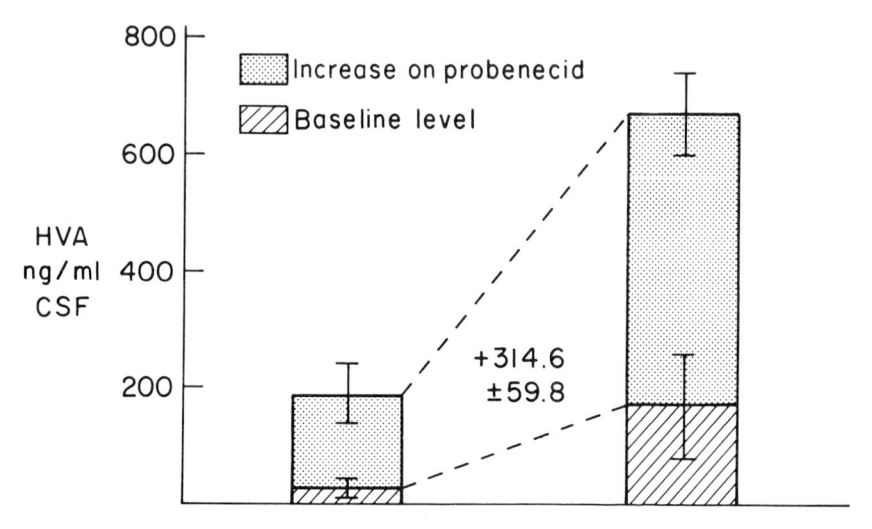

FIG. 3. Changes in HVA accumulation with probenecid during treatment with L-DOPA. (From Goodwin, Dunner, and Gershon, 1971.)

by the administration of L-DOPA or L-DOPA plus a peripheral decarboxylase inhibitor. The question of L-DOPA's effect on brain NE is more complex. As noted above, most of the animal data would suggest that L-DOPA administration does not produce an increase in brain NE; however, under conditions where this brain amine has previously been depleted, L-DOPA can produce an increase in NE (Seiden and Petersen, 1968).

The overall clinical results of L-DOPA in depressed patients are illustrated in Table 2. Note that of the 26 patients there was a small subgroup of six patients who showed some therapeutic response. It is of interest that this small subgroup was comprised of depressed patients with past histories of hypomania, that is, bipolar II patients. In the unipolar patients (no history of mania) and in the bipolar I patients (past history of hospitalization for mania), there was no beneficial effect of these large doses of

---

* Plasma L-DOPA studies (Dunner, Brodie, and Goodwin, 1971) indicate that this lower dose, in combination with the peripheral decarboxylase inhibitor (MK485), yielded a plasma L-DOPA level which was equivalent to that with L-DOPA alone at 100 mg/kg.

L-DOPA. Another way to analyze the data relates to the parameter of retardation-agitation: All the responders were patients with predominant psychomotor retardation. From analysis of the individual patient data, it appeared that the clinical response in this subgroup represented an initial activation of preexisting psychomotor retardation followed by improvement in other depressive symptoms (Goodwin et al., 1970).

TABLE 2. *Clinical response to L-DOPA in subtypes of depressed patients*

	Unipolar	Bipolar II	Bipolar I
Responders	0	5	1
Nonresponders	7	3	10
Totals	7	8	11

Of considerable interest, particularly in relation to the theme of this symposium, was the fact that in all those patients who failed to show a therapeutic response to L-DOPA (that is, 20 of 26), the compound did cause some activating effects—effects which appear to represent an exacerbation of preexisting psychopathology or vulnerability to psychopathology. For example, those depressed patients with preexisting psychotic features evidenced an intensification of psychosis on L-DOPA (Bunney, Brodie, Murphy, and Goodwin, 1971) (Fig. 4). The most striking "activation" effect produced by L-DOPA in these patients was in relation to hypomania; of the 11 depressed patients classified as bipolar I, that is, prior histories of hospitalization for mania, 10 experienced episodes of hypomania during

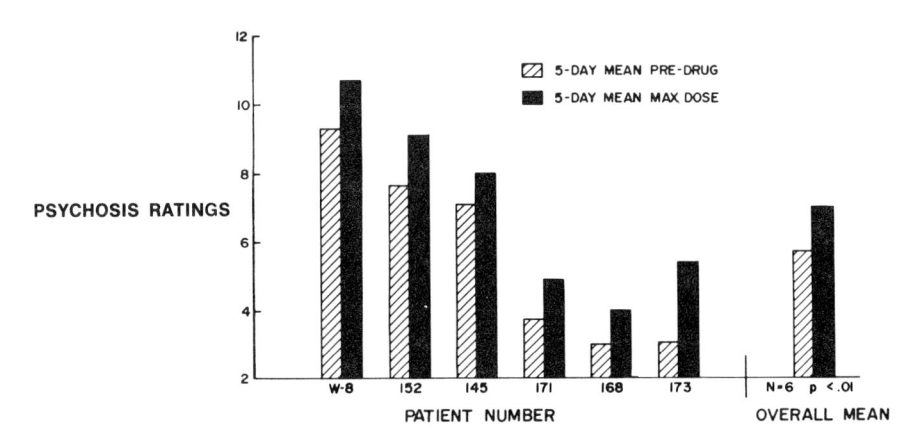

FIG. 4. Average nurses' psychosis ratings for 5 days prior to administration and 5 days at maximal dosage of L-DOPA in a group of depressed patients showing intensification of existing psychotic symptoms with L-DOPA in nonresponders. (From Bunney et al., 1971.)

TABLE 3. *Incidence of hypomania during L-DOPA administration in bipolar versus unipolar depressed patients*

	Hypomanic episode	No hypomanic episode
Bipolar patients	10	1
Unipolar patients	1	12

$\chi^2 = 11.1$, $p < 0.001$.

the period of L-DOPA administration (see Table 3) (Murphy, Brodie, Goodwin, and Bunney, 1971). In general, the periods of hypomanic behavior were of only several days duration, and specifically included increased motor and verbal activity with pressured speech, increased social involvement, intrusiveness, some increase in expressed anger, grandiosity, sleeplessness, some agitated behavior, but little or no true euphoria except in one patient (one of the responder group). More detailed discussions of the activating effects of L-DOPA administration are contained in reviews by Goodwin (1971) and Murphy (1972).

B. Studies with Alpha-methyl-para-tyrosine (AMPT) in Mania

AMPT is a potent and specific inhibitor of tyrosine hydroxylase, the rate-limiting enzyme in the synthesis of DA and NE; when administered in sufficiently high doses, it markedly reduces the levels of these catecholamines in all tissues including brain (Spector et al., 1965). When administered to manic

TABLE 4. *Patient characteristics, parameters of drug administration and clinical response in a group of manic patients treated with the tyrosine hydroxylase inhibitor AMPT*

Patient	Diagnosis	Age	Sex	Days on AMPT	Maximum daily dose (g)	No. of courses	Clinical response
N.R.	Bipolar-manic	30	F	15	2	1	Improvement
R.J.	Bipolar-manic	37	M	5	3	1	Improvement
R.M.	Bipolar-manic	43	F	8	4	1	Improvement
R.E.	Bipolar-manic	52	F	35	4	3	Improvement[a]
M.J.	Bipolar-manic	44	F	48	4	3	Improvement[a]
W.F.	Bipolar-manic	47	M	29	3	1	No change
K.M.	Bipolar-manic	47	F	9	4	1	Worse
M.L.	Bipolar-manic	26	M	18	4	1	No change

[a] Two of the patients who improved showed relapse with placebo substitution.
From Gershon et al., 1971.

patients in doses up to 4 g/day, five of eight patients showed significant improvement (see Table 4) (Brodie, Murphy, Goodwin, and Bunney, 1971; Gershon, Bunney, Goodwin, Murphy, Dunner, and Henry, 1971). Two of the five responders showed a consistent pattern of relapse following placebo substitution (Brodie et al., 1971). In the apparent responders, the antimanic effect was reflected in parallel decreases in the global ratings for both mania and psychosis (see Table 1). The levels of catecholamine metabolites in the CSF were markedly reduced in patients on AMPT, verifying that the brain amines had in fact been reduced.

Since decreases in both DA and NE occur in patients treated with AMPT, ambiguity remains as to whether the clinical effect observed in these patients was secondary to a reduction in one or both catecholamine neurotransmitters.

C. Studies with Fusaric Acid, a Dopamine-beta-hydroxylase Inhibitor

Fusaric acid (5-butylpicolinic acid) (Fig. 5) has been reported to be a potent, nontoxic inhibitor of DBH both *in vivo* and *in vitro*, resulting in a decrease in both the synthesis and levels of central NE (Hidaka, Nagatsu, Takeya, Takeuchi, Suda, Kojiri, Matsuzaki, and Umezawa, 1969; Suda, Takeuchi, Nagatsu, Matsuzaki, Matsumoto, and Umezawa, 1969; Nagatsu et al., 1970; Matta and Wooten, 1973). In order to test the hypothesis that a selective decrease in central NE (rather than the decrease in both catecholamines as produced by AMPT) would decrease manic behavior, we conducted a trial of fusaric acid in a group of eight manic and one depressed patient.

The inhibition produced by fusaric acid, unlike that of disulfiram and related compounds, is noncompetitive and does not involve chelation of copper (Nagatsu et al., 1970). Since the drug can produce inhibition of DBH *in vitro* at a concentration of 10^{-8} M, it is considerably more potent than the copper chelating inhibitors, whose *in vitro* potency is of the order of 10^{-6} M (Green, 1964). Its high degree of specificity is suggested by the fact that at concentrations up to 10^{-4} M it has no effect *in vitro* on tyrosine hydroxylase, monoamine oxidase, or aldehyde dehydrogenase, nor does it appear to effect either amine uptake or release (Nagatsu et al., 1970). Single doses of 100 mg/kg in rats reduced brain NE levels by 40%; although DA levels appear to be unchanged in fusaric acid-treated animals (Nagatsu et al., 1970), the

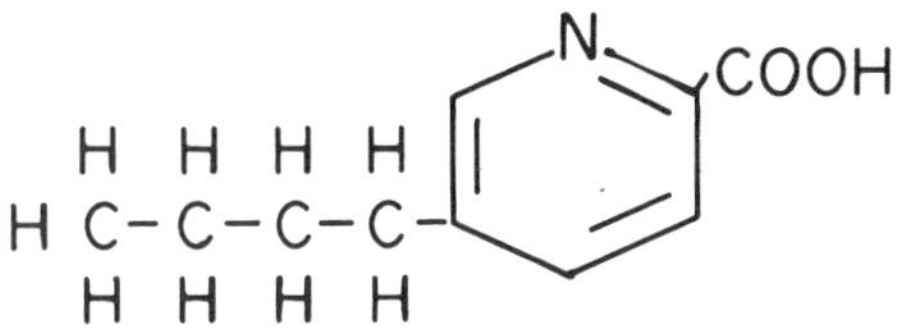

FIG. 5. Fusaric acid (5-butylpicolinic acid).

decreased conversion of ^{3}H-DA to ^{3}H-NE following ^{3}H-DA injection in rats is accompanied by accumulation of labeled DA (Hidaka, Nagasaka, and Takeda, 1973), suggesting some precursor accumulation proximal to the inhibited enzyme.

In man, a single oral dose in the range of 200 to 300 mg reduces plasma DBH by 95% with an apparent biological half-life in plasma of about 20 hr (Nagatsu et al., 1972; Matta and Wooten, 1973); chronic treatment at 600 mg/day maintained inhibition of plasma DBH at 95 to 98% (Matta and Wooten, 1973). Although the drug has been used in more than 200 hypertensive patients in Japan (Hidaka et al., 1973) and in two small trials in Parkinson's disease (Mena, Court, and Cotzias, 1971; T. N. Chase, *personal communication*, 1973), this represents the first trial in psychiatric patients. In the patients reported here, the average maximum dose was 1.12 g/day (range 0.6 to 1.8 g/day), administered over a period of 10 to 20 days.

Studies of CSF amine metabolites demonstrated that chronic administration of fusaric acid caused changes in central amine metabolism that are predicted from its reported mechanism of action (see Fig. 6). Specifically,

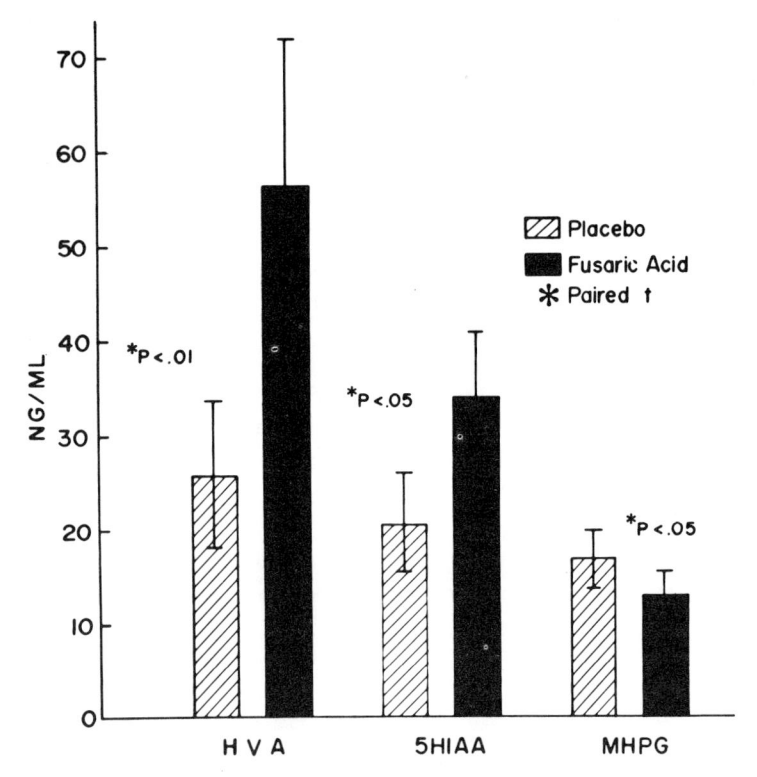

FIG. 6. The effect of fusaric acid on CSF amine metabolites. (MHPG data obtained in collaboration with T. N. Chase, E. Gordon, and L. K. Y. Ng.)

levels of 3-methoxy-4-hydroxy-phenylglycol (MHPG), the major metabolite of NE in brain (Maas and Landis, 1968; Shanberg, Schildkraut, Breese, and Kopin, 1968), were significantly lower when patients on fusaric acid were compared to the same patients on placebo (mean reduction 25%, $p = <0.05$). Note also that the mean CSF concentration of HVA was nearly doubled, a finding which presumably reflects an accumulation of central DA as a result of DBH inhibition. The modest increase in 5-hydroxyindoleacetic acid (5-HIAA) in the CSF is consistent with reports of increased serotonin turnover (Hidaka, 1971), perhaps secondary to an increase in brain tryptophan as has been reported with several other DBH inhibitors (Johnson and Kim, 1973).

In order to assess the effects of fusaric acid in manic patients, the twice-daily nurses' global ratings for mania and for psychosis were averaged separately for the 7 placebo days prior to fusaric acid, and again for the 7 days of maximal drug dosage. Figures 7 and 8 illustrate the effect of fusaric

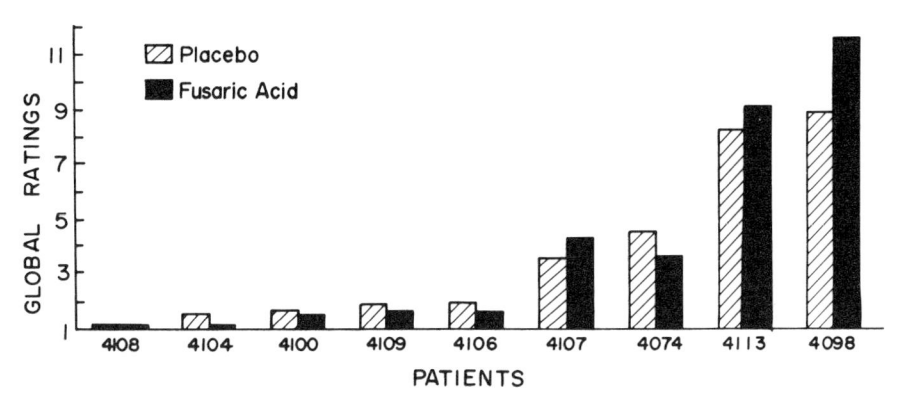

FIG. 7. Effect of fusaric acid on nurses' global ratings for mania in nine manic-depressive patients. Each bar represents the average of twice-daily ratings for 7 days. Placebo ratings obtained just before initiating fusaric acid. Fusaric acid ratings obtained at maximal dose.

acid on mania and psychosis ratings for the nine individual patients. For both mania and psychosis ratings, the effect of the drug appeared to be related to the preexisting clinical state; thus, with the mildly hypomanic patients (mania ratings less than 3) there was a slight decrease in ratings in each case. Of the two patients with intermediate predrug ratings (moderate mania), one became somewhat more manic and the other somewhat less, while the two patients with severe mania both became worse on fusaric acid (Fig. 7).

Baseline clinical state appeared to be even more important with regard to the effects of fusaric acid on psychosis ratings (Fig. 8). In patients without initial psychosis ratings there was no change; that is, the drug was not psychotogenic. However, patients with some (even slight) psychosis ratings

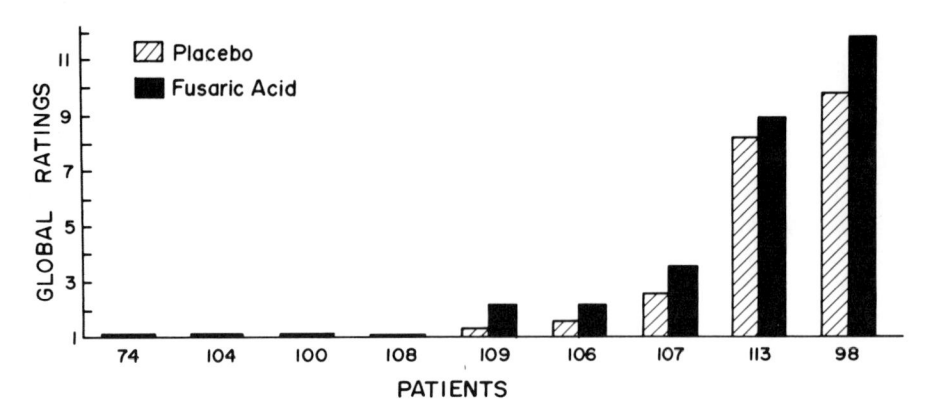

FIG. 8. Effect of fusaric acid on nurses' global ratings for psychosis in nine manic-depressive patients. Each bar represents the average of twice-daily ratings for 7 days. Placebo ratings obtained just before initiating fusaric acid. Fusaric acid ratings obtained at maximal dose.

prior to initiation of drug therapy consistently showed increases in psychosis as reflected by the ratings.

In summary, fusaric acid did not have any consistent beneficial effect on mania; rather, the effects of the treatment appeared to depend on the preexisting clinical state. Patients with Stage III mania (see Fig. 2), or with evidence of preexisting psychotic features, became consistently worse; however, a few patients with mild hypomanic symptoms showed slight improvement in ratings. Conventional antimanic therapies (lithium, phenothiazines) initiated after the fusaric acid protocol, were associated with improvement in the expected time period in all the patients.

D. Subgroup Differences in Central Dopamine "Turnover"

Central dopamine "turnover" was estimated in a group of hospitalized depressed patients by measuring the accumulation of HVA in the CSF following the administration of high doses of probenecid (100 mg/kg over 18 hr) as previously described (Goodwin et al., 1973). At the time of the study all patients were at least moderately depressed and had been off all drugs for 2 weeks or more. The patients were divided into bipolar or unipolar according to the presence or absence of a prior history of mania or hypomania. As can be seen in Fig. 9, the probenecid-induced accumulation of HVA was approximately 40% greater in the bipolar than in the unipolar group ($p = <0.05$). In these patients there were no significant unipolar-bipolar differences in sex ratio, age, duration of illness, or severity of depression at the time of the probenecid study.

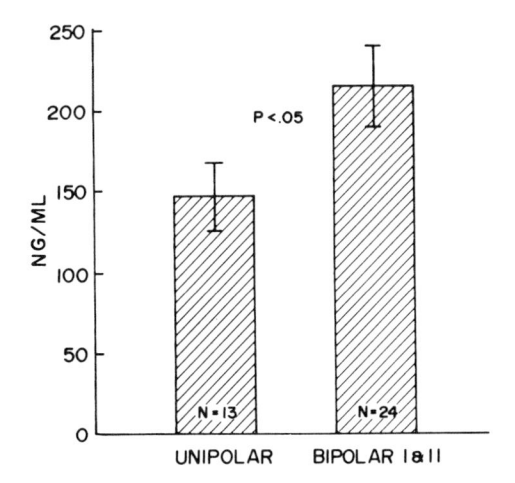

FIG. 9. Central dopamine "turnover." Differences in the probenecid-induced accumulation of homovanillic acid (HVA) in unipolar versus bipolar depressed patients. For details of probenecid methodology, see Goodwin et al., 1973.

IV. DISCUSSION

With regard to the possible role of central DA function in affective illness, what are some of the implications of the clinical-pharmacological data reviewed here? First, the failure of L-DOPA in high doses to produce antidepressant effects in the majority of patients, in spite of the evidence of substantial increases in brain DA, would seem to weigh against the hypothesis that depression *per se* is etiologically related to a depletion of brain DA. Thus, reserpine-induced "depression," which can be reversed by L-DOPA in this dose range both in animals and in man (reviewed in Goodwin et al., 1970), does not provide a good model for naturally occurring depressions in the majority of patients. On the other hand, our data as well as that from studies of L-DOPA administration to schizophrenic (Yaryura-Tobias, Diamond, and Merlis, 1970; Angrist, Sathananthan, and Gershon, 1973) and to parkinsonian patients (Godwin-Austen, Tomlinson, Frears, and Kok, 1969; Celesia and Barr, 1970; Jenkins and Groh, 1970; McDowell, 1970) make it clear that this DA precursor produces an increase in activation or arousal, with the qualitative expression of this activation appearing to be at least partially dependent on the preexisting clinical state (Goodwin, 1971; Murphy, 1972). Thus, within this group of depressed patients we observed a variety of response patterns, including no improvement in depression with some activation of preexisting psychosis in the unipolar patients, activation of hypomania without improvement in depression in the bipolar I patients, and improvement in depression in five of eight of the bipolar II patients—improvement which seemed to represent (at least initially) an activation of a retarded psychomotor state.

A possible schema for relating these differential effects of L-DOPA in depressed patients is outlined in Table 5. Murphy and Weiss (1972), studying the activity of platelet monoamine oxidase (MAO) in subgroups of depressed patients, have found a low MAO in bipolar I patients compared to unipolar or bipolar II patients or normal controls. On the other hand, as noted above we have found unipolar-bipolar differences in the probenecid-induced accumulation of HVA, with bipolar I and II patients evidencing higher central DA "turnover" when compared to unipolar patients.

TABLE 5. *Differential response to L-DOPA in depressive patients: possible relationship to differences in neurochemical substrate*

	Clinical response	Neurochemical differences	
Unipolar	No improvement in depression Activation of psychosis and anger	Normal MAO	Lower DA turnover
Bipolar II	Some improvement in depression (5 of 8)	Normal MAO	Higher DA turnover
Bipolar I	No improvement in depression Activation of hypomania	Low MAO	Higher DA turnover

The differential clinical response to L-DOPA in the three subgroups can reasonably be represented as an increasing degree of "activation" as one goes from unipolar to bipolar II to bipolar I. Considering the subgroup differences in MAO and in DA "turnover" together, a possible explanation for the differential clinical response to L-DOPA might be formulated as follows: Since the major route of DA metabolism involves MAO, the lower activity of this enzyme in bipolar I patients could result in larger L-DOPA-induced increases in dopaminergic activity (assuming that the platelet MAO differences reflect similar differences in the mitochondrial MAO in central dopaminergic neuronal systems). With regard to the DA "turnover" differences, there is at least indirect data to suggest that the central effects of L-DOPA are increased in situations where endogenous DA turnover is already elevated, such as with chlorpromazine treatment (Angrist et al., 1973). Thus, the bipolar I patients, who clinically were most activated with L-DOPA, have alterations in both MAO and DA turnover in the direction that should potentiate the action of L-DOPA; the bipolar II patients, who were intermediate in activation, have a higher DA turnover but do not have low MAO; and the unipolar patients have neither. This schema is offered not as a definitive explanation of the differential effects of L-DOPA in these subgroups, but rather as one example of an experimental approach to the understanding of differential drug-substrate interactions.

With regard to the role of central DA in mania, the data from L-DOPA,

AMPT, and fusaric acid do not lend themselves to any simple formulations. As we have noted previously (Goodwin and Sack, 1974), the clinical-pharmacological correlation between mania and functionally increased central catecholamines appears to be more consistent than the correlation between depression and decreased amines. However, with most of these correlations it is not possible to distinguish DA from NE. In this regard fusaric acid may offer an advantage, since unlike tyrosine hydroxylase inhibition with AMPT, the inhibition of DBH results in a decreased formation of NE without a decrease in DA; in fact the animal data indicate that a wide variety of DBH inhibitors produce an increase in brain DA and/or its metabolites (Goldstein and Kazuhiko, 1967; Johnson, Baukma, and Kim, 1969; Von Voigtlander and Moore, 1970)—findings consistent with the effect of fusaric acid on CSF HVA in our patients. From a functional point of view, it is important to keep in mind that any increase in DA following DBH inhibition would be occurring *in noradrenergic neurons* and not in dopaminergic neurons, since the latter do not contain DBH. This selective effect on the NE/DA ratio in noradrenergic neuronal systems might be compared to the effects of L-DOPA: This catecholamine precursor can also reduce NE and increase DA in noradrenergic neurons (Ng, Chase, Colburn, and Kopin, 1970; Ng, Colburn, and Kopin, 1971), presumably by a flooding mechanism in which the amino acid is converted to DA in the noradrenergic neuron by the nonspecific aromatic L-amino acid decarboxylase creating a DA concentration exceeding the capacity of DBH to convert it to NE (i.e., a relative DBH deficiency). Evidence has also been presented to suggest that this flooding mechanism, producing a "false transmitter" situation, can also occur in serotonergic neurons following large doses of L-DOPA (Ng et al., 1970).

Thus, in the noradrenergic neuron the effects of fusaric acid and of L-DOPA could be regarded as similar; on the other hand, while fusaric acid should have no direct effect on dopaminergic neurons, L-DOPA administration produces large increases in DA synthesis in these cells. Thus, one might speculate that similarities between the behavioral effects of fusaric acid and L-DOPA might be traceable to alterations in noradrenergic neurons—that is, an increase in the DA/NE ratio—while dissimilar effects of the two drugs might reside in dopaminergic (and/or serotonergic) systems. AMPT and fusaric acid have dissimilar effects on both dopaminergic and noradrenergic neurons: In noradrenergic neurons NE is decreased by both drugs, but with fusaric acid this decrease is accompanied by an increase in DA.

The differential catecholaminergic effects of these drugs and their possible relationship to behavioral changes in patients with affective illness is outlined in Table 6.

Earlier experience with psychiatric complications accompanying the clinical use of disulfiram (a DBH inhibitor) in alcoholics (Child, 1952), as

TABLE 6. *The effect of L-DOPA, AMPT, and fusaric acid on catecholamine neurons: possible relationship to behavioral effects*

Drug	Direct effects on catecholamine systems		Behavioral effects in patients with affective illness		
	Dopaminergic	Noradrenergic	Mania	Psychosis	Depression
L-DOPA	↑↑DA	↓NE ↑DA	↑(ppt. in	↑	↑or↓
AMPT	↓DA	↓NE	↓B.P. I)	↓	↑?
Fusaric acid	—	↓NE ↑DA	↓?	↑	—?

well as reports of mental changes accompanying carbon disulfide intoxication (Braceland, 1942), had led to speculation about the relationship between behavioral changes and the presumed inhibition of brain DBH. As noted by Kane (1970), the more severe reactions to disulfiram can involve both depressive and manic features, but more often the picture is dominated by a nonspecific psychotic state accompanied by signs of an organic brain syndrome ("toxic psychosis"). Interpretation of these data is complicated by the relative lack of specificity of disulfiram (i.e., it is also a potent inhibitor of aldehyde dehydrogenase) and by its associated potential for inducing nonspecific organic brain syndromes. Thus, in the report by Heath, Nesselhof, Bishop, and Byers (1965) of worsening of schizophrenic symptoms by disulfiram, the high incidence of associated organic symptoms makes it plausible that the observed increase in psychosis could be a nonspecific reaction to a drug-induced organic brain impairment, rather than a direct result of DBH inhibition. Of the available DBH inhibitors, on the other hand, fusaric acid appears to be the most specific, particularly in that it does not inhibit aldehyde dehydrogenase. Also of importance is the fact that fusaric acid has not been associated with any signs of an organic brain syndrome in any of its clinical trials. In light of these considerations, it seems reasonable to hypothesize that the behavioral changes accompanying fusaric acid administration represent the effects of DBH inhibition.

Although the major focus on the clinical significance of central catecholamine functioning has been in the area of affective illness, there has recently been renewed interest in the possibility that the pathophysiology of schizophrenia may involve disturbances in central DA or NE or both. On the basis of the apparently selective effects on brain DA systems of drugs which either induce a schizophrenic-like psychosis (amphetamines) or which alleviate schizophrenic symptoms (phenothiazines), it has been hypothesized that excess functional DA may be involved in schizophrenia, at least in its acute symptoms (Snyder, 1972). On the other hand, using animal data derived from a self-stimulation model, Stein and Wise (1971) have hypothesized that chronic schizophrenia may represent a degeneration of noradrenergic neurons; recently these authors have reported on a postmortem study of human brains in which they found significantly lower DBH activity in the material from chronic schizophrenics compared to controls (Wise and Stein, 1973). They cite this as evidence in support of the noradrenergic deficit hypothesis, since DBH can be considered a marker for NE neurons.

Our data on drug-catecholamine relationships in affective illness would appear to have some bearing on the role of these amines in schizophrenic psychosis. The L-DOPA- and fusaric acid-related increases in preexisting psychosis, which occur in association with an apparent increase in DA and decrease in NE in noradrenergic neurons, could be consistent with both the DA-psychosis correlations emphasized by Snyder and the noradrenergic hypothesis of Stein and Wise. Finally, the data reviewed here further underline the symptomatic overlap between affective illness and schizophrenia, and emphasize the usefulness of attempts to correlate alterations in specific neurotransmitters with specific symptom profiles rather than simply with broad diagnostic categories.

REFERENCES

Angrist, B., Sathananthan, G., and Gershon, S. (1973): Behavioral effects of L-DOPA in schizophrenic patients. *Psychopharmacologia*, 31:1-12.
Axelrod, J. (1972): Dopamine-β-hydroxylase: Regulation of its synthesis and release from nerve terminals. *Pharmacological Reviews*, 24:233-243.
Bartholini, G., Constantinidis, J., Tissot, R., and Pletscher, A. (1971): Formation of monoamines from various amino acids in the brain after inhibition of extracerebral decarboxylase. *Biochemical Pharmacology*, 20:1243-1247.
Beigel, A., and Murphy, D. L. (1971): Assessing clinical characteristics of the manic state. *American Journal of Psychiatry*, 128:688-699.
Braceland, F. J. (1942): Mental symptoms following carbon disulphide absorption and intoxication. *Annals of Internal Medicine*, 16:246-261.
Brodie, H. K. H., Murphy, D. L., Goodwin, F. K., and Bunney, W. E., Jr. (1971): Catecholamines and mania: The effect of alpha-methyl-para-tyrosine on manic behavior and catecholamine metabolism. *Clinical and Pharmacological Therapeutics*, 12:218-224.
Bunney, W. E., Jr., Brodie, H. K. H., Murphy, D. L., and Goodwin, F. K. (1971): Studies of alpha-methyl-para-tyrosine, L-DOPA, and tryptophan in depression and mania. *American Journal of Psychiatry*, 127:872-881.
Bunney, W. E., Jr., and Hamburg, D. A. (1963): Methods for reliable longitudinal observation of behavior. *Archives of General Psychiatry*, 9:280-294.
Bunney, W. E., Jr., and Davis, J. M. (1965): Norepinephrine in depressive reactions. *Archives of General Psychiatry*, 13:483-494.
Butcher, L. L., and Engel, J. (1969): Behavioral and biochemical effects of L-DOPA after peripheral decarboxylase inhibition. *Brain Research*, 15:233-242.
Carlson, G. A., and Goodwin, F. K. (1973): The stages of mania: A longitudinal analysis of the manic episode. *Archives of General Psychiatry*, 28:221-228.
Carlsson, A., Lindqvist, M., and Magnusson, T. (1960): On the biochemistry and possible functions of dopamine and noradrenaline in brain. In: *Ciba Symposium on Adrenergic Mechanisms*, edited by J. R. Vane, G. E. W. Wolstenholme, and M. O'Connor. J. and A. Churchill, London, pp. 432-439.
Carlsson, A., Rosengren, E., Bertler, A., and Nilsson, J. (1957): Effect of reserpine on the metabolism of catecholamines. In: *Psychotropic Drugs*, edited by S. Garattini and V. Ghetti. Elsevier, Amsterdam.
Celesia, G. G., and Barr, A. N. (1970): Psychosis and other psychiatric manifestations of levodopa therapy. *Archives of Neurology*, 23:193-200.
Child, J., and Crump, C. (1952): Disulfiram. *Acta Pharmacologica*, 8:305-314.

Coppen, A. (1967): The biochemistry of affective disorders. *British Journal of Psychiatry*, 113:1237-1264.
Dunner, D. L., Brodie, H. K. H., and Goodwin, F. K. (1971): Plasma DOPA response to L-DOPA administration in man: Potentiation by a peripheral decarboxylase inhibitor. *Clinical and Pharmacological Therapeutics*, 12:212-217.
Everett, G. M., and Borcherding, J. W. (1970): L-DOPA, effect on concentrations of dopamine, norepinephrine, and serotonin in brains of mice. *Science*, 168:849-850.
Feighner, J. P., Robins, E., Guze, S. B., Woodruff, R. A., Winokur, G., and Munoz, R. (1972): Diagnostic criteria for use in psychiatric research. *Archives of General Psychiatry*, 26:57-63.
Gershon, E. S., Bunney, W. E., Jr., Goodwin, F. K., Murphy, D. L., Dunner, D. L., and Henry, G. M. (1971): Catecholamines and affective illness: Studies with L-DOPA and alpha-methyl-para-tyrosine. In: *Brain Chemistry and Mental Disease*, Vol. 1, edited by B. T. Ho and W. M. McIsaac. Plenum Press, New York,
Glassman, A. (1969): Indoleamines and affective disorders. *Psychosomatic Medicine*, 31:107-114.
Godwin-Austen, R. B., Tomlinson, E. B., Frears, C. C., and Kok, H. W. L. (1969): Effects of L-DOPA in Parkinson's disease. *Lancet*, 2:165-168.
Goldstein, M., and Kazuhiko, N. (1967): The effect of disulfiram on catecholamine levels in the brain. *Journal of Pharmacology and Experimental Therapeutics*, 157:96-102.
Goodwin, F. K. (1971): Psychiatric side effects of levodopa in man. *Journal of the American Medical Association*, 218:1915-1920.
Goodwin, F. K., Brodie, H. K. H., Murphy, D. L., and Bunney, W. E., Jr. (1970): L-DOPA, catecholamines and behavior: A clinical and biochemical study in depressed patients. *Biological Psychiatry*, 2:341-366.
Goodwin, F. K., Dunner, D. L., and Gershon, E. S. (1971): Effect of L-DOPA treatment on brain serotonin metabolism in depressed patients. *Life Sciences*, 10:751-759.
Goodwin, F. K., Murphy, D. L., Dunner, D. L., and Bunney, W. E., Jr. (1972): Lithium response in unipolar vs. bipolar depression. *American Journal of Psychiatry*, 129:44-47.
Goodwin, F. K., Post, R. M., Dunner, D. L., Gordon, E. K., and Bunney, W. E., Jr. (1973): Cerebrospinal fluid amine metabolites in affective illness: The probenecid technique. *American Journal of Psychiatry*, 130:73-79.
Goodwin, F. K., and Sack, R. L. (1974): The catecholamine hypothesis revisited. In: *Frontiers in Catecholamine Research*, edited by E. Usdin and S. Snyder, Pergamon Press, Elmsford, New York, pp. 1157-1164.
Green, A. L. (1964): The inhibition of dopamine-β-oxidase by chelating agents. *Biochimica et Biophysica Acta*, 81:394-397.
Heath, R. G., Nesselhof, W., Bishop, M. P., and Byers, L. W. (1965): Behavioral and metabolic changes associated with administration of tetraethylthiuram disulfide (antabuse). *Diseases of the Nervous System*, 26:99-104.
Hidaka, H. (1971): Fusaric (5-butylpicolinic) acid, an inhibitor of dopamine β-hydroxylase, affects serotonin and noradrenalin. *Nature*, 231:54-55.
Hidaka, H., Nagasaka, A., and Takeda, A. (1973): Fusaric (5-butylpicolinic) acid: Its effect on plasma growth hormone. *Journal of Clinical Endocrinology*, 37:145-147.
Hidaka, H., Nagatsu, T., Takeya, K., Takeuchi, T., Suda, H., Kojiri, K., Matsuzaki, M., and Umezawa, H. (1969): Fusaric acid, a hypotensive agent produced by fungi. *Journal of Antibiotics*, 22:228-230.
Iversen, L. L. (1967): The catecholamines. *Nature*, 214:8-14.
Jenkins, R. B., and Groh, R. H. (1970): Mental symptoms in Parkinsonian patients treated with L-DOPA. *Lancet*, 2:177-180.
Johnson, G. A., Boukma, S. J., and Kim, E. G. (1969): Inhibition of dopamine-β-hydroxylase by aromatic and alkyl thioureas. *Journal of Pharmacology and Experimental Therapeutics*, 168:229-234.
Johnson, G. A., and Kim, E. G. (1973): Increase of brain levels of tryptophan induced by inhibition of dopamine-β-hydroxylase. *Journal of Neurochemistry*, 20:1761-1763.
Kane, F. J. (1970): Carbon disulfide intoxication from overdosage of disulfiram. *American Journal of Psychiatry*, 127:690-694.
Kotin, J., and Goodwin, F. K. (1972): Depression during mania: Clinical observations and theoretical implications. *American Journal of Psychiatry*, 129:679-686.

Lapin, T. P., and Oxenkrug, G. F. (1969): Intensification of the central serotonergic processes as a possible determinant of the thymoleptic effect. *Lancet,* 1:132–136.
Maas, J. W., and Landis, D. H. (1968): In vivo studies of the metabolism of norepinephrine in the central nervous system. *Journal of Pharmacology and Experimental Therapeutics,* 163:147–162.
Matta, R. J., and Wooten, G. F. (1973): The pharmacology of fusaric acid in man. *Clinical Pharmacology and Therapeutics,* 14:541–546.
McDowell, F. H. (1970): Changes in behavior and mentation. In: L-*DOPA and Parkinsonism,* edited by A. Barbeau and F. H. McDowell. F. A. Davis, Philadelphia, pp. 321–324.
Mena, I., Court, J., and Cotzias, G. (1971): Levodopa, involuntary movements and fusaric acid. *Journal of the American Medical Association,* 218:1829–1830.
Murphy, D. L. (1972): L-DOPA, behavioral activation and psychopathology. In: *Neurotransmitters.* Res. Publ. A.R.N.M.D., Vol. 50, pp. 472–493.
Murphy, D. L., Brodie, H. K. H., Goodwin, F. K., and Bunney, W. E., Jr. (1971): L-DOPA: Regular induction of hypomania in bipolar manic-depressive patients. *Nature,* 229:135–136.
Murphy, D. L., and Weiss, R. (1972): Reduced monoamine oxidase activity in blood platelets from bipolar depressed patients. *American Journal of Psychiatry,* 128:1351–1357.
Nagatsu, T., Hidaka, H., Kuzuya, H., Takeya, K., Umezawa, H., Takeuchi, T., and Suda, H. (1970): Inhibition of dopamine-β-hydroxylase by fusaric acid (5-butypicolinic acid) in vitro and in vivo. *Biochemical Pharmacology,* 19:35–44.
Nagatsu, T., Kato, T., Kuzuya, H., Okada, T., Umezawa, H., and Takeuchi, T. (1972): Inhibition of human serum dopamine-β-hydroxylase after the oral administration of fusaric acid. *Experientia,* 28:779–780.
Neff, N. H., Tozer, T. N., and Brodie, B. B. (1967): Application of steady-state kinetics to studies of the transfer of 5-hydroxyindoleacetic acid from brain to plasma. *Journal of Pharmacology and Experimental Therapeutics,* 158:214–218.
Ng, K. Y., Chase, T. N., Colburn, R. W., and Kopin, I. J. (1970): L-DOPA induced release of cerebral monoamines. *Science,* 170:76–77.
Ng, K. Y., Colburn, R. W., and Kopin, I. J. (1971): Effects of L-DOPA on efflux of cerebral monoamines from synaptosomes. *Nature,* 230:331–332.
Schildkraut, J. J. (1965): The catecholamine hypothesis of affective disorders: A review of supporting evidence. *American Journal of Psychiatry,* 122:509–522.
Schneider, K. (1959): *Clinical Psychopathology.* Grune and Stratton, New York.
Seiden, L. S., and Petersen, D. D. (1968): Blockade of L-DOPA reversal of reserpine-induced conditioned avoidance response suppression by disulfiram. *Journal of Pharmacology and Experimental Therapeutics,* 163:84–90.
Shanberg, S. M., Schildkraut, J. J., Breese, G. R., and Kopin, I. J. (1968): Metabolism of normetanephrine-H³ in rat brain—identification of conjugated 3-methoxy-4-hydroxyphenylglycol as the major metabolite. *Biochemical Pharmacology,* 17:247–254.
Snyder, S. H. (1972): Catecholamines in the brain as mediators of amphetamine psychosis. *Archives of General Psychiatry,* 27:169–179.
Spector, S., Sjoerdsma, A., and Udenfriend, S. (1965): Blockade of endogenous norepinephrine synthesis by alpha-methyl-tyrosine, an inhibitor of tyrosine hydroxylase. *Journal of Pharmacology and Experimental Therapeutics,* 147:86–95.
Stein, L., and Wise, C. D. (1971): Possible etiology of schizophrenia: Progressive damage to the noradrenergic reward system by 6-hydroxydopamine. *Science,* 171:1032–1036.
Suda, H., Takeuchi, T., Nagatsu, T., Matsuzaki, M., Matsumoto, I., and Umezawa, H. (1969): Inhibition of dopamine-β-hydroxylase by 5-alkylpicolinic acid and their hypotensive effects. *Chemical and Pharmacological Bulletin (Tokyo),* 17:2377–2380.
Terasawa, F., and Kameyama, M. (1971): The clinical trial of a new hypotensive agent, "fusaric acid (5-butypicolinic acid)": The preliminary report. *Japan Circulation Journal,* 35:339–357.
Von Voigtlander, P. F., and Moore, K. E. (1970): Behavioral and brain catecholamine depleting actions of U-14,624, an inhibitor of dopamine-β-hydroxylase. *Proceedings of the Society for Experimental Biology and Medicine,* 133:817–820.
Wise, D. C., and Stein, L. (1973): Dopamine-beta-hydroxylase deficits in the brains of schizophrenic patients. *Science,* 181:344–347.
Yaryura-Tobias, J. A., Diamond, B., and Merlis, S. (1970): The action of L-DOPA on schizophrenic patients (a preliminary report). *Current Therapeutic Research,* 12:528–531.

Neuropsychopharmacology of Monoamines and Their Regulatory Enzymes, edited by E. Usdin.
Raven Press, New York © 1974.

Behavioral and EEG Changes in the Amphetamine Model of Psychosis

Everett H. Ellinwood, Jr.

Behavioral Neuropharmacology Section, Department of Psychiatry, Duke University Medical Center, Durham, North Carolina 27706

I. INTRODUCTION

The psychoses associated with chronic amphetamine or cocaine intoxication and that associated with temporal lobe epilepsy more closely mimic schizophrenia, especially paranoid schizophrenia, than any other organic condition (Connell, 1958; Fish, 1962; Slater et al., 1963; Ellinwood, 1967, 1968). These conditions have not infrequently been diagnosed as schizophrenia. Other stimulants can, of course, produce a similar picture of schizophrenia, although the number of case studies is less extensive than for amphetamines; these other stimulants include phenmetrazine, diethylpropion, and methylphenidate (Bethel, 1957; McCormick and McNeil, 1963; Mandels, 1964; Jones, 1968). That an epileptic and a pharmacological condition produce a similar psychological syndrome presents conceptual difficulties, since (1) most current theoretical positions state that excessive dopamine activity is responsible for the amphetamine effect (Fog et al., 1967) and (2) the epileptic effect is secondary to a focal abnormal neuronal responsiveness to which dopamine makes no known contribution. Examination of the similarities and differences of these two conditions will contribute to our understanding of the mechanisms common to these two states as well as to the functional psychoses.

Perhaps a main reason that the psychoses associated with chronic stimulant intoxication and temporal lobe epilepsy closely relate to functional psychoses and to each other is that the duration of hyperarousal and distorted perceptual mechanisms is sufficiently extended (over weeks, months, and even years) to permit the development of cognitive processes and their underlying neuronal substrata to become reorganized or distorted. In many other drug-induced psychoses and organic conditions, the state of abnormal arousal is an acute one, and delusional belief systems do not establish themselves. The psychosis associated with temporal lobe epilepsy has features that are strikingly similar to those of the amphetamine psychosis, including ideas of influence; thought disorders; paranoid ideas, often colored with religiosity or philosophical concerns; auditory, visual, and olfactory hallucinations; distorted body images, especially distorted facial percepts; depersonalization; *deja vu;* and bruxism (Ellinwood, 1967, 1968). Another feature common to both syndromes is that the behaviors involved not infrequently become stereotyped (Ellinwood, 1967; Kramer et al., 1967; Oppenheimer, 1971). However, the most significant common feature is emotional arousal, perhaps best characterized by Conrad's term "apophany," which describes the intense meaning and significance attached to both thoughts and percepts in the initial stages of schizophrenia (Conrad, 1958). With both amphetamine and temporal lobe psychosis, patients express the same feelings of portentousness and heightened significance to their thoughts, especially what they describe as "eureka thoughts," and also to the meaning of "signifying" percepts. Statements from epileptic patients quoted from Slater and Beard (1963) illustrate the common similarities in thought between the two conditions:

> I had two thoughts side by side and then realized the untruth of Christianity. It all falls into a pattern. Things have some kind of connection. People's Christian names have a significance.

Increased general arousal also occurs in the functional psychotic states. The observation that many schizophrenics, both acute and chronic, function at a sustained and even abnormally high level of arousal has been developed by several investigators (Venables and Wing, 1962; Goldstein et al., 1963; Venables, 1963; Kornetsky and Mirsky, 1966). Related to the overarousal in schizophrenics are deficits noted in experiments on reaction times (Shakow, 1946) and distractibility based on overconclusive attention or a propensity to give equal weight to relevant and irrelevant stimuli (McGhie et al., 1965; Orzack and Kornetsky, 1966; Silverman, 1968).

Venables and Wing (1962) and Venables (1963) have demonstrated a linear relationship between the state of physiological arousal and the degree of withdrawal in schizophrenics. Venables (1963) also presents evidence that increased arousal adversely affects perceptual selectivity and functioning. This is consistent with the concept of Malmo (1959) and of Kornetsky and Mirsky (1966) that the relationship between levels of arousal and performance can be represented as an inverted U when one uses a continuum from sleep to states of hyperarousal. Schizophrenics have a basal arousal level near or at the peak of the inverted U; thus, any subsequent arousal leads to a deterioration of performance (Kornetsky and Mirsky, 1966). In keeping with these observations on arousal, the electroencephalogram of a schizophrenic is usually manifested as a low-voltage, desynchronized record. In more discriminative analysis, the low-voltage, fast activity has been repeatedly demonstrated to be comprised of very fast frequencies of 26 to 50 cycles/sec (Davis, 1942; Finley, 1944; Kennard and Schwartzman, 1957; Itil, 1972). Itil (1972) found the most significant differences between schizophrenics and normal subjects in the fast frequencies, especially in the 24- to 33-cycles/sec range and in the 40- to 50-cycles/sec range. Findings by Grinker (1973) indicate a specific increase in 29-cycles/sec activity in his schizophrenic patients. This cortical activity has been hypothesized to be reflective of subcortical mechanisms (Lester and Edwards, 1966). Since the peak frequency effect is often in the 28- to 29-cycles/sec range, one might wonder whether the amygdala or limbic spindle activity may contribute to these effects, since the frequency of the primate limbic spindle is most often approximately 28 to 30 cycles/sec. More recently, Hanley et al. (1972) reported the strong predictive correlation between schizophrenic stereotyped rituals and bandwidth narrowing in the 25- to 28-hertz range from the septum-amygdala. Although this represented only a single case report, Hanley et al. also demonstrated in another patient that parameters in these higher frequencies successfully achieved pre- and postspike (epileptic) discrimination from the septal region. In animals the amygdala spindle has been reported to be a much more sensitive and, specifically, a more predictive indicator of the upper levels of normal arousal than cortical changes from slow-wave activity to low-voltage, desynchronized activity (Paganno and Gault, 1964). For several years now, we have used chronic amphetamine intoxication in experimental animals as a model of psychosis (Ellinwood and Escalante, 1970; Ellinwood and Sudilovsky, 1973). In our studies we have also found that the spindle parameters in the olfactory system accurately reflect both the normal and abnormal states of hyperarousal in the form of an inverted U. In addition, there is often an evolution of slow-wave activity associated with the development of stereotyped behavior. In this chapter we will describe some of these electrical changes and their relationship to attention and arousal behaviors.

II. AMPHETAMINE MODEL PSYCHOSIS

Our model psychosis in cats is based on a 2-week period of chronic intoxication with increasing doses of amphetamine (15 mg/kg per day to 28 mg/kg per day) (Ellinwood, 1971; Ellinwood and Sudilovsky, 1973). Although longer periods of intoxication produce more pronounced bizarre behaviors, the 2-week period has certain logistic advantages. Samples of behavior and EEG were simultaneously recorded on analogue and videotape periodically before and after the amphetamine injections. Behavior during 10-sec intervals is rated according to an extensive behavioral rating system, details of which have been presented elsewhere (Ellinwood, Sudilovsky, and Nelson, 1972). Briefly, the rating chart consists of 28 behavioral categories such as head movement, arousal level, and attitude. The EEG is analyzed by computer techniques in a variety of ways, including a spindle identification and quantification program (Grabowy and Ellinwood, 1971; Ellinwood, Sudilovsky, and Grabowy, 1973). The reference level is defined for each cat based on the average value of the spindle parameter at arousal level 3 in preinjection recordings. All spindles are presented as a percent of this reference level.

In general, behaviors on day 1 are characterized by patterns of "investigative" stereotypies. Over the days of intoxication the behavior becomes more autonomous, fragmented, and disorganized. At times, behavior evolves into periods of repeated startle reactiveness in the absence of any adequate stimuli. Although even on day 1 the animals become hyperaroused, the behavior is more cohesive at that time than on the later days of intoxication. The reactive attitude and attitudes we have called "focused" (the animal's attentive attitude is frozen in a focus on one narrow aspect of the environment) appear to be the most bizarre behavior states and are usually seen on the last days of the 2-week cycle. In this stage, one also observes many adventitious movements and dyskinesias (Ellinwood and Sudilovsky, 1973).

III. SPINDLES AND AROUSAL

Spindle amplitude, frequency, and duration demonstrate a remarkable U-shaped function in relation to the level of arousal of the olfactory bulb parameters, as can be demonstrated in Fig. 1. In this figure, arousal level 1 represents sleep, arousal level 2 is awake but in a different level of arousal, 3 is the normal level of peak arousal in the cat, 4 is an abnormal state of arousal, and 5 is a hyperexcitable reactive or frozen state of arousal. Figure 1 demonstrates two studies, one a repetition of the other. As can be noted, amplitude, frequency, and duration all reach a peak at normal arousal level 3 and then fall off. Changes from arousal level 3 to the abnormal states 4 and 5 are reflected most sensitively in the frequency and duration of spindles. During the experiments, observation of EEG readout would demonstrate changes from an average spindle frequency of 40 to 45 cycles/sec to one in the range of 50, 55, or even 70 cycles/sec as the behavioral arousal level increased. In addition, the envelope of the spindle would be reduced in duration. The one exception to the U-shaped function of the parameters is the change in the amplitude from arousal level 4 to 5. At arousal level 5 the amplitude shows a tendency to increase or at least to reach a plateau.

There is a large variability in an arousal 5 behavioral condition which we have designated as reactive. Some cats will have high-voltage spindles during periods in which they are in the reactive attitude; others will have practically no spindles at this time. The reactive attitude is one in which the cat suddenly reacts disproportionately to restricted stimuli with a jumpy, nervous, or jittery quality (Ellinwood, Sudilovsky, and Nelson, 1972). It is highlighted by startle reactions, sudden jumps, reaching out

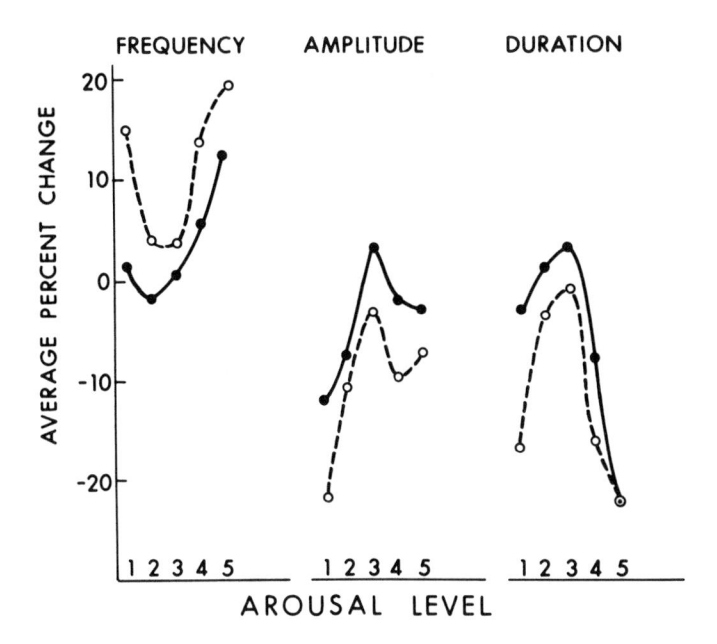

FIG. 1. Average spindle parameters in the olfactory bulb. Solid lines: original study (cats 63–74); broken lines: replication study (cats 75–82).

with a paw or shaking the paw as if there were something stuck to it, and sudden head shakes. Most of the behaviors indicated an apparent hyperreactivity to mainly "imaginary" stimuli. For example, an animal might appear to be intensely watching a nonexistent bug crawling on the cage wall and might then either reach out for it or suddenly jump back. The reactive behaviors appear in the later days of intoxication, especially at the end of the second week of intoxication. The reactive attitude is quite different from the investigatory attitude, which is intimately associated with the more intense stereotyped activity during which the animal focuses on one or a restricted number of areas in his environment. In the investigatory animal there is little if any reaction to extraneous stimuli; usually the animal continues to investigate the particular area involved in his stereotypy and does not react to such stimuli as catnip, mice, or food.

The spindling associated with the reactive attitude frequently appeared to initiate from the amygdala, olfactory tubercle, or accumbens rather than from the usual pacemaking role of the olfactory bulb. Spindles will at times immediately precede startle reactions, as can be seen in Fig. 2, where a spindle is shown leading into the movement artifact.

At times the spindles appear in sustained paroxysms in several subcortical nuclei in what we have called "spindle showers" (Ellinwood, Sudilovsky, and Grabowy, 1973). At such times spindles will appear in the interpeduncular nucleus and substantia nigra, whereas for the most part

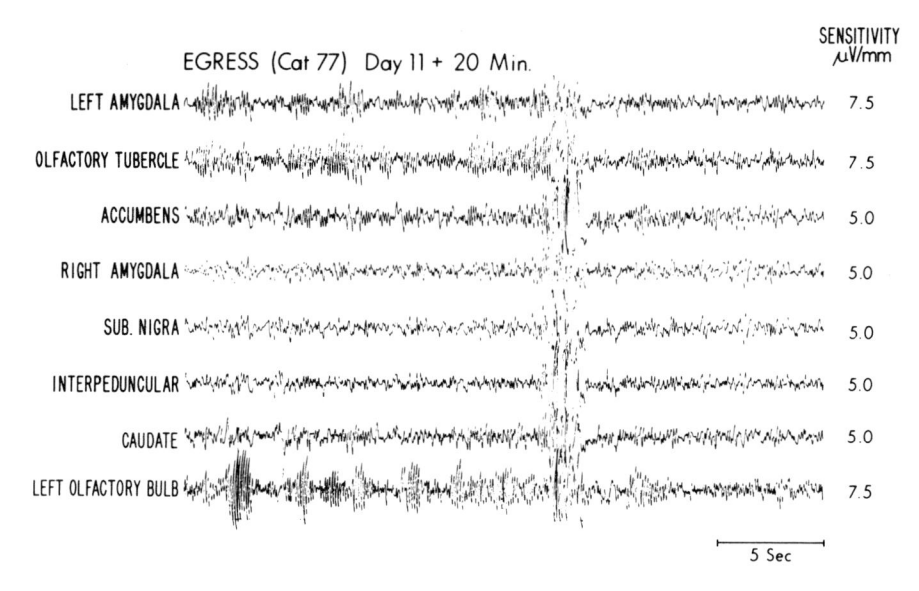

FIG. 2. Spindle activity preceding reactive startle movement.

they normally do not appear in these nuclei. The change toward high-voltage spindles produced by amphetamine can be enhanced and the process speeded up by simultaneously administering disulfiram, a dopamine-beta-hydroxylase inhibitor (Ellinwood, Sudilovsky, and Grabowy, 1973). Although disulfiram is not a "clean" drug and causes multiple metabolic changes, we were specifically interested in using this compound to block norepinephrine synthesis because it has been reported to produce psychosis (Liddon and Satran, 1967). In addition, Heath et al. (1965) also demonstrated that disulfiram (1 to 2 g/day) either produced hallucinations and delusions *de novo* or enhanced these symptoms in schizophrenics. The spindle burst in the amphetamine- and disulfiram-treated animals was also associated with an increase in aroused frozen or reactive behaviors. These behavioral states could be produced in half the time and at much lower doses of amphetamine when the combination of drugs was administered.

IV. NOREPINEPHRINE MECHANISMS AND HYPERSYNCHRONY

In addition, the amphetamine- and disulfiram-treated cats developed forebrain seizures that were associated with a marked increase in spindle amplitude and a decrease in spindle frequency. Figure 3 demonstrates the increase in spindle voltage and decrease in frequency associated with the onset of seizures. Spindles in the early stages of intoxication usually preceded the discharges in a regular manner and were often phase-locked with

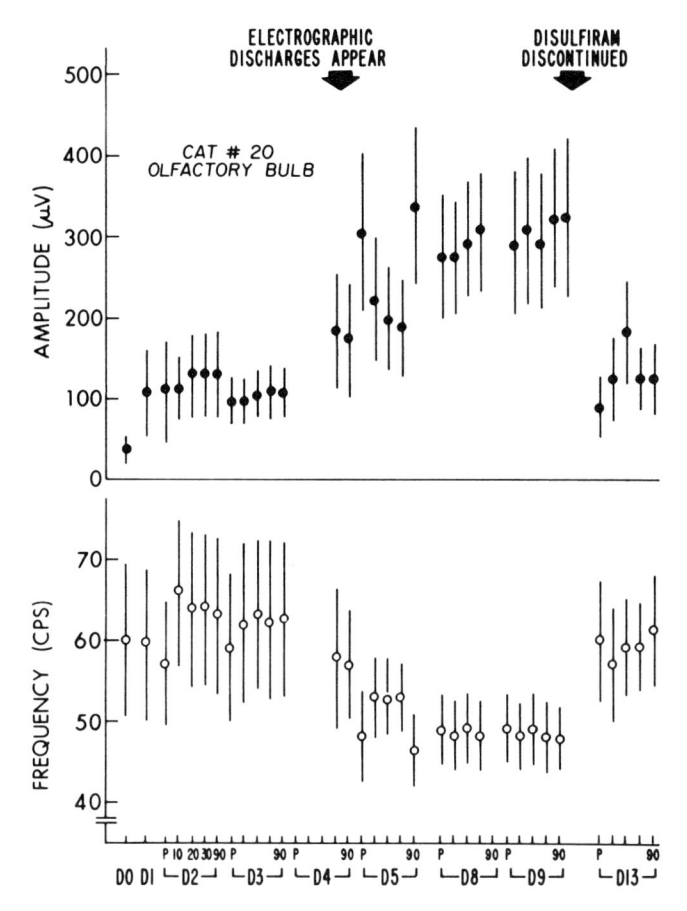

FIG. 3. Spindle amplitude and frequency change associated with onset of amphetamine and disulfiram seizures.

the discharge (Ellinwood, Sudilovsky, and Grabowy, 1973). The discharges at this stage appeared to be paced by respiratory mechanisms. Often the decreased frequency in the olfactory bulb tubercle, amygdala, and accumbens was a 20- to 25-cycles/sec harmonic frequency of the regular spindle and was at times admixed with the regular spindle frequency.

The electrograph discharges subsequently spread to the other electrodes. Either previously or at about the same time, a 10- to 20-cycles/sec rhythm often increased in voltage and regularity in the region of the interpeduncular nucleus and substantia nigra. Alternatively, when the epileptic discharge spreads to the interpeduncular nucleus substantia nigra, it may be followed by a 10- to 20-cycles/sec afterdischarge (Fig. 4). When generalized seizures occur, they appear to form as a gradual recruitment of seizure activity into or superimposed upon the afterdischarges of the high-voltage mesencephalic

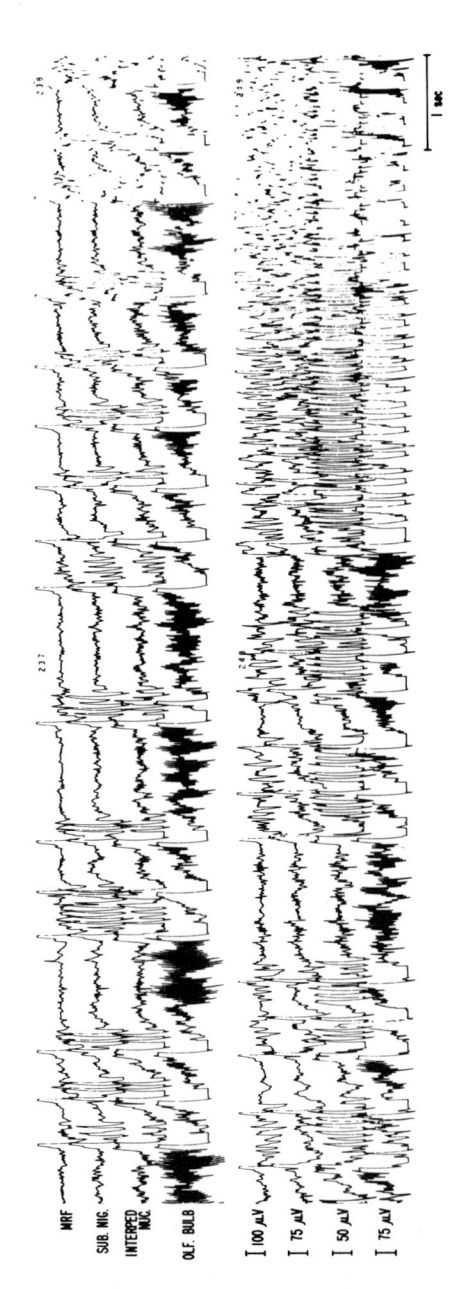

FIG. 4. Seizure discharges and generalized seizure. Note (A) regular association of spindles with discharges, (B) 12/sec afterdischarge in interpeduncular nucleus leading to generalized seizures.

rhythms. We (Ellinwood, Sudilovsky, and Grabowy, 1973) have previously presented the hypothesis that amphetamine and disulfiram combination produces a stimulating effect in the relative absence of norepinephrine and reflects the same condition seen in chronic amphetamine intoxication, where there is a relatively greater depletion of norepinephrine than dopamine (Gunne and Lewander, 1967). Although generalized seizures are not regularly encountered either in experimental animals or in human amphetamine abusers, they may occur after administration of large or especially chronic doses of either amphetamine or cocaine to experimental animals (Eidelberg et al., 1961, 1963; Ellinwood, Sudilovsky, and Grabowy, 1973). We have been interested in this electrical activity leading to seizures because it may be related to the abnormal behavioral states noted in chronic stimulant intoxication. The mechanisms leading to seizures may provide insight into the underlying conditions of the abnormal behavioral states.

Recently, Spencer and Turner (1969) reported that dextroamphetamine acts as a proconvulsant when combined with leptazol. In addition, they found that an intact dopamine synthesis was necessary for amphetamine potentiation with the convulsant action of leptazol; however, neither intact norepinephrine nor 5-hydroxytryptamine synthesis was necessary. We have found that chronically intoxicated monkeys not infrequently die in status epilepticus when methamphetamine is increased by 0.25 mg/kg per day or more. In direct opposition to these findings, both D-amphetamine and cocaine are noted for their effectiveness in increasing electric shock convulsion thresholds (Tainter et al., 1943). Moreover, amphetamine in substantial doses (20 mg/kg) increases the threshold to electric shock seizures by 90% (DeSchaepdryver et al., 1962). Dextroamphetamine (15 mg/kg) also has a significant protective action against audiogenic seizures (Lehmann, 1967). Lehmann (1970) recently summarized the data on audiogenic seizures, which indicate that all agents with anticonvulsant activity to audiogenic seizures, including dextroamphetamine, have a net effect of increasing the release of norepinephrine onto the receptor site.

Thus, it appears that there may be at least two possibly mutually antagonistic actions of dextroamphetamine: (1) a norepinephrine-mediated inhibition of seizure activity; (2) an activation of olfactory forebrain rhythms followed by seizures which might be related to dopamine action or action of other neural transmitters in the relative absence of norepinephrine as produced by disulfiram or chronic amphetamine intoxication.

Understanding the mechanisms that lead to electrical activity preceding temporal lobe seizures (amygdala), especially the mechanisms related to chronic amphetamine intoxication, would be an exciting advance since, as we mentioned earlier, the psychosis associated with temporal lobe epilepsy has a remarkable correspondence to the psychosis produced in humans with chronic amphetamine intoxication (Ellinwood, 1968). The story is even more exciting when one considers that cocaine, even in acute doses, produced the same type (Eidelberg et al., 1961, 1963) of amygdala seizure as that associated with the high-voltage, slow spindles seen with amphetamine and disulfiram. But, alas, the story is not complete without mentioning that almost all local anesthetics produce the amygdala seizure pattern (Wagman et al., 1967; Tuttle and Elliot, 1969; Riblet and Tuttle, 1970); yet cocaine is the only local anesthetic thought to have an effect on catecholamine mechanisms. More recently we have, however, demonstrated *(unpublished results)* that amygdala seizure activity produced by cocaine and lidocaine has a lower threshold when the animals are pretreated with FLA 63, another dopamine-beta-hydroxylase inhibitor.

Another surprising observation in the cocaine- and lidocaine-induced seizures is that the effects can be localized or unilateral. Thus high-voltage spindle activity leading to seizures can begin in one amygdala with associated high-voltage theta activity in the hippocampus, and the entire spindle activity in the forebrain appears to be paced from this amygdala. The other amygdala might show slight response or at times no response to the activity in the contralateral amygdala.

In contrast to this rather spectacular spindle pacemaking with local anesthetics, in amphetamine intoxication even with quite high doses one does not observe this type of electrophysiological organization unless norepinephrine mechanisms are blocked. We have been particularly interested in pacemaker-type activities with amphetamine intoxication because of their possible explanatory role in the very constricted behaviors known as stereotypies. The most consistent repeated EEG pattern that spreads or could have a pacing function with other nuclei is the olfactory bulb slow wave (see Fig. 5). Most frequently the olfactory slow wave increases in frequency and then is noted almost always in the olfactory tubercle and accumbens but in some cases spreads to the interpeduncular nucleus and caudate as well as the amygdala. This olfactory wave is phase-related to stereotyped sniffing movements (Ellinwood and Escalante, 1970), but the olfactory wave is also frequently present in the absence of stereotyped sniffing. In some cats one observes a gradually organizing slow activity that often appears first in the amygdala and also in the olfactory tubercle and accumbens. Slow activity then appears to spread to the dopamine-containing nuclei in high-voltage synchronous waves that are often, but not invariably, associated with the stereotyped motor pattern (Fig. 5).

V. DISCUSSION

In their book on experimental psychosis, Weil-Malherbe and Szara (1971) present the case that theories of psychosis using arousal concepts may provide heuristic and integrative functions which bridge the gap between molecular and behavioral hypotheses as well as between animal and human observations. Perhaps the most important bridging function is between the molecular level, where drugs act, and the more complex symbolic functions of the human mind, which are most often the data observed with pharmacologically induced psychoses (Weil-Malherbe and Szara, 1971). We (Ellinwood, and Sudilovsky, 1973; Ellinwood, Sudilovsky and Nelson, 1973) have argued that sufficient periods of emotional hyperarousal with associated distorted percepts are necessary if pharmacological models of psychosis are to be extended to human counterparts in the functional psychoses. This type of concept was originally presented by Fish (1961); that is, hyperarousal produced by overactivity of the mesencephalic reticular activating systems, if of sufficient duration, may cause new reverberating associative circuits or "parallel processes" in the cortical and subcortical regions (Hebb's cell assemblies), and these can become autonomous and remain even though the arousal overactivity eventually subsides. The psychological and neuronal reorganization represents the chronic stages of psychosis. In our work with chronic amphetamine intoxication, such phenomena do appear in the latter stages of intoxication in what we have called autonomous behavioral configurations (Ellinwood and Sudilovsky, 1973). These autonomous behavioral units are enhanced by the disulfiram potentiation of amphetamine effect. In the original stereotypies (at the initial stage of intoxication) the body posture, head movements, and attending movements all formed a behavioral configuration with a central vector of organization. In the end stages of amphetamine intoxication, one notes that autonomous units of behavior appear to have little or no relationship with one another. This is best noted in the relative autonomy of activity between the various segments of the body, including the hip-hindleg region, the shoulder-foreleg region, and the head-neck region.

Most of the stereotypies are comprised primarily of the postural-motor components of patterns of attending and examining (Ellinwood, 1971). We have proposed that these components represent the ground plan of perceptual-motor behaviors, perhaps as an intrinsic scaffolding of behaviors that are programmed in the limbic extrapyramidal catecholamine systems. For example, we would say that the stereotypy associated with sniffing is made up of the motor as well as the postural components of sniffing behavior, and that this behavioral complex represents the postural-motor underpinnings of olfactory perception. These postural-motor base components can be activated by drugs that stimulate dopamine mechanisms. Thus, the so-called reverberating associative circuits may represent overactivity in one of the postural subsets of attention.

We have been interested in the limbic spindle and olfactory slow wave not only as a harbinger of arousal but because it may take part in pacemaking or gating mechanisms associated with arousal in the forebrain as well as overactivity in certain circuits. Our studies have shown that there is a spread of the olfactory slow waves associated with stereotypy (Ellinwood and Escalante, 1970) and that seizures produced by amphetamine as well as by

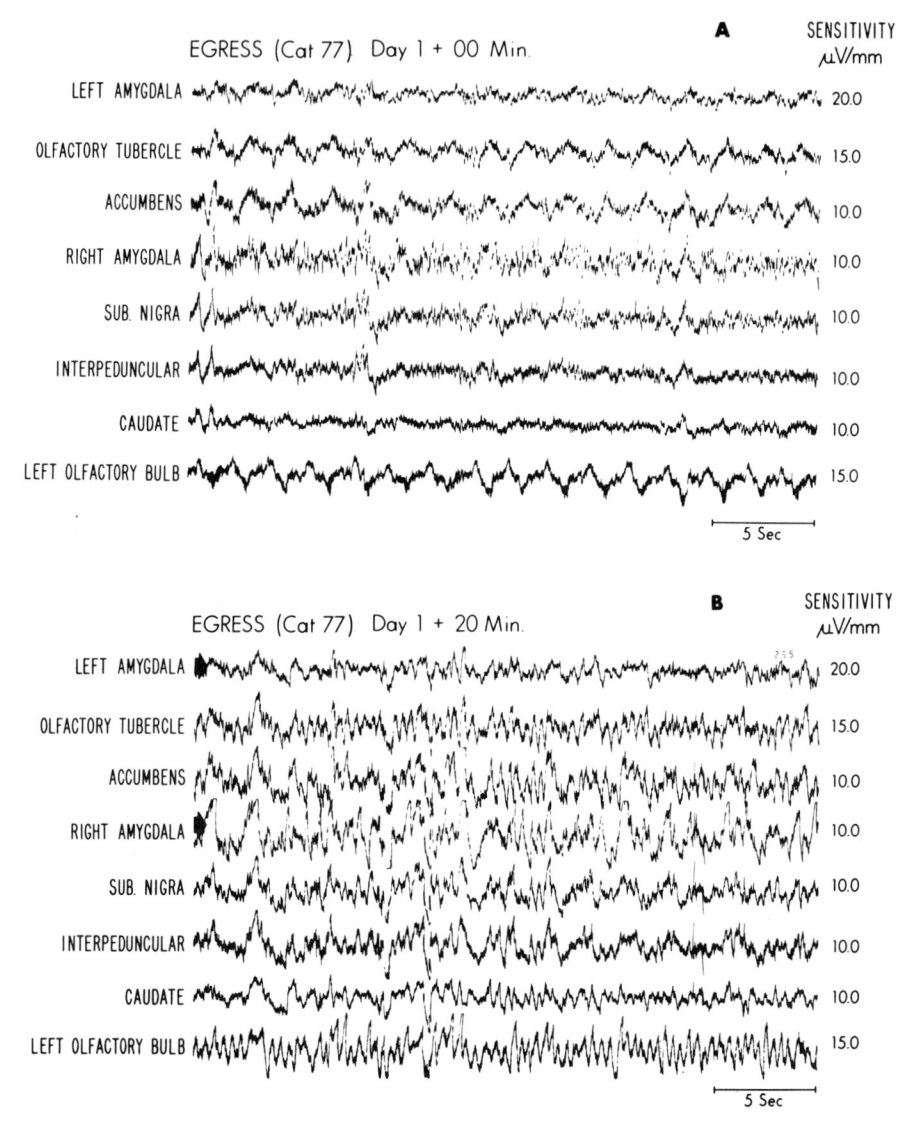

FIG. 5. **A.** Electrographic recording at "00" min (actually 2½ min) after injection. Note slow-wave activity in olfactory bulb and tubercle and accumbens that is beginning to become more regular and faster. **B.** Slow-wave now at 2½ cycles/sec and phase related to tiny sniffing movements; high-voltage slow activity primarily in right amygdala. **C.** High-

local anesthetics are dependent on nasal air flow (Tuttle and Elliot, 1969). In addition, the epileptic discharge-spindle complex is often synchronous with nasal breathing (Wagman et al., 1964; Ellinwood, Sudilovsky, and Grabowy, 1973). These preseizure activities may provide clues as to the basis of the pacemaking mechanisms underlying the autonomous behavioral configurations. The movements and posture associated with coordination of voltage slow has now spread and predominates in accumbens, substratia nigra, interpeduncular, and caudate. This activity is in phase with a stereotyped head movement shown by mark under amygdala. *D.* Olfactory waves have now spread throughout the leads.

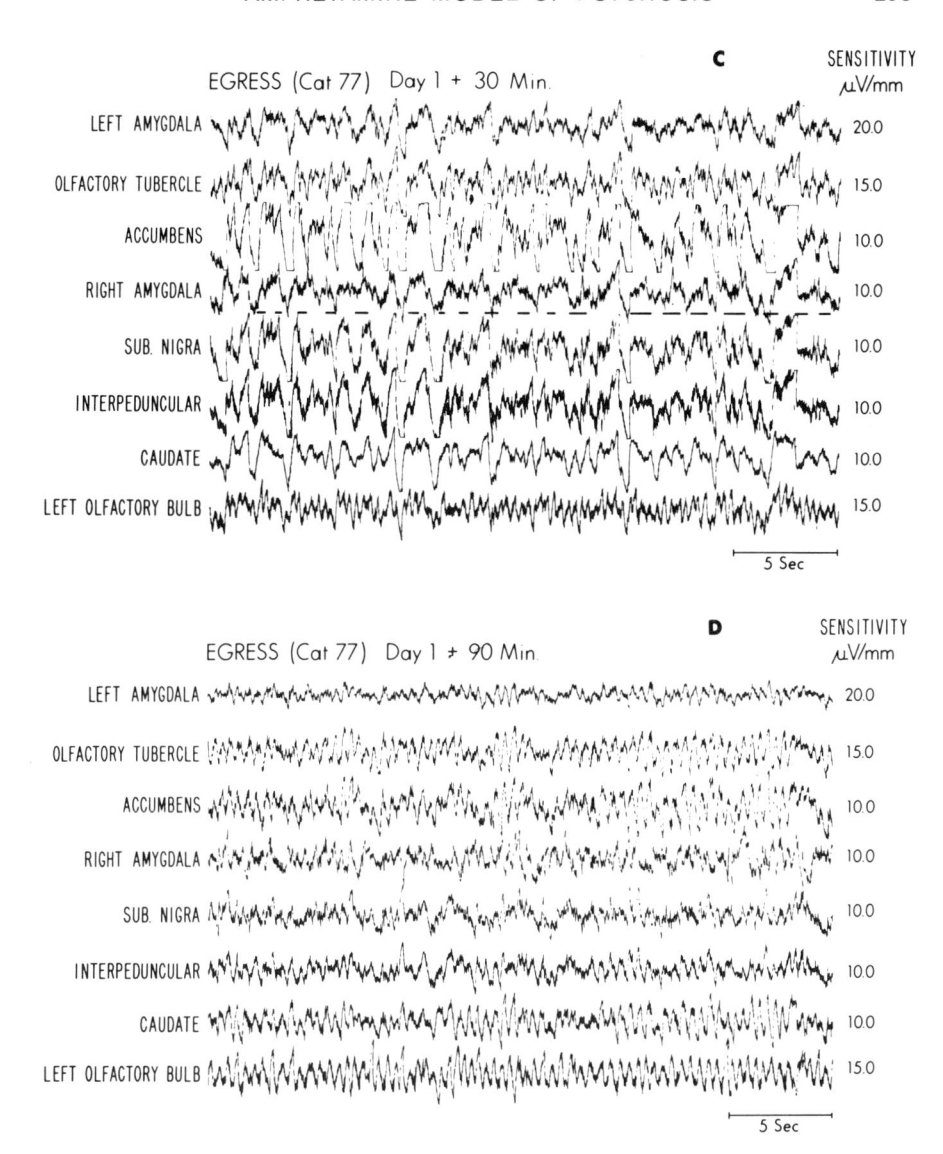

breathing with behavior (including sniffing), and especially with attention to olfactory inputs, may be one postural subset of attention. In a similar manner, eye-hand coordinated attention patterns represent another postural base of attention in primates that becomes part of the stereotypy complex in both monkeys and man (Ellinwood, 1971).

Stimulation of postural stereotypies in the initial stage of intoxication also provides clues to the nature of parakinesias or dyskinesias noted in the end stages of amphetamine intoxication as well as in schizophrenia. With amphetamine intoxication, the movements involved in initial stereotype behavior not infrequently become fragmented and autonomous, and one of these autonomous fragments is often represented by the end-stage dyskinesia. For example, a cat that was initially involved in a stereotypy of grooming the groin region subsequently developed a leg-raising movement that occurred any time she turned her head 20° toward the side of the grooming reaction. The grooming reaction was not completed at this stage of the stereotypy. Later one could note a leg jerk as an isolated behavior. Similarly, a monkey that began with a hand-to-mouth examining pattern stereotypy subsequently, over a period of 4 to 5 months of intoxication, developed a dystonic neck turn or dyskinetic mouthing movement, with the associated hand movement long since having dropped out of the picture (Ellinwood, 1971). The original manner in which many of these behaviors seem to be elicited is based on what appear to be prepotent attending behaviors which became active with the arousal effects of amphetamine and then later become more autonomous. In keeping with these observations, amphetamine is known to stimulate circuits that are already active. This local circuit gradient of neuronal overactivity may have exhausted certain catecholamine neurons, leading to neuronal chromatolysis such as we have previously observed in chronic amphetamine intoxication (Escalante and Ellinwood, 1970, 1972). Neuronal chromatolysis is found primarily in areas with catecholamine-containing neurons. Neuronal chromatolysis of catecholamine neurons could result in circuits supersensitive to catecholamines especially produced by subsequent states of arousal not induced by amphetamines. After periods sufficient for recovery (1 to 2 months), many of our chronic intoxication animals will manifest stereotypies or dyskinesias on reentry into the experimental cage; even the sight of the experimenter can trigger these responses. This type of response can be explained both on the basis of conditioning or on states of arousal triggering these behaviors.

In summary, it is important to conceive of psychosis and experimental models of psychoses as developing systems that require time and noting that there are several stages to the development, each characterized by its own type of behavior and organization. In our studies we have been interested in the electrophysiological changes associated with these behavioral states. Although the data are quite preliminary, there are indications that there may be changes in the organization gradients of electrical activity of the forebrain that may be related to pacemaking activity. Preseizure activity in the olfactory forebrain indicates that electrical waveforms associated with respiration are involved in the elicitation of the responses. There are dramatic changes in the limbic spindle complex associated not only with the states of abnormal arousal but also with the onset of this seizure activity. Arousal concepts may provide the intermediate conceptual bridge for integrating electrophysiological behavioral and pharmacological observations in model psychoses.

ACKNOWLEDGMENTS

Many of the studies reviewed here were done in collaboration with Dr. Abraham Sudilovsky.

This work was supported in part by National Institute of Mental Health grants MH 07073 and MH 18904.

REFERENCES

Bethel, M. F. (1957): Toxic psychosis caused by preludin. *British Medical Journal*, 1:30.
Connell, P. H. (1958): *Amphetamine Psychosis,* Institute of Psychiatry, Maudsley Monograph 5. Oxford University Press, London.
Conrad, K. (1958): *Die deginnende Schizophrenia*. Thieme, Stuttgart.
Davis, P. A. (1942): Evolution of electroencephalograms of schizophrenic patients. *American Journal of Psychiatry*, 96:851.
De Schaepdryver, A. F., Piette, Y., and Delaunois, A. L. (1962): Brain amines and electroshock threshold. *Archives Internationales de Pharmacodynamie et de Thérapie*, 140:358.
Eidelberg, E., Lesse, H., and Gault, F. P. (1961): Convulsant effects of cocaine. *Federation Proceedings*, 20:322.
Eidelberg, E., Lesse, H., and Gault, F. P. (1963): Experimental model of temporal lobe epilepsy: Studies of convulsant properties of cocaine. In: *EEG and Behavior,* edited by G. H. Glass. Basic Books, New York, p. 272.
Ellinwood, E. H., Jr. (1967): Amphetamine psychosis I: Description of the individuals and process. *Journal of Nervous and Mental Diseases*, 44:273–283.
Ellinwood, E. H., Jr. (1968): Amphetamine psychosis I: Theoretical implications. *International Journal of Neuropsychiatry*, 4:45–54.
Ellinwood, E. H., Jr. (1971): Comparative methamphetamine intoxication in experimental animals. *Pharmakopsychiatrica*, 24:4, 351–61.
Ellinwood, E. H., Jr., and Escalante, D. O. (1970): Chronic amphetamine effect on the olfactory forebrain. *Biological Psychiatry*, 2:189–203.
Ellinwood, E. H., Jr., and Sudilovsky, A. (1973): Chronic amphetamine intoxication: Behavioral model of psychoses. In: *Psychopathology and Psychopharmacology,* edited by J. Cole et al. Johns Hopkins Press, Baltimore, pp. 51–70.
Ellinwood, E. H., Jr., Sudilovsky, A., and Grabowy, R. S. (1973): Olfactory forebrain seizures induced by methamphetamine and disulfiram. *Biological Psychiatry*, 7(2):89–99.
Ellinwood, E. H., Jr., Sudilovsky, A., and Nelson, L. (1972): Behavioral analysis of chronic amphetamine intoxication. *Biological Psychiatry*, 4(3):215–229.
Ellinwood, E. H., Jr., Sudilovsky, A., and Nelson, L. (1973): Evolving behavior in the clinical and experimental amphetamine (model) psychosis. *American Journal of Psychiatry*, 130(10): 1088–1093.
Escalante, D. O., and Ellinwood, E. H., Jr. (1970): Central nervous system cytopathological changes in cat with chronic methedrine intoxication. *Brain Research*, 21:555.
Escalante, D. O., and Ellinwood, E. H., Jr. (1972): Effects of chronic amphetamine intoxication on adrenergic and cholinergic structures. In: *Current Concepts on Amphetamine Abuse,* edited by E. H. Ellinwood and S. Cohen. U.S. Government Printing Office, Washington, D.C., pp. 97–106.
Finley, K. H. (1944): On the occurrence of rapid frequency potential changes in the human electroencephalogram. *American Journal of Psychiatry*, 101:194.
Fish, F. (1961): A neurophysiological theory of schizophrenia. *Journal of Mental Science*, 107:828–38.
Fish, F. K. (1962): *Schizophrenia.* Williams and Wilkins Co., Baltimore, p. 128.

Fog, R. L., Randrup, A., and Pakkenberg, H. (1967): Adrenergic mechanisms in corpus striatum and amphetamine-induced stereotyped behavior. *Psychopharmacologia*, 11:179.

Goldstein, L., Murphree, H. B., Sugarman, A. A., Pfeiffer, C. C., and Jenney, E. H. (1963): Quantitative electroencephalographic analysis of naturally occurring schizophrenia in drug induced psychotic states in human males. *Clinical and Pharmacological Therapeutics*, 4:10.

Grabowy, R. S., and Ellinwood, E. H., Jr. (1971): On-line detection of EEG spindle activity. *Decus* Spring Symposiums, pp. 42–45.

Grinker, R. R., Holzman, P., and Kayton, L. A. (1973): Unpublished results as quoted in *Schizophrenia Bulletin*, Issue 7, pp. 36–37.

Gunne, L. M., and Lewander, T. (1967): Long term effects of some dependence producing drugs of the brain monoamines. In: *Molecular Basis of Some Aspects of Mental Activity*, Vol. 2, edited by O. Wahaas. Academic Press, New York.

Hanley, J., Rickles, W. R., Crandall, P. H., and Walter, R. D. (1972): Automatic recognition of EEG correlates of behavior in a chronic schizophrenic patient. *American Journal of Psychiatry*, 128:12.

Heath, R. G., Nesselhof, W., Bishop, N. P., and Byers, L. W. (1965): Behavior and metabolic changes associated with administration of tetraethylthiuram disulfide (Antabuse). *Diseases of the Nervous System*, 26:99.

Itil, T. M., Saletu, B., and Davis, J. (1972): EEG findings in chronic schizophrenics based on digital computer period analysis and analog power spectrum. *Biological Psychiatry*, 5:1.

Jones, H. S. (1968): Diethylpropion dependence. *Medical Journal of Australia*, 1:267.

Kennard, M. A., and Schwartzman, A. F. (1957): A longitudinal study of electroencephalographic frequency patterns in mental hospital patients and normal controls. *Electroencephalography and Clinical Neurophysiology*, 9:263.

Kornetsky, G., and Mirsky, A. F. (1966): On certain pharmacological and physiological differences between schizophrenic and normal persons. *Psychopharmacologia* (Berlin), 8:309.

Kramer, J. C., Fischman, V. S., and Littlefield, D. C. (1967): Amphetamine abuse—pattern and effects of high doses taken intravenously, *Journal of the American Medical Association*, 201:305–309.

Lehmann, A. (1967): Audiogenic seizures data in mice supporting new theories of biogenic amines mechanisms in the central nervous system. *Life Sciences*, 6:1421.

Lehmann, A. (1970): Psychopharmacology of the response to noise, with special reference to audiogenic seizure in mice. In: *Physiological Effects of Noise*, edited by B. L. Welch and A. S. Welch. Plenum Press, New York.

Lester, B. K., and Edwards, R. J. (1966): EEG fast activity in schizophrenic and control subjects. *International Journal of Neuropsychiatry*, 2:143.

Liddon, S., and Satran, R. (1967): Disulfiram (and abuse) psychosis. *American Journal of Psychiatry*, 123:1284.

Malmo, R. B. (1959): Activation: A neuropsychological dimension. *Psychological Reviews*, 66:367–386.

Mandels, J. (1964): Paranoid psychosis associated with phenmetrazine addiction. *British Journal of Psychiatry*, 110:865.

McCormick, T. C., and McNeil, T. W. (1963): Acute psychosis in Ritalin abuse. *Texas Medicine*, 59:99–100.

McGhie, A., Chapman, J., and Lawson, J. S. (1965): The effect of distraction on chronic schizophrenic performance. I: Perception and immediate memory. *British Journal of Psychiatry*, 11:383.

Oppenheimer, H. (1971): *Clinical Psychiatry*. Harper & Row, New York, pp. 167–168.

Orzack, M. H., and Kornetsky, C. (1966): Attention disfunction in chronic schizophrenia. *Archives of General Psychiatry*, 14:323.

Pagano, R. R., and Gault, F. P. (1964): Amygdala activity: A central measure of arousal. *Electroencephalography and Clinical Neurophysiology*, 17:255.

Riblet, L. A., and Tuttle, W. W. (1970): Investigation of the amygdaloid and olfactory electrographic response in the cat after toxic dosage of lidocaine. *Electroencephalography and Clinical Neurophysiology*, 28:601–608.

Shakow, D. (1946): The nature of deterioration of schizophrenic condition. *Nervous Diseases Monograph Series*, 70:1.

Silverman, J. (1968): A paradigm for the study of altered states of consciousness. *British Journal of Psychiatry*, 114:1201.

Slater, E., Beard, A. W., and Glithero, E. (1963): The schizophrenic-like psychosis of epilepsy. *British Journal of Psychiatry,* 109:95–150.
Spencer, P. S. J., and Turner, T. A. R. (1969): Blockade of biogenic amine synthesis: Its effect on the responses to leptazol and dextroamphetamine in rats. *British Journal of Pharmacology,* 37:94.
Tainter, M. L., Tainter, E. G., Lawrence, W. S., Neuru, E. N., Lackey, R. M., Luduena, F. P., Kirtland, H. B., and Gonzalez, R. I. (1943): Influence of various drugs on the threshold for electrical convulsions. *Journal of Pharmacology and Experimental Therapeutics,* 79:42.
Tuttle, W. W., and Elliot, H. W. (1969): Electrographic and behavioral study of convulsants in the cat. *Anesthesiology,* 30:48.
Venables, P. H. (1963): Selectivity of attention withdrawal and cortical activation. Studies in chronic schizophrenia. *Archives of General Psychiatry,* 9:74.
Venables, P. H., and Wing, K. K. (1962): Level of arousal and the sub-classification of schizophrenia. *Archives of General Psychiatry,* 7:114.
Wagman, I. H., De Jong, R. H., and Prince, D. (1964): Effects of lidocaine upon spontaneous and evoked activity within the limbic system. *Electroencephalography and Clinical Neurophysiology,* 17:453.
Wagman, I. H., De Jong, R. H., and Prince, D. (1967): Effects of lidocaine upon the nervous system. *Anesthesiology,* 28:155.
Weil-Malherbe, H., and Szara, S. I. (1971): *The Biochemistry of Functional and Experimental Psychoses.* Charles C Thomas, Springfield, Ill.

Serum Prolactin Levels in Newly Admitted Psychiatric Patients

H. Y. Meltzer,* E. J. Sachar,** and A. G. Frantz†

*Department of Psychiatry, University of Chicago Pritzker School of Medicine and the Illinois State Psychiatric Institute, Chicago, Illinois 60637 **Department of Psychiatry, Albert Einstein College of Medicine, Bronx, New York 10461 †Department of Medicine, Columbia University College of Physicians and Surgeons and the Presbyterian Hospital, New York, New York 10032

I. INTRODUCTION

During the last two decades there has been a great deal of interest in the possibility that an abnormality in one or more of the biogenic amines, e.g., serotonin, norepinephrine, and dopamine, is importantly involved in the etiology of at least some types of schizophrenia. There have been myriad experimental approaches to support or refute these theories. The aim of this chapter is to report the initial results of a neuroendocrinologic approach to the study of the role of dopamine in the etiology of schizophrenia. We will first briefly review the evidence for hyperactivity of some dopaminergic neurons as an etiologic factor in schizophrenia. We will then briefly present some of the evidence for the dopaminergic control of prolactin levels in serum which, if it were the predominant influence on serum prolactin levels, would then permit the serum prolactin level to be an indirect measure of the dopaminergic activity of the tuberoinfundibular tract of the brain. As will be apparent, it remains to be established how significant are the effects of the release of dopamine in the median eminence to the ultimate secretion of prolactin from the anterior hypothalamus in comparison with other influences on prolactin secretion.

The hypothesis that the pathogenesis of some forms of schizophrenia is associated with hyperactivity of brain dopaminergic neurons — or else underactivity of cholinergic or serotonergic neurons which may normally antagonize the effects of dopamine — is based on a variety of observations, only some of which will be reviewed here. The antipsychotic drugs, such as the phenothiazines, butyrophenones, and pimozide, are believed to exert their antipsychotic action by blocking dopamine receptors (Carlsson and Lindquist, 1963; Andén, Butcher, Corrodi, Fuxe, and Ungerstedt, 1970). Horn and Snyder (1971) have demonstrated by X-ray crystallography a similar conformation of chlorpromazine and dopamine, which would provide a molecular mechanism for dopamine receptor competition and blockade by the phenothiazines. Clinically effective phenothiazines and butyrophenones in low doses can inhibit a dopamine-stimulated adenylate cyclase from rat caudate nucleus, whereas phenothiazines such as promethazine, which are not effective in treating schizophrenia, have no effect on this enzyme (Kebabian, Petzold, and Greengard, 1972). York (1972) found that chlorpromazine blocked the excitant and inhibitory effect of iontophoretically applied dopamine on the frequency of discharge of cells in the rat putamen. Aghajanian and Bunney (1973) have reported an excellent correlation between the clinical efficacy of a series of phenothiazines and butyrophenones and their capacity to block the inhibitory effect of dopamine, administered by iontophoresis, on the discharge frequency of olfactory tubercle cells.

Further evidence for the dopamine hypothesis of schizophrenia is supplied by the human and rat studies with amphetamine. These have been extensively reviewed by Snyder (1972, 1973). Briefly, D-amphetamine can induce a paranoid psychosis in normal volunteers without previous history of psychosis as well as in individuals with a previous history of psychosis (Connell, 1958; Davis and Janowsky, 1973). The amphetamine psychosis in man has some features in common with amphetamine-induced stereotyped behavior in laboratory animals, which has been attributed to release of dopamine by amphetamine (Snyder, Taylor, Coyle, and Meyerhoff, 1970). The recent claims that the approximately 10-fold greater potency of D-amphetamine over L-amphetamine in producing a paranoid psychosis in volunteers is evidence that the psychosis is due to dopaminergic effects of amphetamine (Angrist, Shopsin, and Gershon, 1971; Davis and Janowsky, 1973) is disputed by the results of other investigators (Svensson, 1971; Ferris, Tang, and Maxwell, 1972; Harris and Baldessarini, 1973; Thornburg and Moore, 1973), who do not support the conclusions of Taylor and Snyder (1971) that there is a differential effect of D- and L-amphetamine on the uptake of norepinephrine and dopamine by rat brain synaptosomes.

L-DOPA, the administration of which can raise brain dopamine, can in high doses produce psychotic symptoms in about 20% of parkinsonian patients (Barbeau, 1969; Jenkins and Groh, 1970; Brogden, Speight, and Avery, 1971) and can exacerbate symptoms in schizophrenic patients (Yaryura-Tobias, Diamond, and Merlis, 1970; Angrist, Sathananthan, and Gershon, 1973) but has few psychotomimetic effects in nonpsychotic psychiatric patients (Sathananthan, Angrist, and Gershon, 1973). On the other hand, small doses of L-DOPA have been reported to be of considerable clinical benefit to schizophrenic patients who were receiving phenothiazines with only slight or moderate clinical improvement (Inanaga, Inoue, Tachibana, Oshima, and Kotorii, 1972). Alpha-methyl-paratyrosine, which inhibits catecholamine synthesis, has been reported to potentiate the antipsychotic effect of phenothiazines (Carlsson, Persson, Roos, and Walinder, 1972) but has no clinical effect on schizophrenics on its own (Charalampous and Brown, 1967; Gershon, Hekimian, Floyd, and Hollister, 1967). The recent reports of decreased brain dopamine-β-hydroxylase (DBH) activity (Wise and Stein, 1973) and decreased platelet monoamine oxidase (MAO) activity in some schizophrenic patients (Murphy and Wyatt, 1972; Meltzer and Stahl, 1974; Wyatt, Murphy, Belmaker, Cohen, Donnelly, and Pollin, 1973) could provide a basis for excessive central dopaminergic activity in some schizophrenic patients. However, the findings of low DBH activity may well be an artifact, and the platelet MAO may not be indicative of brain MAO activity toward dopamine as substrate.

Some of the evidence that the balance between dopaminergic, cholinergic, and serotonergic activity is the critical factor in the development of schizophrenia is summarized by Friedhoff and Alpert (1973). Recently, Bowers (1973) has reported that, following probenecid loading, homovanillic acid (HVA), the major metabolite of dopamine, 5-hydroxyindoleacetic acid (5-HIAA), the major metabolite of serotonin, and 5-HIAA/HVA ratios in the spinal fluid of a group of 18 unmedicated psychotic patients were not significantly different from those of two control groups. The seven psychotic subjects who lacked Schneider first-rank symptoms of schizophrenia did, however, have increased HVA and a decreased 5-HIAA/HVA ratio. The 5-HIAA/HVA ratio in the 11 Schneider-positive patients was increased, whereas the 5-HIAA and HVA levels were not themselves abnormal. Some aspects of thought disorder were better correlated with 5-HIAA levels than with HVA levels. However, there are many problems associated with the probenecid loading method, particularly the poor tolerance of most subjects for large doses of probenecid, which makes the procedure variably stressful, and the variable levels of probenecid in the brain after a standard loading dose, with resultant variation in the blockade of transport of acid metabolites from brain. This requires monitoring probenecid levels in the spinal fluid, which was not done in Bowers' study. We must therefore interpret the results of this important study with caution. In any case, increased activity of a particular dopaminergic tract, particularly one of the smaller ones, or a portion of any of the tracts, might not be reflected in overall levels of metabolites in cerebrospinal fluid because of the greater contribution of other dopaminergic neurons.

There are at least four major dopaminergic neuronal tracts in the mammalian brain: the nigrostriatal tract, the retinal tract, the tuberoinfundibular tract, and the mesolimbic tract (Ungerstedt, 1971). Through consideration of the known anatomy and physiology of these tracts, several investigators have concluded that the mesolimbic dopaminergic tract may be the most likely candidate for the dopaminergic tract whose hyperactivity may be most relevant to the schizophrenic process itself (Matthysse, 1973; Stevens, 1973). There is as yet no direct or indirect measure of the activity of this system in man.

There is, however, good reason to believe that the dopaminergic activity of the tuberoinfundibular neurons may be assessed indirectly. There is impressive evidence that the dopaminergic terminals of this tract control the release of prolactin-inhibiting factor (PIF), which is stored in the median eminence. After release from the median eminence, PIF enters the pituitary portal vessels, which carry it to the anterior pituitary where it exerts an inhibitory effect on the synthesis and release of prolactin and thus lowers serum prolactin levels. The literature prior to 1972 which supports this hypothesis has been reviewed extensively (Meites, Lu, Wuttke, Welsch, Nagasawa, and Quadri, 1972). There is also strong evidence for a prolactin-releasing factor (PRF) of hypothalamic origin which can promote the release of prolactin from the anterior pituitary (see Meites et al., 1972, for references), but the influence of this factor on prolactin secretion is believed to be much less than that of PIF (Schally, Arimura, and Kastin, 1973).

Early studies of the role of biogenic amines in the release of prolactin are reviewed by Meites et al., (1972). Brief mention of some of the major findings in that review and the more recent findings, with particular emphasis on those most relevant to biological psychiatry, follows. A single injection of reserpine, chlorpromazine, α-methyl-paratyrosine, or α-methyl-metatyrosine (AMMT), all of which reduce the amount of catecholamines at postsynaptic receptor sites, produce a rapid increase in serum prolactin levels and, with the exception of AMMT, a decrease in the amount of prolactin in the anterior pituitary (Fuxe, Hökfelt, and Nilsson, 1969; Lu, Amenomori, Chen, and Meites, 1970; Donoso, Bishop, Fawcett, Krulich, and McCann, 1971). L-DOPA, which increases brain dopamine, decreases serum prolactin levels in rats (Donoso et al., 1971; Lu and Meites, 1971); and man (Kleinberg, Noel, and Frantz, 1971; Malarkey, Jacobs, and Daughaday, 1971). MAO inhibitors decrease serum prolactin levels and potentiate the effect of L-DOPA on prolactin levels (Lu and Meites, 1971). L-DOPA and the MAO inhibitors all increase hypothalamic PIF activity (Lu and Meites, 1971).

The previous studies implicate a catecholamine in the control of PIF release and serum prolactin levels but do not clearly distinguish between the role of norepinephrine and dopamine. Kamberi, Mical, and Porter (1971a) found that injection of minute amounts of dopamine into the third ventricle of rats produced a marked drop in serum prolactin levels in 10 min. Injections of similar amounts of norepinephrine and epinephrine had no effect. When the three catecholamines were given into the pituitary portal circulation or the arterial supply, which would deliver it directly to the anterior pituitary, none had any effect on serum prolactin levels. Donoso et al. (1971) found that the decrease in serum prolactin in rats after L-DOPA was not blocked by pretreatment with diethyldithiocarbamate, which blocks the synthesis of norepinephrine by inhibiting dopamine-β-hydroxylase but does not interfere with the synthesis of dopamine. Recently, Smythe and Lazarus (1973) have shown that dimethoxyphenylethylamine (DMPEA), the so-called pink spot, can elevate plasma prolactin levels in rats after intraperitoneal administration. It can also block the decrease in plasma prolactin produced by L-DOPA. These results have been interpreted to indicate that dopamine normally inhibits the secretion of prolactin and that DMPEA can interfere with this inhibitory effect by competing for receptor sites. Finally, there have been numerous demonstrations that, in man, phenothiazines, which have strong antidopaminergic properties and weak antiadrenergic properties, can markedly raise serum prolactin levels (Kleinberg et al., 1971). Thus, there is powerful evidence for a tonic-inhibitory dopaminergic effect on serum prolactin levels.

There is, however, considerable evidence of numerous other influences, including the effects of other bioamines, on serum prolactin levels in both laboratory animals and man. Intraventricular serotonin, melatonin, and N-acetylserotonin can elevate plasma prolactin (Kamberi, Mical, and Porter, 1971b; Porter, Mical, and Cramer, 1971). 5-Hydroxytryptophan can markedly increase serum prolactin levels in rats, tryptophan less so (Lu and Meites, 1973). Lu and Meites (1973) have proposed that serotonin may promote the secretion of prolactin by stimulating release of PRF, which opposes the action of PIF. Libertun and McCann (1973) have presented evidence that cholinergic pathways are also involved in the central control of prolactin secretion. Atropine, given subcutaneously or intravenously, was found to block the afternoon increases in serum prolactin in rats and the proestrous rise in serum prolactin. In addition to these amines, nicotine (Terkel, Blake, Hoover, and Sawyer, 1973) has been shown to inhibit the release of prolactin; on the other hand, estradiol and testosterone (Kalra, Fawcett, Krulich, and McCann, 1973), infusions of the amino acids arginine, leucine, and phenylalanine (Davis, 1972), and thyrotropin-releasing hormone (Jacobs, Snyder, Wilber, Utiger, and Daughaday, 1971) markedly stimulate serum prolactin levels in laboratory animals or man. Plasma osmolarity also affects prolactin secretion, although the mechanism is unknown (Buckman and Peake, 1973).

It must be pointed out, however, that these studies were all done with either single injections of chemicals or at most a few days of treatment. The effect of chronic increase or decrease in central dopaminergic activity, cholinergic activity, estradiol, and so on has not yet been assessed.

Plasma prolactin levels have been shown to be related to the sleep cycle in man. Plasma prolactin concentration rises shortly after sleep onset and continues to rise throughout sleep. If the time of day when sleep is begun shifts, the time of the prolactin increase shifts. Plasma prolactin levels reach their lowest level within 1 to 3 hr after awakening (Sassin, Frantz, Kapen, and Weitzman, 1973).

It is well established that stress can elevate plasma prolactin in rats (Neill, 1970; Ajika, Kalra, Fawcett, Krulich, and McCann, 1972). Noel, Suh, Stone, and Frantz (1972) have shown that human plasma prolactin increases markedly in a number of stress situations. The effects of stress on plasma prolactin were generally greater in females. Preoperative levels of prolactin in females were nearly four times greater than normal but were not increased in males. At their peak during surgery, prolactin levels were increased 17-fold in females, sevenfold in males. They were still increased 24 hr after surgery in females, but not in males. Gastroscopy, proctoscopy, strenuous exercise, and insulin-induced hypoglycemia in both sexes, and sexual intercourse in females, particularly if orgasm occurred, raise serum prolactin levels. Sexual intercourse had no effect on serum prolactin levels in males. The biochemical system(s) that mediate this stress response have not as yet been elucidated.

We were interested in studying serum prolactin levels in newly admitted, unmedicated schizophrenic patients. Despite all the potential influences on serum prolactin levels cited above, we believed it was still possible that serum prolactin levels might serve as an index of the activity of the major system which controls serum prolactin levels, the dopaminergic influence on the release of PIF. If the hypothesized increase in dopaminergic activity in schizophrenia was a general phenomenon affecting all dopaminergic tracts and thus affecting the neurons regulating prolactin secretion, one would expect reduced prolactin concentration in schizophrenic patients; on the other hand, the psychological stress of acute psychosis and hospitalization might produce increased prolactin concentration, particularly in females; finally, these influences might cancel each other out.

II. METHODS

Thirty patients admitted to a research unit at the Illinois State Psychiatric Institute were included in this study. The serum prolactin levels in 26 patients in samples obtained between 8 A.M. and 9 A.M. were studied at least once prior to medication; 13 were studied twice prior to medication. The serum prolactin levels 1 to 6 days after initiating phenothiazines were studied in 20 patients. Oral medication was given at 9 A.M. and 9 P.M. in equal doses on this unit, so that the postmedication samples were obtained 12 hr after the last oral dose. Diagnosis was made by two psychiatrists and a social worker who had no knowledge of any biological data. All the patients diagnosed as schizophrenic satisfied the criteria for schizophrenia of the New Haven Schizophrenic Index (Astrakhan, Harrow, Adler, Brauer, Schwartz, Schwartz, and Tucker, 1972). The patients were fasting prior to obtaining serum samples. No sleeping medication was utilized for these patients during the initial period of study. No attempt was made to monitor the relationship between time of awakening and time of obtaining the serum sample. The behavior and mental status of all patients were rated daily with a modification of the rating scale of Hargreaves (1968). Two experienced raters rated each patient daily on a 15-item scale, with a 1 to 9 rating possible for each item. We will report only on two ratings: Anxiety and Psychiatric Illness. The interrater reliability for these items were 0.778 and 0.938, respectively. All the psychotic patients, except for the individual diagnosed as latent schizophrenic, were in a state of florid psychosis when their premedication blood samples were collected, which was generally within 48 hr of admission. Their Psychiatric Illness ratings, on a scale of 1 to 9, ranged from 5 to 9 (mean 6.54 ± SD 1.50). The onset of psychotic symptoms or the major exacerbation of symptoms was generally within 1 week of admission.

After clotting, the serum was removed, frozen, and assayed on a blind basis by radioimmune assay (Frantz, Kleinberg, and Noel, 1972). Split samples differed by less than 10%. Samples from 12 normal males were also included with the patient samples and were also determined on a blind basis. These will be referred to as Chicago controls. Serum prolactin levels in 54 normal males and 43 normal females with this method have been previously reported (Frantz et al., 1972). These will be referred to as New York controls.

III. RESULTS

The mean serum prolactin levels for the controls and the unmedicated patients are given in Table 1. Individual values for the controls and schizophrenic patients are given in Fig. 1. For the patients who had more than one serum prolactin determination during this period, the initial prolactin level was used. The levels of serum prolactin in our small group of controls was

TABLE 1. *Serum prolactin levels in newly admitted psychiatric patients and controls*

Group	Number	Mean prolactin level[a] (ng/ml)	Range (ng/ml)
Chicago controls, male	12	6.1 ± 2.5	2.8–11.0
New York controls, male	54	9.2 ± 5.7	0.6–29.0
New York controls, female	43	10.3 ± 8.0	1.0–39.0
All schizophrenics	22	6.0 ± 3.2	<1.0–14.2
Acute schizophrenia	16	5.4 ± 2.8	<1.0– 9.8
Chronic schizophrenia	5	8.3 ± 3.8	5.1–14.2
Latent schizophrenia	1	3.7	3.7
Paranoid schizophrenia	12	5.3 ± 2.5	<1.0– 8.2
Nonparanoid schizophrenia	9	7.3 ± 3.8	0.6–14.2
Male schizophrenics	10	4.5 ± 2.7	<1.0– 8.2
Female schizophrenics	12	6.8 ± 2.8	2.3–14.2
Manic-depressives, manic phase	2	14.3 ± 1.9	12.9–15.6
Unipolar psychotic depression	1	4.3	4.3
Anxiety neurosis	1	17.0	17.0

[a] Mean ± SD.

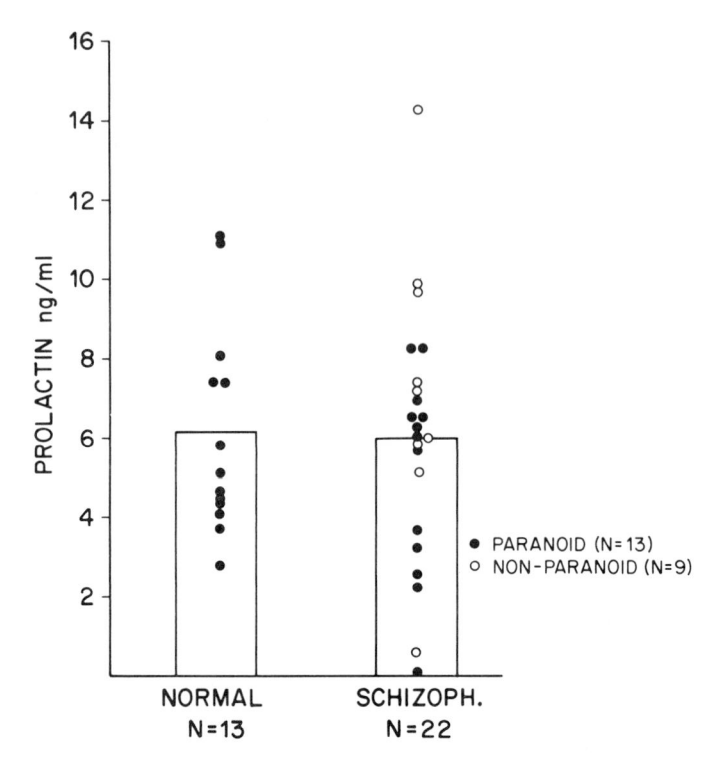

FIG. 1. Serum prolactin levels in controls and schizophrenic patients at admission to the hospital.

slightly but not significantly less than that of the New York controls, who had a broader range of prolactin levels. The difference is probably due to batch-to-batch variations common to radioimmune assay. The mean serum prolactin levels of the 22 schizophrenic patients were not significantly different from those of the control groups. There was no significant effect of acuteness or chronicity, paranoia versus nonparanoia, or sex on serum prolactin levels in the schizophrenic patients. The two female manic-depressive patients, manic phase, had higher serum prolactin levels than any of the 12 male controls from Chicago and all but one of the 22 schizophrenic patients. The highest serum prolactin level was in the patient diagnosed as an anxiety neurosis. The patient diagnosed as a unipolar psychotic depression had low normal levels of prolactin, which might have been influenced by 10 mg of diazepam which she had received from the referring physician 1 day prior to admission.

Since stress has been shown to influence serum prolactin levels in man (Noel et al., 1972), and since ratings of Anxiety and Psychiatric Illness are two measures related to stress, it was of interest to evaluate whether the stress-serum prolactin relationship was present in the psychiatric patients.

The initial premedication prolactin level and the Anxiety and Psychiatric Illness ratings on the day of the sample are given in Table 2. Anxiety and Psychiatric Illness ratings from the day before discharge are also provided to demonstrate that these patients were relatively disturbed at the time the blood for the serum prolactin determination was obtained. The Anxiety and Psychiatric Illness ratings were significantly less on the day before discharge than on the initial ratings ($p < 0.001$ for both, paired t test). The Pearson correlation coefficient, $r = -0.276$ for prolactin and the concomitant Anxiety rating, and for prolactin and the concomitant Psychiatric Illness rating, $r = -0.236$, were not significant. The negative signs indicate a trend toward an inverse relationship between these behavioral parameters and

TABLE 2. *Admission prolactin levels in relation to initial and discharge anxiety ratings and psychiatric illness ratings*

Subject	Diagnosis[a]	Prolactin Level (ng/ml)	Anxiety rating Initial	Anxiety rating Discharge	Psychiatric illness rating Initial	Psychiatric illness rating Discharge
1	AS-NP	7.1	7	4	8	4
2	AS-NP	5.9	5	3	7	4
3	AS-NP	0.6	5	3	7	2
4	AS-NP	7.2	5	2	6	4
5	AS-NP	9.8	5	3	5	4
6	AS-P	6.5	4	3	6	3
7	AS-P	8.2	6	3	8	4
8	AS-P	8.2	4	3	6	4
9	AS-P	0.1	7	2	8	3
10	AS-P	5.7	4	3	8	3
11	AS-P	6.5	3	2	5	3
12	AS-P	3.3	4	3	9	3
13	AS-P	6.0	3	3	6	4
14	AS-P	2.6	3	3	5	4
15	AS-P	2.3	4	5	6	6
16	AS-P	7.0	7	2	7	4
17	CS-NP	9.7	5	3	9	4
18	CS-NP	5.1	6	2	7	4
19	CS-NP	14.2	4	4	6	4
20	CS-NP	6.0	4	3	5	5
21	CS-P	6.3	4	3	6	4
22	LS	3.7	5	2	5	2
23	M-D,M	12.9	2	3	3	4
24	M-D,M	15.6	4	3	8	3
25	PD	4.3	6	2	9	2
26	Anx.	17.0	4	2	5	3
Mean ± SD		6.99± 4.16	4.62± 1.27	2.85± 0.72	6.54± 1.50	3.62± 0.88

[a] AS-NP = acute schizophrenic, nonparanoid; AS-P = acute schizophrenic, paranoid; CS-NP = chronic schizophrenic, nonparanoid; CS-P = chronic schizophrenic, paranoid; LS = latent schizophrenic; MD-M = manic-depressive, manic phase; PD = psychotic depression; Anx. = anxiety.

serum prolactin. There was no evidence of a stress-related increase in prolactin levels, despite the fact that most of the Psychiatric Illness ratings were in the severe range (mean 6.54 ± 1.50) while the Anxiety ratings were in the moderate range (mean 4.62 ± SD 1.27) in the first week of hospitalization.

As another test of a possible relationship between serum prolactin levels and the Anxiety and Psychiatric Illness ratings, we compared the second behavior ratings and serum prolactin levels prior to medication to the data of the initial sample (Table 3). The prolactin levels from the 10 schizophrenic patients are given in Fig. 2. It may be seen that in eight of the 10 schizophrenic patients, and one of three cases of affective psychoses, the second serum prolactin level was greater than the first. Prolactin levels increased four- to 70-fold in five of the eight schizophrenic subjects. The mean of the second determination serum prolactin levels for the 10 schizophrenic patients was significantly greater than that of the first (Mann-Whitney $U = 20$, $p < 0.025$). Nevertheless, the second prolactin levels are still within normal limits for all but one of the patients. The change, if any, in Anxiety Ratings and Psychiatric Illness ratings between the two serum prolactin determinations was not significantly correlated with the change in prolactin levels

TABLE 3. *Changes in serum prolactin levels, anxiety, and psychiatric illness prior to medication*

Dx	Prolactin levels (ng/ml) 1st	2nd	Anxiety ratings 1st	2nd	Psychiatric illness ratings 1st	2nd
CS	6.3	10.0	4	4	6	6
CS	14.2	17.0	4	4	6	4
AS	6.0	5.9	3	3	6	6
AS	8.2	7.3	6	7	8	8
AS	2.6	10.4	3	—	5	—
AS	0.6	3.3	5	6	7	8
AS	5.9	40.0	5	5	7	6
AS	1.4	11.0	5	5	7	7
AS	0.1	7.1	7	8	8	9
LS	3.7	6.3	6	7	8	8
MD	12.9	10.4	2	5	3	7
MD	15.6	5.0	4	4	8	8
UPD	4.3	6.4	6	6	9	8
Mean ± SD						
All patients	6.3± 5.1	10.8± 9.5	4.5± 1.4	5.2± 1.47	6.5± 1.6	6.7± 1.5
All AS + CS	4.9± 4.2	11.8± 10.6	4.8± 1.3	5.4± 1.7	6.8± 1.0	6.9± 1.5

CS = chronic schizophrenia; AS = acute schizophrenia; LS = latent schizophrenia; MD = manic-depressive; UPD = unipolar psychotic depressive.

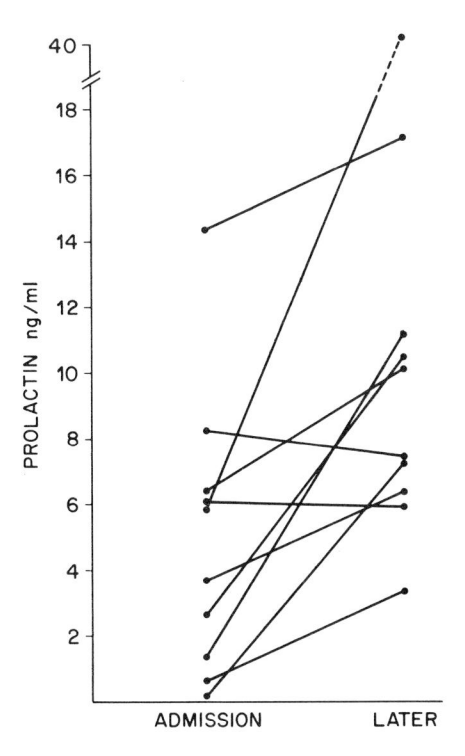

FIG. 2. Initial and later serum prolactin levels in unmedicated acutely schizophrenic patients. Median number of days between samples was 1 to 3.

during this time. In general, both ratings tended to show little change between the first and second determinations.

The effect of phenothiazines on serum prolactin levels in 20 schizophrenic patients is given in Table 4. The postmedication levels of prolactin were obtained from the first day after medication to 6 days after beginning medication. The median was 1.5 days. The range of prolactin levels postmedication was 4 to > 80 ng/ml. The median was 26.5 ng/ml. Seven of the 20 subjects had levels which were within normal limits. Three of these had supposedly received chlorpromazine, three had received trifluoperazine, and one, thioridazine for 1 to 4 days prior to obtaining the samples. Six of the seven patients without increases had been prescribed oral medication for the first time 12 hr prior to the blood sample. It is conceivable that they did not swallow their medication, that there was a delay in onset of action at dopamine receptor sites, or that serum prolactin levels had returned to normal. Six of eight patients given oral medication for the first time 1 day prior to the postdrug sample had no increase, whereas only one of nine patients prescribed oral drugs 2 or more days before had normal prolactin levels. The three patients given parenteral medication all had increased

TABLE 4. Effect of phenothiazines on prolactin levels

Medication (mg)	Duration (days)	Prolactin (ng/ml) Predrug	Postdrug
Chlorpromazine, 300	2	7.3	42.0
Chlorpromazine, 150	6	—	60.0
Chlorpromazine, 300	5	—	19.0
Chlorpromazine, 200	1	7.1	9.0
Chlorpromazine, 400	1	3.3	11.5
Chlorpromazine, 300	1	5.1	50.0
Chlorpromazine, 400	5	2.3	32.0
Chlorpromazine, 100	1	6.5	5.1
Trifluperazine, 10	2	6.5	25.0
Trifluperazine, 10	4	8.2	6.1
Trifluperazine, 10	1	5.7	4.0
Trifluperazine, 10	1	10.4	12.6
Trifluperazine, 10	5	7.2	20.0
Thioridazine, 400	1	3.3	11.0
Thioridazine, 400	1	5.9	80.0
Thioridazine, 200	4	6.0	38.0
Thioridazine, 75	6	—	33.0
Prolixin enanthate, 25	1	7.1	28.0
Prolixin enanthate, 25	1	—	50.0
Prolixin enanthate, 25	6	9.8	>80.0

prolactin levels. There was no apparent difference in the magnitude of the increases which resulted from chlorpromazine, trifluoperazine, thioridazine, or fluphenazine enanthate. There was also no significant correlation between the drug dose when expressed as chlorpromazine equivalent and the level of prolactin.

IV. DISCUSSION

Serum prolactin levels were within normal limits for the unmedicated, severely disturbed schizophrenic patients in this study. Slight elevations were present in two manic patients and one anxiety patient. The interpretation of these results is not easy. As was indicated in the Introduction, there are numerous influences on the secretion of prolactin by the anterior pituitary. While the interrelationships of these influences are not known, e.g., if they have one or at most a few final common pathways, there is general agreement that in man the major influence affecting the secretion of prolactin is the inhibitory effect of PIF, which is released into the portal circulation by dopamine-sensitive cells (Schally et al., 1973). Based on this, it is probably valid to conclude that had there been recent onset of a greatly increased activity of the dopaminergic neurons in the tuberoinfundibular tract in the absence of influences strongly tending to elevate serum prolactin levels, serum prolactin levels should have fallen to very low levels in the schizophrenic patients, as they do after relatively small doses of L-DOPA in normals (Kleinberg et al., 1971).

The question as to whether there were strong influences tending to increase serum prolactin requires consideration next. The factor most likely to elevate serum prolactin in these patients is stress. The schizophrenic patients were acutely disturbed, with very recent onset of marked psychotic symptoms and with a considerable anxiety component to their illness. In comparison with their discharge ratings of Anxiety and Psychiatric Illness, these patients were significantly more disturbed at the time of obtaining the first prolactin sample, which, if this is an equivalent state of stress to that found in patients undergoing diagnostic procedures or facing surgery, could be expected to produce elevated prolactin levels. The answer to whether these states are equivalent is, of course, unknown. The fact that the two manic patients and the patient with an acute anxiety state did have the highest serum prolactin levels in this group of patients does suggest that, at least in nonschizophrenic, newly admitted psychiatric patients, the serum prolactin level may be increased by stress-related phenomena. The normal serum prolactin level in the psychotic depressive patient could have been due to receiving a minor tranquilizer, diazepam. Thus, *if* the stress-related mechanism which leads to increased serum prolactin levels is intact in schizophrenic patients, one would have expected increased serum prolactin levels in the samples obtained after admission. Unmedicated schizophrenic patients similar to those in our study typically show a marked hormonal stress response during hospital admission, and during periods of emotional turmoil (Sachar, Mason, Kolmer, and Artiss, 1963; Sachar, Kanter, Buie, Engle, and Mehlman, 1970; Wohlberg, Knapp, and Vachon, 1970). If the stress-related increase in prolactin secretion did occur, the most likely explanation for the normal serum prolactin levels would be an increase in dopamine-dependent release of PIF, which would block the stress-induced increase. Exhaustion of supplies of prolactin in the anterior pituitary is ruled out by the brisk response of the prolactin levels of most patients to phenothiazines.

However, the increase in prolactin which occurs with stress in normals may not occur in schizophrenic patients. There was a trend toward an *inverse* relationship between serum prolactin and Anxiety Ratings in these patients. Stress-sensitive hormones such as cortisol and urinary epinephrine in normals and psychiatric patients decline after adaptation to hospitalization (Sachar et al., 1963; Mason, Sachar, Fishman, Hamburg, and Handlon, 1965; Sachar, 1967; Sachar et al., 1970; Wohlberg et al., 1970). Serum prolactin levels significantly *increased* between the admission sample and that obtained 2 to 25 days later (median 3 days) in eight of 10 schizophrenic patients. If the schizophrenic patients experienced less stress in the first few days of hospitalization, which our ratings of Anxiety and Psychiatric

Illness were not sensitive enough to record, these results suggest a paradoxical suppressive effect of stress on prolactin secretion in schizophrenics. If this is the case, then there is no evidence for increased dopaminergic activity in the tuberoinfundibular tract in schizophrenics. Future studies will need to address themselves to the elucidation of the response of serum prolactin levels in unmedicated schizophrenic patients to known stressors. The simultaneous measurement of as many of the known factors which influence serum prolactin levels, e.g., plasma osmolarity, gonadal steroids, thyrotropin-releasing factor, is a much more difficult approach. Assay of PIF levels in the general circulation, which is not yet possible, could also yield more direct information about dopaminergic activity in the tuberoinfundibular dopaminergic neurons.

The magnitude of the increase in prolactin levels in most of the schizophrenic patients after all four phenothiazines was consistent with the increases previously reported in normal volunteers given chlorpromazine (Kleinberg et al., 1971; Turkington, 1972). The variability in levels could reflect failure to take the drug, differences in degree of absorption of drug, rates of metabolism, sensitivity to the phenothiazines of dopaminergic neurons or other systems which influence prolactin secretion, and clearance of prolactin from serum. It is of interest that thioridazine produced effects on prolactin secretion that were comparable to those produced by the other phenothiazines. Thioridazine produced much less parkinsonian symptoms in schizophrenic patients than the other phenothiazines and had little effect on dopamine turnover in monkeys (Matthysse, 1973). Either it has an effect on dopaminergic receptors in the median eminence equal to those of the other phenothiazines or the increase in prolactin levels produced by these drugs is based on some effect other than blockade of dopamine receptors. If the first interpretation is correct, it would indicate that the failure of thioridazine to affect dopamine receptors in the corpus striatum may not be a serious inconsistency in the theory that the antipsychotic effect of the phenothiazines is based on dopamine receptor blockade.

ACKNOWLEDGMENTS

This research was supported in part by U.S. Public Health Service grants MH 16, 127; MH 18, 396; and 5 K02 MH 47, 808, and Department of Mental Health of the State of Illinois grant 431-13-RD (Dr. Meltzer), U.S. Public Health Service grants MH-25133 and 5 K2-NIH 22613 (Dr. Sachar), AM-11294, CA 11704, and American Heart Association grant 68-111 (Dr. Frantz).

REFERENCES

Aghajanian, G. K., and Bunney, B. S. (1973): Central dopaminergic neurons: Neurophysiological identification and response to drugs. In: *Frontiers in Catecholamine Research,* edited by E. Usdin and S. H. Snyder. Pergamon Press. Elmsford. New York.

Ajika, K., Kalra, S. P., Fawcett, C. P., Krulich, L., and McCann, S. M. (1972): The effect of stress and nembutal on plasma levels of gonado tropius and prolactin in ovariectomized rats. *Endocrinology*, 90:707-715.

Andén, N.-E., Butcher, S. G., Corrodi, H., Fuxe, K., and Ungerstedt, U. (1970): Receptor activity and turnover of dopamine and noradrenaline after neuroleptics. *European Journal of Pharmacology*, 11:303-314.

Angrist, B., and Gershon, S. (1970): The phenomenology of experimentally induced amphetamine psychoses. *Biological Psychiatry*, 2:95-107.

Angrist, B., Sathananthan, G., and Gershon, S. (1973): Behavioral effects of L-DOPA in schizophrenic patients. *Psychopharmacologia* (Berlin), 31:1-12.

Angrist, B. M., Shopsin, B., and Gershon, S. (1971): Comparative psychotomimetic effects of stereoisomers of amphetamine. *Nature*, 234:152-153.

Astrachan, B. M., Harrow, M., Adler, D., Brauer, L., Schwartz, A., Schwartz, C., and Tucker, G. (1972): A checklist for the diagnosis of schizophrenia. *British Journal of Psychiatry*, 121:529-539.

Barbeau, A. (1969): L-DOPA therapy in Parkinson's disease: A critical review of nine years experience. *Canadian Medical Association Journal*, 101:791-800.

Bowers, M. B., Jr. (1973): 5-Hydroxyindoleacetic acid (5-HIAA) and homovanillic acid (HVA) following probenecid in acute psychotic patients treated with phenothiazines. *Psychopharmacologia* (Berlin), 28:309-318.

Brogden, R. N., Speight, T. M., and Avery, G. S. (1971): Levodopa: A review of its pharmacological properties and therapeutic uses with particular reference to parkinsonism. *Drugs*, 2:257-408.

Buckman, M. T., and Peake, G. T. (1973): Osmolar control of prolactin secretion in man. *Science*, 181:755-757.

Carlsson, A., and Lindquist, M. (1963): Effect of chlorpromazine or haloperidol on formation of 3-methoxytyramine and normetanephrine in mouse brain. *Acta Pharmacologica et Toxicologica*, 20:140-144.

Carlsson, A., Persson, T., Roos, B. E., and Wålinder, J. (1972): Potentiation of phenothiazines by α-methyltyrosine in treatment of chronic schizophrenia. *Journal of Neural Transmissions*, 33:83-90.

Charalampous, K. D., and Brown, S. (1967): A clinical trial of α-methylparatyrosine in mentally ill patients. *Psychopharmacologia* (Berlin), 11:422-429.

Connell, P. H. (1958): *Amphetamine Psychosis*. Chapman and Hall, London.

Davis, J. M., and Janowsky, D. S. (1973): Amphetamine and methylphenidate psychosis. In: *Frontiers in Catecholamine Research*, edited by E. Usdin and S. H. Snyder. Pergamon Press, New York.

Davis, S. L. (1972): Plasma levels of prolactin, growth hormone, and insulin in sheep following the infusion of arginine, leucine and phenylalanine. *Endocrinology*, 91:549-555.

Donoso, A. O., Bishop, W., Fawcett, C. P., Krulich, L., and McCann, S. M. (1971): Effect of drugs that modify brain monoamine concentrations on plasma gonadotropin and prolactin levels in the rat. *Endocrinology*, 89:774-784.

Ferris, R. M., Tang, F. L. M., and Maxwell, R. A. (1972): A comparison of the capacities of isomers of amphetamine, deoxypipradrol and methylphenidate to inhibit the uptake of tritiated catecholamines into rat cerebral cortex slices, synaptosomal preparations of rat cerebral cortex, hypothalamus and striatum and into adrenergic nerves of rabbit aorta. *Journal of Pharmacology and Experimental Therapeutics*, 181:407-416.

Frantz, A. G., Kleinberg, D. L., and Noel, G. L. (1972): Prolactin studies in man. *Recent Progress in Hormone Research*, 28:527-573.

Friedhoff, A. J., and Alpert, M. (1973): A dopaminergic-cholinergic mechanism in production of psychotic symptoms. *Biological Psychiatry*, 6:165-169.

Fuxe, K., Hökfelt, T., and Nilsson, O. (1969): Factors involved in the control of the activity of the tubero-infundibular dopamine neurons during pregnancy and lactation. *Neuroendocrinology*, 5:107-120.

Gershon, S., Hekimian, L. J., Floyd, A., and Hollister, L. E. (1967): Methyl-p-tyrosine (AMT) in schizophrenia. *Psychopharmacologia* (Berlin), 11:189-194.

Hargreaves, W. A. (1968): Systematic nursing observation of psychopathology. *Archives of General Psychiatry*, 18:518-531.

Harris, J. E., and Baldessarini, R. J. (1973): The uptake of [^{3}H] dopamine by homogenates of rat corpus striatum: Effects of cations. *Life Sciences*, 13:303-312.

Horn, A. S., and Snyder, S. H. (1971): Chlorpromazine and dopamine: Conformational similarities that correlate with the antischizophrenic activity of phenothiazine drugs. *Proceedings of the National Academy of Sciences* (U.S.), 68:2325–2328.

Inanaga, K., Inoue, K., Tachibana, H., Oshima, M., and Kotorii, R. (1972): Effect of L-DOPA in schizophrenia. *Folin Psychiatrica Neurologica Japonica*, 26:145–157.

Jacobs, L. S., Snyder, C. J., Wilber, J. F., Utiger, R. D., and Daughaday, W. H. (1971): Increased serum prolactin after administration of synthetic thyrotropin releasing hormone (TRH) in man. *Journal of Clinical Endocrinology*, 33:996–988.

Jenkins, R. B., and Groh, R. H. (1970): Mental symptoms in parkinsonian patients treated with L-DOPA. *Lancet*, 2:177–179.

Kalra, P. S., Fawcett, C. P., Krulich, L., and McCann, S. M. (1973): The effects of gonadal steroids on plasma gonadotropins and prolactin in the rat. *Endocrinology*, 92:1256–1268.

Kamberi, I. A., Mical, R. S., and Porter, J. C. (1971a): Effect of anterior pituitary perfusion and intraventricular injection of catecholamines on prolactin release. *Endocrinology*, 88:1012–1020.

Kamberi, I. A., Mical, R. S., and Porter, J. C. (1971b): Effects of melatonin and serotonin on the release of FSH and prolactin. *Endocrinology*, 88:1288–1293.

Kebabian, J. W., Petzold, G. L., and Greengard, P. (1972): Dopamine-sensitive adenylate cyclase in caudate nucleus of rat brain, and its similarity to the "dopamine receptor." *Proceedings of the National Academy of Sciences* (U.S.), 69:2145–2149.

Kleinberg, D. L., Noel, G. L., and Frantz, A. G. (1971): Chlorpromazine stimulation and L-DOPA suppression of plasma prolactin in man. *Journal of Clinical Endocrinology and Metabolism*, 33:873–876.

Libertun, C., and McCann, S. M. (1973): Blockade of the release of gonadotropins and prolactin by subcutaneous or intraventricular injection of atropine in male and female rats. *Endocrinology*, 92:1714–1724.

Lu, K-H., Amenomori, Y., Chen, C-L., and Meites, J. (1970): Effect of central acting drugs on serum and pituitary prolactin levels in rats. *Endocrinology*, 87:667–672.

Lu, K-H., and Meites, J. (1971): Inhibition of L-DOPA and monoamine oxidase inhibitors of pituitary prolactin release, stimulation by methyldopa and D-amphetamine. *Proceedings of the Society for Experimental Biology and Medicine*, 137:480–483.

Lu, K-H., and Meites, J. (1973): Effects of serotonin precursors and melatonin on serum prolactin release in rats. *Endocrinology*, 93:152–155.

Malarkey, W. B., Jacobs, L. S., and Daughaday, W. H. (1971): Levodopa supression of prolactin in nonpeurperal galactorrhea. *New England Journal of Medicine*, 285:1160–1163.

Mason, J. W., Sachar, E. J., Fishman, J. R., Hamburg, D. A., and Handlon, J. H. (1965): Corticosteroid responses to hospital admission. *Archives of General Psychiatry*, 13:1–8.

Matthysse, S. (1973): Antipsychotic drug actions: A clue to the neuropathology of schizophrenia? *Federation Proceedings*, 32:200–205.

Meites, J., Lu, K-H., Wuttke, W., Welsch, C. W., Nagasawa, H., and Quadri, S. K. (1972): Recent studies on functions and control of prolactin secretion in rats. *Recent Progress in Hormone Research*, 28:471–526.

Meltzer, H. Y., and Stahl, S. M. (1974): Platelet monoamine oxidase activity and substrate preferences in schizophrenic patients. *Research Communications in Chemical Pathology and Pharmacology*, 7:419–431.

Murphy, D. L., and Wyatt, R. J. (1972): Reduced monoamine oxidase activities in blood platelets of schizophrenic patients. *Nature*, 238:225–226.

Neill, J. D. (1970): Effect of "stress" on serum prolactin and lutenizing hormone levels during the estrous cycle of the rat. *Endocrinology*, 87:1192–1197.

Noel, G. L., Suh, H. K., Stone, G. J., and Frantz, A. G. (1972): Human prolactin and growth hormone release during surgery and other conditions of stress. *Journal of Clinical Endocrinology and Metabolism*, 35:840–851.

Porter, J. C., Mical, R. S., and Cramer, O. M. (1971): Effect of serotonin and other indoles on the release of LH, FSH, and prolactin. *Hormones, Antagonists, and Gynecological Investigations*, 2:13–22.

Sachar, E. J. (1967): Corticosteroids in depressive illness II. A longitudinal psychoendocrine study. *Archives of General Psychiatry*, 17:554–567.

Sachar, E. J., Kanter, S. S., Buie, D., Engle, R., and Mehlman, R. (1970): Psychoendocrinology of ego disintegration. *American Journal of Psychiatry*, 126:1067–1078.

Sachar, E. J., Mason, J. W., Kolmer, H. S., and Artiss, K. L. (1963): Psychoendocrine aspects of acute schizophrenic reactions. *Psychosomatic Medicine*, 25:510–537.

Sassin, J. F., Frantz, A. G., Kapen, S., and Weitzman, E. D. (1973): The nocturnal rise of human prolactin is dependent on sleep. *Journal of Clinical Endocrinology and Metabolism*, 37:436–440.

Sathananthan, G., Angrist, B. M., and Gershon, S. (1973): Response threshold to L-DOPA in psychiatric patients. *Biological Psychiatry*, 7:139–146.

Schally, A. V., Arimura, A., and Kastin, A. J. (1973): Hypothalamic regulatory hormones. *Science*, 179:341–350.

Smythe, G. A., and Lazarus, L. (1973): Blockade of the dopamine-inhibitory control of prolactin secretion in rats by 3,4-dimethoxyphenylethylamine (3,4-di-O-methyldopamine). *Endocrinology*, 93:147–151.

Snyder, S. H. (1972): Catecholamines in the brain as mediators of amphetamine psychosis. *Archives of General Psychiatry*, 27:169–179.

Snyder, S. H. (1973): Amphetamine psychosis: A "model" of schizophrenia mediated by catecholamines. *American Journal of Psychiatry*, 130:61–66.

Snyder, S. H., Taylor, K. M., Coyle, J. T., and Meyerhoff, J. L. (1970): The role of brain dopamine in behavioral regulation and actions of psychotropic drugs. *American Journal of Psychiatry*, 127:199–207.

Stevens, J. R. (1973): An anatomy of schizophrenia? *Archives of General Psychiatry*, 29:177–189.

Svensson, T. H. (1971): Functional and biochemical effects of D- and L-amphetamine or central catecholamine neurons. *Naunyn-Schmiedeberg's Archives of Pharmacology*, 271:170–180.

Taylor, K. M., and Snyder, S. H. (1971): Differential effects of D- and L-amphetamine on behavior and on catecholamine disposition in dopamine and norepinephrine containing neurons of rat brain. *Brain Research*, 28:295–309.

Terkel, J., Blade, C. A., Hoover, V., and Sawyer, C. H. (1973): Pup survival and prolactin levels in nicotine-treated lactating rats. *Proceedings of the Society for Experimental Biology Medicine*, 143:1131–1135.

Thornburg, J. E., and Moore, J. E. (1973): Dopamine and norepinephrine uptake by rat brain synaptosomes: Relative inhibitory potencies of L- and D-amphetamine and amantadine. *Research Communications in Chemical Pathology and Pharmacology*, 5:81–89.

Turkington, R. W. (1972): The clinical endocrinology of prolactin. *Advances in Internal Medicine*, 18:363–387.

Ungerstedt, U. (1971): Sterotaxic mapping of the monoamine pathways in the rat brain. *Acta Physiologica et Scandinavica*, 367 (Suppl.):1–49.

Wise, C. D., and Stein, L. (1973): Dopamine β-hydroxylase deficits in the brains of schizophrenic patients. *Science*, 181:344–347.

Wohlberg, G. W., Knapp, P. H., and Vachon, L. (1970): A longitudinal investigation of adrenocortical function in acute schizophrenia. *Journal of Nervous and Mental Diseases*, 151:245–265.

Wyatt, R. J., Murphy, D. L., Belmaker, R., Cohen, S., Donnelly, C. H., and Pollin, W. (1973): Reduced monoamine oxidase activity in platelets: A possible genetic marker for vulnerability to schizophrenia. *Science*, 179:916–918.

Yaryura-Tobias, J. E., Diamond, B., and Merlis, S. (1970): The action of L-DOPA on schizophrenic patients. *Current Therapeutic Research*, 12:528–533.

York, D. H. (1972): Dopamine receptor blockade – A central action of chlorpromazine on striatal neurones. *Brain Research*, 37:91–99.

Neuropsychopharmacology of Monoamines and
Their Regulatory Enzymes, edited by E. Usdin.
Raven Press, New York © 1974.

Dopamine, Psychomotor Stimulants, and Schizophrenia: Effects of Methylphenidate and the Stereoisomers of Amphetamine in Schizophrenics

David S. Janowsky* and John M. Davis**

*Department of Psychiatry, University of California at San Diego, School of Medicine, San Diego, California 92037 and **Department of Psychiatry, University of Chicago School of Medicine and The Illinois State Psychiatric Institute, Chicago, Illinois 60612.

I. INTRODUCTION

Considerable indirect evidence has been accumulated suggesting that schizophrenia may be related to changes in central dopamine (Snyder, Taylor, Cole, and Meyerhoff, 1970; Taylor and Snyder, 1970; Janowsky, El-Yousef, Davis, and Sekerke, 1973a). For example, all antipsychotic drugs have been found to cause extrapyramidal symptoms, probably through their ability to block central dopamine receptors. Also, all effective antipsychotic compounds have been found to have a molecular configuration similar to that of dopamine (Horn and Snyder, 1971). L-DOPA, a precursor of dopamine that is preferentially converted to dopamine rather than to norepinephrine (Everett and Borcherding, 1970), has been found to activate schizophrenic symptoms (Yaryura-Tobias, Diamond, and Merlis, 1970; Angrist, Sathananthan, and Gershon, 1973).

Centrally active psychostimulants have been used as important research tools to understand the etiology of schizophrenia (Scheel-Kruger, 1971). In animals, amphetamine-induced stereotyped behavior, generally accepted to be due to increased dopaminergic activity in the extrapyramidal system and thought by some to be an animal model of psychotic behavior in man (Randrup and Munkrad, 1967), is effectively blocked by administration of those antipsychotic agents which are effective in the treatment of schizophrenia. In man, amphetamine and other related psychostimulants, such as methylphenidate, administered chronically in large amounts, have been found to produce a psychosis that is clinically indistinguishable from paranoid schizophrenia (Angrist and Gershon, 1970; Griffith, Cavanaugh, Held, and Oates, 1972). This "model psychosis" can occur in patients who have not been psychotic or prepsychotic. For example, Griffith et al. (1972) and Angrist and Gershon (1970) noted that when they gave amphetamine abusers with no evidence of preexisting psychotic symptoms large, frequent doses of D-amphetamine, all subjects developed a paranoid psychosis within days of beginning the regimen. This "amphetamine psychosis" cleared upon discontinuation of amphetamine intake. Since amphetamine psychosis markedly resembles paranoid schizophrenia, it may be an excellent drug-induced model of schizophrenia.

Further evidence in support of a "dopamine hypothesis" of schizophrenia has been presented by Janowsky et al. (1970a). These authors observed that another, quantitatively different behavioral phenomenon can occur following acute psychostimulant administration to schizophrenics. This phenomenon consists of an activation or worsening of preexisting psychotic symptoms in patients who have active psychotic symptoms by a small, single intravenous dose of methylphenidate. The phenomenon does not occur in remitted patients or in normals. In actively ill schizophrenic patients, following methylphenidate injection, psychotic symptoms become more florid, in the direction of the preexisting psychosis. For example, catatonic patients become more rigid and exhibit waxy flexibility without necessarily showing paranoid symptoms. This differs from the predominantly paranoid psychosis that follows high-dose amphetamine administration. We have suggested that this psychosis activation, caused by methylphenidate, is more likely due to a dopaminergic mechanism rather than a noradrenergic mechanism, since there is evidence that methylphenidate selectively causes central dopamine release in preference to norepinephrine release (Ferris, Tang, and Maxwell, 1973).

Another way in which psychostimulants have been utilized in the understanding of psychotic illness has taken advantage of the differential effects of D- and L-amphetamine on behavior on the one hand, and central dopamine and norepinephrine activity on the other (Snyder et al., 1970; Taylor and Snyder, 1970; Angrist and Gershon, 1971; Angrist, Shopsin, and Gershon, 1971). In animals, D-amphetamine is approximately 10 times as potent as L-amphetamine in causing increased locomotor activity, generally thought to be a noradrenergic phenomenon. In contrast, D-amphetamine is only two times as potent as L-amphetamine in causing stereotyped sniffing and gnawing behavior in animals, thought to be a dopaminergic phenomenon (Snyder et al., 1970; Taylor and Snyder, 1970), and considered to be an animal model for human psychosis. These observations concerning the differential effects of D- and L-amphetamine can be applied to human behavioral phenomena. Thus, if in man, D-amphetamine is also 10 times as potent as L-amphetamine in producing behaviors mediated by norepinephrine, but only two times as potent as L-amphetamine in causing dopaminergically mediated behaviors, the relative potencies of the two isomers in eliciting a given behavior may indicate whether the behavior is under dopaminergic or noradrenergic control. Angrist et al. (1971) have produced a chronic "amphetamine psychosis" in volunteers by administering repeated doses of

D- or L-amphetamine over a period of days. They noted that D-amphetamine was approximately two times as potent as L-amphetamine in causing paranoid psychotic symptoms, thus suggesting a dopaminergic etiology to this phenomenon.

D-Amphetamine and L-amphetamine have not previously been used as investigative tools to determine whether the acute activation of psychotic symptoms by psychostimulants, observed (Janowsky et al., 1973a) in schizophrenic patients with preexisting symptoms, is under dopaminergic or noradrenergic control. Therefore, in the current study we have used D- and L-amphetamine, as well as methylphenidate, as research tools to study whether psychostimulant-induced psychosis activation, occurring in actively ill schizophrenic patients following a single, low-dosage injection, is more likely a dopaminergic or a noradrenergic phenomenon.

II. METHOD

A total of 17 actively ill schizophrenic inpatients in good health, without cardiovascular or other physical illness, were administered active drug by a single intravenous injection, preceded and followed by placebo injections every 5 min. Equimolar doses of D-amphetamine sulfate solution, L-amphetamine succinate solution, or methylphenidate hydrochloride solution in random order on different days were administered. A blind rater observed each patient for changes in a number of variables every 10 min using a five-point rating scale. The items rated included: (1) psychosis, (2) conceptual disorganization, (3) unusual thoughts, (4) anger, (5) irritability, (6) interactions, (7) talkativeness, as well as (8) "activation" and (9) "inhibition" as defined previously (Janowsky, El-Yousef, Davis, and Sekerke, 1973b).

In each experiment, the rater and patient were blind to the time during an intravenous injection sequence when active drug was substituted for placebo, and to which active drug was given. Blood pressures and pulses were monitored every 5 min. Data were analyzed by comparing average change scores between the baseline-placebo phase and the 10- to 20-min and 20- to 30-min phases after active drug administration for each of the experimental drugs. Thus, each patient served as his own control. Patients received D-amphetamine and L-amphetamine, 20 mg (0.11 mM) and 28 mg (0.11 mM), respectively; methylphenidate, 29 mg (0.11 mM); as well as D-amphetamine, 4 mg (0.027 mM) and 10 mg (0.055 mM).

For clinical reasons, it was sometimes not possible to give each subject all the injections in the series. All comparisons between drug effects were made in those subjects who received injection of two experimental drugs using a Student's *t* test for paired results. Comparison was made between the baseline-placebo phase and the average of the 10- to 20-min and 20- to 30-min postinjection phase ratings.

III. RESULTS

Table 1 compares the relative abilities of intravenously administered D-amphetamine, L-amphetamine, and methylphenidate to increase the "combined psychosis score" (the sum of the global psychosis, conceptual disorganization, and unusual thoughts scores) as well as the "activation" scale score (sum of the interaction plus talkativeness scores). The values represent an increase in the score from the baseline value in all those patients receiving a specific drug, expressed as a percentage of the ability of L-amphetamine (0.11 mM) to increase these scores from baseline levels. The table indicates that methylphenidate (0.11 mM) is 300% and D-amphetamine (0.11 mM) 207% as potent as L-amphetamine in increasing psychosis activation. Significantly, D-amphetamine (0.055 mM – half the number of moles of L-amphetamine) is only 41% as active as L-amphetamine (0.11 mM) in increasing psychotic symptoms. Thus, D-amphetamine appears to be about two times as potent as L-amphetamine when given in equimolar doses. Similar ratios exist in relation to the abilities of the psychostimulants to increase activation. Thus, methylphenidate is 321%, and D-amphetamine 220%, as potent as L-amphetamine in causing an increase in the sum of the talkativeness and interaction scales.

TABLE 1. *Comparison of the abilities of various intravenously administered psychostimulants to increase psychosis and activation scores in psychiatric patients expressed as a percentage of the effects of L-amphetamine (28 mg)*

	Psychosis (combined psychosis score)	Activation (talkativeness + interaction)	Number of subjects (N)
Methylphenidate (29 mg)[a]	300%	321%	10
D-Amphetamine (20 mg)	207%	220%	18
L-Amphetamine (28 mg)	100%	100%	14
D-Amphetamine (10 mg)	41%	80%	6

[a] Methylphenidate (29 mg), D-amphetamine (20 mg), and L-amphetamine (28 mg) are equimolar.

For clinical reasons, it was impossible to have the patients receive all doses of all drugs. Therefore, the comparison of the potency of drugs in our study was a comparison of the effects in those same patients receiving a given pair of drugs administered in random order, using a crossover design.

In Table 2, data are presented indicating that D-amphetamine is significantly more effective in causing behavioral activation than is L-amphetamine given in equimolar doses to the same patients. Although D-amphetamine caused a greater degree of psychosis activation than did L-amphetamine, the difference was not quite statistically significant. L-Amphetamine was significantly more effective than D-amphetamine in causing an increase in irritability scores and anger scores. Thus, D-amphetamine was more active in increasing interaction and talkativeness, but less effective in inducing anger and irritability than was L-amphetamine.

TABLE 2. *Comparison of change scores between average baseline-placebo phase scores and the average of ratings done at 10 and 20 min after injection of L-amphetamine (28 mg) and D-amphetamine (20 mg) in 13 patients*[a]

	L-Amphetamine (28 mg)	D-Amphetamine (20 mg)	Difference between change scores
Conceptual disorganization	0.22 ± 0.16	0.61 ± 0.21	0.39 ± 0.24
Unusual thoughts	0.25 ± 0.11	0.42 ± 0.17	0.17 ± 0.15
Combined psychosis score	0.69 ± 0.32	1.43 ± 0.51	0.74 ± 0.52
Anger	0.35 ± 0.19	−0.05 ± 0.25	−0.40 ± 0.22[b]
Irritability	0.23 ± 0.18	−0.16 ± 0.14	−0.38 ± 0.20[c]
Activation (interaction + talkativeness)	0.56 ± 0.21	1.31 ± 0.25	0.76 ± 0.23[d]

[a] L-Amphetamine (28 mg) and D-amphetamine (20 mg) are equimolar.
[b] $p < 0.05$; [c] $p < 0.02$; [d] $p < 0.01$. One-tailed paired t test comparing differences between change scores.

IV. DISCUSSION

The data indicate that D-amphetamine worsened psychotic symptoms but was less potent than methylphenidate, and that L-amphetamine was the least potent of the three drugs studied.

The data indicate that D-amphetamine is approximately two times rather than 10 times as potent as an equimolar dose of L-amphetamine in activating preexisting psychotic behavior. The above results are substantiated by the fact that D-amphetamine (0.055 mM), which is half the molar dose of L-amphetamine given, had a lesser psychosis-activating effect than did L-amphetamine. Since the ratio of the effectiveness of D-amphetamine and L-amphetamine in activating dopaminergically mediated stereotyped behavior in rats is approximately 2:1, it would seem likely that, by analogy, the acute psychosis activation noted in patients following psychostimulant administration would be mediated by dopamine rather than by norepinephrine (Snyder et al., 1970; Taylor and Snyder, 1970). This finding is consistent with that of Angrist and Gershon (1971), who noted that the paranoid psychosis occurring in nonpsychotic volunteers receiving chronic doses of amphetamine was probably dopaminergically mediated (Angrist et al., 1971).

The above information may augment another set of data. It has recently been demonstrated that platelet monoamine oxidase is decreased in schizophrenia (Murphy and Wyatt, 1972). If brain monoamine oxidase is decreased in schizophrenics' brains as it is in platelets, it is tempting to speculate that psychostimulant-induced psychosis activation might be related in part to a central monoamine oxidase deficit. Specifically, central catecholamines released by methylphrenidate could be expected to remain in active form, since they would not be effectively metabolized intraneuronally by monoamine oxidase.

It is most important in considering the use of D-amphetamine and L-amphetamine as investigative tools to stress that many of the assumptions and conclusions presented here are based on neurochemical and behavioral experiments in animals. Animals are not humans, and it is thus quite possible that D-amphetamine and L-amphetamine have effects on central dopamine and norepinephrine in man which are different from those in rats. Also, a "drug challenge" experiment obviously involves administration of exogenous agents, and these events may not accurately reflect endogenous events. It is also important to stress that dopaminergic hypothesis of schizophrenia is at present quite speculative, based predominantly on indirect information. Thus, obvious caution must be used in applying the results of these or other psychostimulant experiments to the understanding of endogenous psychotic disorders. Nevertheless, the results presented in this paper are most consistent with the notion that psychostimulant-induced psychosis activation and generalized activation in man is more likely to be a dopaminergic than a noradrenergic phenomenon.

V. SUMMARY

The psychostimulants methylphenidate, D-, and L-amphetamine offer pharmacologic tools for the indirect evaluation of the role of norepinephrine and dopamine in schizophrenia. In actively ill schizophrenic patients, intravenous methylphenidate causes a dramatic intensification of preexisting symptoms such as hallucinations and delusions. In the current study, intravenous methylphenidate, D-, and L-amphetamine, in equimolar doses, were administered to schizophrenic patients. The results indicate that methylphenidate is a more potent activator of psychotic behavior than is D-amphetamine. If one assumes that a ratio of 2:1 between D- and L-amphetamine represents a dopaminergic phenomenon, it would seem that the general activating and psychosis-activating effects of D- and L-amphetamine represent a dopaminergic rather than a noradrenergic phenomenon. Investigations by others have shown that oral L-DOPA also produces a worsening of schizophrenic symptoms. These findings augment a body of information suggesting that dopamine may play a role in the schizophrenic process.

ACKNOWLEDGMENTS

This research was supported in part by grant GM 15431 from the National Institutes of Health, grant MH 11468 from the National Institute of Mental Health, and by research support from the State of Tennessee Department of Mental Health.

REFERENCES

Angrist, B. M., and Gershon, S. (1970): The phenomenology of experimentally induced amphetamine psychosis, preliminary observations. *American Journal of Psychiatry,* 126:95–107.
Angrist, B. M., and Gershon, S. (1971): A pilot study of pathogenic mechanisms in amphetamine psychosis utilizing differential effects of D- and L-amphetamine. *Pharmakopsychiatrie,* 4:64–75.
Angrist, B. M., Sathananthan, G., and Gershon, S. (1973): Behavioral effects of L-DOPA in schizophrenic patients. *Psychopharmacologia (in press).*
Angrist, B. M., Shopsin, B., and Gershon, S. (1971): Comparative psychotomimetic effects of stereoisomers of amphetamine. *Nature,* 234:152–153.
Everett, G. M., and Borcherding, J. W. (1970): L-DOPA, effect on concentrations of dopamine, norepinephrine, and serotonin in brains of mice. *Science,* 168:849–850.
Ferris, R. M., Tang, F., and Maxwell, R. A. (1973): A comparison of the capacities of isomers of amphetamine, deoxypipradrol, and methylphenidate to inhibit the uptake of tritiated catecholamines into rat cerebral cortex slices, synaptosomal preparation of rat cerebral cortex, hypothalamus and striatum and into adrenergic nerves of rabbit aorta. *Journal of Pharmacology and Experimental Therapeutics (in press).*
Griffith, J. J., Cavanaugh, J., Held, J., and Oates, J. A. (1972): Dextroamphetamine: Evaluation of psychomimetic properties in man. *Archives of General Psychiatry,* 26:97–100.
Horn, A. S., and Snyder, S. H. (1971): Chlorpromazine and dopamine: Conformational similarities that correlate with the antischizophrenic activity of phenothiazine drugs. *Proceedings of the National Academy of Sciences* (U.S.), 68:2325–2328.
Janowsky, D. S., El-Yousef, M. K., Davis, J. M., and Sekerke, H. J. (1973*a*): Provocation of schizophrenic symptoms by intravenous administration of methylphenidate. *Archives of General Psychiatry,* 28:185–191.
Janowsky, D. S., El-Yousef, M. K., Davis, J. M., and Sekerke, H. J. (1973*b*): Parasympathetic suppression of manic symptoms of physostigmine. *Archives of General Psychiatry,* 28:542–547.
Murphy, D. L., and Wyatt, R. J. (1972): Reduced monoamine axidase activity in blood platelets of schizophrenic patients. *Nature,* 238:225–226.
Randrup, A., and Munkrad, I. (1967): Behavioral stereotypes induced by pharmacological agents. *Pharmakopsychiatrie,* 1:8–26.
Scheel-Kruger, J. (1971): Comparative studies of various amphetamine analogues demonstrating different interactions with the metabolism of the catecholamines in the brain. *European Journal of Pharmacology,* 14:47–59.
Snyder, S. H., Taylor, J. T., Coyle, J. L., and Meyerhoff, J. L. (1970): The role of brain dopamine in behavioral regulation and the actions of psychotropic drugs. *American Journal of Psychiatry,* 127:199–207.
Taylor, K. M., and Snyder, S. H. (1970): Amphetamine: Differentiation of D- and L-isomers of behavior involving brain norepinephrine or dopamine. *Science,* 168:1487–1489.
Yaryura-Tobias, J. A., Diamond, B., and Merlis, S. (1970): The action of L-DOPA on schizophrenic patients (a preliminary report). *Current Therapeutic Research,* 12:528–531.

Neuropsychopharmacology of Monoamines and Their Regulatory Enzymes, edited by E. Usdin. Raven Press, New York © 1974.

Catecholamines, Drugs, and Behavior: Mutual Interactions

Lewis S. Seiden and A. Bruce Campbell

Departments of Pharmacological and Physiological Sciences, Psychiatry, and Psychology, University of Chicago, Chicago, Illinois 60637

Progress in understanding the mechanisms by which drugs influence behavior has stemmed from studies on the neurochemical basis of drug action and the function of catecholamines in the maintenance of behavior. Insight into the role of neurochemical substrates in maintenance of behavior is fundamental to the attempt to modify behavior through the use of drugs. In this context the primary question is: What role do monoamines play in the development and/or the maintenance of a certain behavior? Answers to this question are not simple and depend on a multitude of factors. According to one research strategy, endogenous monoamines may be altered by drugs and subsequent alterations in behavior observed; however, as has frequently been pointed out, covariance between two events does not provide proof of a causal relationship in the absence of supporting evidence. Such evidence has been provided by antagonizing the effects of a drug with another drug or chemical precursor.

Research over the past 15 years has shown that the effects of drugs on behavior depend not only on the type of drug and the dose employed, but also on the ongoing behavior of the animal. The work of Dews (1958) demonstrated that amphetamine decreased the rate of an operant pecking response in pigeons under an FR 50 schedule of reinforcement, but that the same dose of amphetamine increased response rate when the pigeon was pecking under an FI 5-min schedule. This type of drug-behavior interaction appears to be general and has been extended to other drugs, species, and schedules of reinforcement (Dews, 1958; Kelleher and Morse, 1968; Schoenfeld and Seiden, 1969).

This chapter will review some experimental evidence which shows that (1) catecholamines play a functional role in the maintenance of behavior under stimulus control, (2) behavior can modify the metabolism of brain catecholamines, and (3) mutual interactions between drug effects on catecholamines and behavioral or stimulus effects on catecholamines may account for the fact that drug effects depend on the ongoing behavior and the environmental setting in which the drug is administered. The model which has been evolved from this research is presented in Fig. 1.

According to the model, behavior is a function of brain biochemistry and environmental contingencies, and brain biochemistry is a function of drug administered, ongoing behavior, and the environment. Evidence for this model stems from research in a number of laboratories.

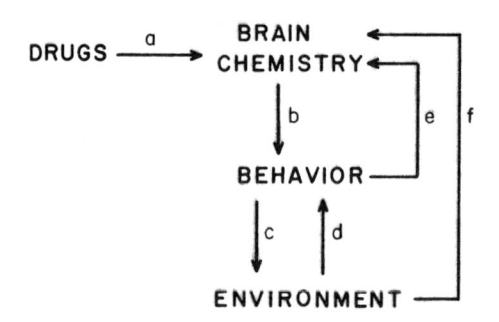

FIG. 1. Modification of behavior through the internal and external milieu. Many, if not all, drugs which modify behavior act in some manner to alter brain chemistry (a). This chapter discusses the alteration of a conditioned avoidance response through manipulation of biogenic amines (b) (Seiden and Carlsson, 1964), as well as the modification of brain norepinephrine by ongoing operant behavior (e) (Schoenfeld and Seiden, 1969; Lewy and Seiden, 1972). The environment itself may influence behavior (d) by providing the conditions necessary for the occurrence of behavior (Skinner, 1938; Reynolds, 1961; Terrace, 1963), or by providing a stimulus which directly elicits the behavior (reflex behavior). Operant behavior, by definition, acts on the environment to produce a consequence (c). Further, the environment may directly alter behavior by modifying brain chemistry (f) (Rosenzweig, Krech, Bennett, and Diamond, 1962; Moore, 1963).

The functional role of neurochemicals in the maintenance of behavior is well exemplified in studies of the catecholaminergic system and behavior. L-Dihydroxyphenylalanine (L-DOPA) can antagonize many of the behavioral actions of reserpine, and the available evidence strongly suggests that L-DOPA effects are mediated through its metabolites, dopamine (DA) and norepinephrine (NE). Several workers have shown that L-DOPA administration can antagonize the reserpine-induced suppression of locomotor activity (Carlsson et al., 1957; Blaschko and Chrusciel, 1960; Smith and Dews, 1962), and this evidence, along with the clinical pathology and treatment of Parkinson's disease, strongly implicates DA in motor function. Seiden and Carlsson (1963, 1964) have demonstrated that DOPA can antagonize the reserpine-induced suppression of a conditioned avoidance response. When mice had reached criterion performance in a shuttle-box avoidance situation they were injected with reserpine, which abolished the maintenance of avoidance behavior and caused a reduction in escape behavior (Fig. 2). Injection of L-DOPA partially and temporarily antagonized this reserpine effect. Further, work has indicated that the L-DOPA effect is mediated by the formation of DA and to some extent NE in the central nervous system (Seiden and Peterson, 1968a, 1968b; Seiden and Martin,

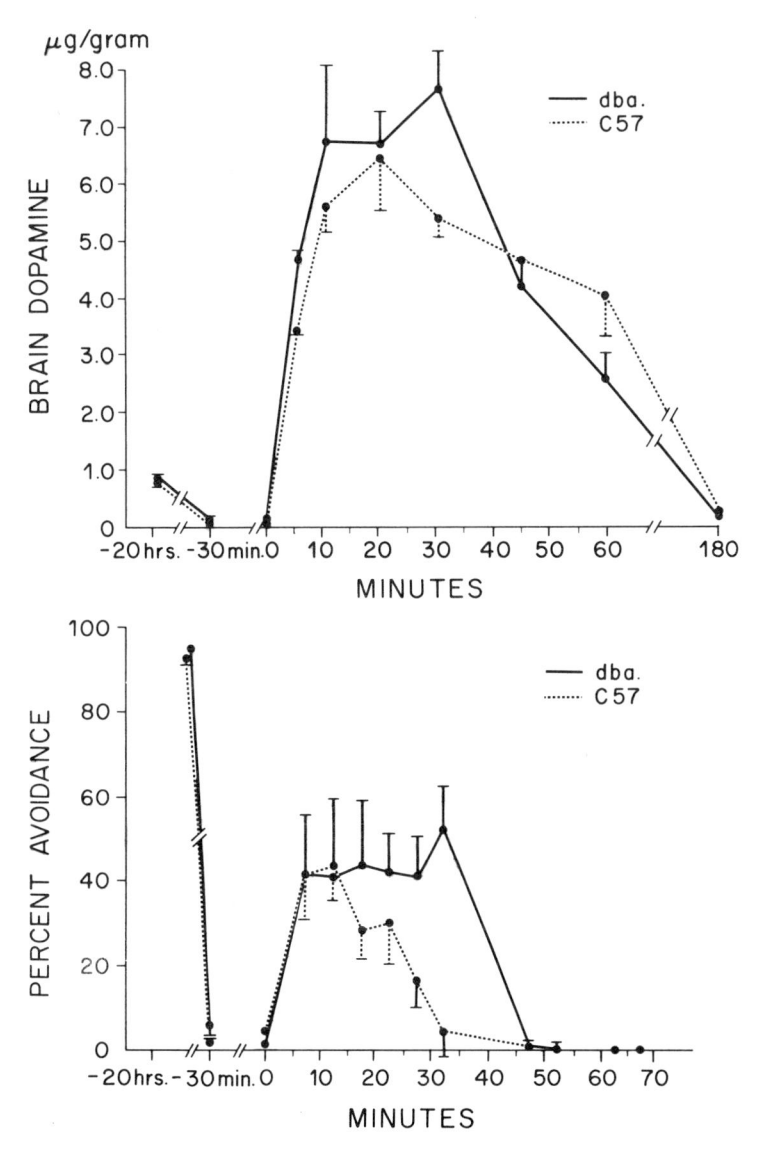

FIG. 2. Top: Brain dopamine levels in mice treated with reserpine (2.5 mg/kg) followed by L-DOPA (400 mg/kg). Vertical lines represent SEM. Bottom: The effect of reserpine and L-DOPA on the CAR. −20 hr represents the level of avoidance responding before any drug treatment. Immediately after this control session, the mice were injected with reserpine (2.5 mg/kg, i.p.) and tested at −30 min. Mice were injected with L-DOPA (400 mg/kg, i.p.) at time 0. No mouse was tested for more than 10 min after DOPA administration. Vertical lines represent SEM. From Seiden and Peterson (1968a); reproduced with permission of the publisher.

1971). Other investigators have obtained similar results using different species and different drugs to lower concentrations of catecholamines in brain (Wada et al., 1963; Seiden and Hanson, 1964; Moore and Rech, 1967; Ahlenius and Engel, 1971; Breese et al., 1973). It is appropriate to generalize that DA and NE play a role in the maintenance of a conditioned avoidance response in several species of animals.

Catecholamines also appear to play a role in the maintenance of operant behavior under appetitive control (Schoenfeld and Seiden, 1969). Using water as a reinforcing stimulus, groups of rats were trained to press a lever under one of several fixed-ratio (FR) or fixed-interval (FI) schedules of reinforcement: FR 1, FR 5, FR 10, FI 30-sec, FI 60-sec, and FI 90-sec. When lever-pressing performances had stabilized, rats were treated with alpha-methyl-para-tyrosine (AMT) in 75-mg/kg doses given 8 and 4 hr prior to testing. AMT treatment caused a decrease in response rates under all schedules (Fig. 3). The response decrement, however, depended on the

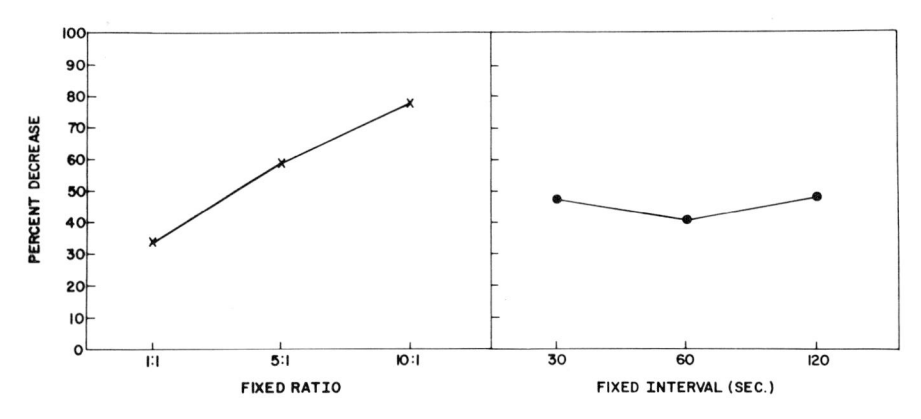

FIG. 3. The effect of different schedules of reinforcement on the response to AMT (two doses, 75 mg/kg).

schedule. Rats pressing under the fixed-interval schedule showed a 30 to 40% decrease in response rate that was constant for all parameter values of the schedule. On the other hand, the response rate decrease depended on the schedule parameter in rats pressing under a fixed-ratio schedule. Rats pressing under FR 1 showed less of a decrement than those pressing under FR 5 which in turn showed less decrement than rats pressing on FR 10. The schedule (or rate) dependency for the effect of AMT is consistent with the finding by Dews (1958) and Kelleher and Morse (1968), who showed similar drug-behavior interactions using different drugs, schedules, and species. In addition, it was found that FI performances were maintained at a constant, but slightly decreased rate under AMT. On the other hand, response rate was dramatically altered throughout the session in rats pressing under FR schedules (Fig. 4). This continuous decrease in FR rates over the session

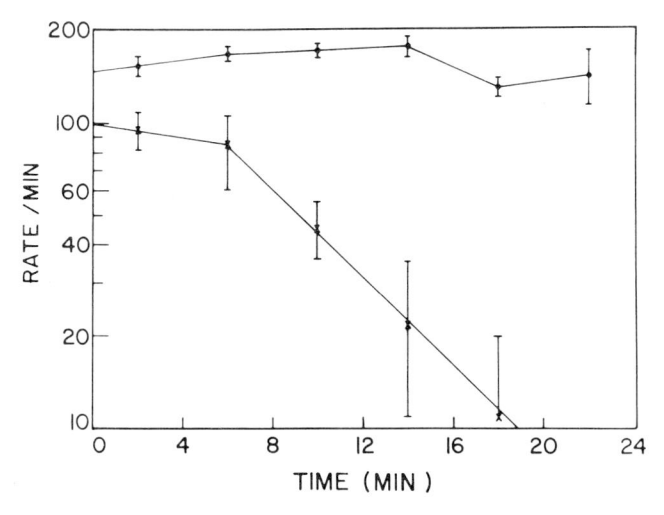

FIG. 4. Effect of AMT (two doses, 75 mg/kg) on rate of lever pressing during FR-10 schedule of reinforcement. Each value represents the mean rate per minute over consecutive 4-min periods. N = 5. Circle: control session; ×: AMT-treated rats.

suggests that the behavior itself continuously modified the effect of AMT. The fact that the effect of AMT could be antagonized in a dose-dependent fashion by L-DOPA indicated that the effects of AMT were mediated through catecholamines (Fig. 5).

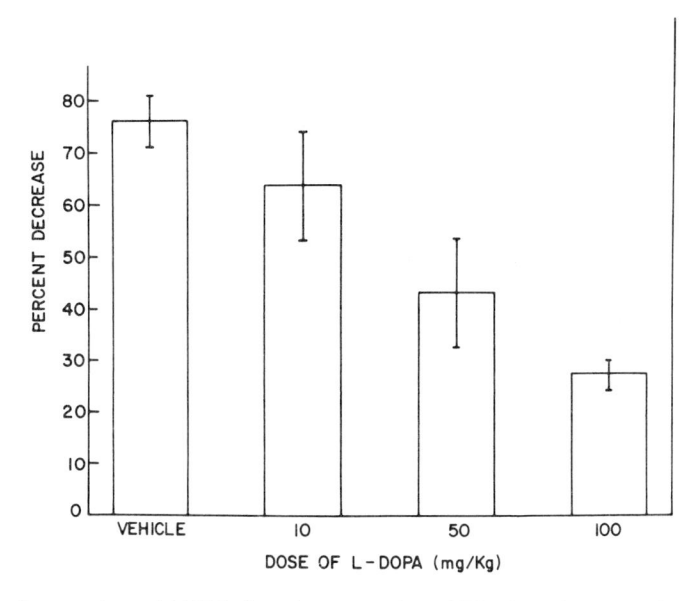

FIG. 5. Antagonism of AMT-induced suppression of FR-10 performance by L-DOPA.

Proceeding on the evidence that there was a drug-behavior interaction, and that the drug effect was mediated by catecholamines, it seemed logical to explore the possibility that ongoing behavior also interacted with brain catecholamines. An experiment was designed in which rats were placed into the operant chamber $\frac{1}{2}$ hr after a large dose of AMT (200 mg/kg), lever pressed under a VI 30-sec schedule for 2 hr, and were then killed and their brains analyzed for DA and NE. Another group of rats was injected with the same dose of AMT, but was neither water-deprived nor trained to press the lever. The rats in the latter group were also killed after 2 hr, and the relative depletion of CAs between the two groups was compared. It was found that the lever-pressing group showed about a 15% greater depletion than the group remaining in their cages. Therefore, it appeared that some aspect of the operant situation accelerated depletion of brain catecholamines. Such factors might include water deprivation, operant training, lever pressing, intermittent water consumption, or stimulation occurring in the operant chamber.

In order to isolate some of the variables responsible for increased CA depletion, additional experiments employed tritiated NE to measure changes in NE metabolism to eliminate the possibility that the changes observed in the above experiments with AMT were due to a difference in the distribution of the drug that was dependent on the behavior. A tracer amount of tritiated NE was administered intraventricularly to rats that had been trained under a VI 30-sec schedule. The hypothalamus-midbrain and pons-medulla were analyzed for total NE and tritiated NE. It was found that rats pressing under the VI 30-sec schedule had the same amount of endogenous NE as naive water-deprived and ad-lib controls. The total tritiated NE, however, was lower in the responding group than in either control group; the specific activity of the behaving group was lowered to a similar extent (Fig. 6). In separate experiments it was found that neither training nor motor activity influenced the metabolism of NE in the brainstem. Although it is still not clear what aspect of the operant situation is responsible for the observed changes in NE metabolism, it appears that deprivation, motor activity, and training are not critical variables.

The experiments discussed above strongly suggest that drug behavior interactions noted by several investigators can be accounted for by the differential actions of behavior and/or environmental stimuli on brain catecholamines. In order to test this notion, a series of studies were recently completed. The drugs employed in this study were amphetamine and AMT, since there is substantial evidence that the behavioral actions of these compounds are mediated by catecholamines. Rats were trained to press a lever on a VI 20-sec schedule. Separate groups of rats were reinforced with different amounts of water (0.01 ml, 0.04 ml, 0.10 ml). Each group of rats had the same rate of lever pressing irrespective of the amount of water reinforcement. The effects of amphetamine, however, depended on the rein-

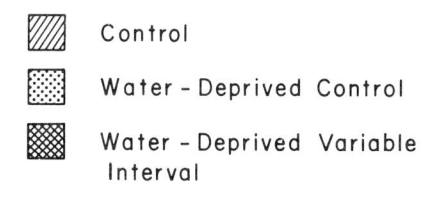

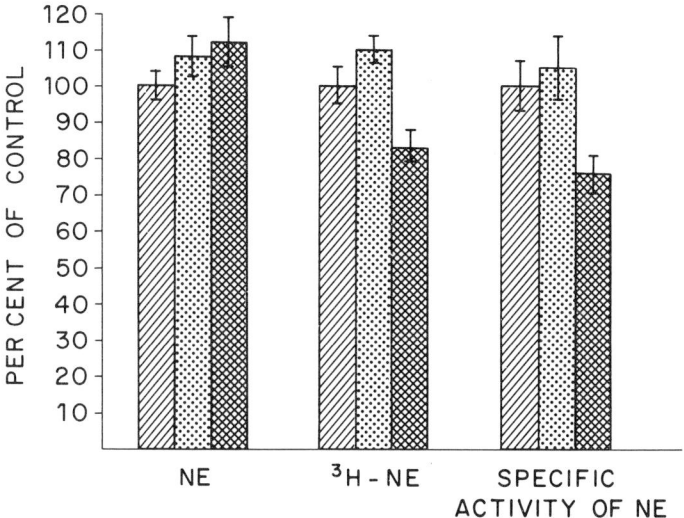

FIG. 6. The effect of operant behavior and water deprivation on the concentrations of NE, and the specific activity of NE in the rat brainstem-diencephalon. Values represent the mean of five animals ±SEM. Absolute values for the control group are 0.555 ± 0.0222 γ/g of NE, 92.7 ± 5.3 mµC/g of ^{3}H-NE, and 168 ± 13 mµC/µg specific activity of NE.

forcement magnitude (Fig. 7). Rats reinforced with small amounts of water showed the greatest amphetamine-induced decrease in response rate, with the amount of behavioral depression being a function of the dose of amphetamine. The group reinforced with the largest amount of water showed no such dose-dependent decrease in response rate, and at the highest dose of amphetamine approximately half of the animals showed an amphetamine-induced increase in response rate. Since the rates of responding prior to drug were the same in all groups, we have concluded that the size of reinforcement is a critical determinant of the action of amphetamine. Analysis of interresponse times (IRT) revealed that amphetamine caused a dose-dependent increase in the short IRTs (defined as IRTs falling within three standard deviations of the mean IRT) for both the groups reinforced with 0.01 and 0.04 ml of water, but did not affect short IRTs in the group reinforced with 0.10 ml of water (Fig. 8).

When AMT was administered in two doses (75 mg/kg) 8 and 4 hr prior

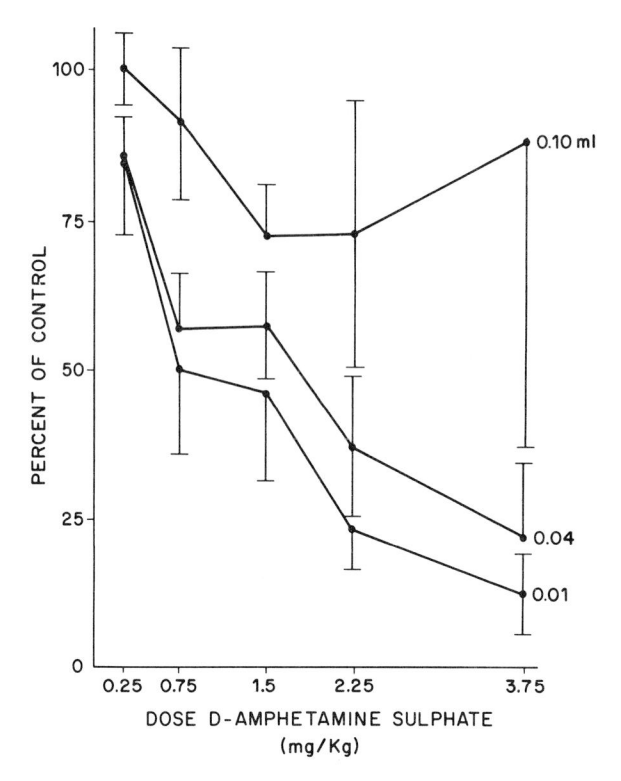

FIG. 7. Response rate (as percent of control) during a variable-interval 20-sec schedule, for the three reinforcement magnitude groups, after five different doses of D-amphetamine sulfate. All values (ml) are expressed as group (N = 6) means ±SE.

to the testing session, a small decrease in response rates was seen in all animals. The rate-depressant effect of AMT did not depend on the size of reinforcement. With only one dose of AMT administered 2 hr before the session, no rate-decreasing effects of AMT were noted. Since there is evidence to suggest that the newly synthesized catecholamine pool is necessary for producing amphetamine effects, experiments were undertaken in which the combined effects of AMT and amphetamine were examined. AMT (75 mg/kg) was given 2 hr before the experimental session and amphetamine was administered $\frac{1}{2}$ hr before the experimental session. The results of the drug interactions showed that the rate-decreasing effects of amphetamine were potentiated by AMT pretreatment. Furthermore, unlike the effect of amphetamine given alone, reinforcement magnitude did not influence the rate-decreasing effect of AMT plus amphetamine.

In view of these results and the postulated mechanism of catecholamine-amphetamine interactions, an extensive series of neurochemical studies

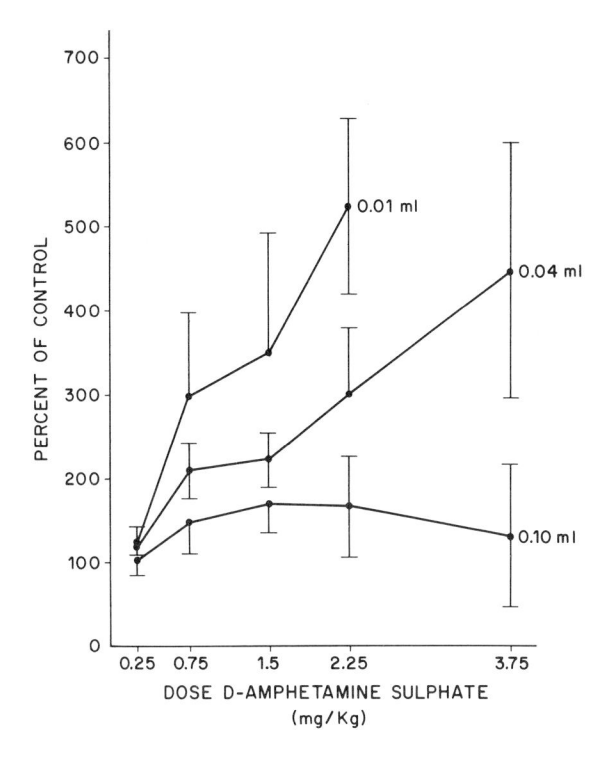

FIG. 8. Mean short IRT values, plotted as a percent of control, of the three reinforcement magnitude groups after five different doses of D-amphetamine sulfate. The 0.01-ml group was not plotted at 3.75 mg/kg due to insufficient data for computer analysis at this dose point. All values are expressed as group means ±SE.

was undertaken to determine the possible correlations between the effects of amphetamine, AMT, and ongoing behavior, with a view toward understanding the neurochemical basis of these drug-behavior interactions.

In the neurochemical studies to be discussed below, the concentration of brain catecholamines was measured in several areas of the brain under differing drug conditions and in behaving rats reinforced with different amounts of water. In the latter case the catecholamine concentrations were measured after the animals had performed in the operant chamber. These rats had been pretreated with AMT (in order to inhibit the formation of newly synthesized amines). Therefore differences in CA concentrations among the three groups reflect differences in the rate of CA metabolism (probably release and utilization). Although several brain areas were analyzed in all experiments, the pons-medulla results correspond most closely to the behavioral results; therefore the following data will show mainly the results of this area, although it should be pointed out that there were interesting qualitative differences between various areas of the brain examined.

Amphetamine did not alter the concentration of amines in brain except

TABLE 1. Effect of amphetamine on pons-medulla NE concentration

Amphetamine (mg/kg)	0.75	2.25	3.75
NE (% of control)	100%	100%	115%[a]

[a] Differs from control, $p < 0.05$.

at the highest dose, which produced a slight increase (Table 1). Increased NE concentration following amphetamine has not been reported before, but rather amphetamine has been reported to decrease NE concentrations (e.g., Moore et al., 1970). In the latter studies, however, the range of doses employed exceeded those in this study, and the increase that we observed may be related to the dose of amphetamine.

AMT, when administered in two separate doses 8 and 4 hr prior to sacrifice, decreased catecholamines. Neither AMT (75 mg/kg, 2 hr pretreatment) nor amphetamine administered alone had effects on NE concentrations in the pons-medulla. However, when the same doses and pretreatment times were used and the two drugs were administered together, NE concentrations in the pons-medulla were substantially reduced (Table 2). The amount

TABLE 2. Effect of AMT followed by amphetamine on pons-medulla NE concentration

AMT administered 2 hr before killing (mg/kg)	75	75	75
Amphetamine administered 20 min before killing (mg/kg)	0.75	2.25	3.75
NE (% of control)	95%	95%	65%[a]

[a] Differs from control, $p < 0.05$.

of NE reduction was proportional to the dose of amphetamine. AMT-induced NE depletion in the pons-medulla depended upon the magnitude of reinforcement among rats pressing on a VI 20-sec schedule of reinforcement. The NE depletion induced by AMT was inversely proportional to the magnitude of reinforcement, suggesting that low magnitudes of reinforcement engender a faster rate of turnover in the pons-medulla than higher magnitudes of reinforcement (Table 3).

The above results suggest that the differential effects of amphetamine on behavior maintained by different quantities of reinforcement might be explained in the following way. We assume that amphetamine acts by causing release of NE onto postsynaptic receptor sites. Under most circumstances, no drug-induced changes in concentration are detected, because increased NE synthesis can compensate for release. On the other hand, we have observed that with low reinforcement magnitude there appears to be greater NE release. If the NE stores are necessary for the maintenance of constant release of NE, it can be seen that when the small reinforcement magnitudes

TABLE 3. *Effect of reinforcement magnitude and AMT on pons-medulla NE concentration*

Reinforcement (ml)	Ad lib. (control)	0.01	0.04	0.10
AMT administered 2 hr before beginning of session (mg/kg)	75	75	75	75
NE concentration 0.5 hr after beginning of session (% of ad lib. control)	100%	80%	80%	92%

are combined with amphetamine the effects on the stores will be additive. As a result, the supply of NE in the presynaptic terminal will be decreased, and behavior will be depressed. However, with the larger reinforcement magnitude the turnover of NE will not be stimulated to the extent that it is with smaller reinforcement magnitudes. Therefore, amphetamine and behavior do not add up to the same effect on NE as they did when reinforcement magnitude was high; in the latter case turnover is not being driven by the behavior at a high rate, and therefore the displaced NE will either stimulate responding, as it did with about half of the rats, or do nothing at all, as it did with the other half.

A model of the type of situation of drug-biochemical-behavioral interactions is shown in Fig. 9. Consider two cases. In the first, NE is turning over at a high rate. Accordingly, behavior is occurring at a high rate. If amphetamine is added to the already high turnover situation, the amount of NE in the presynaptic membrane is further depleted as a result of the additive effect of the drug and the ongoing behavior. Therefore the synaptic cleft becomes flooded with NE, and the excess transmitter causes a block. According to the model, the conditions produce a rate decrease. In the second case, NE turnover occurs at a low rate. Under these circumstances amphetamine increases the NE available for transmission and thereby increases the rate of behavior. It has been shown that reinforcement magnitude engenders different NE turnover rates in pons-medulla. In addition, other data support the notion that various aspects of the operant situation also engender different turnover rates of NE in brain. The different baseline of NE secreted from the presynaptic nerve may well account for the dependence of the effect of amphetamine on the environment and the ongoing behavior.

Other classes of drugs, including barbiturates, anticholinergics, phenothiazines, and imipramine types of compounds, have been shown to have effects on behavior that are dependent on the stimuli that maintain the behavior and the behavior itself. Since many of the drugs mentioned interact with other neurochemical transmitters (e.g., acetycholine, serotonin, histamine) it is likely that the environmental stimuli and ongoing behavior itself will modulate the metabolism, distribution, or turnover rate of these various transmitter substances.

336   CATECHOLAMINES, DRUGS, AND BEHAVIOR

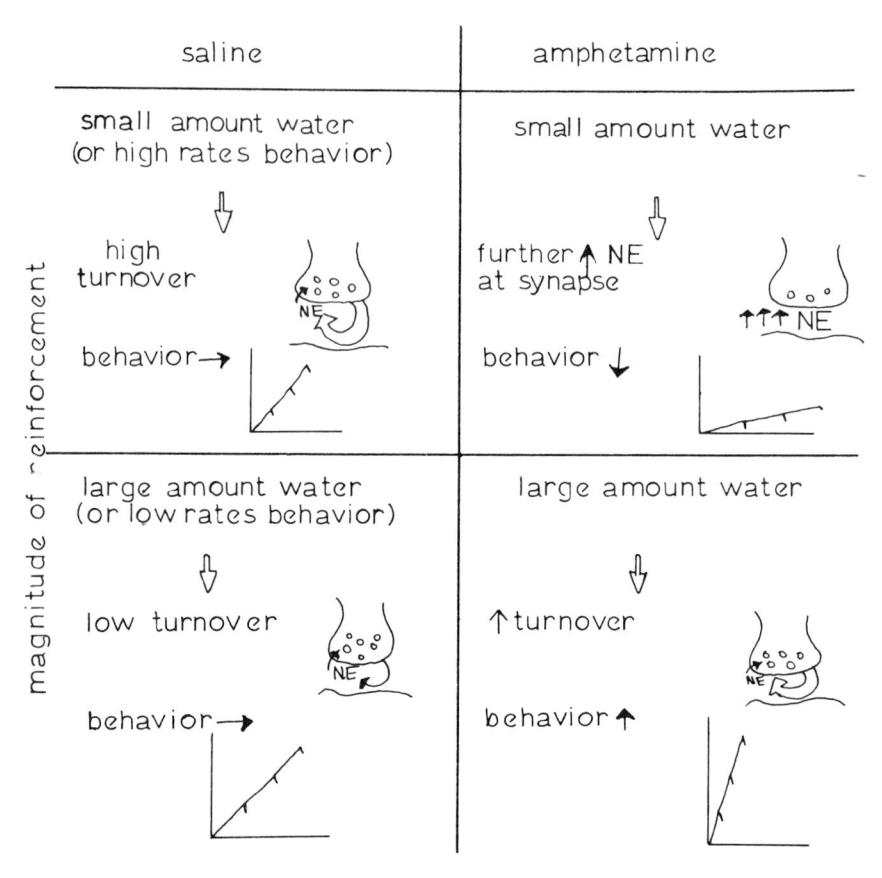

	saline	amphetamine
magnitude of reinforcement	small amount water (or high rates behavior) ⇩ high turnover behavior→	small amount water ⇩ further ↑ NE at synapse behavior ↓
	large amount water (or low rates behavior) ⇩ low turnover behavior→	large amount water ⇩ ↑ turnover behavior ↑

FIG. 9. A model of events at the synaptic level which underlie the interactions between amphetamine, reinforcement magnitude, and operant behavior. Note that the small inserted graphs in each of the four figures are cummulative response records in which rate of response is proportional to the scope of the function. For further details see text.

SUMMARY

Mutual interaction between conditioned behavior and brain catecholamines were studied using pharmacological, neurochemical, and analytical behavioral techniques. Drug-induced modifications of the metabolism or concentration of brain catecholamines altered behavior in a conditioned avoidance response procedure as well as in operant procedures which were maintained by appetitive reinforcement. L-Dihydroxyphenylalanine (L-DOPA) antagonized these drug effects, and further experimental work indicated that the L-DOPA effect depended on the conversion of L-DOPA to its catecholamine metabolites, dopamine (DA) and norepinephrine (NE). The effect of the drugs depended on both the degree of catecholamine modification and the nature of the ongoing behavior at the time of testing. In a series of studies where behavior was the independent variable, it was found that catecholamine metabolism was altered as a function of behavior; changes in NE metabolism were estimated using alpha-methyltyrosine or tritiated NE. Other experiments demonstrated that changes in the effects of amphetamine in rats pressing a lever on a variable interval schedule depended on the magnitude of reinforcement. Furthermore, there was an interaction between the magnitude of reinforcement and the metabolism of brain NE. It was concluded from these studies that the interaction between amphetamine and the behavioral situation can be understood in neurochemical terms if changes in NE metabolism are measured on a regional neuroanatomical basis in behaving animals. These findings suggest a general biochemical basis for drug effects which are dependent on the ongoing behavioral situation.

ACKNOWLEDGMENTS

This work was supported by a research grant from the National Institute of Mental Health, USPHS MH-11191; a Research Scientist Development Award from the National Institute of Mental Health, PHS 5K02-MH 10, 562 (L.S.S); and Mental Health Training Grant MH-07083 (A.B.C.).

REFERENCES

Ahlenius, S., and Engel, J. (1971): Behavioral and biochemical effects of L-DOPA after inhibition of dopamine-β-hydroxylase in reserpine pretreated rats. *Naunyn-Schmiedeberg's Archives of Pharmacology*, 270:349–360.

Blaschko, H., and Chrusciel, T. L. (1960): The decarboxylation of amino acids related to tyrosine and their awakening effect in reserpine-treated mice. *Journal of Physiology* (London), 151:272–284.

Breese, G. R., Cooper, B. R., and Smith, R. D. (1973): Biochemical and behavioral alterations following 6-hydroxydopamine administration into brain. In: *Frontiers in Catecholamine Research: Third International Catecholamine Symposium*, edited by E. Usdin and S. Snyder. Pergamon Press, Elmsford, New York.

Carlsson, A., Lindquist, M., and Magnusson, T. (1957): 3,4-Dihydroxyphenylalanine and 5-hydroxytryptophane as reserpine antagonist. *Nature*, 180:1200.

Dews, P. B. (1958): Analysis of effects of psychopharmacological agents in behavioral terms. *Federation Proceedings*, 17:1024–1030.

Kelleher, R. T., and Morse, W. H. (1968): Determinants of the specificity of behavioral effects of drugs. *Reviews of Physiology, Biochemistry and Experimental Pharmacology*, 60:1–56.

Lewy, A. J., and Seiden, L. S. (1972): Operant behavior changes norepinephrine metabolism in rats. *Science*, 175:454–456.

Moore, K. E. (1963): Toxicity and catecholamine releasing actions of D- and L-amphetamine in isolated and aggregated mice. *Journal of Pharmacology and Experimental Therapeutics*, 142:6–12.

Moore, K. E., Carr, L. A., and Dominic, J. A. (1970): Functional significance of amphetamine-induced release of brain catecholamines. In: *Amphetamines and Related Compounds. Proceedings of the Mario Negri Institutes for Pharmacological Research*, edited by E. Costa and S. Garattini. Raven Press, New York, pp. 371–384.

Moore, K. E., and Rech, R. H. (1967): Reversal of α-methyltyrosine-induced behavioral depression with dihydroxyphenylalanine and amphetamine. *Journal of Pharmacy and Pharmacology*, 19:405–407.

Reynolds, G. S. (1961): Behavioral contrast. *Journal of the Experimental Analysis of Behavior,* 4:57–71.

Rosenzweig, M. R., Krech, D., Bennett, E. L., and Diamond, M. C. (1962): Effects of environmental complexity and training on brain chemistry and anatomy: A replication and extension. *Journal of Comparative Physiology and Psychology,* 55:429–437.

Schoenfeld, R., and Seiden, L. S. (1969): Effect of α-methyltyrosine on operant behavior and brain catecholamine levels. *Journal of Pharmacology and Experimental Therapeutics,* 167:319–327.

Seiden, L. S., and Carlsson, A. (1963): Temporary and partial antagonism by L-DOPA of reserpine induced suppression of a conditioned avoidance response. *Psychopharmacologia,* 4:418–423.

Seiden, L. S., and Carlsson, A. (1964): Brain and heart catecholamine levels after L-DOPA administration in reserpine treated mice: Correlations with a conditioned avoidance response. *Psychopharmacologia,* 5:178–181.

Seiden, L. S., and Hanson, L. C. F. (1964): Reversal of the reserpine induced suppression of the conditioned avoidance response in the cat by L-DOPA. *Psychopharmacologia,* 6:239–244.

Seiden, L. S., and Martin, T. W. (1971): Potentiation of effects of L-DOPA on conditioned avoidance behavior by inhibition of extracerebral DOPA-decarboxylase. *Physiology and Behavior,* 6:453–458.

Seiden, L. S., and Peterson, D. D. (1968a): Reversal of the reserpine-induced suppression of the conditioned avoidance response by L-DOPA: Correlation of behavioral and biochemical differences in two strains of mice. *Journal of Pharmacology and Experimental Therapeutics,* 159:422–428.

Seiden, L. S., and Peterson, D. D. (1968b): Blockade of L-DOPA reversal of the reserpine induced CAR suppression by disulfiram. *Journal of Pharmacology and Experimental Therapeutics,* 163:85–90.

Skinner, B. F. (1938): *The Behavior of Organisms: An Experimental Analysis.* Appleton-Century-Crofts, New York.

Smith, C. B., and Dews, P. B. (1962): Antagonism of locomotor suppressant effects of reserpine in mice. *Psychopharmacologia,* 3:55–59.

Terrace, H. S. (1963): Discrimination learning with and without "errors." *Journal of the Experimental Analysis of Behavior,* 6:1–27.

Wada, J. A., Wrinch, J., Hill, D., McGeer, P. L., and McGeer, E. G. (1963): Central aromatic amine levels and behavior. *Archives of Neurology,* 9:69–89.

Neuropsychopharmacology of Monoamines and Their Regulatory Enzymes, edited by E. Usdin. Raven Press, New York © 1974.

The Effect of Catechol-O-methyl Transferase Inhibitors on Behavior and Dopamine Metabolism

G. M. McKenzie

Department of Pharmacology, Wellcome Research Laboratories, Research Triangle Park, North Carolina 27709

I. INTRODUCTION

It is currently thought that apomorphine is a specific dopaminergic agonist in the central nervous system. This conclusion is supported by the following experimental findings. Apomorphine produces stereotyped behavior in a number of species (Dhawan and Saxena, 1960; Ernst, 1967; Scheel-Krüger, 1970; Willner, Samach, Angrist, Wallach, and Gershon, 1970; McKenzie, 1973). In addition, apomorphine lowers homovanillic acid (HVA) concentrations in the striatum of rodents (Roos, 1969) and decreases the rate of dopamine synthesis (Nybäck, Shubert, and Sedvall, 1970) without producing marked changes in dopamine levels. Both the behavioral and biochemical changes induced by apomorphine are effectively antagonized by neuroleptic agents which are thought to block dopamine receptors (Andén, Rubenson, Fuxe, and Hökfelt, 1967; Ernst, 1969; Nybäck et al., 1970; Lahti, McAllister, and Wozniak, 1972; Rostrosen, Wallach, Angrist, and Gershon, 1972; McKenzie, Viik, and Boyer, 1973). Furthermore, the depletion of dopamine following α-methyltyrosine is retarded by apomorphine (Andén et al., 1967). Lastly, animals with unilateral 6-hydroxydopamine lesions in the substantia nigra, medial forebrain bundle, or nucleus caudatus putamen respond to apomorphine treatment by circling contralateral to the lesion (Ungerstedt, 1971; Von Voigtlander and Moore, 1973).

It has been proposed that apomorphine-induced changes in dopamine turnover are due to the action of a negative feedback mechanism brought into play by dopamine receptor activation, thus producing a decrease in the firing rate of dopaminergic neurons (Carlsson and Lindqvist, 1963; Andén, Corrodi, Fuxe, and Ungerstedt, 1971). Conversely, neuroleptics would produce a compensatory increase in firing rates due to blockade of dopamine receptors and thus increase dopamine turnover. It has been recently demonstrated that the systemic administration of apomorphine does result in a decrease in the firing rate of striatal dopaminergic neurons (Bunny, Walters, Roth, and Aghajanian, 1973). How changes in neuron activity signal appropriate changes in the rate of transmitter synthesis is unknown; however, a trans-synaptic messenger has been proposed to explain the changes in dopamine metabolism following apomorphine and neuroleptics (Kehr, Carlsson, Lindqvist, Magnusson, and Atack, 1972).

The fact that apomorphine is a catechol suggested that this agent should be metabolized by catechol-O-methyl transferase (COMT) and thus would be a competitive inhibitor of this enzyme. Furthermore, it has been reported recently that tropolone, a known competitive inhibitor of COMT, lowered striatal HVA levels and blocked the chlorpromazine-induced increase in HVA (Roth, 1973). These considerations led to the idea that changes in dopamine turnover following apomorphine could be a result of an interaction between apomorphine and the enzyme COMT.

The purpose of these experiments was to study (1) the possible interactions between apomorphine and COMT and (2) how such an interaction might be involved in the mechanism of action of apomorphine.

II. POTENTIATION OF APOMORPHINE BY COMT INHIBITORS

Long Evans rats, pretreated with COMT inhibitors, were given apomorphine s.c. and scored under blind conditions according to the method described by McKenzie (1972). COMT inhibitors were administered i.p. according to the following schedule: pyrogallol, 250 mg/kg and 150 mg/kg; tropolone, 100 mg/kg and 50 mg/kg; and 8-hydroxyquinoline, 100 mg/kg and 100 mg/kg; 30 min before and 1 hr after apomorphine, respectively. The effects of these drugs on stereotyped behavior were tested against a wide range of doses of apomorphine in the case of pyrogallol or against standard doses of 1 mg/kg and 5 mg/kg of apomorphine in animals pretreated with tropolone and 8-hydroxyquinoline.

Pretreatment with COMT inhibitors potentiated markedly the stereotyped behavior induced by apomorphine (Fig. 1). Potentiation by pyrogallol was indicated by a reduction in the threshold dose of apomorphine from 0.5 mg to 0.1 mg/kg (Fig. 1A). In addition, mean behavioral scores at all doses of apomorphine were approximately doubled (Fig. 1B, C). Lastly, the second-hour mean scores after 5 and 10 mg/kg of apomorphine were increased from 4 ± 0.08 and 14.7 ± 1.4 to 19.8 ± 1.0 and 23.1 ± 0.9, respectively (Fig. 1B).

Tropolone and 8-hydroxyquinoline also produced marked prolongation of stereotyped behavior and significant increases in mean scores during the second-hour scoring period (Fig. 1B, C).

In the absence of apomorphine, none of the COMT inhibitors produced stereotyped behavior.

Missala, Lal, and Sourkes (1973) also demonstrated that pyrogallol prolonged apomorphine-induced stereotyped behavior. These *in vivo* data clearly suggested that methylation by COMT may be an important metabolic pathway for the inactivation of apomorphine.

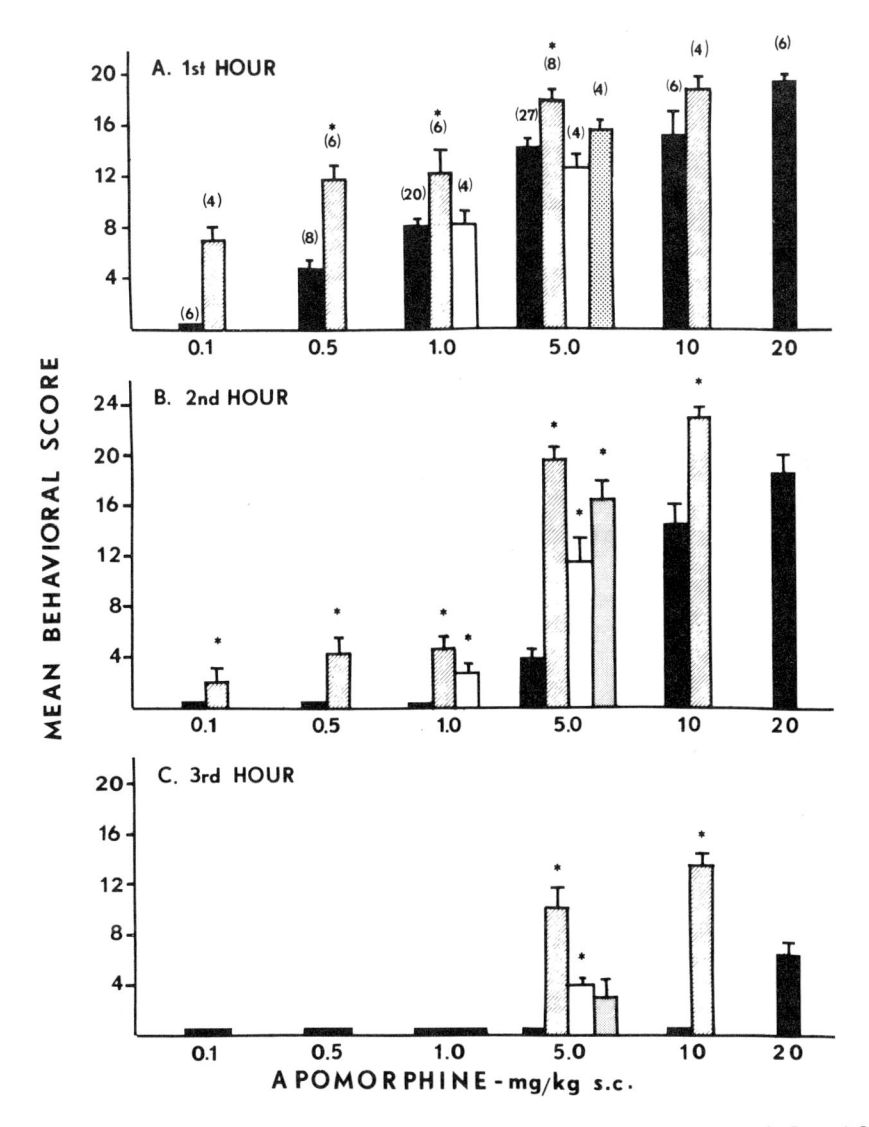

FIG. 1. Mean stereotyped behavior scores following apomorphine treatment. A, B, and C show the mean scores during the first, second, and third hour, respectively. Solid bars, control animals receiving only apomorphine; hatched bars, pyrogallol pretreated; open bars, tropolone pretreated; stippled bars, 8-hydroxyquinoline pretreated. *indicates significantly different ($p < 0.05$) from control animals for that dose of apomorphine. p values were calculated using Student's t test.

III. *IN VITRO* METHYLATION OF APOMORPHINE

Using purified COMT from either rat liver or brain and ^{14}C-S-adenosyl-L-methionine as the methyl donor, White and McKenzie (1971) and McKenzie and White (1973) demonstrated that apomorphine was a substrate for

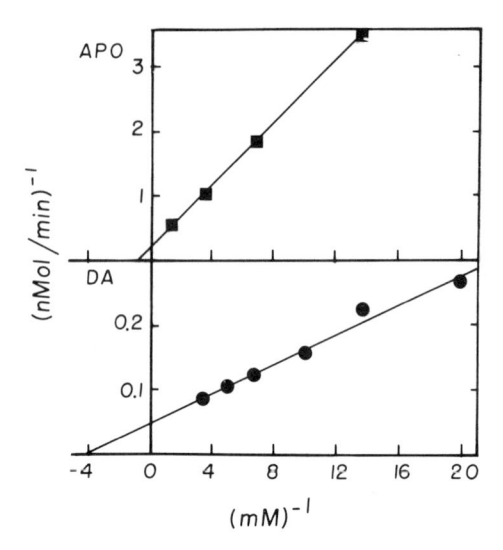

FIG. 2. Substrate kinetics for apomorphine or dopamine methylation by rat liver COMT. Ordinate is reciprocal of initial velocity expressed as nanomoles of O-methylated product formed per minute. Abscissa represents reciprocal of substrate concentrations: boxes, apomorphine; circles, dopamine. Each assay contained approximately 4 mg of extract protein. Assays were performed in duplicate with maximum variation of <3%. K_m for apomorphine = 1.4×10^{-3} M; K_m for dopamine = 2.6×10^{-4} M.

COMT with an apparent K_m of 1.4×10^{-3} M (Fig. 2). The formation of this ^{14}C-methylated product was competitively inhibited by the addition of either tropolone or pyrogallol to the incubation mixture. The K_i values are shown in Table 1 and agree closely with the K_i values reported by Belleau and Burba (1963) for the inhibition of the O-methylation of norepinephrine.

TABLE 1. *Competitive inhibition of apomorphine (Apo) methylation by known COMT inhibitors*

Inhibitor	K_i(Apo)[a] μM	K_i(NE)[b] μM
Tropolone	10	12
Pyrogallol	11	17

[a] K_i was calculated from slopes of reciprocal plots obtained by varying apomorphine concentration in the absence and presence of inhibitors.

[b] NE data from Belleau and Burba (1963). (Taken from McKenzie and White, 1973.)

Cannon, Smith, Modiri, Sood, Borgman, and Aleem (1972) and Missala et al. (1973) also demonstrated the *in vitro* methylation of apomorphine and determined that the methylated product was apocodeine.

IV. THE EFFECTS OF APOMORPHINE AND TROPOLONE ON DOPAMINE SYNTHESIS AND UTILIZATION IN THE CAT CAUDATE NUCLEUS

The effects of apomorphine and tropolone on dopamine synthesis and utilization were determined in cats using labeled tyrosine. Unanesthetized cats, immobilized with succinylcholine and artificially ventilated, received ^{3}H-tyrosine (sp. act. 30 to 50 C/mmole) at an infused rate of 1.66 μC/kg/min, for 2 hr, through a femoral cannula. At selected times after infusion the animals were deeply anesthetized with halothane, and both caudate nuclei were quickly dissected and frozen in ethanol and CO_2. Tissues were analyzed for ^{3}H-dopamine, ^{3}H-acidic catechols, and total tritium, using alumina column separation and liquid scintillation.

Figure 3 shows the time course for the accumulation and disappearance of ^{3}H-dopamine, ^{3}H-acidic catechols, and total tritium in the caudate nucleus of control animals. The time course for the accumulation and disappearance of labeled dopamine is quite similar to that found in rodents by other authors (Nybäck, 1971).

To determine the effects of drugs on the accumulation of ^{3}H-dopamine, animals were pretreated with either apomorphine or tropolone and sacrificed immediately following the 2-hr infusion of ^{3}H-tyrosine. Tropolone was

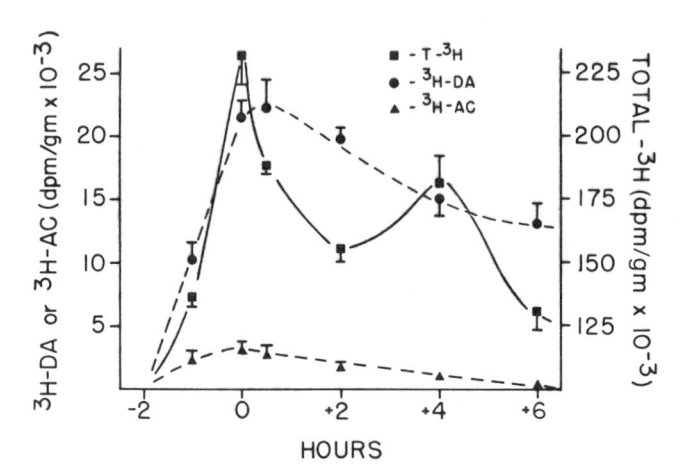

FIG. 3. Time course for the accumulation and disappearance of ^{3}H-dopamine, ^{3}H-acidic catechols, and total tritium in the cat caudate nucleus. Animals received ^{3}H-tyrosine (200 μC/kg body wt) by continuous i.v. infusion at the rate of 1.66 μC/kg/min. Total infusion time was 2 hr. With exception of points at −1 hr, 0 hr represents the end of the ^{3}H-tyrosine infusion.

TABLE 2. *Effects of drugs on the accumulation of tritiated dopamine, acidic catechols, and total tritium in the cat caudate nucleus*

Drug (mg/kg)	(N)	³H-DA	p[a]	dpm/g (mean ± SE) ³H-AC	p[a]	T-³H	p[a]
Control	(10)	21,550 ± 1,342	—	3,307 ± 450	—	231,897 ± 11,169	—
Apomorphine							
1 × 3	(4)	19,160 ± 3,614	N.S.	7,088 ± 1,982	N.S.	246,028 ± 6,830	N.S.
1.5 × 3	(8)	10,655 ± 1,290	<0.001	2,727 ± 605	N.S.	230,143 ± 20,659	N.S.
10 × 3	(4)	4,030 ± 189	<0.001	1,825 ± 155	<0.01	239,138 ± 21,917	N.S.
Tropolone							
10	(4)	18,370 ± 1,600	<0.1	2,520 ± 822	N.S.	360,301 ± 8,509	<0.001
20	(4)	12,532 ± 1,557	<0.001	2,571 ± 478	N.S.	280,931 ± 10,055	<0.01
40	(6)	6,570 ± 521	<0.001	1,175 ± 122	<0.001	259,234 ± 15,317	N.S.

Cats received ³H-tyrosine (200 μC/kg body wt) as a continuous 2-hr i.v. infusion. Infusion rate was 1.66 μC/kg/min. Animals were killed immediately after the infusion and ³H-dopamine (³H-DA), ³H-acidic catechols (³H-AC), and total tritium (T-³H) in the caudate nucleus determined immediately after the end of infusion. Tropolone was administered i.p. 1 hr prior to the start of ³H-tyrosine infusion. Apomorphine was administered s.c., three times, commencing 1 hr before ³H-tyrosine infusion and thereafter every hour. Controls received a single saline injection.

[a] p Values were calculated using Student's t test.

administered as a single dose, i.p., 1 hr prior to the ³H-tyrosine infusion, whereas apomorphine was administered at 1-hr intervals commencing 1 hr before infusion.

Both apomorphine and tropolone decreased markedly the conversion of ³H-tyrosine to ³H-dopamine in the caudate nucleus (Table 2). This effect was dose related for both drugs. At the highest dose tested (3 × 10 mg/kg), apomorphine reduced ³H-dopamine levels from a control value of 21,550 ± 1,342 dpm/g to 4,030 ± 189 dpm/g. Tropolone at 40 mg/kg lowered ³H-dopamine values to 6,570 ± 521 dpm/g. Both apomorphine and tropolone at high doses significantly decreased ³H-acidic catechols. Tropolone tended to elevate total tritium, but not at all doses.

The disappearance of ³H-dopamine from the caudate was studied by administering drugs 10 min before the end of the ³H-tyrosine infusion and sacrificing the animals 4 hr after infusion. The results of these experiments are shown in Table 3. Both apomorphine and tropolone retarded the disappearance of ³H-dopamine in the caudate nucleus. However, in contrast to the data obtained in the accumulation study, the effects of apomorphine did not appear to be dose related. The levels of ³H-acidic catechols tended to be elevated by apomorphine, and were significantly elevated by tropolone. Neither drug consistently affected total tritium levels.

The fact that apomorphine and tropolone affected similarly the metabolic disposition of dopamine again suggests that the biochemical effects of apomorphine may involve an interaction with COMT. Indeed, it appears that COMT inhibition following tropolone alone can produce similar changes in dopamine metabolism.

TABLE 3. *Effects of drugs on the disappearance of tritiated dopamine, acidic catechols, and total tritium in the cat caudate nucleus*

Drug (mg/kg)	(N)	³H-DA	p^a	dpm/g (mean ± SE) ³H-AC	p^a	T-³H	p^a
Control	(12)	15,666 ± 1,359	—	1,087 ± 147	—	181,156 ± 11,257	—
Apomorphine							
1 × 4	(2)	23,360 ± 1,371	<0.01	797 ± 15	N.S.	151,008 ± 364	<0.05
2.5 × 4	(6)	24,204 ± 2,839	<0.02	828 ± 126	N.S.	141,078 ± 8,195	<0.02
5 × 4	(4)	24,153 ± 1,428	<0.001	1,061 ± 324	N.S.	213,829 ± 23,384	N.S.
10 × 4	(6)	17,712 ± 2,775	N.S.	1,716 ± 291	<0.01	183,673 ± 10,984	N.S.
20 × 4	(2)	19,445 ± 480	<0.05	2,730 ± 2,685	N.S.	144,609 ± 1,968	<0.02
Tropolone							
40	(6)	24,738 ± 9,853	N.S.	2,753 ± 750	<0.05	170,125 ± 25,291	N.S.
50	(2)	21,879 ± 90	<0.01	3,877 ± 181	<0.001	176,480 ± 2,422	N.S.

Animals received ³H-tyrosine (200 µC/kg body wt) as a continuous 2-hr i.v. infusion. Infusion rate was 1.66 µC/kg/min. Animals were sacrificed 4 hr after the end of infusion. ³H-dopamine (³H-DA), ³H-acidic catechols (³H-AC), and total tritium (T-³H) were determined in the caudate nucleus. Tropolone was administered in a single i.p. dose 10 min before the end of infusion. Apomorphine was given s.c., four times, commencing 10 min before the end of infusion and thereafter every hour. Controls received a single saline injection.

^{a} p Values calculated using Student's t test.

It is well known that at high concentrations most catechols, including apomorphine (Goldstein, 1970), are capable of inhibiting the enzyme tyrosine hydroxylase, and it may be argued that the foregoing results could be due to such inhibition. Furthermore, it is also possible that apomorphine and tropolone may affect the uptake of tyrosine into dopaminergic nerve terminals. To test these possibilities, cats were treated with either apomorphine (in multiple doses) or tropolone (single dose) and endogenous dopamine concentrations in the caudate determined 3 hr after pretreatment. Doses which reduced profoundly the conversion of ³H-tyrosine to ³H-dopamine

TABLE 4. *Effects of tropolone or apomorphine on dopamine concentrations in the cat caudate nucleus*

Drug (mg/kg)	DA (µg/g) mean ± SE	N	Percent decrease
Vehicle	13.84 ± 0.65	6	—
Tropolone			
20	11.82 ± 0.27	2	11
40	10.29 ± 0.47	2	23
Apomorphine			
10 × 3	10.24 ± 1.2	2	23

Apomorphine, 10 mg/kg, was administered every hour beginning 3 hr prior to sacrificing the animals. Pretreatment time for tropolone was 3 hr.

(Table 2) produced only minimal (10 to 23%) reductions in endogenous dopamine levels (Table 4).

It is concluded from these studies that changes in dopamine turnover induced by apomorphine or tropolone cannot be attributed to either direct tyrosine hydroxylase inhibition or to inhibition of tyrosine uptake into nerve tissue.

V. DRUG EFFECTS ON DOPAMINE SYNTHESIS IN THE RAT STRIATUM

Further studies on the possible regulatory role of COMT on dopamine synthesis were conducted in Long Evans, hooded rats. ^{3}H-Tyrosine (500 µC/kg body wt) was injected through the tail vein and ^{3}H-dopamine, ^{3}H-acidic catechols, and total tritium in the nucleus caudatus putamen (striatum) were determined 1 hr after ^{3}H-tyrosine injection using the same methods described previously for the cat.

Both apomorphine and tropolone reduced markedly the accumulation of ^{3}H-dopamine in the rat striatum (Table 5). Tropolone, 100 mg/kg, reduced ^{3}H-dopamine levels from 48,310 ± 1,377 dpm/g in control animals to 11,928 ± 526 dpm/g. At the highest apomorphine dose (2 × 10 mg/kg), ^{3}H-dopamine content was reduced to 10,056 ± 3,106 dpm/g. Both apomorphine and tropolone tended to increase ^{3}H-acidic catechols. Neither agent consistently affected total tritium content.

In an attempt to differentiate between the effects of apomorphine and those of tropolone, these two agents were tested against chlorpromazine-induced changes in dopamine metabolism. Chlorpromazine, 10 to 20 mg/kg, i.p., pretreated 1 hr prior to ^{3}H-tyrosine injection, produced dose-related

TABLE 5. *The effects of apomorphine and tropolone on the accumulation of ^{3}H-dopamine, ^{3}H-acidic catechols, and total tritium in the rat striatum*

Drug (mg/kg)	N	^{3}H-DA	p	^{3}H-AC	p	^{3}H-T	p
Control	13	48,310 ± 1,377	—	5,914 ± 698	—	363,443 ± 11,913	—
Apomorphine							
2.5 × 2	3	18,157 ± 1,875	<0.001	2,199 ± 402	<0.01	294,633 ± 13,987	<0.01
5 × 2	3	15,755 ± 3,645	<0.001	4,013 ± 978	N.S.	342,967 ± 39,382	N.S.
10 × 2	3	10,056 ± 3,106	<0.001	7,882 ± 3,065	N.S.	382,670 ± 54,884	N.S.
Tropolone							
20	3	43,646 ± 1,622	<0.05	5,056 ± 1,149	N.S.	300,649 ± 26,000	N.S.
40	3	34,672 ± 1,870	<0.001	5,571 ± 321	N.S.	427,227 ± 26,272	0.05
60	3	31,275 ± 4,754	<0.001	7,659 ± 1,625	N.S.	466,641 ± 53,120	N.S.
100	2	11,928 ± 526	<0.001	8,027 ± 326	<0.02	358,748 ± 6,636	N.S.

Rats received ^{3}H-tyrosine (500 µC/kg body wt) injected through the tail vein and ^{3}H-dopamine (^{3}H-DA), ^{3}H-acidic catechols (^{3}H-AC), and total tritium (^{3}H-T) in the striatum determined 1 hr after label. Apomorphine was administered 1 hr prior to and again immediately after ^{3}H-tyrosine. Tropolone was administered 1 hr before ^{3}H-tyrosine. p Values were calculated using Student's *t* test.

increases in ³H-dopamine levels and ³H-acidic catechols (Table 6). Following 20 mg/kg of chlorpromazine, ³H-dopamine content was increased from 48,310 ± 1,377 dpm/g (mean ± SE) to 78,597 ± 8,420 dpm/g. Both apomorphine and tropolone effectively antagonized this chlorpromazine-induced increase in dopamine synthesis.

TABLE 6. *Antagonism of the chloropromazine-induced increase in dopamine turnover by apomorphine and tropolone in the rat striatum*

Drug (mg/kg)		N	³H-DA	p	³H-AC	p	³H-T	p
Control		13	48,310 ± 1,377	—	5,914 ± 698	—	363,443 ± 11,913	—
Chlorpromazine								
10		3	45,651 ± 7,165	N.S.	12,885 ± 3,428	N.S.	347,769 ± 50,905	N.S.
15		3	55,018 ± 2,347	<0.05	16,126 ± 1,488	<0.001	311,852 ± 40,877	N.S.
20		4	78,597 ± 8,420	<0.001	23,457 ± 1,832	<0.001	382,339 ± 16,195	N.S.
CPZ + Apo								
20	2.5 × 2	4	97,288 ± 3,195	N.S.[a]	5,926 ± 1,314	<0.001[a]	335,884 ± 4,264	<0.05[a]
20	5 × 2	3	29,751 ± 7,860	<0.01[a]	6,239 ± 1,810	<0.001[a]	260,620 ± 35,017	<0.05[a]
20	10 × 2	3	16,454 ± 4,723	<0.001[a]	5,929 ± 1,298	<0.001[a]	285,707 ± 29,262	<0.05[a]
CPZ + Trop								
20	20	3	67,644 ± 1,210	N.S.[a]	14,512 ± 310	<0.01[a]	333,553 ± 4,285	<0.05[a]
20	40	3	49,748 ± 2,390	<0.05[a]	10,495 ± 2,296	<0.02[a]	262,445 ± 31,623	<0.05[a]
20	60	3	31,547 ± 1,244	<0.01[a]	10,251 ± 3,064	<0.02[a]	368,703 ± 16,106	N.S.[a]
20	100	3	18,957 ± 1,030	<0.001[a]	8,128 ± 340	<0.01[a]	362,116 ± 11,412	N.S.[a]

Rats received ³H-tyrosine (500 µC/kg body wt) injected through the tail vein and ³H-dopamine (³H-DA), ³H-acidic catechols (³H-AC), and total tritium (³H-T) in the striatum determined 1 hr after label. Apomorphine (Apo) was administered 1 hr prior to and again immediately after ³H-tyrosine. Chlorpromazine (CPZ) and tropolone (Trop) were administered 1 hr before ³H-tyrosine. *p* Values were calculated using Student's *t* test. [a] Compared to CPZ, 20 mg/kg.

³H-acidic catechols were increased markedly by pretreatment with chlorpromazine (Table 6). Both apomorphine and tropolone effectively antagonized this increase.

VI. CHLORPROMAZINE ANTAGONISM OF APOMORPHINE-INDUCED STEREOTYPED BEHAVIOR

The foregoing results would suggest that chlorpromazine on the one hand, and either apomorphine or tropolone on the other, are mutual antagonists at some central receptor site which influences dopamine metabolism. It therefore seemed of interest to determine dose-response relationships for the antagonism of stereotyped behavior induced by apomorphine.

Apomorphine was administered at three dose levels, 1, 5, and 10 mg/kg, s.c. The dose of chlorpromazine, pretreated 30 min prior to apomorphine, was varied from 1 to 30 mg/kg, i.p. Stereotyped behavior was scored for

1 hr using the scoring system described by McKenzie (1972). The results are shown in Table 7. The mean stereotypy score following 1 mg/kg of apomorphine was reduced to 50% of the control score by pretreatment with 2 mg/kg of chlorpromazine. When the dose of apomorphine was increased to 5 and 10 mg/kg, the doses of chlorpromazine required to reduce mean behavioral scores by approximately 50% were 10 and 20 mg/kg, respectively. Thus, it appears that as regards stereotyped behavior chlorpromazine is a competitive antagonist of apomorphine, presumably at central dopamine receptors.

TABLE 7. *Antagonism of apomorphine (Apo)-induced stereotyped behavior by chlorpromazine*

Chlorpromazine (mg/kg)	Apo (1 mg/kg)	Apo (5 mg/kg)	Apo (10 mg/kg)
—	9.0 ± 0.6 (26)	18.8 ± 1.0 (20)	19.4 ± 0.08 (28)
1	6.0 ± 0.5[b] (6)	14.6 ± 2.2 (6)	20.2 ± 1.9 (10)
2	4.5 ± 0.5[c] (6)	15.1 ± 2.6 (6)	19.5 ± 1.5 (10)
5	1.8 ± 0.4[c] (6)	15.8 ± 1.8 (6)	19.0 ± 2.0 (10)
10	0.5 ± 0.2[c] (6)	8.6 ± 2.3[b] (6)	16.1 ± 2.0 (10)
20	0 (2)	6.5 ± 1.5[c] (6)	12.8 ± 1.0[c] (6)
30	0.5 ± 0.5[c] (2)	3.0 ± 0.3[c] (6)	15.8 ± 1.5[a] (6)

Chlorpromazine was administered i.p., 30 min prior to apomorphine, s.c. Animals were scored for 1 hr following apomorphine. [a] $p < 0.05$, [b] $p < 0.01$, [c] $p < 0.001$ compared to the appropriate control dose of apomorphine. p Values were calculated using Student's t test. (N) = number of animals.

Despite the fact that tropolone mimics the effects of apomorphine on dopamine metabolism and on chlorpromazine-induced changes in dopamine metabolism, tropolone neither induces stereotyped behavior nor antagonizes chlorpromazine-induced behavioral depression. Therefore, it seems that either tropolone is affecting dopamine metabolism via a mechanism different from that of apomorphine or there is no direct connection between the behavioral changes induced by apomorphine and its effects on dopamine turnover.

VII. EFFECTS OF AGROCLAVINE ON DOPAMINE METABOLISM

To test the hypothesis that the biochemical changes induced by apomorphine may be the result of COMT inhibition, we sought a dopaminergic agonist which was neither a substrate for COMT nor an inhibitor of the enzyme. Agroclavine and ergocornine, a structurally related ergot drug, have been reported to be possible dopaminergic agonists, based on the finding that (1) they induce stereotyped behavior, (2) they induce contralateral headturning in 6-hydroxydopamine-lesioned rats, and (3) they retard the depletion of dopamine following α-methyltyrosine treatment (Corrodi, Fuxe, Hökfelt, Lidbrink, and Ungerstedt, 1973; Stone, 1973).

In our hands, agroclavine induced stereotyped behavior in rodents and proved to be a potent emetic in dogs (McKenzie, *unpublished observation*). Furthermore, chlorpromazine antagonized the stereotypy induced by agroclavine (McKenzie, *unpublished*). However, unlike apomorphine, agroclavine was neither a substrate for COMT nor an inhibitor *in vitro* (H. L. White, *personal communication*).

When agroclavine was tested for its effects on dopamine metabolism, it was found that this agent did not reduce the incorporation of ^{3}H-tyrosine to ^{3}H-dopamine in the rat striatum (Table 8). In fact, at doses which caused stereotyped behavior (i.e., 5 to 10 mg/kg), agroclavine produced significant increases in ^{3}H-dopamine levels. This is in direct contrast to the effects observed with apomorphine. However, similar to apomorphine, agroclavine partially antagonized the chlorpromazine-induced increase in ^{3}H-dopamine and ^{3}H-acidic catechols.

TABLE 8. *Effects of agroclavine on the accumulation of ^{3}H-dopamine, ^{3}H-acidic catechols, and total tritium in the rat striatum*

Drug (mg/kg)	N	^{3}H-DA	p	^{3}H-AC	p	^{3}H-T	p
Control	13	48,310 ± 1,377	—	5,915 ± 698	—	363,443 ± 11,913	—
Agroclavine							
1 × 2	3	35,710 ± 6,487	N.S.	5,392 ± 552	N.S.	263,477 ± 44,984	N.S.
2.5 × 2	3	61,646 ± 11,010	N.S.	7,632 ± 2,592	N.S.	418,505 ± 76,530	N.S.
5 × 2	3	79,307 ± 6,887	<0.001	6,476 ± 2,386	N.S.	476,883 ± 41,306	<0.05
10 × 2	3	59,389 ± 9,956	N.S.	5,977 ± 388	N.S.	404,520 ± 24,050	N.S.
Chlorpromazine							
20	4	78,597 ± 8,420	—	23,457 ± 1,832	—	382,339 ± 16,195	—
CPZ + Agro							
20 5 × 2	4	57,198 ± 2,931	<0.05[a]	4,895 ± 1,591	<0.001[a]	349,192 ± 45,372	N.S.[a]

Animals received ^{3}H-tyrosine (500 μC/kg body wt) through the tail vein and ^{3}H-dopamine (^{3}H-DA), ^{3}H-acidic catechols (^{3}H-AC), and total tritium (^{3}H-T) determined 1 hr after. Agroclavine (Agro) was administered 1 hr prior to and immediately after ^{3}H-tyrosine infusion. Chlorpromazine (CPZ) was administered 1 hr prior to ^{3}H-tyrosine. p Values were calculated using Student's t test. [a] Compared to CPZ, 20 mg/kg.

VIII. CONCLUSIONS

Apomorphine is a substrate for COMT and therefore a competitive inhibitor of the enzyme. The finding that tropolone, a COMT inhibitor, mimics the effects of apomorphine on dopamine synthesis and utilization in the cat and dopamine synthesis in the rat raises the question of whether the biochemical effects of apomorphine are mediated by dopamine receptor activation or COMT inhibition or both. The fact that tropolone does not induce stereotyped behavior can be interpreted to mean that, unlike apomorphine, tropolone has no postsynaptic receptor agonistic properties. However, the chlorpromazine-induced increase in dopamine turnover, which is considered to be due to dopamine receptor blockade, was antagonized by both tropolone and apomorphine. The only known common denominator that can explain the close similarities between apomorphine and tropolone on dopamine turnover is COMT inhibition.

Agroclavine fulfills many of the criteria for a dopaminergic agonist; that is, (1) it produces contralateral turning in unilateral-strio-nigral-lesioned rats, (2) it induces stereotyped behavior which is blocked by chlorpromazine, (3) it retards the α-methyltyrosine-induced depletion of striatal dopamine, and (4) it causes emesis in the dog. Yet, agroclavine did not reduce the conversion of ^{3}H-tyrosine to ^{3}H-dopamine in the rat striatum but, instead, enhanced this phenomenon. In addition, in contrast to apomorphine, agroclavine does not inhibit COMT *in vitro*. These considerations support the concept that the effects of apomorphine on dopamine metabolism may not be due entirely to an interaction between apomorphine and the postsynaptic dopamine receptor, but may also involve COMT inhibition.

It is interesting that tropolone completely antagonized the increase in dopamine synthesis produced by chlorpromazine. This finding may suggest some utility for COMT inhibitors in the management of tardive dyskinesias, a syndrome currently thought to be the result of an increased dopamine turnover following chronic treatment with neuroleptic agents.

REFERENCES

Andén, N. E., Corrodi, H., Fuxe, K., and Ungerstedt, U. (1971): Importance of nervous impulse flow for the neuroleptic induced increase in amine turnover in central dopamine neurons. *European Journal of Pharmacology,* 15:193–199.

Andén, N. E., Rubenson, A., Fuxe, K., and Hökfelt, T. (1967): Evidence for dopamine receptor stimulation by apomorphine. *Journal of Pharmacy and Pharmacology,* 19:627–629.

Belleau, B., and Burba, J. (1963): Occupancy of adrenergic receptors and inhibition of catechol O-methyl transferase by tropolones. *Journal of Medicinal Chemistry,* 6:755–759.

Bunney, B. S., Walters, J. R., Roth, R. H., and Aghajanian, G. K. (1973): Dopaminergic neurons: Effect of antipsychotic drugs and amphetamine on single cell activity. *Journal of Pharmacology and Experimental Therapeutics,* 185:560–571.

Cannon, J. G., Smith, R. V., Modiri, A., Sood, S. P., Borgman, R. J., Aleem, M. A., and Long, J. P. (1972): Centrally acting emetics. 5. Preparation and pharmacology of 10-hydroxy-11-methoxyaporphine (isoapocodeine). *In vitro* enzymatic methylation of apomorphine. *Journal of Medicinal Chemistry,* 15:273–276.

Carlsson, A., and Lindqvist, M. (1963): Effect of chlorpromazine or haloperidol on formation of 3-methoxytyramine and normetanephrine in mouse brain. *Acta Pharmacologica et Toxicologica,* 20:140–144.

Corrodi, H., Fuxe, K., Hökfelt, T., Lidbrink, P., and Ungerstedt, U. (1973): Effect of ergot drugs on central catecholamine neurons: Evidence for a stimulation of central dopamine neurons. *Journal of Pharmacy and Pharmacology,* 25:409–411.

Dhawan, B. N., and Saxena, P. N. (1960): Apomorphine-induced pecking in pigeons. *British Journal of Pharmacology*, 15:285-289.
Ernst, A. M. (1967): Mode of action of apomorphine and dexamphetamine on gnawing compulsion in rats. *Psychopharmacologia*, 10:316-323.
Ernst, A. M. (1969): The role of biogenic amines in the extra-pyramidal system. *Acta Physiologica et Pharmacologia Neerlander*, 15:141-154.
Goldstein, M., Freedman, L. S., and Backstrom, T. (1970): The inhibition of catecholamine biosynthesis by apomorphine. *Journal of Pharmacy and Pharmacology*, 22:715-717.
Kehr, W., Carlsson, A., Lindqvist, M., Magnusson, T., and Atack, C. (1972): Evidence for a receptor-mediated feedback control of striatal tyrosine hydroxylase activity. *Journal of Pharmacy and Pharmacology*, 24:744-747.
Lahti, R. A., McAllister, B., and Wozniak, J. (1972): Apomorphine antagonism of the elevation of homovanillic acid induced by antipsychotic drugs. *Life Sciences*, 11:605-613.
McKenzie, G. M. (1972): Role of the tuberculum olfactorium in stereotyped behavior induced by apomorphine in the rat. *Psychopharmacologia*, 23:212-219.
McKenzie, G. M. (1973): *The Pharmacology of Apomorphine*. Futura Publishers, Mt. Kisco, New York, *(in press)*.
McKenzie, G. M., Viik, K., and Boyer, C. E. (1973): Selective blockade of apomorphine-induced aggression and gnawing following fenfluramine or raphe-lesions in the rat. *Federation Proceedings*, 32:248.
McKenzie, G. M., and White, H. L. (1973): Evidence for the methylation of apomorphine by catechol-O-methyl transferase *in vivo* and *in vitro*. *Biochemical Pharmacology*, 22:2329-2336.
Missala, K., Lal, S., and Sourkes, T. L. (1973): O-Methylation of apomorphine and the metabolic prolongation of apomorphine-induced stereotyped behavior. *European Journal of Pharmacology*, 22:54-58.
Nybäck, H. (1971): Regional disappearance of catecholamines formed from ^{14}C-tyrosine in rat brain: Effect of synthesis inhibitors and of chlorpromazine. *Acta Pharmacologica et Toxicologica*, 30:372-384.
Nybäck, H., Schubert, J., and Sedvall, G. (1970): Effect of apomorphine and pimozide on synthesis and turnover of labelled catecholamines in mouse brain. *Journal of Pharmacy and Pharmacology*, 22:622-624.
Roos, B. E. (1969): Decrease in homovanillic acid as evidence for dopamine receptor stimulation by apomorphine in the neostriatum of the rat. *Journal of Pharmacy and Pharmacology*, 21:263-264.
Roth, R. H. (1973): Inhibition by γ-hydroxybutyrate of chlorpromazine-induced increase in homovanillic acid. *British Journal of Pharmacology*, 47:408-414.
Rotrosen, J., Wallach, M. B., Angrist, B., and Gershon, S. (1972): Antagonism of apomorphine-induced stereotypy and emesis in dogs by thioridazine, haloperidol and pimozide. *Psychopharmacologia*, 26:185-194.
Scheel-Krüger, J. (1970): Central effects of anticholinergic drugs measured by the apomorphine gnawing test in mice. *Acta Pharmacologica et Toxicologica*, 28:1-16.
Stone, T. W. (1973): Studies on the central nervous system effects of agroclavine, an ergot alkaloid. *Archives of International Pharmacodynamics and Therapeutics*, 202:62-65.
Ungerstedt, V. (1971): Postsynaptic supersensitivity after 6-hydroxydopamine induced degeneration of the nigro-striatal dopamine system. *Acta Physiologica Scandinavica*, Suppl. 367, 69-93.
Von Voigtlander, P. F., and Moore, K. E. (1973): Turning behavior of mice with unilateral 6-hydroxydopamine lesions in the striatum: Effects of apomorphine, L-DOPA, amantadine, amphetamine and other psychomotor stimulants. *Neuropharmacology*, 12:451-462.
Wilner, J., Samach, M., Angrist, B., Wallach, M., and Gershon, S. (1970): Drug induced stereotyped behavior and its antagonism in dogs. *Communications in Behavioral Biology*, 5:135-141.

Neuropsychopharmacology of Monoamines and Their Regulatory Enzymes, edited by E. Usdin.
Raven Press, New York © 1974.

Relationship of Dopamine Neural Systems to the Behavioral Alterations Produced by 6-Hydroxydopamine Administration into Brain

Barrett R. Cooper and George R. Breese

Departments of Psychiatry and Pharmacology, Biological Sciences Research Center, Child Development Institute, University of North Carolina School of Medicine, Chapel Hill, North Carolina 27514

I. INTRODUCTION

Administration of 6-hydroxydopamine into brain results in a prolonged depletion of brain catecholamine content (Breese and Traylor, 1970; Uretsky and Iversen, 1970), a chronic reduction of brain tyrosine hydroxylase activity (Breese and Traylor, 1970), and ultrastructural damage in brain regions rich in adrenergic terminals (Bloom, Algeri, Groppetti, Revuetta, and Costa, 1969). These findings support the view that 6-hydroxydopamine produces a destruction of catecholamine neural processes in the central nervous system.

Recent studies of the pharmacology of 6-hydroxydopamine and its interaction with other compounds known to alter metabolism and uptake of endogenous catecholamines led to the development of methods utilizing 6-hydroxydopamine to produce relatively selective destruction of either noradrenergic or dopaminergic fibers in brain (Breese and Traylor, 1970, 1971; Evetts and Iversen, 1970; Smith, Cooper, and Breese, 1973*b*). These procedures to reduce the content of brain norepinephrine, dopamine, or both catecholamines with 6-hydroxydopamine were subsequently employed by several laboratories to examine the functional significance of central catecholamine fibers in various pharmacological, physiological, and behavioral responses (Breese, Cooper, and Smith, 1973*a*; Thoenen and Tranzer, 1973; Ungerstedt and Ljungberg, 1973; Breese, 1974). In this chapter we will review the acute and chronic deficits produced by administering 6-hydroxydopamine into brain as well as present current evidence which relates deficits in 6-hydroxydopamine-treated rats to destruction of either noradrenergic or dopaminergic neural systems.

II. ACUTE EFFECTS OF 6-HYDROXYDOPAMINE ON TEMPERATURE REGULATION AND INGESTIVE BEHAVIOR

After intracisternal injection of 6-hydroxydopamine, rats appear sedated during the first few hours and, to some extent, resemble animals treated with reserpine (Laverty and Taylor, 1970; Breese et al., 1973a). During this period, it has also been observed that, like norepinephrine, 6-hydroxydopamine administration into brain reduces body temperature (Simmonds and Uretsky, 1970; Breese and Howard, 1971; Nakamura and Thoenen, 1971; Breese, Moore, and Howard, 1972). This hypothermia produced by 6-hydroxydopamine can be potentiated if rats are first pretreated with pargyline and then placed in the cold 30 min after treatment (Breese et al., 1972). Since the reduction of body temperature after administration of 6-hydroxydopamine can be prevented by prior destruction of catecholamine-containing fibers in brain, it would appear that the hypothermia produced by 6-hydroxydopamine requires the release of endogenous catecholamines (Simmonds and Uretsky, 1970; Breese et al., 1972). The acute hypothermic response after 6-hydroxydopamine administration has been attributed to release of brain dopamine (Simmonds and Uretsky, 1970) as well as to the release of brain norepinephrine (Breese and Howard, 1971; Nakamura and Thoenen, 1971; Breese et al., 1972). Further work will be required to resolve the apparent contradiction in these proposals.

In addition to the acute alteration of temperature regulation produced by 6-hydroxydopamine, several recent studies indicate that 6-hydroxydopamine treatment will also acutely disrupt ingestive behavior (Ungerstedt, 1971; Fibiger, Lonsbury, Cooper, and Lytle, 1972; Breese et al., 1973a). Ungerstedt (1971) has reported that bilateral injection of 6-hydroxydopamine into the substantia nigra causes acute aphagia and adipsia, which disappears after several days. This acute syndrome has been proposed to be similar to the aphagia and adipsia observed after lesions of the lateral hypothalamus, suggesting that interruption of eating and drinking may be related to catecholamine neural destruction. Support for this view has been provided by studies demonstrating that lateral hypothalamic lesions, which disrupt eating and drinking, are also accompanied by depletion of norepinephrine and dopamine in brain (Oltsman and Harvey, 1972). Acute aphagia and adipsia have also been shown to accompany intracisternal injection of 6-hydroxydopamine, if given in combination with pargyline (Zigmond and Stricker, 1972; Breese, Smith, Cooper, and Grant, 1973b) or other monoamine oxidase inhibitors (Fibiger et al., 1972).

III. CHRONIC EFFECTS OF 6-HYDROXYDOPAMINE ON INGESTIVE BEHAVIOR

Since the acute effects of 6-hydroxydopamine on eating and drinking resembled those produced by lesions of the lateral hypothalamus (Teitelbaum and Epstein, 1962), the possibility was examined that 6-hydroxydopamine-treated rats might have chronic deficits in ingestive behavior similar to those produced by lateral hypothalamic lesions (see Ungerstedt, 1971). In accord with this view, recent studies have shown that "recovered"

6-hydroxydopamine-treated rats have a reduced feeding response to injection of insulin (Breese et al., 1973b) and to the administration of 2-deoxyglucose (Zigmond and Stricker, 1972). In addition, treated rats have also been found to consume less saline than control animals after desoxycorticosterone acetate treatment (Breese et al., 1973b). One additional deficit in ingestive behavior observed in 6-hydroxydopamine-treated rats was the failure of treated animals to increase fluid intake, as controls do, when sucrose solution was substituted for water. Although it was first thought that an inability to taste was responsible for this observation, it was soon established that treated animals actually preferred drinking the sucrose solution if offered a choice between sucrose solution and water. This result suggests that taste is not a factor in this reduced intake of sucrose solution in 6-hydroxydopamine-treated rats (Breese et al., 1973b).

Unlike rats recovered from the acute effects of lateral hypothalamic lesions, animals treated with 6-hydroxydopamine were found to drink in the absence of food and to consume water like control rats after injection of hypertonic saline solution (Breese et al., 1973b). These latter results suggest that alterations in ingestive behavior produced by 6-hydroxydopamine are not, in every respect, like those observed after lateral hypothalamic lesions. This could be due to the possibility that certain aspects of the "lateral hypothalamic syndrome" arise from interruption of pathways other than the catecholamine neural systems in the lateral hypothalamus (Breese et al., 1973b). However, Fibiger, Zis, and McGeer (1973) have recently reported that 6-hydroxydopamine-treated animals do not drink water in the absence of food, nor do they drink water after injection of a hypertonic saline solution. A possible explanation for these contradictory findings is that insufficient reduction of brain catecholamine content was obtained in previous studies to reveal all deficits in the "recovered" 6-hydroxydopamine-treated animals (Breese et al., 1973b). However, it is also possible that the differences in the way data were presented account for the discrepancy in interpretation. Breese et al. (1973b) found that if fluid intake of control and 6-hydroxydopamine-treated rats was based upon body weight, apparent deficits in water consumption after injection of hypertonic saline or in the absence of food were no longer present. Fibiger et al. (1973) did not correct fluid intake in 6-hydroxydopamine-treated rats for their reduced body weight (Zigmond and Stricker, 1972; Breese et al., 1973b).

IV. EFFECTS OF 6-HYDROXYDOPAMINE ON OPERANT BEHAVIOR

Initial investigations of the effects of 6-hydroxydopamine on operant behavior showed that, following an acute period of behavioral depression, performance in a variety of tasks progressively improved until responding reached levels similar to those observed prior to treatment (Laverty and Taylor, 1970; Scotti de Carolis, Ziegler, Del Basso, and Longo, 1971;

Schoenfeld and Zigmond, 1972; Cooper, Breese, Howard, and Grant, 1972a). Therefore, even though brain catecholamine-containing fibers were irreversibly destroyed, it was not possible to correlate a chronic behavioral change with depletion of brain catecholamine content.

This apparent dissociation of the neurochemical and behavioral effects following the administration of 6-hydroxydopamine, however, is not unique to this treatment. For example, Harvey (1965) found no evidence of chronic depression of a continuously reinforced schedule of responding for water reward in rats with bilateral lesions of the medial forebrain bundle which reduced brain catecholamine and serotonin content. In subsequent experiments, these rats with medial forebrain bundle lesions were found to be several times more sensitive to the depressant effects of reserpine (Harvey, 1965). Similar to the recovery observed in 6-hydroxydopamine-treated rats, the depressed performance observed after injection of reserpine has also been found to return to levels observed prior to treatment, even though depletion of brain monoamines persisted (Pirch, Rech, and Moore, 1967). Such "recovered" reserpinized rats showed an enhanced sensitivity to the depressant effects of α-methyltyrosine (Rech, Carr, and Moore, 1968). In view of these latter findings, perhaps it is not surprising that 6-hydroxydopamine-treated rats which responded normally in an active avoidance task, a continuously reinforced bar-press task for food reward, and a T-maze task, also displayed an enhanced sensitivity to low doses of reserpine as well as α-methyltyrosine (Cooper et al., 1972a). Since the 6-hydroxydopamine-treated rats had brain catecholamine content reduced by approximately 75 to 80%, it would appear that the catecholamine neural systems in brain have a remarkable ability to compensate for a considerable loss of integrity, either by mechanisms of enhanced receptor sensitivity (e.g., Ungerstedt, 1971; Breese et al., 1972, 1973a; Smith, Strohmayer, and Reis, 1972), or by increased turnover of catecholamines in surviving neurons (Nakamura and Thoenen, 1972), or both. Regardless of the mechanism responsible for recovery of function, it was predicted that chronic behavioral deficits would be produced if sufficient reduction of catecholamine content could be obtained following treatment with 6-hydroxydopamine (Cooper et al., 1972a; Breese, et al., 1973a).

In accord with this proposal, performance of rats in a shuttle-box avoidance task was reduced after multiple injections of 6-hydroxydopamine, the first in combination with pargyline and the second without (Cooper, Breese, Grant, and Howard, 1973b). This treatment, which reduces brain catecholamine content by approximately 90%, has also been shown to block acquisition and performance of a position habit in a T-maze (Howard, Grant, and Breese, 1973) and to depress electrical self-stimulation of the brain (Breese, Howard, and Leahy, 1971). The similarity of the deficits produced by α-methyltyrosine and 6-hydroxydopamine in the acquisition of the shuttle-box avoidance response is shown in Fig. 1. The fact that acquisition of the

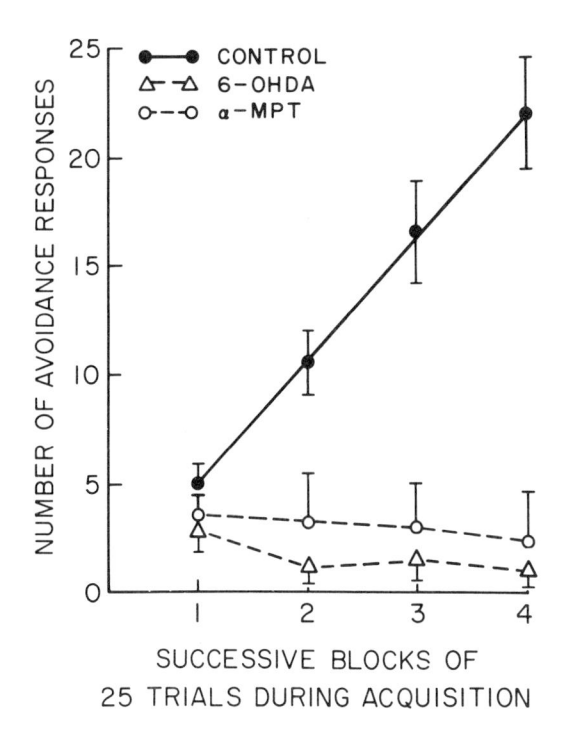

FIG. 1. Effects of 6-hydroxydopamine (6-OHDA) or α-methyltyrosine (α-MPT) treatments on the acquisition of a shuttle-box avoidance response. "6-OHDA" refers to rats injected intracisternally with 6-hydroxydopamine (200 μg) 30 min after pargyline (50 mg/kg), followed by a second 6-hydroxydopamine treatment (200 μg) 1 week later. Animals were tested approximately 3 months after treatment. "α-MPT" refers to rats given 50 mg/kg of L-α-methyltyrosine intraperitoneally 4 hr before testing.

avoidance response was determined 3 months after 6-hydroxydopamine treatment suggests that the deficit in active avoidance acquisition is permanent (Cooper et al., 1973b). In spite of the severe reduction of brain catecholamine content, neither motor activity nor the ability to acquire a passive-avoidance response has been shown to be altered by this 6-hydroxydopamine treatment schedule.

It would appear that 6-hydroxydopamine must destroy large numbers of catecholamine-containing fibers before chronic deficits are obtained in the rat after 6-hydroxydopamine treatment. Recently, the ability of catecholamine neural systems to compensate for destruction has also been suggested from work with brains from patients with Parkinson's disease. Hornykiewicz (1973) has demonstrated that depletion of dopamine in caudates from humans has to reach a rather marked degree (approximately 80%) before any clinical symptoms of Parkinson's disease are manifested. Patients with severe symptoms differ from those with mild symptoms in that caudate dopamine is reduced by only an additional 10% (Hornykiewicz, 1973).

V. THE ROLE OF NOREPINEPHRINE AND DOPAMINE IN THE BEHAVIORAL DEFICITS OF 6-HYDROXYDOPAMINE-TREATED RATS

The observations that chronic deficits in behavior would result if sufficient destruction of brain catecholaminergic fibers occurred after 6-hydroxydopamine treatment led to efforts to relate these deficits to the absence of noradrenergic or dopaminergic neural systems. Similar to the earlier work with 6-hydroxydopamine, rats treated to deplete norepinephrine or dopamine in brain (see Breese and Traylor, 1971) failed to display chronic behavioral deficits (Cooper, Breese, Howard, and Grant, 1972b; Breese et al., 1973a). Nevertheless, subtle differences between groups were noted. Rather than a reduction in acquisition, rats in which brain norepinephrine was preferentially reduced displayed enhanced acquisition of a shuttle-box avoidance response and had elevated activity levels. Any interpretation of these findings, however, must take into account the small reduction of serotonin found after this treatment (Breese et al., 1973a). Acquisition of the avoidance task by rats treated to deplete dopamine did not differ from control (Cooper et al., 1972b).

Previous work indicated that additional catecholamine amine depletion with α-methyltyrosine produced an enhanced behavioral depression in 6-hydroxydopamine-treated rats (Cooper et al., 1972a). Therefore, it was felt that the administration of α-methyltyrosine to rats in which brain norepinephrine or dopamine had previously been reduced with 6-hydroxydopamine might be used to define the role of each of the catecholamines in the chronic deficits observed after 6-hydroxydopamine treatment. In addition to α-methyltyrosine, a dopamine-β-hydroxylase inhibitor, U-14,624,[*] was included in these experiments so that any changes related to reduced norepinephrine levels in brain could be more clearly specified. Figure 2 shows the results of experiments in which α-methyltyrosine and U-14,624 were given to rats performing a shuttle-box avoidance response or a bar-press response for food reward. In each task, the greatest enhanced depression of behavior after α-methyltyrosine treatment was observed in rats treated with 6-hydroxydopamine to deplete dopamine in brain. Rats treated to reduce brain norepinephrine preferentially displayed only slight decreases in responding when compared with control. Interestingly, U-14,624 failed to produce any changes that could be attributed to functional deficits in noradrenergic neurons (Fig. 2), in spite of *additional* reductions of norepinephrine content in brain of approximately 70% in all groups (see Cooper, Grant, and Breese, 1973a). A similar pattern of results has also been obtained in animals bar pressing for electrical stimulation applied to the lateral hypothalamus (Breese et al., 1973a; Cooper, Smith, and Breese, 1973).

Such findings suggested that reduction of dopamine in brain might be of

---

[*] U-14,624 is 1-phenyl-3-(2-thiazolyl)-2-thiourea.

A. AVOIDANCE RESPONDING

B. OPERANT RESPONDING

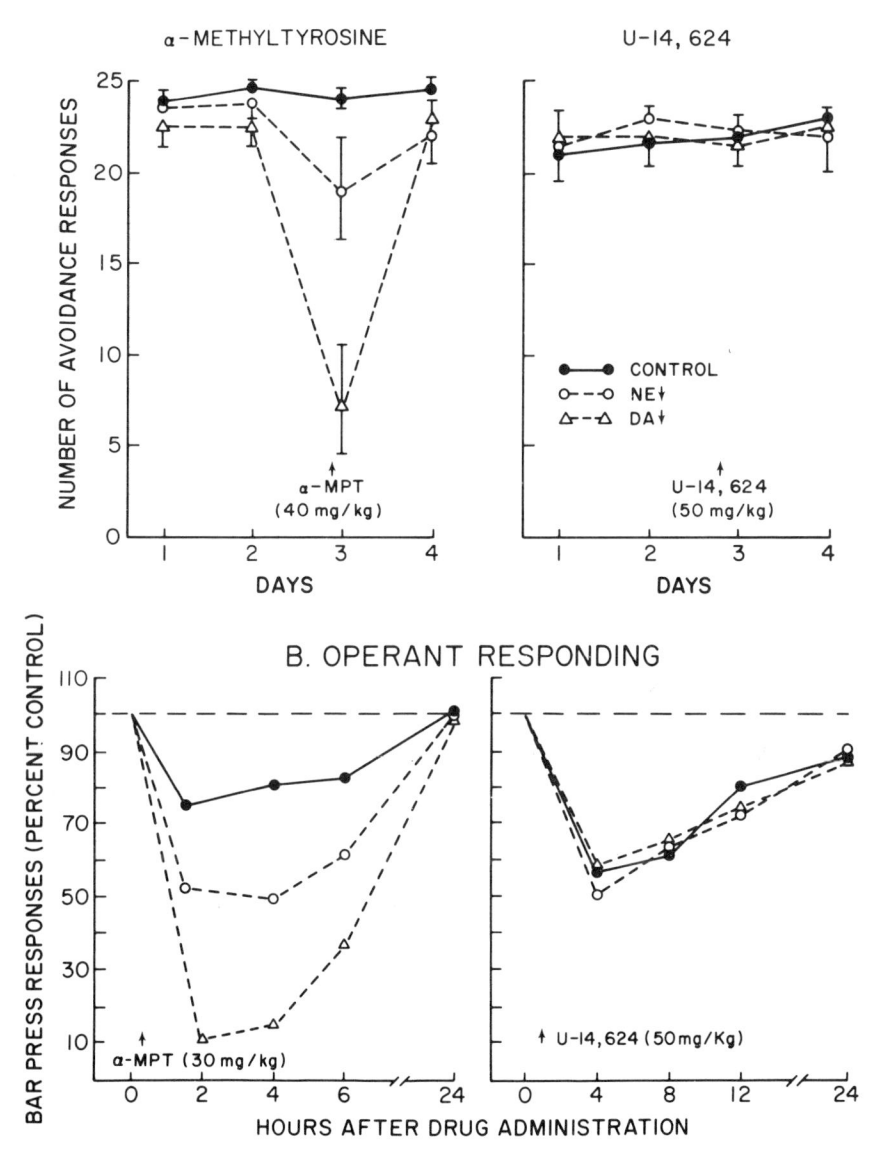

FIG. 2. Effects of α-methyltyrosine (α-MPT) or U-14,624 on the performance of an avoidance response (A) or an operant response (B). In A, performance of a shuttle-box avoidance response was determined daily during a 25-trial session for 7 days before drugs were administered. α-Methyltyrosine (40 mg/kg) was given 4 hr before testing, and U-14,624 (50 mg/kg) was given 6 hr before testing. In B, bar-press responding was assessed for at least 7 days prior to treatment with either α-methyltyrosine (30 mg/kg) or U-14,624 (50 mg/kg). The time designated "0" hour refers to performance prior to drug treatment. NE ↓ designates rats that received 3 × 25 μg of 6-hydroxydopamine. DA ↓ refers to rats given 200 μg of 6-hydroxydopamine 60 min after desipramine (25 mg/kg).

primary importance to the behavioral depression observed after destruction of both noradrenergic and dopaminergic fibers with 6-hydroxydopamine treatments. These results also suggested that additional destruction of brain dopaminergic fibers would produce chronic deficits in behavior similar to those previously described following multiple injections of 6-hydroxydopamine. This possibility was confirmed in rats treated with multiple injections of 6-hydroxydopamine in combination with desipramine, which reduced brain dopamine by 80 to 90% with only a moderate effect on brain norepinephrine (see Breese et al., 1973a; Cooper et al., 1973b). Behaviorally, these animals displayed an acute aphagia and adipsia characteristic of rats treated with either microinjections of 6-hydroxydopamine administered into the substantia nigra (Ungerstedt, 1971) or with 6-hydroxydopamine given intracisternally after pargyline (Fibiger et al., 1972; Zigmond and Stricker, 1972; Breese et al., 1973a, b). Table 1 shows the effect of the different 6-hydroxydopamine treatments that reduced brain norepinephrine or dopamine on the consumption of a sucrose solution substituted for water. Deficits were observed in animals in which dopamine was destroyed but not in rats in which brain norepinephrine was reduced. In addition to disrupting

TABLE 1. *Effect of catecholamine alteration on intake of a sucrose solution*

Treatment[a]	Body weight	Fluid consumption (ml/100 g body weight) Water	Sucrose
Control (N = 18)	413 ± 8	11.9 ± 0.8	37.6 ± 2.5
NE Down (N = 11)	400 ± 11	11.5 ± 0.5	36.8 ± 2.4
DA down (N = 23)	333 ± 10[b]	11.3 ± 0.7	22.5 ± 2.1[b]
6-OHDA (N = 20)	300 ± 9[b]	11.2 ± 0.9	17.3 ± 2.2[b]

[a] Treatments with 6-hydroxydopamine are described by Breese et al., 1973a, b. "NE down" refers to animals treated with 3 × 25 µg of 6-OHDA intracisternally to reduce norepinephrine. "DA down" refers to animals treated with 240 µg of 6-OHDA intracisternally 1 hr after desipramine HCl (30 mg/kg, i.p.) on two occasions. "6-OHDA" refers to animals that received 2 × 200 µg of 6-hydroxydopamine, one dose with pargyline and one without. All animals were allowed to recover for approximately 2 months after treatment with 6-hydroxydopamine before weight and fluid consumption were determined. Numbers in parentheses refer to number of animals in each group.
[b] $p < 0.001$ when compared with control.

ingestive behavior, preferential reduction of brain dopamine was also shown to block the acquisition of an active avoidance response (Cooper et al., 1973b) as well as the acquisition and performance of a position habit in a T-maze (Grant, Howard, and Breese, *unpublished observations*).

VI. EFFECTS OF DOPAMINE ALTERATIONS ON GROWTH AND DEVELOPMENT

With the view that destruction of central catecholamine-containing fibers might assist in understanding the physiological and behavioral significance of these systems, 6-hydroxydopamine has also been administered to developing rats (Breese and Traylor, 1971, 1972; Lytle, Shoemaker, Cottman, and Wurtman, 1972). Intracisternal injection of 6-hydroxydopamine into rat pups on or before 7 days of age not only caused a marked destruction of catecholamine-containing fibers as observed in adult rats, but also caused a striking reduction of body growth (Breese and Traylor, 1972; Table 2). In subsequent experiments, rats were treated with 6-hydroxydopamine using methods that result in the preferential depletion of brain norepinephrine or selective reduction of brain dopamine. Only those rats in which brain dopamine was reduced failed to grow normally, implicating the absence of dopaminergic fibers in the growth deficits observed in 6-hydroxydopamine-treated animals (Smith et al., 1973b; Table 2).

Several possibilities have been suggested to explain the growth deficits in developing rats after 6-hydroxydopamine treatment. These include possible alterations in the neural regulation of hormonal function and changes in ingestive behavior (Breese and Traylor, 1972; Lytle et al., 1972). At the present time, there is some experimental support for the view that chronic alterations of ingestive behavior may be responsible for the growth deficit observed in rats treated with 6-hydroxydopamine when immature. Smith, Breese, and Cooper (1973a) found that adult rats treated during infancy with 6-hydroxydopamine to produce severe reductions of both catecholamines in brain, or to reduce only brain dopamine, failed to eat after insulin treatment, failed to drink large amounts of saline after treatment with desoxycorticosterone, and failed to consume large amounts of sucrose when this solution was substituted for water. These findings provide further evidence that the chronic deficits in ingestive behavior described earlier in adult rats treated with 6-hydroxydopamine are related to a disruption of dopaminergic fibers in brain (Breese et al., 1973b).

The chronic alterations of ingestive behavior observed in rats treated with 6-hydroxydopamine during development suggested that the treatments might be interfering with normal food intake, producing a state of malnutrition. In order to determine if the growth deficits observed after 6-hydroxydopamine might be reproduced by an acute period of malnutrition, nursing mothers and then weaned pups were fed a diet deficient in phenyl-

TABLE 2. *Effects of neonatal administration of 6-hydroxydopamine on brain catecholamines and behavior when adult*

Treatments[a]	Body weight (g)	Brain catecholamines (ng/g) Norepinephrine	Dopamine	Sucrose consumption (ml/24 hr/100 g b.w.)	Avoidance acquisition (avoidance responses/100 trials)
Control	403 ± 6	340 ± 10	620 ± 40	34 ± 4	47 ± 10
NE down	408 ± 11	160 ± 20[b]	490 ± 50	30 ± 5	70 ± 8
DA down	200 ± 10[b]	350 ± 20	49 ± 20[b]	20 ± 7[c]	20 ± 6[c]
6-OHDA	180 ± 16[b]	30 ± 20[b]	50 ± 20[b]	21 ± 2[c]	23 ± 10[c]
Malnourished	207 ± 8[b]	390 ± 50	790 ± 114	32 ± 5	37 ± 10

[a] Treatments with 6-hydroxydopamine are described by Smith et al. (1973) and were given to immature rats on or prior to 7 days of age. Malnourished rats were fed a diet deficient in tyrosine and phenylalanine for 3 weeks starting at 5 days of age.
[b] $p < 0.01$ when compared with control.
[c] $p < 0.05$ when compared with control.

alanine and tyrosine from day 5 to 3 weeks of age. This treatment, as well as the 6-hydroxydopamine treatments, was found to produce a deficit in growth (Table 2). However, malnourishment did not alter the content of norepinephrine or dopamine in brain. Furthermore, no chronic deficits in ingestive behavior or in avoidance behavior could be associated with this method of producing neonatal malnutrition.

Thus, whereas the growth deficits in the 6-hydroxydopamine-treated rats may in part be related to an acute disruption of food intake, the chronic behavioral effects produced by reducing brain dopamine during development cannot be attributed to malnourishment.

VII. EVIDENCE FOR THE INVOLVEMENT OF DOPAMINERGIC FIBERS IN THE STIMULANT EFFECTS OF AMPHETAMINE

Since treatment with 6-hydroxydopamine was found to alter the behavioral depressant effects of α-methyltyrosine and reserpine, it was of interest to determine if 6-hydroxydopamine treatment would also affect the stimulant properties of amphetamine (Breese et al., 1973a; Hollister, Cooper, and Breese, 1973). This was of particular interest, since noradrenergic fibers as well as dopaminergic fibers have been implicated in the stimulant actions of this compound (Taylor and Snyder, 1971).

Supporting the view that the actions of amphetamine are dependent upon catecholamine neural systems (Hanson, 1966). Hollister et al. (1973) recently reported that treatment with 6-hydroxydopamine decreased amphetamine-induced motor activity as well as reduced the stereotypies produced by this drug. Following this work, experiments were designed to define whether the actions of amphetamine were related to noradrenergic or dopaminergic fibers. Contrary to previous proposals implicating noradrenergic fibers in the locomotor stimulation produced by amphetamine (Randrup and Scheel-Kruger, 1966), preferential reduction of brain dopamine, but not norepinephrine, was found to antagonize amphetamine-induced motor activity (Fig. 3A; Breese, Cooper, and Hollister, 1974). In keeping with previous work indicating that intact dopaminergic fibers are essential for amphetamine-induced stereotypies (Taylor and Snyder, 1971), destruction of dopaminergic fibers was found to reduce stereotyped behavior after the administration of amphetamine.

In other studies, the response to L-DOPA was examined in 6-hydroxydopamine-treated rats (Breese et al., 1973a; Hollister et al., 1973). In contrast to the reduction of amphetamine-induced motor activity, 6-hydroxydopamine treatment produced a significant potentiation of DOPA-stimulated motor activity. These findings are comparable to those reported by Schoenfeld and Uretsky (1973). In other work, it was found that this enhanced activity was evident in rats with brain dopamine preferentially reduced, but not in animals in which brain norepinephrine was depleted

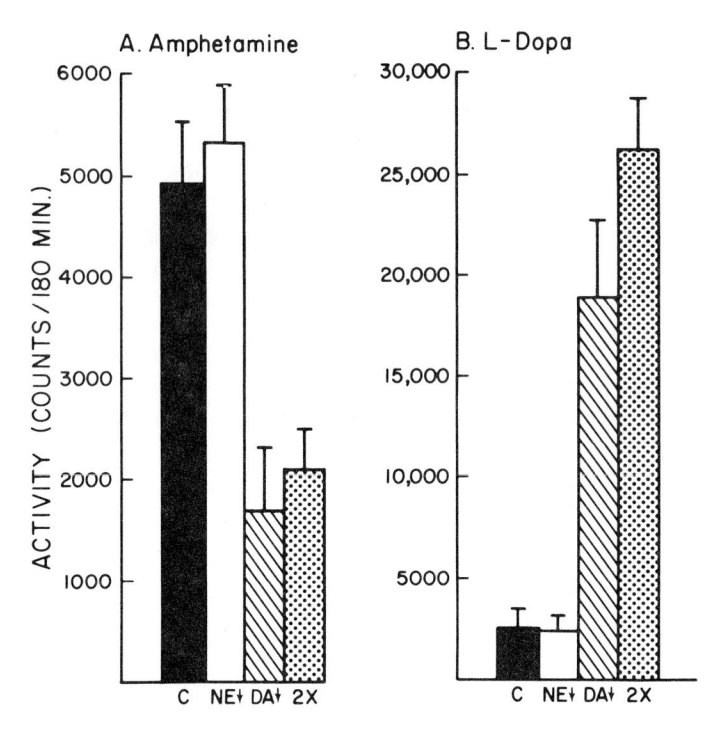

FIG. 3. Effects of various 6-hydroxydopamine treatments on *(A)* amphetamine- and *(B)* L-dihydroxyphenylalanine (L-DOPA)-induced locomotor activity. NE ↓ refers to animals that received 3 × 25 µg of 6-hydroxydopamine. DA ↓ treatment consisted of administering 2 × 240 µg of 6-hydroxydopamine intracisternally 60 min after desipramine (30 mg/kg). 2X refers to rats that were injected with 2 × 200 µg of 6-hydroxydopamine, the first dose with pargyline and the second without. C refers to vehicle-treated rats. Animals were allowed at least 3 weeks to recover from the surgery before amphetamine (2 mg/kg) or L-DOPA (100 mg/kg) was administered. Rats that received L-DOPA were pretreated with N-(DL-seryl-N'-(2,3,4-trihydroxybenzyl)hydrazine (50 mg/kg) 1 hr before receiving the L-DOPA. Counts reflect the number of photocell beam interruptions obtained in a circular photocell activity cage.

(Fig. 3*B*). The finding that U-14,624 did not reduce the motor response to L-DOPA in the animals treated to deplete dopamine in brain suggests that the dopamine formed from L-DOPA is responsible for this stimulation of locomotor activity. Attempts to determine whether the supersensitivity of 6-hydroxydopamine-treated rats to DOPA-induced motor activity is the result of an inhibition of presynaptic uptake of dopamine or is the result of a change in postsynaptic receptor sensitivity to the transmitter have not been resolved.

VIII. SUMMARY

The acute and chronic changes produced following intracisternal administration of 6-hydroxydopamine have been studied in an attempt to relate these alterations to destruction of noradrenergic or dopaminergic neural systems in brain. Whether administered to adult or to developing rats, results indicate that destruction of dopaminergic fibers in brain is responsible for the deficits in ingestive and operant behavior that occur following 6-hydroxydopamine treatment. The behavioral depressant effects of α-methyltyrosine and reserpine would also appear to be dependent upon a disruption of dopaminergic function in brain. Unexpectedly, the locomotor stimulant action of amphetamine was reduced following the destruction of dopaminergic fibers, but not after the reduction of brain norepinephrine, suggesting that the integrity of dopaminergic fibers is also essential for amphetamine-induced locomotor activity.

With the possible exception of the acute hypothermia produced by 6-hydroxydopamine, noradrenergic fibers could not be implicated in the deficits produced by this compound. Nevertheless, the inability to define a role for brain norepinephrine may be due, in part, to a failure to produce sufficient reduction of brain norepinephrine following 6-hydroxydopamine treatment. It is also possible that noradrenergic fibers do not have a significant role in the maintenance of the behavioral, physiological, and pharmacological functions studied. Should this be the case, alternative avenues of inquiry to define the functional significance of norepinephrine-containing neurons will need to be developed. The inhibitory role of norepinephrine in the cerebellum (Hoffer, Siggins, Oliver, and Bloom, 1973) and its proposed central role in the control of peripheral cardiovascular mechanisms (Bolme, Fuxe, and Lidbrink, 1972) are promising areas for future research. Regardless of the role for brain noradrenergic fibers, results obtained with 6-hydroxydopamine support the view that dopaminergic neural systems are important for the maintenance of a wide variety of behavioral and pharmacological responses.

ACKNOWLEDGMENTS

We are grateful to Marcine Kinkead, Susan Hollister, Edna Edwards, and Joseph Farmer for their excellent assistance. Supported by U.S. Public Health Service grants MH-16522 and HD-03110. Barrett R. Cooper is a Postdoctoral Fellow in Neurobiology (MH-11107); George R. Breese is a U.S. Public Health Service Career Development Awardee (HD-24585).

REFERENCES

Bloom, F. E., Algeri, S., Groppetti, A., Revuetta, A., and Costa, E. (1969): Lesions of central norepinephrine terminals with 6-OH-dopamine: Biochemistry and fine structure. *Science*, 166:1284–1286.

Bolme, P., Fuxe, K., and Lidbrink, P. (1972): On the function of catecholamine neurons—their role in cardiovascular and arousal mechanisms. *Research Communications in Chemistry, Pathology, and Pharmacology,* 4:657–697.

Breese, G. R. (1974): Chemical and immunochemical lesions by specific neurotoxic substances and antisera. In: *Handbook of Psychopharmacology,* edited by L. L. Iversen, S. D. Iversen, and S. H. Snyder. Plenum Press, New York *(in press).*

Breese, G. R., Cooper, B. R., and Hollister, A. S. (1974): Relationship of biogenic amines to behavior. *Journal of Psychiatric Research (in press).*

Breese, G. R., Cooper, B. R., and Smith, R. D. (1973a): Biochemical and behavioral alterations following 6-hydroxydopamine administration into brain. In: *Frontiers in Catecholamine Research,* edited by E. Usdin and S. Snyder. Pergamon Press, Elmsford, N.Y., pp. 701–706.

Breese, G. R., and Howard, J. L. (1971): Effect of central catecholamine alterations on the hypothermic response to 6-hydroxydopamine in desipramine-treated rats. *British Journal of Pharmacology,* 43:671–674.

Breese, G. R., Howard, J., and Leahy, P. (1971): Effect of 6-hydroxydopamine on electrical self-stimulation of brain. *British Journal of Pharmacology,* 43:255–257.

Breese, G. R., Moore, R. A., and Howard, J. L. (1972): Central actions of 6-hydroxydopamine and other phenylethylamine derivatives on body temperature in the rat. *Journal of Pharmacology and Experimental Therapeutics,* 180:591–602.

Breese, G. R., Smith, R. D., Cooper, B. R., and Grant, L. D. (1973b): Alterations in consummatory behavior following intracisternal injection of 6-hydroxydopamine. *Pharmacology, Biochemistry, and Behavior,* 1:319–328.

Breese, G. R., and Traylor, T. D. (1970): Effect of 6-hydroxydopamine on brain norepinephrine and dopamine: Evidence for selective degeneration of catecholamine neurons. *Journal of Pharmacology and Experimental Therapeutics,* 174:413–420.

Breese, G. R., and Traylor, T. D. (1971): Depletion of brain noradrenaline and dopamine by 6-hydroxydopamine. *British Journal of Pharmacology,* 42:88–99.

Breese, G. R., and Traylor, T. D. (1972): Developmental characteristics of brain catecholamines and tyrosine hydroxylase in the rat. Effects of 6-hydroxydopamine. *British Journal of Pharmacology,* 44:210–222.

Cooper, B. R., Breese, G. R., Grant, L. D., and Howard, J. L. (1973b): Effects of 6-hydroxydopamine treatments on active avoidance responding: Evidence for involvement of brain dopamine. *Journal of Pharmacology and Experimental Therapeutics,* 185:358–370.

Cooper, B. R., Breese, G. R., Howard, J. L., and Grant, L. D. (1972a): Sensitivity to the behavioral depressant effects of reserpine and α-methyltyrosine after 6-hydroxydopamine treatment. *Psychopharmacologia,* 27:99–110.

Cooper, B. R., Breese, G. R., Howard, J. L., and Grant, L. D. (1972b): Effect of central catecholamine alterations by 6-hydroxydopamine on shuttle-box avoidance acquisition. *Physiology and Behavior,* 9:727–731.

Cooper, B. R., Grant, L. D., and Breese, G. R. (1973a): Comparison of the behavioral depressant effects of biogenic amine depleting and neuroleptic agents following various 6-hydroxydopamine treatments. *Psychopharmacologia,* 31:95–109.

Cooper, B. R., Smith, R. D., and Breese, G. R. (1973): Evidence for a role for dopamine in electrical self-stimulation of brain. *Pharmacologist,* 15:163.

Evetts, K. D., and Iversen, L. L. (1970): Effects of protriptyline on the depletion of catecholamines induced by 6-hydroxydopamine in the brain of the rat. *Journal of Pharmacy and Pharmacology,* 22:540–542.

Fibiger, H. C., Lonsbury, B., Cooper, H. P., and Lytle, L. D. (1972): Early behavioral effects of intraventricular administration of 6-hydroxydopamine in rat. *Nature New Biology,* 236:209–211.

Fibiger, H. C., Zis, A. P., and McGeer, E. G. (1973): Feeding and drinking deficits after 6-hydroxydopamine administration in the rat: Similarities to the lateral hypothalamic syndrome. *Brain Research,* 53:135–148.

Hanson, L. C. F. (1966): Evidence that the central action of amphetamine is mediated via catecholamines. *Psychopharmacologia,* 9:78–80.

Harvey, J. A. (1965): Comparison between the effects of hypothalamic lesions on brain amine levels and drug action. *Journal of Pharmacology and Experimental Therapeutics,* 147:244–251.

Hoffer, B. J., Siggins, G., Oliver, A., and Bloom, F. (1973): Activation of the pathway from locus coeruleus to rat cerebellar purkinje neurons: Pharmacological evidence of noradrenergic central inhibition. *Journal of Pharmacology and Experimental Therapeutics,* 184:553–569.

Hollister, A. S., Cooper, B. R., and Breese, G. R. (1973): Evidence for a role of dopamine in amphetamine-induced motor activity and stereotypy. *Pharmacologist,* 15:254.

Hornykiewicz, O. (1973): Parkinson's disease: From brain homogenate to treatment. *Federation Proceedings,* 32:183–190.

Howard, J. L., Grant, L. D., and Breese, G. R. (1974): Effects of intracisternal 6-hydroxydopamine treatment on acquisition and performance of rats in a double T-maze. *Journal of Comparative Physiology and Psychology (in press).*

Laverty, R., and Taylor, L. (1970): Effects of intraventricular 2,4,5-trihydroxyphenylethylamine (6-hydroxydopamine) on rat behavior and brain catecholamine metabolism. *British Journal of Pharmacology,* 40:836–846.

Lytle, L. D., Shoemaker, M. J., Cottman, K. E., and Wurtman, R. J. (1972): Long term effects of postnatal 6-hydroxydopamine treatment on tissue catecholamine levels. *Journal of Pharmacology and Experimental Therapeutics,* 183:56–64.

Nakamura, K., and Thoenen, H. (1971): Hypothermia induced by intraventricular administration of 6-hydroxydopamine in rats. *European Journal of Pharmacology,* 16:46–54.

Nakamura, K., and Thoenen, H. (1972): Increased irritability: A permanent behavior change induced in rat by intraventricular administration of 6-hydroxydopamine. *Psychopharmacologia,* 24:359–372.

Oltmans, G. A., and Harvey, J. A. (1972): LH syndrome and brain catecholamine levels after lesions of the nigrostriatal bundle. *Physiology and Behavior,* 8:69–78.

Pirch, J. H., Rech, R. H., and Moore, K. E. (1967): Depression and recovery of the electrocorticogram, behavior and brain amines in rats treated with reserpine. *International Journal of Neuropharmacology,* 6:375–385.

Randrup, A., and Scheel-Kruger, J. (1966): Diethyldithiocarbamate and amphetamine stereotype behavior. *Journal of Pharmacy and Pharmacology,* 18:752.

Rech, R. H., Carr, L. A., and Moore, K. E. (1968): Behavioral effects of α-methyltyrosine after prior depletion of brain catecholamines. *Journal of Pharmacology and Experimental Therapeutics,* 160:326–335.

Schoenfeld, R. I., and Uretsky, N. J. (1973): Enhancement by 6-hydroxydopamine of the effects of DOPA upon the motor activity of rats. *Journal of Pharmacology and Experimental Therapeutics,* 180:616–624.

Schoenfeld, R., and Zigmond, M. (1970): Effect of 6-hydroxydopamine on fixed ratio performance. *Pharmacologist,* 12:227.

Scotti de Carolis, A., Ziegler, H., Del Basso, P., and Longo, V. G. (1971): Central effects of 6-hydroxydopamine. *Physiology and Behavior,* 7:705–708.

Simmonds, M. A., and Uretsky, N. J. (1970): Central effects of 6-hydroxydopamine on body temperature in the rat. *British Journal of Pharmacology,* 40:630–638.

Smith, R. D., Breese, G. R., and Cooper, B. R. (1973a): Involvement of brain dopamine in growth and consummatory deficits observed after administration of 6-hydroxydopamine (6 OHDA) to immature rats. *Pharmacologist,* 15:163.

Smith, R. D., Cooper, B. R., and Breese, G. R. (1973b): Growth and behavioral changes in developing rats treated intracisternally with 6-hydroxydopamine: Evidence for involvement of brain dopamine. *Journal of Pharmacology and Experimental Therapeutics,* 185:609–619.

Smith, G. P., Strohmayer, A., and Reis, D. J. (1972): Effects of lateral hypothalamic injections of 6-hydroxydopamine on food and water intake in rats. *Nature New Biology,* 235:27–29.

Taylor, K. M., and Snyder, S. H. (1971): Differential effects of D- and L-amphetamine on behavior and on catecholamine disposition in dopamine and norepinephrine containing neurons of rat brain. *Brain Research,* 28:295–309.

Teitelbaum, P., and Epstein, A. N. (1962): The lateral hypothalamic syndrome: Recovery of feeding and drinking after lateral hypothalamic lesions. *Psychological Reviews,* 69:74–90.

Thoenen, H., and Tranzer, J. P. (1973): The pharmacology of 6-hydroxydopamine. *Annual Review of Pharmacology,* pp. 169–180.

Ungerstedt, U. (1971): Striatal dopamine release after amphetamine or nerve degeneration revealed by rotational behavior. *Acta Physiologica Scandinavica,* 367:49–68.

Ungerstedt, U., and Ljungberg, T. (1973): Behavioral-anatomical correlates of brain catecholamines. In: *Frontiers in Catecholamine Research,* edited by E. Usdin and S. Snyder. Pergamon Press, Elmsford, N.Y.

Uretsky, N. J., and Iversen, L. L. (1970): Effects of 6-hydroxydopamine on catecholamine containing neurons in rat brain. *Journal of Neurochemistry,* 17:269–278.

Zigmond, M. J., and Stricker, E. M. (1972): Deficits in feeding behavior after intraventricular injection of 6-hydroxydopamine in rats. *Science,* 177:1211–1214.

Neuropsychopharmacology of Monoamines and Their Regulatory Enzymes, edited by E. Usdin. Raven Press, New York © 1974.

Effects of Alterations in Impulse Flow on Transmitter Metabolism in Central Dopaminergic Neurons

Robert H. Roth, Judith R. Walters, and Victor H. Morgenroth III

Departments of Pharmacology and Psychiatry, Yale University School of Medicine and the Connecticut Mental Health Center, New Haven, Connecticut 06510

INTRODUCTION

It is generally a well-accepted view that alterations in impulse flow in both peripheral and central monoamine-containing neurons lead to predictable changes in the metabolism and turnover of the transmitter associated with the neuronal system under study (Costa, 1970; Weiner, 1970). In fact, this view is now so well accepted that many laboratories routinely employ measures of synthesis, catabolism, and turnover to provide an index of the functional activity of monoamine-containing neurons. This approach seems quite justifiable in the study of peripheral and central noradrenergic neurons and central serotonergic neurons, since increases or decreases in neuronal activity appear to be well correlated with increases or decreases in transmitter metabolism and turnover. Thus, for example, an increase in the activity of the sympathetic input to the vas deferens and salivary gland, as well as to other sympathetically innervated tissues, results in an increase in norepinephrine (NE) synthesis and turnover. A decrease or block in impulse flow leads to a reduction in NE synthesis and turnover (Sedvall, 1969; Weiner, 1970).

Similar results are also obtained in studies on central monoamine systems. Stimulation of noradrenergic neurons originating in the locus coeruleus and projecting to the hippocampus and cerebral cortex leads to an increase in NE turnover (Arbuthnott, Crow, Fuxe, and Ungerstedt, 1970; Korf, Roth, and Aghajanian, 1973b) and a frequency-dependent increase in the accumulation of a major metabolite of NE, 3-methoxy-4-hydroxyphenethylene glycol sulfate (Korf, Aghajanian, and Roth, 1973a; Walters and Eccleston, 1973). A cessation of impulse flow in noradrenergic neurons produced by acute destruction of the locus coeruleus results in a reduction of NE turnover with no significant alteration in the steady state levels of NE in the cerebral cortex (Korf et al., 1973b). In a similar fashion, stimulation of serotonergic neurons in the midbrain raphe results in a frequency-dependent increase in serotonin synthesis (Shields and Eccleston, 1972) and accumulation of 5-hydroxyindoleacetic acid in the forebrain (Sheard and Aghajanian, 1968). If impulse flow in serotonergic neurons is abolished by placement of a lesion in the midbrain raphe, the turnover of serotonin is retarded and no significant alteration in the steady state level of serotonin is produced (Herr and Roth, 1972).

Dopaminergic neurons are similar in some respects to central noradrenergic and serotonergic neurons in their response to alterations in impulse flow, but in other respects dopaminergic neurons are markedly different. Stimulation of the dopamine (DA) neurons originating in the zona compacta of the substantia nigra and projecting to the neostriatum results, as does stimulation of the noradrenergic and serotonergic systems, in a release of transmitter (Von Voigtlander and Moore, 1971) and transmitter metabolites (Vogt, 1969). A frequency-dependent increase in DA synthesis and the rapid accumulation of a DA metabolite (dihydroxyphenylacetic acid, DOPAC) in the neostriatum (Murrin and Roth, 1973) is also observed. An increase in impulse flow in addition results in an increase in the turnover of DA as measured by following the rate of disappearance of DA after inhibition of synthesis with α-methyltyrosine (Corrodi, Fuxe, and Hökfelt, 1967; Andén et al., 1971).

Differences between the DA system and other central monoamine systems become apparent when one begins to investigate the effects of a reduction or cessation of impulse flow in the nigro-neostriatal pathway. A cessation of impulse flow in dopaminergic neurons leads, as might be predicted, to both a reduction in the release of DA and a reduction in the subsequent metabolism of DA to DOPAC and homovanillic acid. In fact, the short-term accumulation of DOPAC in the neostriatum appears to provide a very useful index of changes in functional activity of dopaminergic neurons in the nigro-neostriatal pathway (Table 1). An increase in the activity of these neurons results in an increase in the accumulation of DOPAC, and a decrease in activity causes a reduction in DOPAC accumulation. A decrease in DA turnover is also observed during periods of reduction in functional activity of these neurons.

However, if DA synthesis is followed, a different and rather unusual picture emerges. It has recently become appreciated that dopaminergic neurons respond in a rather curious fashion to a cessation of impulse flow. Pharmacological inhibition of impulse flow caused by systemic administration of the anesthetic, gamma-hydroxybutyrate (GHB), mechanical interruption of impulse flow caused by the placement of an electrolytic lesion in the nigro-neostriatal pathway, or mechanical transection of this pathway all result in a marked increase in the DA content of the dopaminergic terminals in the neostriatum (Aghajanian and Roth, 1970; Andén, Bedard, Fuxe, and Ungerstedt, 1972; Nybäck, 1972; Walters, Aghajanian, and Roth, 1972; Stock, Magnusson, and Andén, 1973; Walters, Roth, and Aghajanian, 1973).

TABLE 1. *Effect of alteration of impulse flow in the nigro-neostriatal pathway on the accumulation of DOPAC in the rat striatum*

Treatment	Effect on unit activity of dopaminergic neurons	N	DOPAC levels (% of control)
None	No change	19	100 ± 3
Sham lesion	No change	4	102 ± 3
30-min SN lesion	Decreased	4	75 ± 4[b]
GBL (750 mg/kg, i.p.)	Decreased	4	76 ± 11[a]
Apomorphine (2 mg/kg, i.p.)	Decreased	6	71 ± 8[b]
Trivastal (10 mg/kg, i.p.)	Decreased	3	72 ± 2[b]
D-Amphetamine* (1.25 mg/kg, i.v.)	Decreased	4	47 ± 2[c]
Chloral hydrate (400 mg/kg, i.p.)	Increased	8	140 ± 5[c]
Halothane	Increased	4	128 ± 4[c]
Haloperidol* (0.1 mg/kg, i.v.)	Increased	4	206 ± 14[c]
Chlorpromazine* (1.25 mg/kg, i.v.)	Increased	4	168 ± 11[c]
Promethazine* (10 mg/kg, i.p.)	No change	4	102 ± 5
Electrical stimulation			
Sham (chloral hydrate anesthesia)	No change	9	105 ± 5
5/sec (chloral hydrate anesthesia)	Increased	3	169 ± 11[c]
10/sec (chloral hydrate anesthesia)	Increased	3	201 ± 4[c]
15/sec (chloral hydrate anesthesia)	Increased	5	331 ± 19[c]
30/sec (chloral hydrate anesthesia)	Increased	3	256 ± 45[c]

Lesions of the nigro-neostriatal pathway (SN-lesion) were performed under halothane anesthesia as described by Walters et al. (1973). In the stimulation experiments the nigro-neostriatal pathway was stimulated for periods of 20 min at the indicated frequency. Results are expressed as a percentage of the unstimulated side. Rats were killed 30 min after drug treatment and DOPAC isolated and analyzed fluorometrically as described previously (Walters and Roth, 1972). All results are expressed as the mean ± SEM. The level of DOPAC in the striatum of untreated rats was 934 ± 27 ng/g. Where indicated, results are significantly different from the appropriate control: [a] $p < 0.01$; [b] $p < 0.002$; [c] $p < 0.001$.

* Indicates that rats were anesthetized with chloral hydrate prior to drug administration in order to duplicate conditions of unit recording studies, and chloral hydrate-treated rats served as the appropriate control. Data taken in part from Bunney et al. (1973), and Murrin and Roth (1973).

Until recently the actual cause for this rapid increase in DA was not appreciated. Recent experiments in our own laboratory now indicate that this rapid increase in the steady state level of DA is due in part to a rapid increase in tyrosine hydroxylase activity triggered by the cessation of impulse flow in the dopaminergic pathway (Roth, Walters, and Aghajanian, 1973; Walters et al., 1973; Walters and Roth, 1974). This was a rather unexpected observation, since one would have predicted that a decrease or cessation of impulse flow and the consequent decrease in utilization of transmitter would evoke homeostatic mechanisms that ultimately would result in a reduction rather than an increase in transmitter synthesis. Because of the unusual nature as well as the reproducibility of this phenomena, we have directed our attention toward determining the possible molecular mechanism by which a cessation of impulse flow results in such a striking activation of tyrosine hydroxylation in dopaminergic neurons but not in noradrenergic neurons.

We were very fortunate in this respect, since earlier studies in our laboratory had clearly indicated that it was possible to inhibit reversibly impulse flow in dopaminergic neurons by systemic administration of a drug, GHB, or its lactone precursor, gamma-butyrolactone (GBL) (Walters et al., 1972). Unit recording studies have demonstrated that this drug, when administered systemically in doses greater than 100 mg/kg, causes a rapid cessation of impulse flow in dopaminergic neurons without having a very significant effect on impulse flow in noradrenergic or serotonergic neurons (Roth et al., 1973). This drug therefore served as a very useful tool with which to study the mechanism by which a cessation of impulse flow leads to an activation of tyrosine hydroxylase.

In order to obtain a useful *in vivo* indication of tyrosine hydroxylase activity, we made use of a technique developed by Carlsson and co-workers (Carlsson, Kehr, Lindqvist, Magnusson, and Atack, 1972), in which the accumulation of dihydroxyphenylalanine (DOPA) following inhibition of DOPA decarboxylase is used to obtain an approximation of the rate of dopamine synthesis. Employing this technique, we observed that the accumulation of DOPA was linear for periods up to 1 hr after administration of a decarboxylase inhibitor (Walters and Roth, 1974). Treatment with GHB or the lactone precursor, GBL, or the placement of electrolytic lesions in the nigro-neostriatal pathway caused, within 35 min, about a 200% increase in DOPA accumulation in the neostriatum (Roth et al., 1973; Walters and Roth, 1974). Ninety minutes after the administration of GBL or after placement of a lesion in the nigro-neostriatal pathway (at a time when endogenous levels of DA have doubled), the 30-min accumulation of DOPA is not significantly different from control. Thus, after the initial period of increased DOPA formation, there appears to be a return to normal synthesis rates, but there does not appear to be an inhibition of synthesis below control levels despite the elevated levels of endogenous DA. On the other hand, when endogenous levels of DA are elevated by pretreatment with monoamine oxidase inhibitors (MAOIs) such as pargyline or iproniazid, the 30-min accumulation of DOPA is markedly decreased (cf. Table 4 and Roth et al., 1973). This decrease in synthesis does not appear to be a direct effect of the MAOIs on tyrosine hydroxylase, because no significant change in DOPA accumulation is observed when the inhibitor is administered 35 min before sacrifice (Walters and Roth, 1974). Since the elevation of DA levels following MAOIs does cause an apparent decrease in tyrosine hydroxylase activity, it seems possible that the increased DA levels observed after cessation of impulse flow might be in a functionally different location that protects it from interacting with tyrosine hydroxylase. Alternatively, cessation of impulse flow might somehow alter the sensitivity of tyrosine hydroxylase to inhibition by endogenous DA.

However, exactly how the inhibition of impulse flow in dopaminergic neurons could trigger an increase in DA synthesis and a protection of tyrosine hydroxylase from interaction with newly synthesized DA, or a change in the sensitivity of tyrosine hydroxylase to inhibition by DA, seems at first glance quite difficult to explain. Perhaps some insight can be gleaned from studies of norepinephrine synthesis and release in the periphery. Here recent experiments have given rise to the concept that there may exist on the presynaptic side of the synapse a receptor sensitive to the amount of transmitter in the synaptic cleft and capable of somehow regulating release of transmitter from the presynaptic terminal (Kirpekar and Puig, 1971; Starke, 1971, 1972; Enero, Langer, Rothlin, and Stefano, 1972). Recently Carlsson's group (Kehr, Carlsson, Lindqvist, Magnusson, and Atack, 1972b) has proposed that such a presynaptic receptor might exist on the dopaminergic nerve terminals and somehow cause an increase in synthesis when impulse flow is interrupted and no DA is released into the cleft.

In an effort to investigate this possibility further, drugs that would stimulate DA receptors either directly, or indirectly by release of DA, were tested to see if they would alter the increase in tyrosine hydroxylase activity observed when impulse flow was blocked in the nigro-neostriatal pathway. Pretreatment with amphetamine, apomorphine, and trivastal was found to antagonize markedly the increase in DOPA accumulation observed after inhibition of impulse flow pharmacologically with GHB or by placement of a lesion in the nigro-neostriatal pathway (Fig. 1, Table 2; see also Walters and Roth, 1974). While it is quite possible that treatment with amphetamine leads to some change in the interneuronal distribution of DA, making more free DA available to act as an endogenous inhibitor of tyrosine hydroxylase, an alternative explanation consistent with the theory of receptor-mediated changes in tyrosine hydroxylase activity is that amphetamine causes an increase in the amount of DA present in the synaptic cleft which then interacts with pre- or postsynaptic DA receptors. The data obtained with apomorphine and trivastal are perhaps more clear cut, since these agents presumably have a direct stimulatory effect on dopaminergic receptors (Andén, Rubenson, Fuxe, and Hökfelt, 1967; Corrodi, Fuxe, and Ungerstedt, 1971).

If an interaction with DA receptors is responsible for the ability of amphetamine, apomorphine, and trivastal to block the increase in tyrosine hydroxylase activity observed during a cessation of impulse flow in DA neurons, one would predict that DA receptor blocking drugs should be able to reverse the inhibitory effects of DA receptor stimulants. Administration of haloperidol (1 mg/kg) and chlorpromazine (10 mg/kg) effectively reverses the inhibitory effects of amphetamine, apomorphine, and trivastal on the increase in tyrosine hydroxylase activity observed after inhibition of impulse flow by GHB (Fig. 1, Table 2; see also Walters and Roth, 1974). Haloperidol was the most effective drug tested. Phenothiazines such as

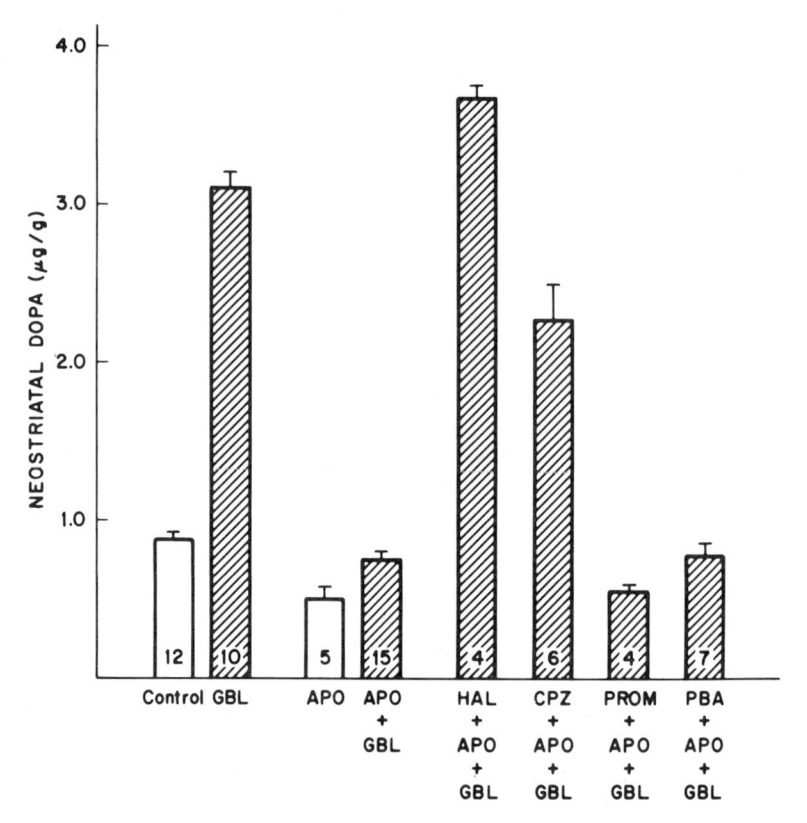

FIG. 1. Effect of apomorphine and other drugs on neostriatal DOPA accumulation following inhibition of firing. The decarboxylase inhibitor [N-(DL-seryl)-N'-(2,3,4-trihydroxybenzyl)]hydrazine (RO4–4602, 800 mg/kg, i.p.) was administered 30 min before sacrifice to all animals. GBL (750 mg/kg, i.p.) was administered 35 min, and apomorphine (APO, 2 mg/kg, i.p.) 40 min before sacrifice. Haloperidol (HAL, 1 mg/kg, i.p.), chlorpromazine (CPZ, 10 mg/kg, i.p.), promethazine (PROM, 10 mg/kg, i.p.), and phenoxybenzamine (PBZ, 25 mg/kg) were administered 45 min before sacrifice where indicated. DOPA determinations were performed by the method of Kehr et al., (1972a), as described in Walters and Roth (1974). The bars represent SEM, and the numbers in the columns indicate the number of individual experiments. APO-treated animals are significantly decreased from control ($p < 0.005$), as are the PROM + APO + GBL-treated rats ($p < 0.01$). GBL, HAL + APO + GBL-, and CPZ + APO + GBL-treated rats are significantly increased over control ($p < 0.001$).

promethazine, devoid of antipsychotic effects and lacking DA receptor blocking capabilities, were ineffective in reversing the inhibitory effects of the DA receptor stimulants. Alpha receptor blockers such as phenoxybenzamine were also ineffective (cf. Fig. 1). These observations provide additional support for the contention that activation of a DA receptor is responsible for the inhibitory effects exerted by amphetamine, apomorphine, and trivastal. These experiments also imply that a presynaptic receptor mechanism may be involved, since the effects of neuronal feedback and sub-

TABLE 2. *Effect of amphetamine, trivastal, and haloperidol on neostriatal DOPA accumulation following inhibition of firing*

Treatment	N	DOPA ± SEM µg/g
Control	13	0.84 ± 0.03
GBL	7	2.96 ± 0.18[a]
Trivastal	9	0.48 ± 0.06[a]
Trivastal + GBL	6	1.08 ± 0.04[a]
Haloperidol + trivastal + GBL	4	3.26 ± 0.18[a]
Amphetamine	6	1.18 ± 0.14[b]
Amphetamine + GBL	4	1.35 ± 0.10[a]
Haloperidol + amphetamine + GBL	6	4.08 ± 0.46[a]

RO4-4602 (800 mg/kg, i.p.) was administered 30 min before sacrifice to all animals. GBL (750 mg/kg, i.p.) was administered 35 min, amphetamine (5 mg/kg, i.p.) and trivastal (10 mg/kg, i.p.) 40 min before sacrifice. Haloperidol (1 mg/kg, i.p.) was administered 45 min before sacrifice where indicated. DOPA determinations were performed by the method of Kehr, Carlsson, and Lindqvist (1972a), as described in Walters and Roth (1974). N equals the number of experiments. Data taken in part from Walters and Roth *(manuscript in preparation)*.
[a] Significantly different from control, $p < 0.001$.
[b] Significantly different from control, $p < 0.005$.

sequent alterations in the firing rate of DA neurons normally produced by haloperidol and chlorpromazine (Bunney, Walters, Roth, and Aghajanian, 1973) are prevented by treatment of the rats with GHB (Walters and Roth, *unpublished data*). However, it is premature to rule out the possibility that a trans-synaptic event mediated by an interaction of these DA receptor stimulants with postsynaptic DA receptors might also be involved.

The idea that a presynaptic receptor may regulate DA synthesis raises certain questions about the possible mechanism(s) involved in this regulation and the nature of the interaction of this regulatory mechanism with end-product inhibition of tyrosine hydroxylase by DA. Iontophoretic application of DA onto the cell body of dopaminergic neurons in the zona compacta has recently been shown to inhibit the firing of these dopaminergic neurons (Aghajanian and Bunney, 1973a). It has been theorized that the DA receptors on the cell body may be responsible for this inhibitory effect of iontophoresed DA (Aghajanian and Bunney, 1973b) and further that these receptors may be similar to the presynaptic receptors on dopaminergic nerve terminals. If this is the case, it seems reasonable to suggest that stimulation of the presynaptic receptors in the synaptic cleft by DA or DA agonists might also affect ionic flow in the presynaptic terminals. Thus, one way in which the presynaptic receptor could control tyrosine hydroxylase activity would be by altering the permeability of the presynaptic membrane to ions or other substances that might influence tyrosine hydroxylase activity. The removal of calcium has already been shown to cause a marked increase in DA synthesis in striatal slices while having no effect on the synthesis of norepinephrine in cortical slices (Goldstein et al., 1970; Harris and Roth, 1970). Two forms of tyrosine hydroxylase with different kinetic properties have also been described in the rat striatum, and calcium has been suggested to cause a change in the distribution of these two forms (Kuczenski and Mandell, 1972). It therefore seemed reasonable to suspect that alterations in the flux of calcium or other ions caused by depolarization or by activation of presynaptic DA receptors could lead to an interconversion between these two forms of enzyme, or could act directly on the enzyme, resulting in an increase in the activity of tyrosine hydroxylase or a change in its sensitivity to feedback inhibition by DA.

For this reason we examined the effects of the presence and absence of calcium ions on the activity of tyrosine hydroxylase in striatal homogenates. Tyrosine hydroxylase was measured by a modification of the method of Shiman and co-workers (Shiman, Akino, and Kaufman, 1971; Coyle, 1972). The results reported below summarize experiments conducted on the soluble supernatant tyrosine hydroxylase assayed in a system containing pteridine reductase and catalase. Removal of calcium by addition of EGTA (10^{-6} M) was observed to cause a significant activation of tyrosine hydroxylase (Fig. 2). The K_m for tyrosine was decreased from 5.39×10^{-5} M to 9.01×10^{-6} M and the K_m for synthetic cofactor 6,7-dimethyl-5,6,7,8-tetrahydropterin (DMPH$_4$) was decreased from 8.95×10^{-4} M to 1.33×10^{-4} M. EGTA did not cause a similar stimulation or change in the kinetic properties of tyrosine hydroxylase prepared from rat pons-medulla or guinea pig vas deferens (Fig. 2). The stimulation produced by EGTA in the striatum was completely reversed by addition of calcium. Removal of calcium also caused a marked increase in the K_i for DA (1.07×10^{-4} M to 7.39×10^{-2} M), indicating that this new form of the enzyme is insensitive to end-product inhibition by DA.

These *in vitro* results suggested the possibility that changes in calcium fluxes in the DA nerve terminal might be responsible for altering tyrosine hydroxylase activity *in vivo*. It would be predicted that a block in impulse flow would prevent depolarization of the DA terminals and the normal influx of calcium. This lack of calcium mobilization could result in an allosteric change in striatal tyrosine hydroxylase, thus increasing its affinity for cofactor and for tyrosine and reducing the inhibitory effects of endogenous DA.

According to the above hypothesis, a cessation of impulse flow leads to a diminished influx of calcium, and the diminished concentration of intercellular calcium causes a change in the kinetics of striatal tyrosine hydroxylase, producing an enzyme that is less susceptible to end-product inhibition by DA.

We tested this hypothesis pharmacologically in the following manner. Rats were pretreated with GBL (750 mg/kg) and killed 30 or 90 min later;

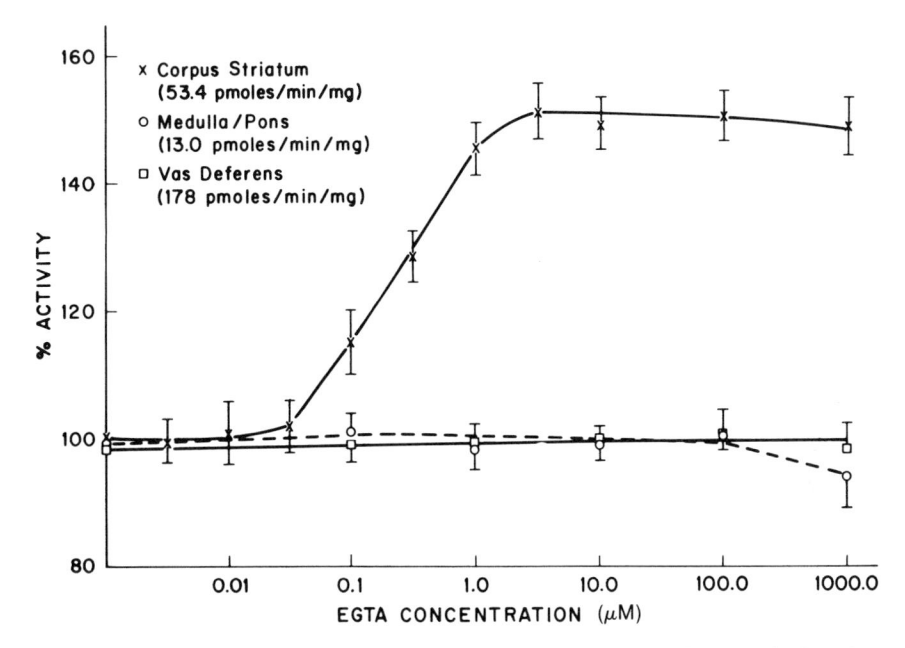

FIG. 2. Effect of calcium removal on tyrosine hydroxylase activity. Tyrosine hydroxylase activity was determined at a tyrosine concentration of 5×10^{-5} M according to a modification of the method of Shiman et al. (1971). Results are expressed as percent of control activity ± SEM. Control values for each tissue are indicated above in terms of pmoles of DOPA/min/mg of protein. The stimulatory effect of EGTA (10^{-6} M) was completely reversed by addition of calcium (10^{-4} M).

the striatum was removed and homogenized in tris-acetate buffer and the kinetics of tyrosine hydroxylase were examined in the supernatant. Pretreatment of rats with GBL caused a significant increase in tyrosine hydroxylase activity and a significant change in the kinetic properties of this enzyme, converting it to a form which was very similar to the EGTA-treated enzyme (Table 3). The K_m for tyrosine and cofactor was decreased, and the K_i for DA significantly increased (Table 3). GHB added directly to the incubation medium in concentrations as high as 10^{-3} M was without effect on tyrosine hydroxylase. These results are consistent with our previous observation that a cessation of impulse flow induced pharmacologically or mechanically in the nigro-neostriatal pathway causes a marked increase in the steady state levels of DA which does not ultimately lead to an inhibition of tyrosine hydroxylase below control levels (Roth et al., 1973; Walters and Roth, 1974). On the other hand, if endogenous levels of DA are increased by pretreatment with a MAOI such as pargyline, a marked inhibition of tyrosine hydroxylase is produced. If a MAOI and GBL are administered to the same animal, resulting in an even larger increase in endogenous DA, less inhibition of tyrosine hydroxylase is observed

TABLE 3

	n	Tyrosine hydroxylase activity (pmoles DOPA/mg/min)	K_m, tyrosine (μM)	K_i, dopamine (mM)
Control	18	23.1 ± 2.5	53.9 ± 5.1	0.11 ± 0.006
EGTA (10⁻⁶M)	10	78.9 ± 7.3[b]	9.0 ± 0.6[c]	74.0 ± 8.0[c]
Gamma-hydroxybutyrate (10⁻³ M)	6	24.6 ± 3.7	54.4 ± 3.8	0.12 ± 0.06
Gamma-hydroxybutyrate (30 min pretreatment)[a]	6	64.9 ± 8.5[b]	18.4 ± 0.3[c]	10.1 ± 0.8[c]
Gamma-hydroxybutyrate (90 min pretreatment)[a]	6	80.5 ± 5.0[b]	8.9 ± 0.6[c]	59.8 ± 8.8[c]
Gamma-hydroxybutyrate[a] (90 min) + apomorphine (40 min)	6	28.7 ± 6.1	50.5 ± 4.3	0.12 ± 0.002
Apomorphine (10⁻⁴ M)	6	31.6 ± 3.1	—	—

[a] Gamma-hydroxybutyrate was administered in the lactone form (750 mg/kg, i.p.) 30 or 90 min prior to sacrifice. Apomorphine (2 mg/kg) was administered 50 min after gamma-hydroxybutyrate and 40 min prior to sacrifice. Results are expressed as the mean ± SEM.

Each kinetic parameter (K_m and K_i) was determined according to the method of Lineweaver-Burk (1934) and Wilkinson (1960). K_m was determined at six different tyrosine concentrations. The K_i was determined at two different dopamine concentrations and over a range of six DMPH₄ concentrations. Protein was determined by the method of Lowry et al. (1951). Tyrosine hydroxylase activity was determined in the 100,000 × g supernatant at a tyrosine concentration of 10⁻⁵ M according to a modification of the method of Shiman et al. (1971). Results are expressed as pmoles of DOPA/mg of protein/min ± SEM.

[b] Significantly different from control, $p < 0.01$.
[c] Significantly different from control, $p < 0.001$.
Data taken in part from Morgenroth and Roth *(manuscript in preparation)*.

than with the MAOI administered alone. Again, this finding could be explained by the assumption that inhibition of impulse flow produced by GBL converts tyrosine hydroxylase to a form insensitive to end-product inhibition by DA (Table 4).

Finally, the speculation that presynaptic DA receptors may regulate tyrosine hydroxylase activity by altering permeability of the presynaptic membrane to calcium receives some indirect support from the observation that stimulation of DA receptors by low doses of apomorphine (2 mg/kg) causes about a 35% inhibition of tyrosine hydroxylase activity *in vivo* (Table 4). However, if endogenous levels of DA are increased to a new steady state by inhibition of impulse flow following administration of GBL (a condition which changes tyrosine hydroxylase to a form which is insensitive to inhibition by endogenous DA), administration of apomorphine to these treated rats now causes an even more potent inhibition of synthesis (85%), perhaps by converting the tyrosine hydroxylase back to a form which is sensitive to the elevated levels of DA (Table 4). In fact, this appears to be the case. If rats are pretreated with GBL (750 mg/kg) and 50 min later administered a DA receptor stimulant such as apomorphine

TABLE 4. *Effect of pargyline and apomorphine on DA and DOPA accumulation in the neostriatum after GBL administration*

Treatment	N	DA (%)	N	DOPA (%)
Control	14	100 ± 4	19	100 ± 4
Pargyline, 3 hr	4	133 ± 8[a]	10	39 ± 6[a]
Apomorphine, 40 min	9	110 ± 5	5	67 ± 9[b]
GBL, 90 min	4	186 ± 13[a]	12	123 ± 10[c]
Pargyline, 3 hr + GBL, 90 min	5	250 ± 10[a]	6	97 ± 6
GBL, 90 min + Apomorphine, 40 min	6	174 ± 16[a]	6	30 ± 3[a]

All DOPA determinations were performed on animals treated with RO4–4602 (800 mg/kg, i.p.) 30 min before sacrifice. DA determinations were performed on a separate group of animals which did not receive RO4–4602. Pargyline (75 mg/kg, i.p.), apomorphine (2 mg/kg, i.p.), and GBL (750 mg/kg, i.p.) were administered before sacrifice at the times indicated. DOPA was measured by the method of Kehr et al. (1972a), as described in Walters and Roth (1974). DA determinations were performed by the method of Laverty and Taylor (1968), as modified by Walters and Roth (1973). N equals the number of experiments.

The data is represented as percent of control ± SEM. Control values for neostriatal DOPA were 795 ± 34 ng/g and for DA, 9.42 ± 0.37 µg/g.

[a] Significantly different from control, $p < 0.001$.
[b] Significantly different from control, $p < 0.005$.
[c] Significantly different from control, $p < 0.05$.
Data taken in part from Walters and Roth (1974).

(2 mg/kg), the tyrosine hydroxylase isolated from the striatum of these rats has kinetic properties almost identical to striatal tyrosine hydroxylase obtained from untreated rats (Table 3). Thus, *in vivo* administration of apomorphine to GBL-treated rats causes a change in the striatal tyrosine hydroxylase from a form which has an increased affinity for substrate and cofactor and a decreased affinity for DA back to a form which has a reduced affinity for substrate and cofactor and an increased affinity for DA. Apomorphine (10^{-4} M) added directly to the incubation medium has no significant effect on tyrosine hydroxylase activity.

Figure 3 illustrates a schematic model of a DA nerve terminal, depicting the possible mechanisms by which alterations in impulse flow and changes in receptor activity might regulate tyrosine hydroxylase activity.

During periods of normal impulse flow, tyrosine hydroxylase is susceptible to end-product inhibition by DA, and a small change in a "strategic pool" of DA could conceivably influence tyrosine hydroxylase activity. This critical pool of DA is probably controlled by a balance between release, reuptake, and synthesis. When increased demands are placed on the neuron and larger amounts of transmitter are utilized, the levels of this

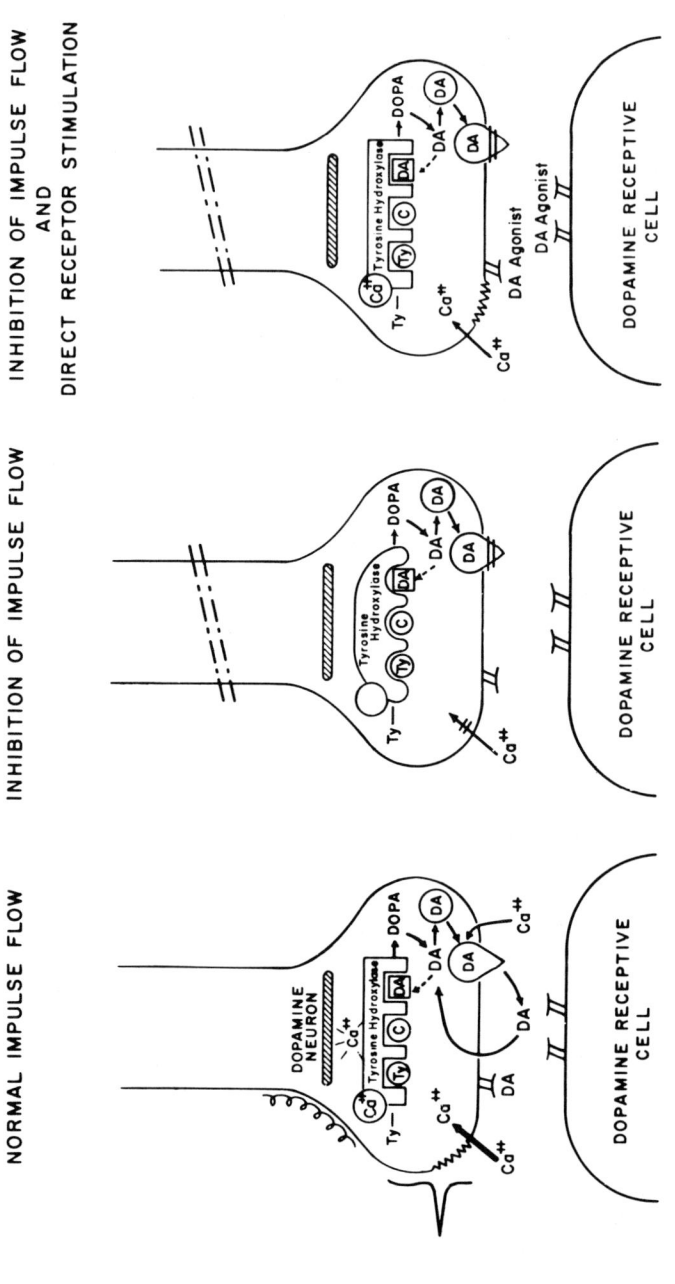

FIG. 3. Schematic model of a dopaminergic nerve terminal illustrating possible mechanisms by which alterations in impulse flow and changes in receptor activity might regulate tyrosine hydroxylase.

"strategic pool" of DA fall and tyrosine hydroxylase is activated. It is also possible that changes in the concentration of intracellular calcium which occur during depolarization may cause an increase in the amount of membrane-bound tyrosine hydroxylase, which has been reported to be a more active form of the enzyme (Kuczenski and Mandell, 1972).

When impulse flow is inhibited in dopaminergic neurons, the nerve terminals are no longer depolarized, calcium influx and DA release are retarded, and the concentration of endogenous DA in the terminal rapidly increases. Due to the lack of calcium, the conformation of tyrosine hydroxylase changes, causing the enzyme to have an increased affinity for tyrosine and synthetic cofactor and a decreased affinity for DA. Thus tyrosine hydroxylase activity is increased and the enzyme has a marked reduction in sensitivity to inhibition by endogenous DA.

Administration of a DA receptor stimulant reverses the increase in tyrosine hydroxylase observed after inhibition of impulse flow, perhaps by interacting with presynaptic dopamine receptors and altering the permeability of the presynaptic membrane to calcium. This partial restoration of intercellular calcium causes tyrosine hydroxylase to revert to its original conformation, in which the enzyme has a reduced affinity for tyrosine and cofactor but an increased affinity for DA. The intercellular DA can now effectively inhibit the tyrosine hydroxylase.

The suggested allosteric nature of the activation of striatal tyrosine hydroxylase observed upon removal of calcium and mediated by an increase in the affinity of the enzyme for substrate (tyrosine) and cofactor ($DMPH_4$) and a decrease in affinity for the end-product inhibitor DA emphasizes the potentially key role calcium fluxes may play in the regulation of tyrosine hydroxylase in the striatum. Recent experiments in both peripheral and central noradrenergic neurons have suggested that here also calcium may play a key role in the control of tyrosine hydroxylase activity (Morgenroth, Boadle-Biber, and Roth, 1974).

CONCLUSIONS

An increase in impulse flow in DA neurons, as in other monoamine-containing neurons, results in an increase in synthesis, catabolism, and turnover of DA. An interruption of impulse flow, on the other hand, results in a decrease in release and catabolism of DA and a paradoxical increase in synthesis, which occurs via an activation of tyrosine hydroxylase. These changes are reflected by a dramatic increase in the steady state levels of DA in the nerve terminals. Once this new steady state level of DA is attained, the rate of synthesis returns to normal but is not inhibited as it is when DA levels are increased by inhibition of monoamine oxidase. Drugs which are capable of stimulating DA receptors block the increase in tyrosine hydroxylase activity and accumulation of DA normally observed after inhibition of impulse flow. DA receptor blocking drugs nullify the ability of DA receptor stimulators to block the increase in tyrosine hydroxylase activity, suggesting that DA synthesis may in part be controlled by alterations in receptor activity. *In vitro* studies on striatal tyrosine hydroxylase indicate that removal of calcium by treatment of the enzyme with EGTA activates the enzyme by increasing its affinity for substrate and cofactor and decreasing its affinity for the end product, DA. Inhibition of impulse flow in the nigro-neostriatal pathway causes striatal tyrosine hydroxylase to behave kinetically like the striatal enzyme treated with EGTA. It is postulated that the increase in tyrosine hydroxylase activity which occurs in the neostriatum during a cessation of impulse flow occurs as a result of a diminished influx of calcium, ultimately resulting in an allosteric activation of tyrosine hydroxylase mediated by an increase in the affinity of the enzyme for substrate and cofactor and a decreased affinity for DA.

ACKNOWLEDGMENTS

This research was supported by National Institute of Mental Health grant MH-14092 and the State of Connecticut.

REFERENCES

Aghajanian, G. K., and Bunney, B. S. (1973a): Central dopaminergic neurons: Neurophysiological identification and responses to drugs. In: *Frontiers in Catecholamine Research*, edited by S. H. Snyder and E. Usdin. Pergamon Press, Elmsford, N.Y.

Aghajanian, G. K., and Bunney, B. S. (1973b): Pre- and postsynaptic feedback mechanisms in central dopaminergic neurons. In: *Neurotransmitters and Brain Function*, edited by P. Seeman. University of Toronto Press, Toronto (in press).

Aghajanian, G. K., and Roth, R. H. (1970): γ-Hydroxybutyrate-induced increase in brain dopamine: Localization by fluorescence microscopy. *Journal of Pharmacology and Experimental Therapeutics*, 175:131–137.

Andén, N. -E., Bedard, P., Fuxe, K., and Ungerstedt, U. (1972): Early and selective increase in brain dopamine levels after axotomy. *Experientia*, 28:300–302.

Andén, N. -E., Corrodi, H., Fuxe, K., and Ungerstedt, U. (1971): Importance of nervous impulse flow for the neuroleptic induced increase in amine turnover in central dopamine neurons. *European Journal of Pharmacology*, 15:193–199.

Andén, N. -E., Rubenson, A., Fuxe, K., and Hökfelt, T. (1967): Evidence for dopamine receptor stimulation by apomorphine. *Journal of Pharmacy and Pharmacology*, 19:627–629.

Arbuthnott, G. W., Crow, T. J., Fuxe, K., Olson, L., and Ungerstedt, U. (1970): Depletion of catecholamines *in vivo* induced by electrical stimulation of central monoamine pathways. *Brain Research*, 24:471–483.

Bunney, B. S., Walters, J. R., Roth, R. H., and Aghajanian, G. K. (1973): Dopaminergic neurons: Effect of antipsychotic drugs and amphetamine on single cell activity. *Journal of Pharmacology and Experimental Therapeutics*, 185:560–571.

Carlsson, A., Kehr, W., Lindqvist, M., Magnusson, T., and Atack, C. F. (1972): Regulation of monoamine metabolism in the central nervous system. *Pharmacological Reviews*, 24:371–384.

Corrodi, H., Fuxe, K., and Hökfelt, T. (1967): The effect of neuroleptics on the activity of central catecholamine neurons. *Life Sciences*, 6:767–774.

Corrodi, H., Fuxe, K., and Ungerstedt, U. (1971): Evidence for a new type of dopamine receptor stimulating agent. *Journal of Pharmacy and Pharmacology*, 23:989–991.

Coyle, J. T. (1972): Tyrosine hydroxylase in rat brain—Cofactor requirements, regional and subcellular distribution. *Biochemical Pharmacology*, 21:1935–1944.

Enero, M. A., Langer, S. Z., Rothlin, R. P., and Stefano, F. J. E. (1972): Role of the α-adrenoceptor in regulating noradrenaline overflow by nerve stimulation. *British Journal of Pharmacology*, 44:672–688.

Goldstein, M., Backstrom, T., Ohi, Y., and Frankel, R. (1970): The effects of Ca^{++} ions on the C^{14}-catecholamine biosynthesis from C^{14} tyrosine in slices from the striatum of rats. *Life Sciences*, 9:919–924.

Harris, J. E., and Roth, R. H. (1970): The effect of potassium on catecholamine biosynthesis and release in rat brain cortical slices. *Federation Proceedings*, 29:941.

Herr, B. E., and Roth, R. H. (1972): Regulation of serotonin biosynthesis in rat hippocampus. *Proceedings of the Fifth International Congress on Pharmacology*, Abs. #600, p. 100.

Kehr, W., Carlsson, A., and Lindqvist, M. (1972a): A method for the determination of 3,4-dihydroxyphenylalanine (DOPA) in brain. *Naunyn Schmiedeberg's Archives of Pharmacology*, 274:273–280.

Kehr, W., Carlsson, A., Lindqvist, M., Magnusson, T., and Atack, C. V. (1972b): Evidence for a receptor-mediated feedback control of striatal tyrosine hydroxylase activity. *Journal of Pharmacy and Pharmacology*, 24:744–747.

Kirpekar, S. M., and Puig, M. (1971): Effect of flow-stop of noradrenaline release from normal spleens and spleens treated with cocaine, phentolamine and phenoxybenzamine. *British Journal of Pharmacology*, 43:359–369.

Korf, J., Aghajanian, G. K., and Roth, R. H. (1973a): Stimulation and destruction of the locus coeruleus: Opposite effects on 3-methoxy-4-hydroxyphenylglycol sulfate levels in the rat cerebral cortex. *European Journal of Pharmacology*, 21:305–310.

Korf, J., Roth, R. H., and Aghajanian, G. K. (1973b): Alterations in turnover and endogenous levels of norepinephrine in cerebral cortex following electrical stimulation and acute axotomy of cerebral noradrenergic pathways. *European Journal of Pharmacology*, 23:276–282.

Kuczenski, R., and Mandell, A. (1972): Regulatory properties of soluble and particulate brain tyrosine hydroxylase. *Journal of Biological Chemistry*, 247:3114–3122.

Laverty, R., and Taylor, K. M. (1968): The fluorometric assay of catecholamines and related compounds: Improvements and extensions to the hydroxyindole technique. *Analytical Biochemistry*, 22:269–279.

Lineweaver, H., and Burk, D. (1934): The determination of enzyme dissociation constants. *Journal of the American Chemical Society*, 56:658–666.

Lowry, O. H., Rosebrough, W. J., Farr, A. L., and Randall, R. J. (1951): Protein measurement with the folin phenol reagent. *Journal of Biological Chemistry*, 193:265–275.

Morgenroth, V. H., III, Boadle-Biber, M. C., and Roth, R. H. (1974): Allosteric activation of tyrosine hydroxylase by calcium. *Federation Proceedings*, 33:535.

Murrin, L. C., and Roth, R. H. (1973): Dopaminergic neurons: Effects of stimulation on dopamine metabolism. *Pharmacologist*, 15:514.

Nybäck, H. (1972): Effect of brain lesions and chlorpromazine on accumulation and disappearance of catecholamines formed *in vivo* from ^{14}C-tyrosine. *Acta Physiologica Scandinavica*, 84:54–64.

Roth, R. H., Walters, J. R., and Aghajanian, G. K. (1973): Effect of impulse flow on the release and synthesis of dopamine in the rat striatum. In: *Frontiers in Catecholamine Research*, edited by S. H. Snyder and E. Usdin. Pergamon Press, Elmsford, N.Y.

Sedvall, G. C. (1969): Effect of nerve stimulation on accumulation and disappearance of catecholamines formed from radioactive precursors *in vivo*. In: *Metabolism of Amines in the Brain*, edited by G. Hooper. Macmillan, New York, pp. 23–28.

Sheard, M. H., and Aghajanian, G. K. (1968): Stimulation of the midbrain raphe: Effect on serotonin metabolism. *Journal of Pharmacology and Experimental Therapeutics*, 163:425–430.

Shields, P. J., and Eccleston, D. (1972): Effects of electrical stimulation of rat midbrain on 5-hydroxytryptamine synthesis as determined by a sensitive radioactive method. *Journal of Neurochemistry*, 19:265–272.

Shiman, R., Akino, M., and Kaufman, S. (1971): Solubilization and partial purification of tyrosine hydroxylase from bovine adrenal medulla. *Journal of Biological Chemistry*, 246:1330–1340.

Starke, K. (1971): Influence of α-receptor stimulants on noradrenaline release. *Naturwissenschaften*, 58:420.

Starke, K. (1972): Influence of extracellular noradrenaline on stimulation-evoked secretion of noradrenaline from sympathetic nerves: Evidence for an α-receptor-mediated feedback inhibition of noradrenaline release. *Naunyn-Schmiedeberg's Archives of Pharmacology*, 275:11–23.

Stock, G., Magnusson, T., and Andén, N. -E. (1973): Increase in brain dopamine after axotomy or treatment with gammahydroxybutyric acid due to elimination of the nerve impulse with gammahydroxybutyric acid. *Naunyn-Schmiedeberg's Archives of Pharmacology*, 278:347–362.

Vogt, M. (1969): Release from brain tissue of compounds with possible transmitter function: Interaction of drugs with these substances. *British Journal of Pharmacology*, 37:325–337.

Von Voigtlander, P. F., and Moore, K. E. (1971): The release of H^3-dopamine from cat brain following electrical stimulation of the substantia nigra and caudate nucleus. *Neuropharmacology*, 10:733–741.

Walters, J. R., Aghajanian, G. K., and Roth, R. H. (1972): Dopaminergic neurons: Inhibition of firing by γ-hydroxybutyrate. *Proceedings of the Fifth International Congress on Pharmacology*, Abs. #1472, p. 246.

Walters, D. S., and Eccleston, D. (1973): Increase of noradrenaline metabolism following electrical stimulation of the locus coeruleus in the rat. *Journal of Neurochemistry*, 21:281–289.

Walters, J. R., and Roth, R. H. (1972): Effect of gamma-hydroxybutyrate on dopamine and dopamine metabolites in the rat striatum. *Biochemical Pharmacology*, 21:2111–2121.

Walters, J. R., and Roth, R. H. (1974): Dopaminergic neurons: Drug-induced antagonism of the increase in tyrosine hydroxylase activity produced by cessation of impulse flow. *Journal of Pharmacology and Experimental Therapeutics (in press)*.

Walters, J. R., Roth, R. H., and Aghajanian, G. K. (1973): Dopaminergic neurons: Similar biochemical and histochemical effects of gamma-hydroxybutyrate and acute lesions of the nigro-neostriatal Pathway. *Journal of Pharmacology and Experimental Therapeutics*, 186:630–639.

Weiner, N. (1970): Regulation of norepinephrine biosynthesis. *Annual Review of Pharmacology*, 10:273–290.

Wilkinson, G. N. (1961): Statistical estimations in enzyme kinetics. *Biochemical Journal*, 80:324–332.

Neuropsychopharmacology of Monoamines and Their Regulatory Enzymes, edited by E. Usdin. Raven Press, New York © 1974.

Ingestive Behavior Following Damage to Central Dopamine Neurons: Implications for Homeostasis and Recovery of Function

Michael J. Zigmond and Edward M. Stricker

Psychobiology Program, Departments of Biology and Psychology, University of Pittsburgh, Pittsburgh, Pennsylvania 15260

I. INTRODUCTION

A. Catecholamines and Homeostasis

One of the prominent features of mammalian life is the constancy of the internal environment. This homeostasis is largely achieved by complementary activities of the autonomic nervous system, the pituitary-adrenal axis, and motivated behavior. It has long been appreciated that catecholamines play a critical role in the physiological contributions to this regulation, functioning within the sympathoadrenal system both as neurotransmitters and as circulatory-borne hormones. We now present evidence for the possible contribution of central catecholamines to motivated behaviors, such as feeding and drinking, that subserve homeostasis.

Like peripheral adrenergic fibers, central catecholamine-containing neural systems are organized into several discrete groups of cell bodies that send axons diffusely throughout wide areas (Ungerstedt, 1971a). In addition, individual fibers are predominantly thin and unmyelinated, and appear to conduct at low velocity and frequency, while postsynaptic effects of nerve stimulation usually far outlast the initial stimulus (e.g., Hillarp, Fuxe, and Dahlström, 1966; Hoffer, Siggins, Oliver, and Bloom, 1973). Gradual activation and persistent responses in multiple projection sites are appropriate to the slow changes and long-term needs of regulated bodily functions and may be contrasted with the specific, rapid, and stimulus-bound communications in most somatic sensory and motor systems. These properties suggested to us that, as in the peripheral nervous system, the catecholamines of the central nervous system are involved in homeostasis. In fact, work in other laboratories already has implicated the ventral norepinephrine bundle in the control of anterior pituitary function (Fuxe and Hökfelt, 1969) and the dorsal norepinephrine bundle in the cortical "arousal" associated with the alert state that is a prerequisite for motivated behavior

(Jouvet, 1972). Our experiments have been focusing on the nigrostriatal dopamine bundle and its contributions to motivated ingestive behaviors.

B. Recovery of Function

The capacity of the brain to recover from neurological damage has both fascinated and perplexed investigators for many years. This capacity to recover has often been taken to indicate "equipotentiality" or "plasticity," and it has been suggested that brain functions could be transferred from one neuronal system to another, at least following damage. However, the accumulating evidence that multiple interrelated processes support synaptic regulation within catecholamine-containing neurons suggests to us that there is a tremendous capacity for recovery of function within catecholamine systems following subtotal damage. For example, following extensive damage to the nigrostriatal dopamine bundle, we observe almost a complete loss of motivated behavior followed by a gradual restoration of function. In the present report we will propose a model for such recovery that is derived from neurochemical investigations of brain catecholamine systems and parallel studies of behavior.

Two lesioning procedures have been used in our studies of brain-damaged rats. In one, we administer 6-hydroxydopamine (6-HDA) by way of the lateral cerebral ventricles, in order to produce permanent depletions of norepinephrine and dopamine with little attendant nonspecific tissue damage (Bloom, Algeri, Groppetti, Revuelta, and Costa, 1969; Breese and Traylor, 1970; Uretsky and Iversen, 1970; see Fig. 1).[1] Selective depletion of dopamine can be obtained by pretreating the animals with desmethylimipramine, an inhibitor of norepinephrine uptake (Breese and Traylor, 1971; see Table 1). Our second procedure involves bilateral electrolytic lesions of the ventrolateral hypothalamus (Fig. 2), which interrupt the nigrostriatal dopamine fibers as well as other neuronal pathways including ascending norepinephrine and serotonin projections (Heller and Moore, 1965; Ungerstedt, 1971a; see Table 1).[2] While each of these procedures would be expected to produce a different spectrum of damage, we believe that the results obtained reflect the damage to ascending dopamine neurons that is a common feature of both of them. In addition, similar effects on behavior

---

[1] Intraventricular 6-HDA appears to be taken up specifically by norepinephrine- and dopamine-containing neurons. Metabolic alterations in some noncatecholamine neurons have been observed following 6-HDA injections (Blondaux, Juge, Sordet, Chouvet, Jouvet, and Pujol, 1973; Kim, 1973), but this may represent compensatory adjustments within other systems rather than nonspecific effects of the drug. Gross nonspecific tissue damage also has been reported (Poirier, Langelier, Roberge, Boucher, and Kitsikis, 1970), but our treatments seldom produce such effects.

[2] Lesions were made with monopolar stainless steel electrodes by passing 1 to 2 mA anodal current for 10 to 15 sec. Each rat was placed in a stereotaxic instrument with its skull positioned flat, and the electrode was placed 6.0 mm anterior to the interaural line, 2.0 mm lateral to the sagittal sinus, and 8.0 mm ventral to the dura of the cortex.

BRAIN DOPAMINE AND INGESTIVE BEHAVIORS 387

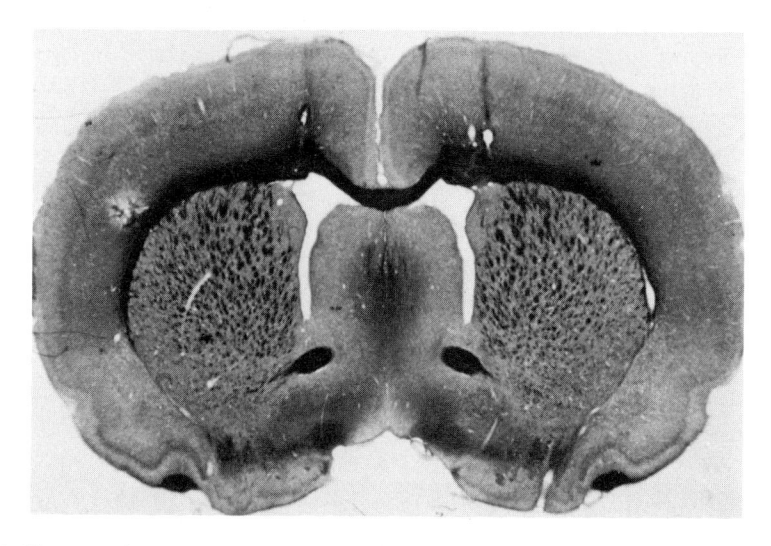

FIG. 1. Absence of nonspecific tissue damage at site of intraventricular injections. Twenty microliters containing 200 μg of 6-HDA was administered to the left lateral ventricle. An equal volume of vehicle was administered to the right ventricle. After 3 days, the rat was perfused with formalin solution and the brain removed and sectioned for histology.

TABLE 1. *Effect of 6-hydroxydopamine treatments or electrolytic lesions on brain catecholamines*

	Control	6-HDA	Pargyline + 6-HDA	DMI, pargyline + 6-HDA	Lateral hypothalamic lesions
Norepinephrine (μg/g)					
Diencephalon	0.76 ± 0.04	0.14 ± 0.02	0.07 ± 0.01	—	—
Brainstem	0.38 ± 0.02	0.13 ± 0.02	0.14 ± 0.01	—	—
Telencephalon	0.28 ± 0.01	0.01 ± 0.01	0	0.25 ± 0.02	0.15 ± 0.04
Dopamine (μg/g)					
Striatum	8.27 ± 0.49	3.44 ± 0.16	0.39 ± 0.13	0.21 ± 0.10	0.38 ± 0.14
Telencephalon	0.80 ± 0.11	0.19 ± 0.03	0.03 ± 0.01	—	—

Drug-tested rats received 6-HDA (200 μg, i.v.), alone or 30 min after pretreatment with pargyline (50 mg/kg, i.p.) or pargyline and desmethylimipramine (DMI) (25 mg/kg, i.p.). Rats given 6-HDA alone or after pargyline pretreatment received two treatments, 48 hr apart; rats given 6-HDA after pargyline and DMI received a single treatment. For lesioning procedure, see text footnote 2. Animals were killed at least 2 weeks after treatment, and tissues were prepared (see Stricker and Zigmond, 1974a) and analyzed spectrophotofluorometrically (Crout, Creveling, and Udenfriend, 1961; Chang, 1964). Values are mean ± standard error of the mean for four to 10 animals, expressed as micrograms of catecholamine per gram of fresh brain weight. There was no significant effect on catecholamine concentrations of any vehicle injections or sham lesions, and control values have been pooled. Zero values indicate an amount below the sensitivity of the assay (approximately 50 ng for dopamine and 20 ng for norepinephrine). Missing values (−) indicate no data available.

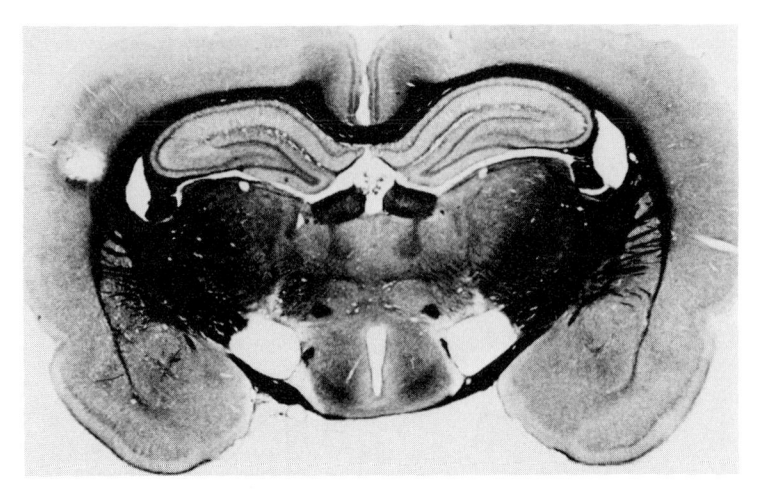

FIG. 2. Photomicrograph of a stained frontal section showing a representative lesion in the ventrolateral hypothalamus. Such large symmetrical lesions produced prolonged aphagia and adipsia.

are observed when the nigrostriatal bundle is interrupted by other procedures (e.g., Albert, Storlien, Wood, and Ehman, 1970; Breese, Smith, Cooper, and Grant, 1973; Fibiger, Zis, and McGeer, 1973; Marshall and Teitelbaum, 1973).

II. SOME EFFECTS OF DAMAGE TO CENTRAL CATECHOLAMINE SYSTEMS

Like many other investigators, we have found that under normal laboratory conditions the general appearance and behavior of rats given intraventricular 6-HDA (2×200 μg) is largely indistinguishable from that of control rats given the vehicle solution, despite the large depletion of central catecholamines (Table 1). These findings are reminiscent of observations that removal of the adrenergic fibers of the sympathetic nervous system, or adrenal demedullation, has only minor effects on animals maintained in a neutral environment. However, the peripherally sympathectomized animals fail to make the appropriate compensatory adjustments to maintain homeostasis when subjected to such stresses as hemorrhage, insulin hypoglycemia, or cold, and often die. In order to determine whether brain-damaged animals had parallel deficits, we examined their capacity to respond behaviorally during more severe regulatory imbalances.

Acute cellular glucopenia. 2-Deoxy-D-glucose produces a "functional hypoglycemia" by inhibiting the cellular uptake and metabolism of glucose. Unlike insulin, it does not compromise various compensatory adjustments, and animals can survive even if they do not increase food intakes. Intact rats show a variety of well-known responses to 2-deoxy-D-glucose, including a rapid elevation of blood sugar, due to increased glycogenolysis resulting from the hypersecretion of epinephrine, increased secretion of gastric acid in the stomach, and increased food intake. In contrast, we find that the brain damage caused by intraventricular 6-HDA treatment disrupts the animals' behavioral responses to 2-deoxy-D-glucose injections, yet leaves the autonomic responses unimpaired (Table 2; Stricker and Zigmond, 1974a; Stricker, Zigmond, Friedman, and Redgate, 1974).

TABLE 2. Effect of 2-deoxy-D-glucose on food intake, blood sugar concentration, and gastric acid secretion of rats with central catecholamine depletions

Group	n	Food intake (g)	n	Blood sugar (mg%)	n	Gastric acid (μEq H$^+$)
Controls	48	6.6 ± 0.3[a]	4	233.0 ± 11.0[a]	10	324.1 ± 18.5[a]
6-HDA	45	2.0 ± 0.2	5	265.6 ± 18.0[a]	6	363.2 ± 34.2[a]
6-HDA, pargyline	22	1.2 ± 0.2	5	272.0 ± 15.8[a]	10	313.0 ± 24.6[a]
6-HDA, pargyline, DMI	9	2.2 ± 0.3	0	—	0	—
LH lesions	7	1.2 ± 0.3	4	298.4 ± 9.6[a]	4	300.5 ± 26.5[a]

All lesioned rats were maintaining themselves on Purina chow and water for at least 2 weeks prior to testing. Rats were given 750 mg/kg of 2-deoxy-D-glucose (i.p.). In one group of animals, the intake of Purina chow was monitored for 6 hr after treatment. In another group, blood was withdrawn 1 to 2 hr after injections (food was not available) for colorimetric determinations of glucose following its reaction with glucose oxidase. In a third group (also food-deprived), gastric secretions were collected from a chronic fistula for 4 hr, their acid contents were determined by titration with 0.01 N NaOH to an end point of pH 7.00, and total acid output was calculated as the accumulated products of titratable acidity and sample volume. The range of nonstimulated control values for all groups was 1.0 to 2.5 g of food, 100 to 120 mg of glucose/100 ml, and 20 to 70 μEq H$^+$. Values represent mean ± standard error of the mean.

[a] $p < 0.001$ in comparison with unstimulated control values. See Table 1 for central catecholamine depletions of comparably treated rats.

Acute cold stress. Exposure to ambient temperatures well below the thermoneutral range of rats elicits a variety of physiological and behavioral responses which help to maintain body temperature by minimizing heat loss or promoting heat gain. Shaving the animals removes important insulation and thereby raises their thermoneutral range, thus making the cold stress more severe. Shaved control rats are able to maintain their body temperatures when exposed continuously to an ambient temperature of 5°C and increase their food intakes significantly. Shaved 6-HDA-treated rats also show good thermoregulation, but do not increase their food intakes during the first 48-hr period of exposure (Stricker et al., 1974).

Acute hypovolemia. Abrupt reduction of circulatory volume can be produced by subcutaneous injection of hyperoncotic colloidal solutions. This procedure reduces plasma volume by gradually withdrawing increasing amounts of isosmotic protein-free plasma fluid into the local interstitium.

Unlike hemorrhage, these losses are not abrupt, anemia is avoided, and compensatory repletion from interstitial fluid reservoirs cannot occur. Control rats show immediate increases in thirst after the loss of plasma fluid in conjunction with decreases in urine volume and sodium concentration. Rats given 6-HDA show similar physiological responses to hypovolemia but have a significant impairment of thirst (Table 3; Stricker, 1973; Stricker and Zigmond, 1974a, b).

TABLE 3. Effect of 20% polyethylene glycol solution on water intakes of rats with central catecholamine depletions

Group	n	Water intake (ml)
Controls	16	5.1 ± 0.5
6-HDA	16	3.5 ± 0.5[a]
6-HDA, pargyline	17	1.5 ± 0.4[a]
6-HDA, pargyline, DMI	9	2.2 ± 0.6[a]
LH lesions	9	0.2 ± 0.1[a]

All lesioned rats were maintaining themselves on Purina chow and water for at least 2 weeks prior to testing. Rats were given 5 ml of 20% polyethylene glycol solution subcutaneously (in the middle of the back) and water intakes were monitored for 4 hr. Food was not available during the test.

[a] $p < 0.01$ in comparison with water intakes of control rats. Sham-injected animals drank 0 to 0.5 ml during a comparable 4-hr test. See Table 1 for central catecholamine depletions of comparably treated rats.

To summarize, 6-HDA-treated rats which had appeared normal fail to behave appropriately during the acute homeostatic imbalances produced by various treatments. In each case, the physiological responses appear intact but the behavioral responses are impaired. We also have studied rats with small lateral hypothalamic lesions and observe similar deficits. Collectively, these results suggest that depleting brain catecholamines impairs the expression of motivated ingestive behaviors when homeostatic imbalances are abrupt and severe.

These effects on ingestive behaviors in 6-HDA-treated rats were associated with almost complete destruction of norepinephrine terminals in the telencephalon but more modest depletion of striatal dopamine. Previous reports have noted prominent behavioral deficits following more complete damage to the nigrostriatal dopamine system (e.g., Anand and Brobeck, 1951; Ungerstedt, 1971c; Oltmans and Harvey, 1972). In order to obtain additional insights into the contribution of brain catecholamines to ingestive behavior, we examined animals in which dopamine depletions were more complete and specific.

In contrast to the effects of 6-HDA when dopamine depletions are more modest, considerable disruption of behavior is apparent even during routine handling of 6-HDA-treated rats when the loss of telencephalic dopamine is almost complete (see Table 2). The animals are extremely hypokinetic and poorly coordinated, have a rigidity of their hindlimbs, and fail to respond normally to sensory stimuli. Most dramatically, they refuse to ingest food pellets or drink water and die unless fed by intragastric intubation (Zigmond and Stricker, 1972, 1973; Stricker and Zigmond, 1974a). These deficits closely resemble those obtained when nigrostriatal damage is produced by lateral hypothalamic lesions (see also Balagura, Wilcox, and Coscina, 1969; Marshall, Turner, and Teitelbaum, 1971).

Many 6-HDA-treated rats resume feeding and drinking behaviors that are adequate for maintenance of body weight (Zigmond and Stricker, 1973; Stricker and Zigmond, 1974a). During the gradual recovery from aphagia and adipsia, rats invariably eat palatable foods and fluids first (such as Pablum and sucrose solution), then accept dry chow, and finally drink water (Fig. 3.). Nevertheless, like animals with smaller depletions of striatal dopamine, these rats do not increase ingestive behaviors during acute glucoprivation, hypovolemia, or (if shaved) exposure to cold (Tables 2 and 3; Zigmond and Stricker, 1972; Stricker and Zigmond, 1974a; Stricker et al., 1974). Both the characteristic pattern of recovery and the residual deficits also are observed following lateral hypothalamic lesions (Tables 2 and 3; Stricker et al., 1974; see also Teitelbaum and Epstein, 1962; Epstein and Teitelbaum, 1967; Stricker and Wolf, 1967).

These results suggest that damage to dopamine-containing neurons is responsible for the observed impairments in motivated ingestive behaviors, since they can be produced in 6-HDA-treated rats when norepinephrine-containing fibers are spared. Furthermore, it would appear that damage to the nigrostriatal dopamine system is responsible for the behavioral deficits, since destruction of the mesolimbic dopamine projections does not produce aphagia and adipsia (Ungerstedt, 1971c). Apparently, the loss of striatal dopamine terminals proportionately disrupts the behavioral response of animals to regulatory imbalances, so that with extreme depletions animals are unable even to survive during the normal privations that occur under neutral laboratory conditions.

Despite the permanent loss of brain catecholamines, there is an impressive recovery from the initial gross incapacitation toward more normal behavior. Three explanations may be suggested for this apparent paradox: (1) the initial aphagia may be due to catecholamine depletions, while resumption of feeding may result from functional recovery of this same system that is not detectable by conventional measurements of amine concentration; (2) the initial aphagia may be due to catecholamine depletions, but resump-

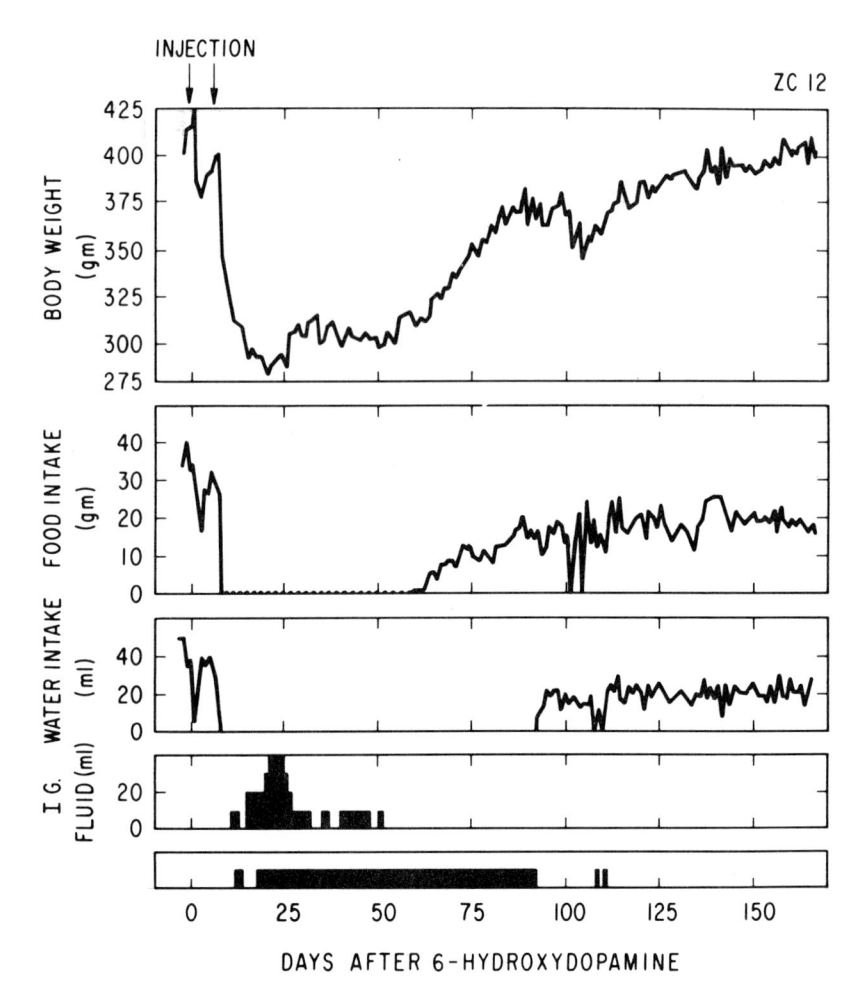

FIG. 3. Recovery of feeding and drinking behaviors in a rat given intraventricular 6-HDA (2 × 200 µg) after pretreatment with pargyline (50 mg/kg, i.p.). Intragastric (IG) feedings and the ingestion of special diets (bottom line) are shown. Although all rats progressed through the same stages of recovery, consideration of this animal (ZC 12), who recovered more slowly than most, permits a more detailed examination of the recovery sequence. At first, this animal ate nothing and had to be maintained by daily intragastric intubations of liquid diet. After 4 days, the animal passed into a second stage in which it would ingest the palatable foods. Gradually, larger amounts of these special foods were consumed and eventually tube feeding was no longer required for body-weight maintenance. After 58 days, the animal entered a third stage in which it would eat dry chow, but only if hydrated. Although it still would not drink water, the rat would accept 5 to 10% sucrose solutions and thereby maintain body fluid hydration. Finally, 92 days after the second 6-HDA treatment, the animal entered a fourth stage in which it maintained body weight on dry chow and tap water, although at a level that was considerably below that of control rats.

Reprinted from Zigmond and Stricker, 1973, with permission.

tion of feeding may result from transfer of the function formerly served by these monoamines to another neurochemical system that is not affected by the lesions: (3) the initial aphagia may not be causally related to observed catecholamine depletions but to another, as yet unknown, effect of the lesions that is more temporary.

Several observations support the first hypothesis. For example, animals with almost complete loss of striatal dopamine that had recovered from the initial effects of 6-HDA treatment or electrolytic lateral hypothalamic lesions become permanently aphagic and adipsic after the intraventricular injection of 200 μg of 6-HDA. In addition, animals recovered from either 6-HDA or electrolytic lesions become aphagic and adipsic for 1 to 2 days after intraperitoneal injections of D,L-α-methyl-p-tyrosine in a dose which inhibits catecholamine synthesis (3 × 75 mg/kg, given every 8 hr) (Zigmond and Stricker, 1973).

If residual catecholamine-containing fibers remain important in the mediation of ingestive behaviors, two basic questions can be raised regarding the phenomena that we have described. First, what are the neurochemical mechanisms by which recovery of function proceeds in the damaged catecholamine system? And second, what are the bases for the deficits in ingestive behaviors that remain after apparent recovery?

III. A POSSIBLE EXPLANATION FOR THE IMPAIRED FUNCTION FOLLOWING DAMAGE TO THE NIGROSTRIATAL DOPAMINE NEURONS

The activity of catecholamines is regulated at the synaptic level by several mechanisms. Although much of the evidence for regulation has been obtained in studies involving increased neural activity, there are reports that adaptive changes also occur in response to decreases in the effective concentration of catecholamine in the synapse. For example, there are compensatory increases in release and synthesis of amine in the central nervous system following the administration of receptor antagonists (Nybäck and Sedvall, 1968), and further increases in biosynthetic capacity through enzyme induction would be expected when elevations in turnover are prolonged (Dairman and Udenfriend, 1970). Comparable changes also are observed when receptor activity is reduced indirectly by chronic reserpine treatment (Rech, Carr, and Moore, 1968; Segal, Sullivan, Kuczenski, and Mandell, 1971). In these animals, as well as in rats given chronic α-methyltyrosine treatment, increased sensitivity of the postsynaptic membrane appears to provide an additional (but less rapidly developing) response to decreases in receptor stimulation, and contributes to the observed recovery of function despite continued catecholamine depletions (Geyer and Segal, 1973).

There is also decreased activity in postsynaptic receptors following subtotal damage to catecholamine-containing fibers, due to a reduction in the number of neurons that are releasing the amine. Assuming that receptors are activated by catecholamine from many adjacent fibers, and that the communication from individual neurons is equivalent, it is possible that the above alterations of turnover, synthesis, and receptor activity also provide the basis for recovery of function within damaged catecholamine systems. The proposed sequence of events is summarized in Fig. 4.[3]

Several lines of evidence support this concept of recovery. First, 6-HDA-

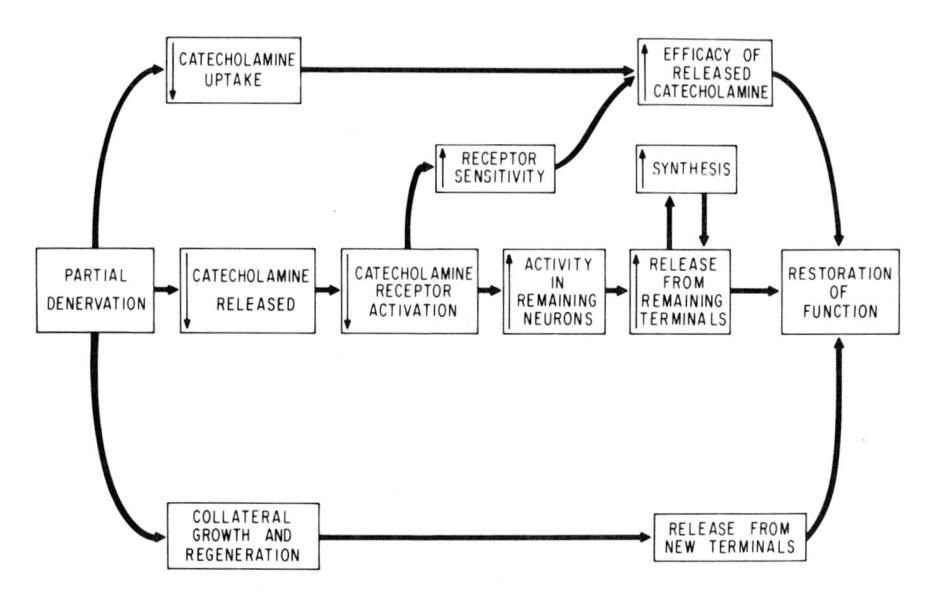

FIG. 4. A model for recovery of function within central catecholamine-containing neurons following subtotal damage. Immediately following the lesion, net catecholamine release will be reduced, reflecting the proportion of undamaged neurons that remain. Even though the loss of uptake sites will increase the efficacy of released amine, net decreases in receptor stimulation should occur and lead to increased catecholamine release from residual neurons. This increased turnover should be accompanied by increases in catecholamine synthesis. Prolonged stimulation should increase the neurons' capacity for catecholamine biosynthesis (e.g., through the induction of tyrosine hydroxylase) and thus progressively raise their capacity for sustained increases in catecholamine turnover, while enhanced sensitivity of the postsynaptic membrane should increase the effectiveness of the released catecholamine and further promote the recovery of function. A decrease in the affinity of residual terminals for catecholamine, collateral growth from intact axons, and regeneration of the damaged fibers might also contribute to recovery of function, although the nature of the stimulus for these processes is unknown.

---

[3] In addition to these processes, recovery of function might be aided by (a) the reversal of certain secondary effects of electrolytic lesions (e.g., vasomotor changes, hemorrhage, edema, the proliferation of glial cells) which might have contributed to the initial deficits, (b) repletion of catecholamines within central neurons that are depleted by the 6-HDA injections in the absence of any ultrastructural damage, and (c) adjustments within systems containing other neurotransmitters that are functionally interrelated with the central catecholamine systems.

treated rats are considerably more sensitive than vehicle-injected controls to the anorexic effects of α-methyltyrosine (Fig. 5), as are rats recovered from lateral hypothalamic lesions (Zigmond and Stricker, 1973). This increased sensitivity is highly correlated with the degree of depletion of striatal

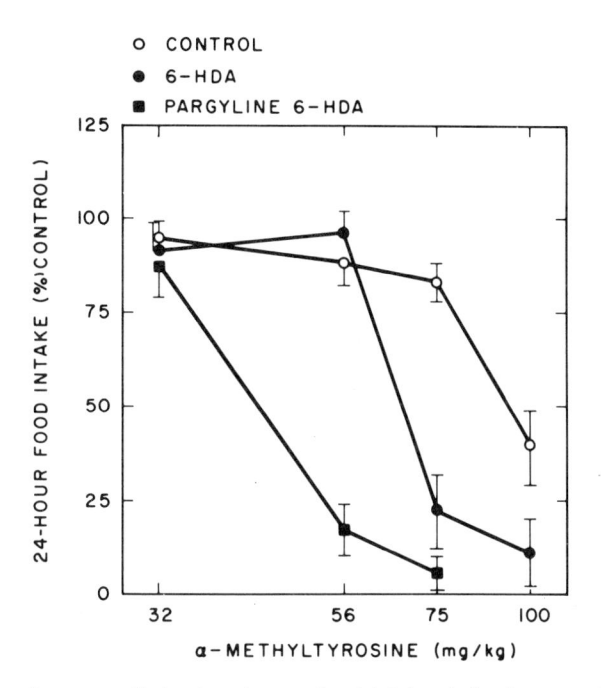

FIG. 5. Effects of D,L-α-methyl-p-tyrosine on food intake. Animals were given intraventricular 6-HDA (200 μg) alone (filled circles) or after pretreatment with pargyline (50 mg/kg, i.p.) (filled squares). After food and water intakes had returned to normal, α-methyltyrosine was given intraperitoneally every 8 hr during a 24-hr test. Food intake was calculated as a percentage of the mean intake during the preceding 3 days and is compared with the effects of the drug on control animals (open circles). Each point represents the mean ± the standard error of the mean for 4 to 12 animals.

dopamine (Fig. 6). Since catecholamine depletion after the inhibition of tyrosine hydroxylase is dependent on the rate of turnover within the neurons, these data suggest that lesion-induced depletions of brain catecholamines are accompanied by an increase in catecholamine turnover and synthesis rates in the remaining neurons. This hypothesis is further supported by recent measurements of catecholamine synthesis in rats following damage to the nigrostriatal dopamine system (Agid, Javoy, and Glowinski, 1973).

Second, damage to central catecholamine-containing fibers is accompanied, *in vitro*, by a decrease in the apparent V_{max} for uptake and an increase in the K_m (Fig. 7). This suggests to us that the reduced uptake of catechola-

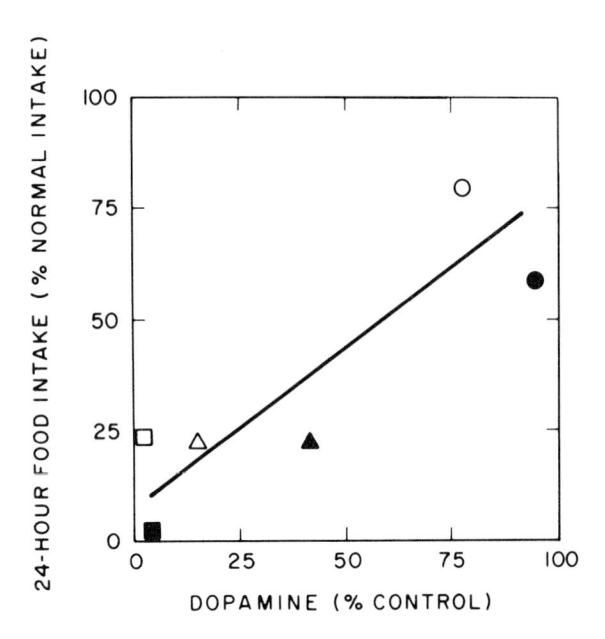

FIG. 6. Effect of α-methyltyrosine (3 × 75 mg/kg, i.p.) on food intake of 6-HDA-treated rats. Rats received 2 × 25 µg (filled circle), 3 × 25 µg (open circle), 2 × 200 µg (filled triangle), or 1 × 400 µg (open triangle) 6-HDA alone, 2 × 200 µg 6-HDA 30 min after pretreatment with pargyline (50 mg/kg, i.p.) (filled square), or 1 × 200 µg 6-HDA 30 min after pretreatment with pargyline and desmethylimipramine (25 mg/kg, i.p.) (open square). All multiple treatments were given 48 hr apart. The test was identical to that described in Fig. 5. Values represent the mean for three to six animals.

mines in brain-damaged rats (Uretsky and Iversen, 1970; Zigmond, Chalmers, Simpson, and Wurtman, 1971) results from a decrease in the affinity for uptake into residual terminals as well as a loss of uptake sites.

Third, we observe that L-DOPA increases the motor activity of 6-HDA-treated rats (pretreated with the peripheral decarboxylase inhibitor, RO4-4602) in a dose range (10 to 100 mg/kg, i.p.) that has little or no effect on control animals, in confirmation of previous findings (Schoenfeld and Uretsky, 1973). This effect is presumably due to increased efficacy of dopamine formed from the DOPA, which results from a decrease in presynaptic uptake (see above) and an increase in the sensitivity of the postsynaptic membrane (Ungerstedt, 1971b; Kalisker, Rutledge, and Perkins, 1973).

Also consistent with the proposed recovery processes are the findings that pretreatments facilitating catecholamine synthesis or receptor sensitivity, such as prior food deprivation (Powley and Keesey, 1970), pretreatment with α-methyltyrosine (Glick, Greenstein, and Zimmerberg, 1972), or prior damage to the frontal cortex (Glick and Greenstein, 1972), reduce the time needed for recovery after lateral hypothalamic damage. Additional support can be obtained by briefly considering how the various compensatory ad-

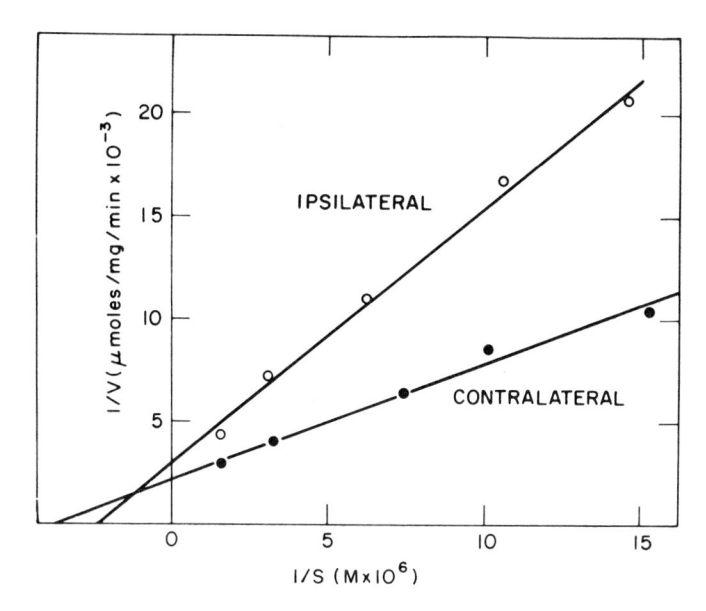

FIG. 7. The effect of lateral hypothalamic lesions on *in vitro* catecholamine uptake. Unilateral lesions were made in the ventrolateral hypothalamus. Synaptosome-rich homogenates were obtained from ipsilateral (open circles) and contralateral (filled circles) telencephalon and incubated with 0.05 µM D,L-norepinephrine-^{3}H for 5 min in the presence of 1.6×10^{-5} mM pheniprazine (see Zigmond et al., 1971). The data are from a representative animal.

justments might provide an explanation for some of the curious functional changes that are observed after brain damage.

1. Rats with greater than 98% depletions of striatal dopamine do not recover ingestive behaviors even after months of tube-feeding maintenance, whereas rats with somewhat smaller depletions (approximately 90 to 98%) do show a gradual recovery from an initial aphagia and adipsia. As might be expected, almost total damage to the ascending dopamine-containing fibers affords insufficient opportunity for recovery within the system. These findings are consistent with the unique, fundamental role which we are postulating for the nigrostriatal dopamine system.

2. Rats in the early stages of recovery do not eat or drink despite considerable nutrient needs unless the added incentive of palatable taste is also present. Since there are relatively few residual neurons, and receptor sensitivity may not be fully developed, it should take an unusually large stimulus (i.e., nutrient deficiency plus incentive) in order to release enough catecholamine to elicit ingestive behavior in these animals. Furthermore, since the gustatory sensations should habituate rather rapidly, it is not surprising that feeding during this period is not sustained.

3. Recovered animals ultimately ingest enough to maintain body weight and hydration in the absence of the incentives provided by highly palatable

foods or fluids. As recovery proceeds, there should be gradual reduction in the stimulus necessary to elicit ingestive behaviors. Nevertheless, note that permanent increases in the thresholds necessary to elicit feeding may explain the reduced body weight maintained by rats after apparent recovery from brain damage (Stricker and Zigmond, 1974a).

4. There are residual deficits in the ability of recovered animals to increase ingestive behaviors during acute and severe stress. This may reflect an inability of the residual catecholamine-containing neurons to increase catecholamine synthesis sufficiently to keep up with turnover during stress. The resultant decrease in receptor stimulation would reestablish the aphagia and adipsia, which would persist until synthesis rates could increase sufficiently to support the catecholamine turnover necessary for ingestive behaviors to reappear (or until the initiating excitatory stimulus was removed). In support of this hypothesis are our recent findings that rats recovered from aphagia and adipsia following intraventricular 6-HDA treatments or lateral hypothalamic lesions ultimately do increase their food intakes during cellular glucopenia or severe cold stress (Stricker et al., 1974) and do increase their water intakes during hypovolemia (Stricker, 1973; Stricker and Zigmond, 1974a) when these stimuli are prolonged.

5. Severe initial deficits are not observed when striatal dopamine depletions are less than 90%. These findings suggest a relatively large capability of this central catecholamine system to adjust its activity rapidly so as to maintain basal function despite extensive and irreversible damage. Increased sensitivity to the effects of α-methyltyrosine and to L-DOPA, and the persistence of residual deficits to acute regulatory imbalances, suggest that the mechanisms responsible for this rapid recovery are among those that also support the gradual recovery from the more severe initial impairments.

IV. SOME CLINICAL IMPLICATIONS

Rats enjoy freedom of movement across a wide range of environments and can function well even under rather stressful conditions. Rats with moderate amounts of damage to the nigrostriatal bundle appear to behave the way rats normally do in the limited environment of the laboratory, but have lost their ability to function properly in a more complex and stressful environment. However, they rarely encounter such a situation, and the consequences of their brain damage might go unnoticed unless there are specific examinations of their function during stress or following certain drug treatments. Rats with more extensive brain damage cannot function properly even in the controlled conditions of the laboratory and are easily discriminated from intact rats, but these differences seem to be more appropriately considered as quantitative rather than qualitative. Thus, the behavioral impairments of brain-damaged rats can be defined by the interaction between the magnitude of their brain damage and the complexity of the environment in which they live.

We believe that this conceptual framework is also relevant to problems of human mental health and cerebral dysfunctions. For example, like our lesioned animals, there is a loss of neostriatal dopamine in patients with Parkinson's disease but an absence of prominent symptoms until depletions reached 70 to 90% (Hornykiewicz, 1973). Since Parkinson's disease appears to be defined in terms of its symptomatology, it would seem that classical neurological tests for this disease are poor indicators of nigrostriatal damage (presumably because of the considerable compensation that is possible within this system). This makes early diagnosis of Parkinson's disease, and related disorders, very difficult and, as a result, neither the physician nor the patient is alerted to the latent problem. This is particularly tragic for conditions in which progressive neural degeneration can be avoided, such as lead poisoning. Chronic lead poisoning ultimately produces many of the behavioral symptoms and much of the neurological damage that is seen in Parkinson's disease. As with parkinsonism, one would expect that in its early stages it would result in few, if any, abnormal responses under neutral conditions. In fact, there are many reports of patients with high blood levels of lead that are discharged without therapy and returned to lead-contaminated environments when neurological tests failed to indicate gross deficits. We suspect that these individuals have suffered brain damage but have partially compensated for it in a way that defied ordinary observation. If so, then tests must be devised that are capable of detecting subclinical damage. Such tests would be based on an understanding of the effects produced by experimental damage to the nigrostriatal dopamine system, and of the processes by which recovery of function occurs, and might include an examination of performance on some complex behavioral task during acute stress (psychological or physiological) or with treatment with drugs that affect catecholamine systems (e.g., L-DOPA, α-methyltyrosine, or reserpine).

V. SUMMARY

1. The general appearance and behavior of rats following 40 to 60% depletions of striatal dopamine by intraventricular injections of 6-HDA is largely indistinguishable from that of control animals under normal laboratory conditions. However, these rats fail to increase ingestive behaviors appropriately when subjected to acute hypovolemia, cellular glucopenia, or low ambient temperatures, although physiological responses remain intact. These observations are analogous to the normal appearance of peripherally sympathectomized animals and their intolerance to similar stresses, and suggest that the nigrostriatal bundle mediates important behavioral contributions to homeostasis.

2. There is a profound aphagia and adipsia when dopamine depletion is

more complete (i.e., greater than 90%). Gradual recovery of ingestive behavior occurs in most animals within a few weeks despite the permanent catecholamine depletions, although in some animals recovery is not attained for many months. These brain-damaged animals ultimately are able to maintain body weight by voluntary consumption of food and water, but, like less severely lesioned animals, do not increase ingestive behavior during marked regulatory imbalances.

3. Several lines of evidence suggest that the partial recovery is due, in part, to compensatory processes within the damaged system. For example, intraventricular 6-HDA, or α-methyltyrosine, reinstates the aphagia and adipsia in recovered animals, suggesting that dopamine was still important in mediating ingestive behaviors. Moreover, recovered animals are sensitive to the anorexic effects of α-methyltyrosine at doses having little effect on control animals, suggesting that the loss of dopamine may be partially compensated by increases in synthesis and release from undamaged neurons. Additional experiments indicate that there is a decreased affinity for uptake into the nerve terminals contributing, in part, to the observed supersensitivity to the effects of L-DOPA.

4. The failure of "centrally sympathectomized" rats to respond appropriately during acute homeostatic imbalances may reflect an inability of the system to compensate fully for the damage. However, some additional compensation must still be possible, since animals ultimately do respond to regulatory challenges when they are prolonged.

5. These and other results may be related to the diagnosis and treatment of Parkinson's disease and related human disorders involving damage to the dopamine-containing neurons of the nigrostriatal bundle.

ACKNOWLEDGMENTS

We are grateful for the colleagueship of our students and associates: Dr. Mark Friedman, Mr. Thomas Heffner, Mr. Gregory Kapatos, Ms. Deborah Levitan, Ms. Debra Rubinstein, Ms. Janet Van Zoeren, Ms. Suzanne Wuerthele, Ms. Jen-shew Yen, and Mr. Mark Zimmerman. Drugs were provided by Abbott Laboratories (pargyline), Hoffmann-LaRoche, Inc. (RO4-4602), and Lakeside Laboratories, Inc. (desmethylimipramine). The research was supported by grants from the U.S. National Institute of Mental Health (MH-20620), the U.S. National Science Foundation (GB-22830), and Eli Lilly and Co.

REFERENCES

Agid, Y., Javoy, F., and Glowinski, J. (1973): Hyperactivity of DA neurons after partial destruction of nigro-striatal DA system. *Nature New Biology*, 245:150–151.
Albert, D. J., Storlien, L. H., Wood, D. J., and Ehman, G. K. (1970): Further evidence for a complex system controlling feeding behavior. *Physiology and Behavior*, 5:1075–1082.

Anand, B. K., and Brobeck, J. R. (1951): Hypothalamic control of food intake in rats and cats. *Yale Journal of Biology and Medicine*, 24:123-140.

Balagura, S., Wilcox, R. H., and Coscina, D. V. (1969): The effect of diencephalic lesions on food intake and motor activity. *Physiology and Behavior*, 4:629-633.

Blondaux, C., Juge, A., Sordet, F., Chouvet, G., Jouvet, M., and Pujol, J.-F. (1973): Modification du metabolisme de la serotonine (5-HT) cerebrale induite chez le rat par administration de 6-hydroxydopamine. *Brain Research*, 50:101-114.

Bloom, F. E., Algeri, S., Groppetti, A., Revuelta, A., and Costa, E. (1969): Lesions of central norepinephrine terminals with 6-OH-dopamine; biochemistry and fine structure. *Science*, 166:1284-1286.

Breese, G. R., Smith, R. D., Cooper, B. R., and Grant, L. D. (1973): Alterations in consummatory behavior following intracisternal injection of 6-hydroxydopamine. *Pharmacology, Biochemistry, and Behavior*, 1:319-328.

Breese, G. R., and Traylor, T. D. (1970): Effect of 6-hydroxydopamine on brain norepinephrine and dopamine: Evidence for selective degeneration of catecholamine neurons. *Journal of Pharmacology and Experimental Therapeutics*, 174:413-420.

Breese, G. R., and Traylor, T. D. (1971): Depletion of brain norepinephrine and dopamine by 6-hydroxydopamine. *British Journal of Pharmacology*, 42:88-99.

Chang, C. C. (1964): A sensitive method for spectrophotofluorometric assay of catecholamines. *International Journal of Neuropharmacology*, 3:643-649.

Crout, R. J., Creveling, C. R., and Udenfriend, S. (1961): Norepinephrine metabolism in rat brain and heart. *Journal of Pharmacology and Experimental Therapeutics*, 132:269-277.

Dairman, W., and Udenfriend, S. (1970): Increased conversion of tyrosine to catecholamines in the intact rat following elevation of tissue tyrosine hydroxylase levels by administered phenoxybenzamine. *Molecular Pharmacology*, 6:350-356.

Epstein, A. N., and Teitelbaum, P. (1967): Specific loss of the hypoglycemic control of feeding in recovered lateral rats. *American Journal of Physiology*, 213:1159-1167.

Fibiger, H. C., Zis, A. P., and McGeer, E. G. (1973): Feeding and drinking deficits after 6-hydroxydopamine administration in the rat: Similarities to the lateral hypothalamic syndrome. *Brain Research*, 55:135-148.

Fuxe, K., and Hökfelt, T. (1969): Catecholamines in the hypothalamus and the pituitary gland. In: *Frontiers in Neuroendocrinology, 1969*, edited by W. F. Ganong and L. Martini. Oxford University Press, New York.

Geyer, M. A., and Segal, D. S. (1973): Differential effects of reserpine and alphamethyl-*p*-tyrosine on norepinephrine and dopamine induced behavior activity. *Psychopharmacologia*, 29:131-140.

Glick, S. D., and Greenstein, S. (1972): Facilitation of recovery after lateral hypothalamic damage by prior ablation of frontal cortex. *Nature New Biology*, 239:187-188.

Glick, S. D., Greenstein, S., and Zimmerberg, B. (1972): Facilitation of recovery by α-methyl-*p*-tyrosine after lateral hypothalamic damage. *Science*, 177:534-535.

Heller, A., and Moore, R. Y. (1965): Effect of central nervous system lesions on brain monoamines in the rat. *Journal of Pharmacology and Experimental Therapeutics*, 150:1-9.

Hillarp, N. A., Fuxe, K., and Dahlström, A. (1966): Demonstration and mapping of central neurons containing dopamine, noradrenaline, and 5-hydroxytryptamine and their reactions to psychopharmaca. *Pharmacological Reviews*, 18:727-741.

Hoffer, B. J., Siggins, G. R., Oliver, A. P., and Bloom, F. E. (1973): Activation of the pathway from locus coeruleus to rat cerebellar Purkinje neurons: Pharmacological evidence of rat noradrenergic central inhibition. *Journal of Pharmacology and Experimental Therapeutics*, 184:553-569.

Hornykiewicz, O. (1973): Parkinson's disease: From brain homogenate to treatment. *Federation Proceedings*, 32:183-190.

Jouvet, M. (1972): The role of monoamines and acetylcholine containing neurons in the regulation of the sleep-waking cycle. *Ergebnisse Physiologie*, 64:166-307.

Kalisker, A., Rutledge, C. O., and Perkins, J. P. (1973): Effect of nerve degeneration by 6-hydroxydopamine on catecholamine-stimulated adenosine 3′,5′-monophosphate formation in rat cerebral cortex. *Molecular Pharmacology*, 9:619-629.

Kim, J. S. (1973): Effects of 6-hydroxydopamine on acetylcholine and GABA metabolism in rat striatum. *Brain Research*, 55:472-475.

Marshall, J. F., and Teitelbaum, P. (1973): A comparison of the eating in response to hypothermic and glucoprivic challenges after nigral 6-hydroxydopamine and lateral hypothalamic electrolytic lesions in rats. *Brain Research,* 55:229–233.

Marshall, J. F., Turner, B. H., and Teitelbaum, P. (1971): Sensory neglect produced by lateral hypothalamic damage. *Science,* 174:523–525.

Nybäck, H., and Sedvall, G. (1968): Effect of chlorpromazine on accumulation and disappearance of catecholamines formed from tyrosine-C^{14} in brain. *Journal of Pharmacology and Experimental Therapeutics,* 162:294–301.

Oltmans, G. A., and Harvey, J. A. (1972): LH syndrome and brain catecholamine levels after lesions of the nigrostriatal bundle. *Physiology and Behavior,* 8:69–78.

Poirier, L. J., Langelier, P., Roberge, A., Boucher, R., and Kitsikis, A. (1972): Non-specific histopathological changes induced by the intracerebral injection of 6-hydroxydopamine (6-OH-DA). *Journal of Neurological Science,* 16:401–416.

Powley, T. L., and Keesey, R. E. (1970): Relationship of body weight to the lateral hypothalamic feeding syndrome. *Journal of Comparative Physiology and Psychology,* 70:25–36.

Rech, R. H., Carr, L. A., and Moore, K. E. (1968): Behavioral effects of α-methyltyrosine after prior depletion of brain catecholamines. *Journal of Pharmacology and Experimental Therapeutics,* 160:326–335.

Schoenfeld, R. I., and Uretsky, N. J. (1973): Enhancement by 6-HDA of the effects of DOPA upon motor activity. *Journal of Pharmacology and Experimental Therapeutics,* 186:616–624.

Segal, D. S., Sullivan, J. L., Kuczenski, R. T., and Mandell, A. J. (1971): Effects of long-term reserpine treatment on brain tyrosine hydroxylase and behavioral activity. *Science,* 173:847–848.

Stricker, E. M. (1973): Thirst, sodium appetite, and complementary physiological contributions to the regulation of intravascular fluid volume. In: *The Neuropsychology of Thirst,* edited by A. N. Epstein, H. R. Kissileff, and E. Stellar. V. H. Winston & Sons, Washington, D.C.

Stricker, E. M., and Wolf, G. (1967): The effects of hypovolemia on drinking in rats with lateral hypothalamic damage. *Proceedings of the Society for Experimental Biology and Medicine,* 124:816–820.

Stricker, E. M., and Zigmond, M. J. (1974a): Effects on homeostasis of intraventricular injection of 6-hydroxydopamine in rats. *Journal of Comparative Physiology and Psychology (in press).*

Stricker, E. M., and Zigmond, M. J. (1974b): Brain catecholamines and thirst. In: *Control Mechanisms of Drinking,* edited by G. Peters and J. T. Fitzsimons. Springer-Verlag, Berlin.

Stricker, E. M., Zigmond, M. J., Friedman, M. I., and Redgate, E. S. (1974): The contribution of central catecholamine-containing neurons to several glucoregulatory mechanisms. *Federation Proceedings,* 33:564.

Teitelbaum, P., and Epstein, A. N. (1962): The lateral hypothalamic syndrome: Recovery of feeding and drinking after lateral hypothalamic lesions. *Psychological Reviews,* 69:74–90.

Ungerstedt, U. (1971a): Stereotaxic mapping of the monoamine pathways in the rat brain. *Acta Physiologica Scandinavica,* Suppl. 367:1–48.

Ungerstedt, U. (1971b): Postsynaptic supersensitivity after 6-hydroxydopamine induced degeneration of the nigro-striatal dopamine system. *Acta Physiologica Scandinavica,* Suppl. 367:69–93.

Ungerstedt, U. (1971c): Adipsia and aphagia after 6-hydroxydopamine induced degeneration of the nigro-striatal dopamine system. *Acta Physiologica Scandinavica,* Suppl. 367:95–122.

Uretsky, N. J., and Iversen, L. L. (1970): Effects of 6-hydroxydopamine on catecholamine containing neurons in the rat brain. *Journal of Neurochemistry,* 17:267–278.

Zigmond, M. J., Chalmers, J. P., Simpson, J. R., and Wurtman, R. J. (1971): Effect of lateral hypothalamic lesions on uptake of norepinephrine by brain homogenates. *Journal of Pharmacology and Experimental Therapeutics,* 179:20–28.

Zigmond, M. J., and Stricker, E. M. (1972): Deficits in feeding behavior after intraventricular injection of 6-hydroxydopamine in rats. *Science,* 177:1211–1214.

Zigmond, M. J., and Stricker, E. M. (1973): Recovery of feeding and drinking by rats after intraventricular 6-hydroxydopamine or lateral hypothalamic lesions. *Science,* 182:717–720.

Behavioral Effects of Direct- and Indirect-Acting Dopaminergic Agonists

K. E. Moore

Department of Pharmacology, Michigan State University, East Lansing, Michigan 48824

I. INTRODUCTION

In 1964 Stein reported that the behavior-stimulating actions of amphetamine were mediated, not by a direct action on central receptors, but indirectly through the local release of brain catecholamines. This indirect action was supported by the findings of Weissman, Koe, and Tenen (1966) that α-methyltyrosine (αMT), which disrupts the synthesis of catecholamines, blocks the central stimulant actions of amphetamine. The results of a variety of other experiments now suggest that amphetamine acts indirectly by facilitating release or by blocking the reuptake of catecholamines at specific nerve terminals in the central nervous system. Until recently it was generally believed that amphetamine interacted primarily with noradrenergic neurons which comprise a "reward" or "activating" system in the central nervous system. Results of recent experiments suggest that amphetamine and a variety of other central stimulant drugs have dopaminergic agonist properties. Direct-acting dopaminergic agonists combine directly with dopamine receptors, while indirect-acting agonists exert characteristic pharmacologic effects by increasing the concentration of dopamine at these same receptors.

A schematic representation of a hypothetical dopaminergic synapse is depicted in Fig. 1. In the nerve terminal dopamine is believed to exist in two separate pools—a large "storage pool" and a smaller "releasable pool." Dopamine is released from the "releasable pool" in response to nerve action potentials or to the administration of certain drugs. The released dopamine diffuses across the synaptic cleft to interact with a receptor on the surface of the postjunctional membrane. Dopamine is then removed from this active site by diffusion and eventual metabolic destruction or by active reuptake into the presynaptic nerve terminal. During intense activity of dopaminergic neurons, or during the actions of "releasing" drugs, the "releasable pool" may in fact be synonymous with a "newly synthesized pool" of dopamine. Selective release of newly synthesized dopamine by amphetamine has been invoked to explain the antiamphetamine effects of αMT (Weissman et al., 1966).

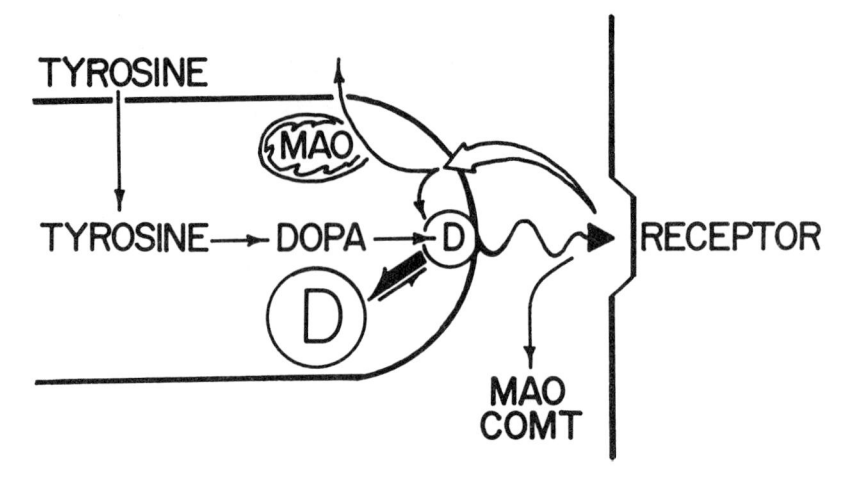

FIG. 1. Schematic diagram of a hypothetical dopaminergic synapse. The small D represents the "releasable pool" and the large D represents the "storage pool" of dopamine.

II. ACTIONS OF AMPHETAMINE FOLLOWING SELECTIVE BLOCKADE OF SYNTHESIS OF NOREPINEPHRINE AND DOPAMINE

αMT disrupts the synthesis of both norepinephrine and dopamine, so it is not possible to relate the antiamphetamine effects of this drug with an exclusive disruption of synthesis of either one of these two amines. Drugs such as U-14,624 or FLA-63 inhibit dopamine-β-hydroxylase and thereby block norepinephrine synthesis without influencing the synthesis of dopamine. If the antiamphetamine effects of αMT result from inhibition of norepinephrine synthesis, then inhibitors of dopamine-β-hydroxylase should mimic the actions of amphetamine. On the other hand, if dopamine-β-hydroxylase inhibitors do not block the stimulant actions of amphetamine, the effect of αMT may result from inhibition of dopamine synthesis. Results of a variety of experiments in our laboratory support the latter hypothesis (Thornburg and Moore, 1973). An example of one such experiment is illustrated in Fig. 2. The continuous activity of mice was determined each day in their home cages during the dark phase of a dark-light cycle. αMT and U-14,624 were added to the diets of mice to avoid nonspecific behavioral depressant actions that accompany the intraperitoneal administration of these insoluble drugs (Thornburg and Moore, 1971). In the experiment depicted in Fig. 2a, mice were placed on a control diet for 4 days and the mean activity on days 3 and 4 was expressed as 100%. On day 5 the mice received the diet containing αMT and a slight reduction of activity was noted. On day 6 a diet containing amphetamine and αMT was presented to the animals and no increase in activity was observed. After 2 days of control diet, a diet containing amphetamine was presented on day 9 which

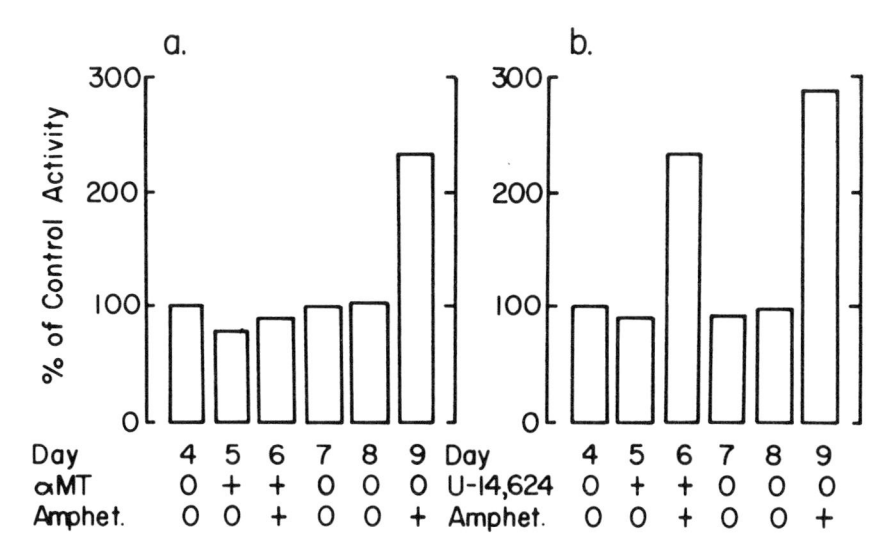

FIG. 2. Effects of diets containing 0.02% D-amphetamine and 0.4% α-methyltyrosine (a) or 0.4% U-14,624 (b) on continuous activity of mice. Groups of four mice received control diets for 4 days during the dark period (17.00 to 08.00 hr), and then beginning on day 5 they received diets containing drugs in the sequence described at the bottom of the figure. Motor activity of mice in their home cage was recorded with a MOTRON motility meter between 22.00 and 08.00 hr and expressed as a percentage of the mean activity of days 3 and 4. [See Thornburg and Moore (1973) for details; reprinted with permission from *Neuropharmacology*.]

caused a marked increase in activity. Thus, the antiamphetamine effects of αMT were clearly demonstrated. In Fig. 2b the same experimental design was utilized except that U-14,624 was substituted for αMT. U-14,624 alone had little effect, but when amphetamine was added to the U-14,624 diet, a marked stimulation of activity was observed. After 2 days of control diet, the addition of amphetamine alone produced stimulation that was approximately equivalent to that seen when amphetamine was added to the U-14,624 diet. Thus, blockade of dopamine-β-hydroxylase with U-14,624 did not influence amphetamine-induced stimulation of motor activity. When added to the diet in this manner, αMT and U-14,624 reduced the brain contents of norepinephrine to about the same extent. Both drugs were equally effective in blocking the conversion of ^{14}C-tyrosine to ^{14}C-norepinephrine, but only αMT reduced the conversion of ^{14}C-tyrosine to ^{14}C-dopamine, and only αMT blocked the stimulant actions of amphetamine. These results suggest that the locomotor stimulant effects of amphetamine involve, at least in part, a dopaminergic mechanism. The antiamphetamine actions of αMT appear related to blockade of dopamine synthesis, since the drug does not have adrenergic blocking properties (Moore and Dominic, 1971) and, in contrast to reports of studies *in vitro* (Enna, Dorris, and Shore, 1973), αMT does not have "antirelease" actions when tested *in vivo* (Chiueh and Moore, 1973).

III. RELEASE OF DOPAMINE FROM BRAIN *IN SITU*

The ability of amphetamine to release dopamine from the brain *in situ* further supports an indirect action of this drug. Using an *in vivo* cerebroventricular perfusing technique, Carr and Moore (1970) demonstrated that amphetamine increases the efflux of exogenously administered ^{3}H-dopamine from cat brain *in situ*. A schematic diagram of the perfusing technique is depicted in Fig. 3. Artificial CSF is introduced into a lateral ventricle of cat brain by means of a stereotaxically placed cannula. The effluent is collected from a catheter inserted into the cerebral aqueduct. Being unable to detect endogenous catecholamines in the perfusate, radioactive catecholamines are introduced into the ventricle before the start of perfusion. The radioactive amines are taken up by tissues lining the ventricles and are subsequently released by drugs or by electrical stimulation of ascending catecholaminergic neurons. In Fig. 3 an electrode is depicted aimed at the dopaminergic nigrostriatal axons in the diencephalon. Electrical stimulation of this site, or of the cell bodies in the substantia nigra or terminals in the caudate nucleus, increases the efflux of ^{3}H-dopamine into the perfusing CSF (Von Voigt-

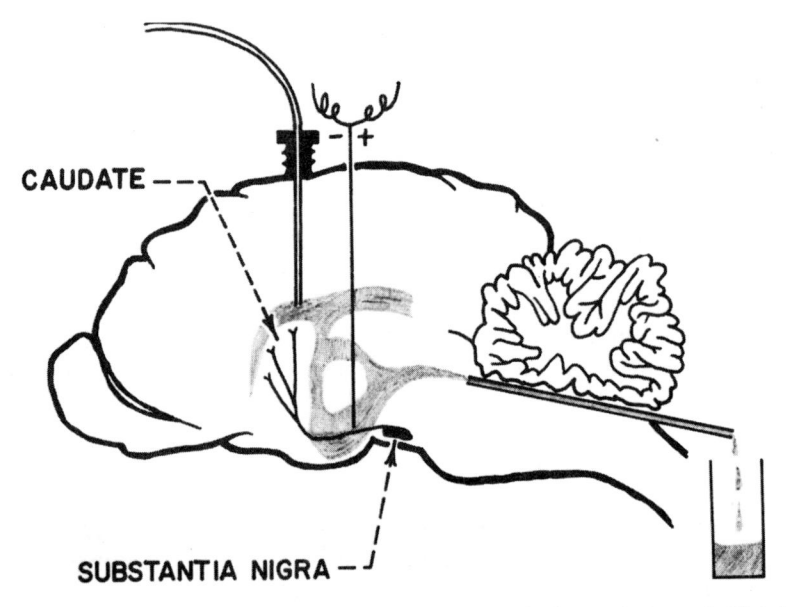

FIG. 3. Schematic diagram of the cerebroventricular perfusion technique. Artificial CSF is introduced into a lateral ventricle of cat brain through a stereotaxically placed cannula and collected from a catheter inserted into the cerebral aqueduct. ^{3}H-Dopamine is injected into the lateral ventricle before the start of perfusion, or ^{3}H-tyrosine is added to the perfusing CSF; the perfusate is analyzed for ^{3}H-norepinephrine and ^{3}H-dopamine by alumina adsorption and ion exchange chromatography. The release of ^{3}H-dopamine can be induced by electrical stimulation of the nigrostriatal pathway or by the administration of drugs (see Fig. 4).

lander and Moore, 1971b). Amphetamine and amantadine facilitate the release of ^{3}H-dopamine from the terminals of the nigrostriatal dopaminergic neurons (Von Voigtlander and Moore, 1971a, 1973a).

When ^{3}H-dopamine is injected into the lateral ventricle in a small volume, the amine accumulates primarily in the caudate nucleus; the amines, however, are probably not taken up by and released exclusively from dopaminergic neurons. Since tyrosine hydroxylase appears to be located only in catecholaminergic neurons, any radioactive dopamine appearing in the cerebroventricular perfusate which was synthesized from radioactive tyrosine must have originated from these neurons. In the experiment illustrated in Fig. 4 the cerebroventricular system was perfused with CSF containing ^{3}H-tyrosine and the perfusate analyzed for ^{3}H-dopamine. The intraventricular infusion or the intravenous injection of D-amphetamine markedly increased the efflux of endogenously synthesized ^{3}H-dopamine.

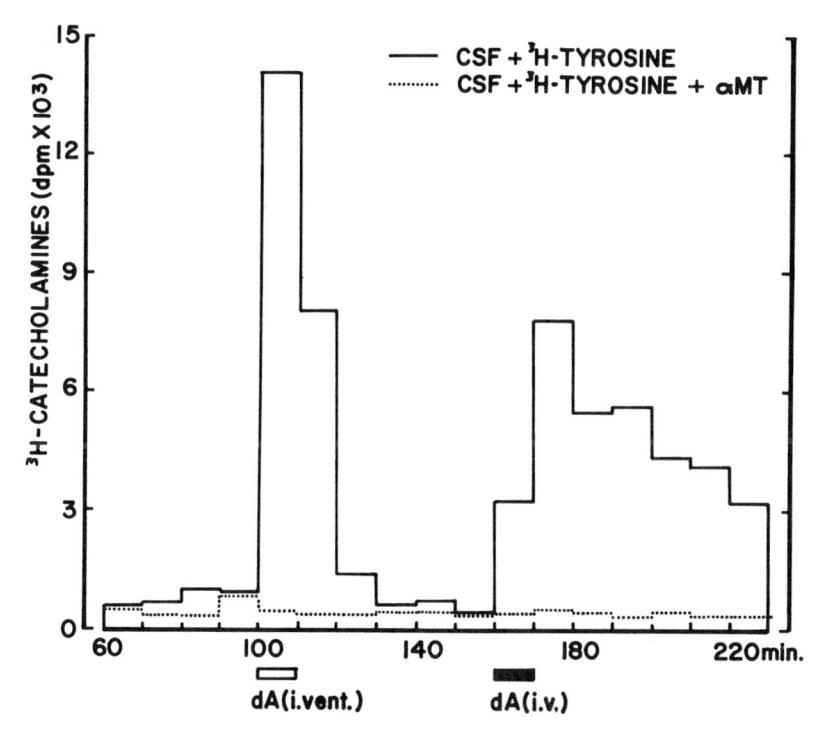

FIG. 4. Amphetamine-induced release of catecholamines from brain *in situ*. CSF containing 20 µC/ml of ^{3}H-tyrosine was infused into the cerebroventricular system at a rate of 0.2 ml/min. Sixty minutes after the start of perfusion samples of perfusate were collected every 10 min and analyzed for ^{3}H-catecholamines; the height of each bar represents the ^{3}H-catecholamine content in each perfusate sample. At 100 min D-amphetamine sulfate (1.1×10^{-5} M) was added to the perfusing CSF for 10 min, and at 160 min 2 mg/kg of D-amphetamine sulfate was injected intravenously. Approximately 95% of the ^{3}H-catecholamines consisted of ^{3}H-dopamine. When α-methyltyrosine (5×10^4 M) was added to the CSF (·······), no ^{3}H-catecholamines could be detected in perfusate.

In similar experiments, apomorphine did not increase the efflux of ^{3}H-dopamine.

IV. EFFECTS OF DIRECT AND INDIRECT DOPAMINERGIC AGONISTS ON CIRCLING BEHAVIOR OF MICE WITH UNILATERAL 6-HYDROXYDOPAMINE (6-HDA) LESIONS IN THE STRIATUM

Activation of central dopamine receptors by both direct (apomorphine) and indirect (D-amphetamine) agonists causes locomotor stimulation (Thornburg and Moore, 1973, 1974) and stereotype gnawing (Ernst, 1967) in rodents. Andén, Dahlström, Fuxe, and Larsson (1966) and Lotti (1971) have made use of the postural asymmetries in rodents induced by mechanical or electrolytic lesions of the striatum to test for dopaminergic agonistic activity; however, unless drugs such as αMT are used concurrently, it is not possible to distinguish between direct- and indirect-acting agonists in these tests. Ungerstedt (1971) utilized direct intracerebral injections of 6-HDA to destroy selectively the dopaminergic nigrostriatal neurons in rats; direct and indirect dopaminergic agonists can be distinguished by the direction in which these lesioned animals rotate. We have developed a similar technique in mice that provides a very rapid and simple means for detecting direct- and indirect-acting dopamine agonists (Von Voigtlander and Moore, 1973b).

Unilateral injections of 6-HDA are made into the striatum of mice anesthetized with methoxyflurane. The site of injection is controlled by a mold which surrounds and immobilizes the mouse head and permits accurate and reproducible microinjections (16 μg of 6-HDA in 4 μl of 0.2% ascorbic acid) into the left or right striatum. The coordinates of the injection site are specified by the location of guide cannulae in the mold; the depth of the injection is controlled by a cuff on the syringe needle.

When examined histologically 10 days after the 6-HDA injection, there was no gross lesion at the injection site; the normal trabecular appearance of the striatum remained unchanged. The dopamine content of the forebrain on the injected side, however, was reduced to 17% of that on the control side. The norepinephrine content was slightly reduced on the injected side (70% of control), while the content of 5-hydroxytryptamine was unchanged. These results suggest a rather selective destruction of dopaminergic nerve terminals in the striatum.

Mice with the unilateral 6-HDA lesions exhibit a characteristic type of circling behavior. Circling is measured by putting a mouse in a 3-liter beaker which is painted white and illuminated from below. The beaker is enclosed in a sound-attenuating box, and the activity of the mouse is viewed through a one-way window in the top of the box. Complete 360° turns are recorded during a 2-min test period. There is a significant correlation between the unilateral loss of dopamine and the rate and direction of circling; mice preferentially circle toward the side of the lesion (ipsilateral circling).

Drugs can influence both the rate and the direction of circling. The effects of a variety of drugs on the circling behavior of mice tested 10 to 20 days after the injection of 6-HDA are summarized in Fig. 5. When tested 30 min after intraperitoneal injections, D- and L-amphetamine and amantadine caused dose-related increases in the normal ipsilateral circling of these mice. That is, they increased the rate of circling toward the side of the lesion. On the other hand, apomorphine and L-DOPA caused the mice to reverse the direction of circling; these drugs caused the mice to circle toward the right, away from the side of the lesion (contralateral circling).

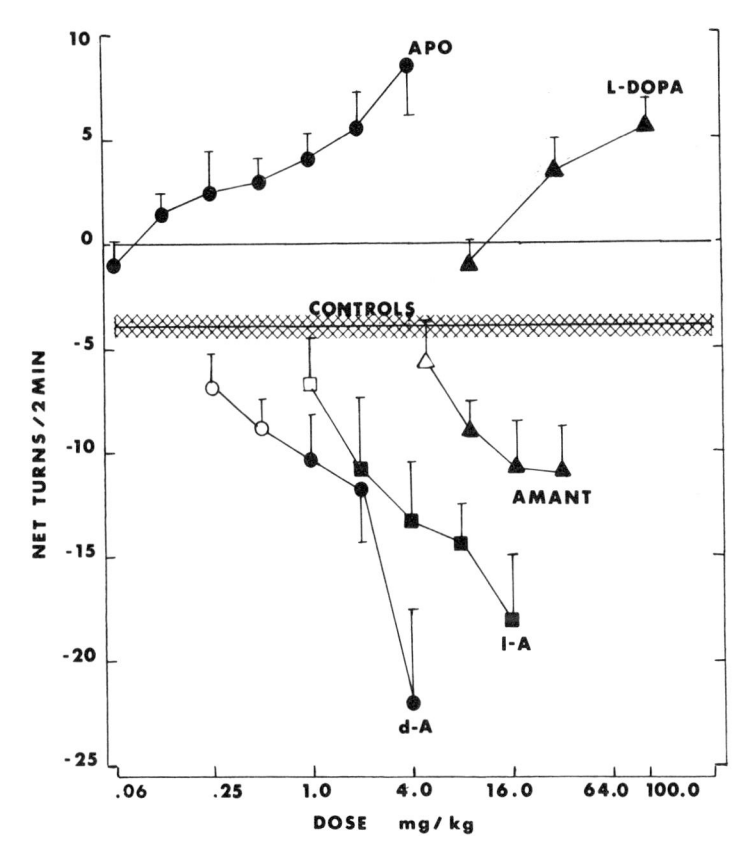

FIG. 5. The effects of direct- and indirect-acting dopaminergic agonists on circling behavior of mice. Ten days or more after the injection of 6-hydroxydopamine (16 μg in 4 μl) into the left striatum, mice were injected i.p. with saline or various doses of D- or L-amphetamine (d-A, l-A), amantadine (AMANT), apomorphine (APO), or L-DOPA and observed for circling behavior 30 min later. The net number of 360° turns in animals injected with saline are represented by the horizontal line and the SE by the hatched area; they circle to the left (toward the side of the lesion; ipsilateral circling), as indicated by the negative numbers and the downward direction in the figure. Circling to the right (away from the side of the lesion; contralateral circling) is indicated by the positive numbers and the upward direction in the figure. Solid symbols indicate those values that are significantly different from control ($p < 0.05$). [See Von Voigtlander and Moore (1973) for details; reprinted with permission from *Neuropharmacology*.]

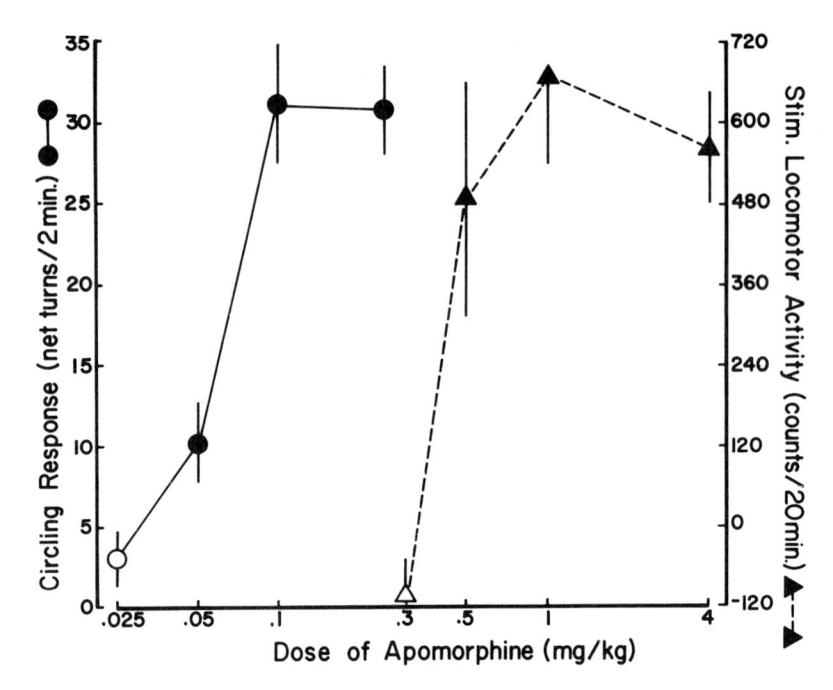

FIG. 6. Effects of apomorphine on circling behavior in 6-hydroxydopamine-injected mice (circles) and on locomotor activity in normal mice (triangles). Symbols represent the means and vertical lines 1 SE from nine to 11 determinations; solid symbols indicate those values that are significantly different from control ($p < 0.05$). Locomotor activity was determined in pairs of mice in Woodard actophotometers for 20 min after injection.

The increased circling obtained after the amphetamines and amantadine occurs with approximately the same doses as are required to increase locomotor activity. On the other hand, doses of apomorphine and L-DOPA required to induce contralateral circling are much less than those required to increase motor activity. This is depicted for apomorphine in Fig. 6. The dose of apomorphine needed to induce circling is approximately one-tenth the dose required to stimulate locomotor activity. A similar ratio is obtained when the doses of L-DOPA required to cause circling and locomotor stimulation are compared.

If both the contralateral circling and the stimulation of locomotor activity in response to L-DOPA and apomorphine result from the direct activation of dopaminergic receptors, the differential doses required to produce these effects may be explained on the basis of induction of "supersensitive" dopamine receptors following the injection of 6-HDA. This proposal is supported by the time course of the development of the circling response to apomorphine (Fig. 7). Dose-response curves for apomorphine-induced circling were determined at various times after the intrastriatal injection of 6-HDA. As the duration between the time of injection and the time of test-

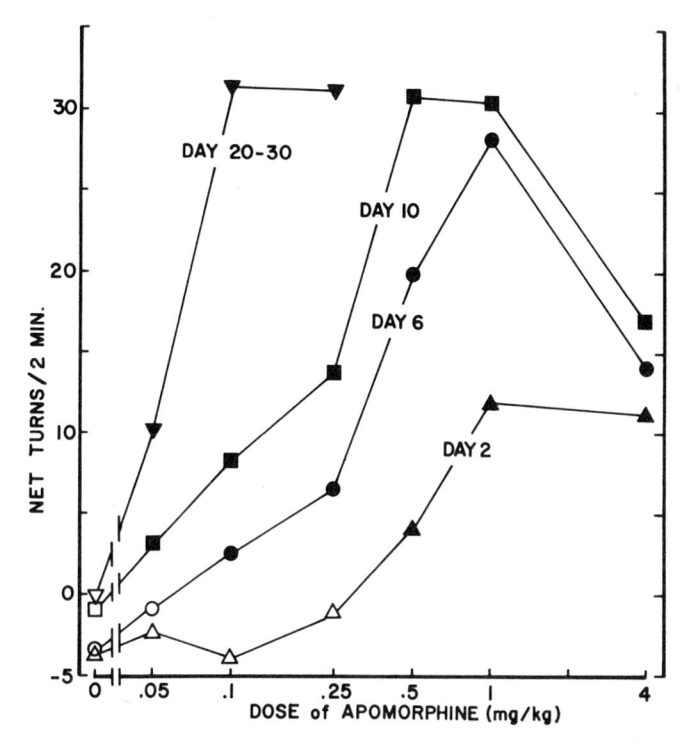

FIG. 7. Dose-response curves to apomorphine-induced circling determined at various times after the intrastriatal injection of 6-hydroxydopamine. Symbols represent the means of nine to 12 determinations; solid symbols indicate those values that are significantly different from saline-injected controls ($p < 0.05$).

ing with apomorphine increased from 2 days to 20 to 30 days, there was a shift in the dose-response curve to the left. There was also an increase in the maximal response. Essentially the same pattern was observed when the temporal development of dose-response curves to L-DOPA was determined. If apomorphine and L-DOPA (or dopamine resulting from the decarboxylation of this amino acid) act directly with dopaminergic receptors, these receptors become progressively more sensitive to the agonists during the first week or so after the intrastriatal injection of 6-HDA.

The mouse circling experiments are summarized in Fig. 8. The bilateral ascending dopaminergic pathways are depicted schematically in the mouse brain. The unilateral intrastriatal injection of 6-HDA selectively destroys the dopaminergic nerve terminals pictured on the left. This is evidenced by the marked reduction in the dopamine but not the 5-hydroxytryptamine content in the mouse forebrain on the injected side. This selective lesion causes an imbalance in the ascending nigrostriatal system, since the dopaminergic synapses, which are depicted on the right, remain functional. As a result of this imbalance the mouse preferentially circles toward the side

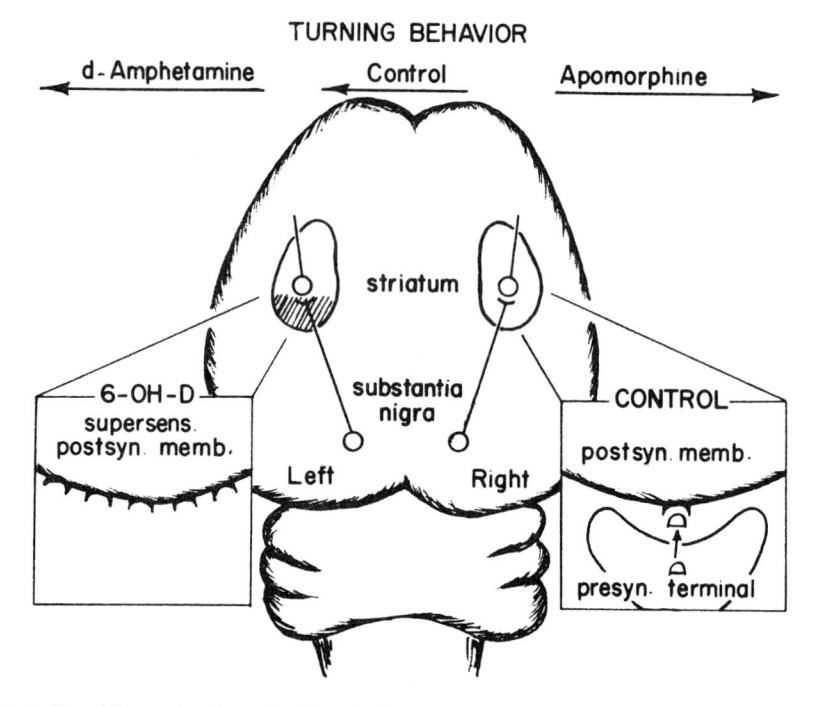

FIG. 8. Possible mechanisms for the circling behavior in mice with unilateral 6-hydroxy-dopamine-induced lesions in the striatum.

of the lesion (ipsilateral circling). If, as we have proposed, D-amphetamine causes locomotor stimulation by releasing dopamine from the terminals of the nigrostriatal neurons (see Fig. 4), then this drug should increase the rate of ipsilateral circling, since only dopaminergic neurons on the right side remain functional. Amphetamine does induce ipsilateral circling (Fig. 5), which appears to be related to the release of newly synthesized dopamine since the effect is blocked by αMT (Von Voigtlander and Moore, 1973b). During the first couple of weeks after the 6-HDA injection, the receptors on the postsynaptic membrane on the injected side become supersensitive to direct-acting dopaminergic agonists. This is depicted schematically on the left side by an increase in the number of dopamine receptors. The administration of a direct-acting dopaminergic agonist, such as apomorphine, has a greater effect on the 6-HDA-injected side. Accordingly, the mice reverse the direction of turning and circle away from the side of the lesion. This simple test, therefore, can be used to detect direct- and indirect-acting dopaminergic agonists.

V. SUMMARY

D-Amphetamine causes locomotor stimulation through a dopaminergic mechanism; it facilitates the release of dopamine from terminals of nigrostriatal neurons. Apomorphine, on the other hand, stimulates postsynaptic dopamine receptors directly. Direct- and indirect-acting dopaminergic agonists can be easily distinguished by determining their effects on the circling behavior of mice with unilateral 6-HDA-induced lesions in the striatum.

ACKNOWLEDGMENTS

Experiments reported in this chapter were conducted by graduate and postdoctoral students (C. C. Chiueh, J. E. Thornburg, P. F. Von Voigtlander) and were supported by U.S. Public Health Service grants NS09174 and MH13174.

REFERENCES

Andén, N.-E., Dahlström, A., Fuxe, K., and Larsson, K. (1966): Functional role of the nigro-striatal dopamine neurons. *Acta Pharmacologica et Toxicologica*, 24:263–274.
Carr, L. A., and Moore, K. E. (1970): Effects of amphetamine on the contents of norepinephrine and its metabolites in the effluent of perfused cerebral ventricles of the cat. *Biochemical Pharmacology*, 19:2361–2374.
Chiueh, C. C., and Moore, K. E. (1973): Failure of α-methyltyrosine to alter D-amphetamine-induced release of dopamine from cat brain *in vivo*. *Pharmacologist*, 15:174.
Enna, S. J., Dorris, R. L., and Shore, P. A. (1973): Specific inhibition by α-methyltyrosine of amphetamine-induced amine release from brain. *Journal of Pharmacology and Experimental Therapeutics*, 184:576–582.
Ernst, A. M. (1967): Mode of action of apomorphine and dexamphetamine on gnawing compulsion in rats. *Psychopharmacologia*, 10:316–323.
Lotti, V. J. (1971): Action of various centrally acting agents in mice with unilateral caudate brain lesions. *Life Sciences*, 10:781–789.
Moore, K. E., and Dominic, J. A. (1971): Tyrosine hydroxylase inhibitors. *Federation Proceedings*, 30:859–870.
Stein, L. (1964): Self-stimulation of the brain and the central stimulant action of amphetamine. *Federation Proceedings*, 23:836–850.
Thornburg, J. E., and Moore, K. E. (1971): Stress-related effects of various inhibitors of catecholamine synthesis in the mouse. *Archives Internationales de Pharmacodynamie et de Thérapie*, 194:158–167.
Thornburg, J. E., and Moore, K. E. (1973): The relative importance of dopaminergic and noradrenergic neuronal systems for the stimulation of locomotor activity induced by amphetamine and other drugs. *Neuropharmacology*, 12:853–866.
Thornburg, J. E., and Moore, K. E. (1974): A comparison of effects of apomorphine and ET495 on locomotor activity and circling behavior in mice. *Neuropharmacology*, 13:189–197.
Ungerstedt, U. (1971): Postsynaptic supersensitivity after 6-hydroxydopamine induced degeneration of the nigro-striatal dopamine system. *Acta Physiologica Scandinavica*, Suppl. 367:69–73.
Von Voigtlander, P. F., and Moore, K. E. (1971a): Dopamine: Release from the brain *in vivo* by amantadine. *Science*, 174:408–410.
Von Voigtlander, P. F., and Moore, K. E. (1971b): The release of H³-dopamine from cat brain following electrical stimulation of the substantia nigra and caudate nucleus. *Neuropharmacology*, 10:733–741.
Von Voigtlander, P. F., and Moore, K. E. (1973a): Involvement of nigro-striatal neurons in the *in vivo* release of dopamine by amphetamine, amantadine and tyramine. *Journal of Pharmacology and Experimental Therapeutics*, 184:542–552.
Von Voigtlander, P. F., and Moore, K. E. (1973b): Turning behavior of mice with unilateral

6-hydroxydopamine lesions in the striatum: Effects of apomorphine, L-DOPA, amantadine, amphetamine and other psychomotor stimulants. *Neuropharmacology,* 12:451–462.

Weissman, A., Koe, B. K., and Tenen, S. S. (1966): Antiamphetamine effects following inhibition of tyrosine hydroxylase. *Journal of Pharmacology and Experimental Therapeutics,* 151:339–352.

//
Neuropsychopharmacology of Monoamines and
Their Regulatory Enzymes, edited by E. Usdin.
Raven Press, New York © 1974.

The Use of a Dopaminergic Receptor Stimulating Agent (Piribedil, ET 495) in Parkinson's Disease

Abraham Lieberman, Yves Le Brun, Dinkar Boal, and Mehdi Zolfaghari

Department of Neurology, New York University School of Medicine, New York, New York 10016

I. INTRODUCTION

Since the initial reports establishing the efficacy of treating Parkinson's disease with large amounts of L-DOPA, most observers have noted improvement in approximately two-thirds of patients (1–6). Recently, however, there have been reports of patients who, while initially responsive to L-DOPA, have after 2 to 3 years of optimal therapy shown progression of their disease (7,8). Additionally, there have been reports of patients who, while appearing to have Parkinson's disease clinically, were unresponsive to L-DOPA therapy (9,10). Further, a number of observers have noted that all of the symptoms of Parkinson's disease do not respond equally well to L-DOPA: Rigidity and bradykinesia respond more than tremor (2,3,6,11,12). Lastly, approximately one-half of the patients treated with L-DOPA develop abnormal involuntary movements (AIM), which limits therapy (1–5,7). Decreased responsiveness with time, inability of some patients to respond, inability of L-DOPA to ameliorate all of the symptoms of parkinsonism, and development of AIM have led to the continued search for new agents.

The development of Parkinson's disease is thought to be secondary to a loss of the normal (inhibitory) dopaminergic input into the corpus striatum. This loss is secondary to a degeneration of the dopaminergic cell bodies in the substantia nigra. L-DOPA probably ameliorates parkinsonism by reestablishing the dopaminergic inhibition of the striatum. Failure to respond to L-DOPA is thought to be related to degeneration of striatal dopaminergic receptors (12). There has thus been a search for agents that stimulate dopaminergic receptors directly.

Apomorphine has been shown to be a dopaminergic receptor stimulating agent (13–16). Apomorphine has been used in the treatment of Parkinson's disease (17-19), and has been reported to be effective in abolishing the tremor but not the rigidity and bradykinesia and, like L-DOPA, can also produce AIMs. The usefulness of apomorphine is limited by the necessity for parenteral administration and by its relatively brief duration of action. The reports by Corrodi and Fuxe (20,21) of a new agent, 1-(2″-pyrimidyl) 4-piperonylpiperazinel, piribedil (ET 495), with a prolonged stimulatory action on dopaminergic receptors, led to its use first in primate models of parkinsonism (monkeys with mesencephalic ventromedial tegmental lesions), where it was shown to be effective in relieving tremor and evoking abnormal involuntary movements (22), and then to its use in human Parkinson's disease.

II. METHODS

A. Use of Piribedil (Oral) in Parkinson's Disease

Seven patients with Parkinson's disease, ranging in age between 34 and 75 years, were selected for the study. Six of the patients had been on, or were being treated with, L-DOPA. All had, at one time, shown a response to L-DOPA, advancing by more than one stage on the scale of Hoehn and Yahr (23). In all patients L-DOPA had been decreased or discontinued because of adverse effects: AIMs in four and confusion in two. At the time of initiation of therapy with piribedil, three patients were still receiving L-DOPA. One of these patients was also on amantadine, and one was on amantadine and trihexyphenidyl (Artane®, an anticholinergic drug). All of the antiparkinsonian drugs were held constant during the trial of piribedil. The purpose of the study was carefully explained to the patients and their families, and informed consent was obtained.

Piribedil was administered in 20-mg tablets beginning on a three-times-a-day schedule. Dosage was titrated upward by 20 mg every other to every third day as tolerated until toxicity (usually nausea). Maximal dose did not exceed 280 mg/day. All of the patients were evaluated at weekly intervals by a physician who was familiar with the patient (herein designated as the "physician therapist") and who was aware of their medication. The patients were also evaluated at weekly intervals by another physician (herein designated the "blind observer") who was unfamiliar with their medication. This physician scored the patient at weekly intervals on a Parkinson's Disease Evaluation Form (Table 1), which is a modification of the Northwestern University Disability Scale. Separate scores for each of the cardinal manifestations of Parkinson's disease (rigidity, tremor, bradykinesia, and gait disturbance) were obtained as well as a total score: The higher the score, the worse the patient (the scores recorded in Table 1 are those of the blind observer).

TABLE 1. *Parkinson's disease evaluation form*

Rigidity (judged on passive movement of major joints with patient relaxed in sitting position; cogwheeling to be ignored)
0 — Absent
1 — Slight or detectable only when activated by contralateral or other movements
2 — Mild to moderate
3 — Marked, but full range of motion easily achieved
4 — Severe, full range of motion achieved with difficulty

	(736) RUE	(737) LUE	(738) NECK	(739) RLE	(740) LLE
0	0	0	0	0	0
10	1 2	1 2	1 2	1 2	1 2
20	3 4	3 4	3 4	3 4	3 4

Tremor
0 — Absent
1 — Slight and infrequently present
2 — Moderate in amplitude but only intermittently present
3 — Moderate and present most of the time
4 — Marked in amplitude and present most of the time

	(731) RUE	(732) LUE	(733) FACE	(734) RLE	(735) LLE
0	0	0	0	0	0
10	1 2	1 2	1 2	1 2	1 2
20	3 4	3 4	3 4	3 4	3 4

Bradykinesia (brought out on four-limb simultaneous movement — scored on worse side)
0   0 — None
10 1 — Minimal slowness giving movement a deliberate character; could be normal for some persons
30 2 — Mild degree of slowness and poverty of movement which is definitely abnormal
40 3 — Moderate slowness with occasional hesitation on initiating movement and arrests of ongoing movement
75 4 — Marked slowness and poverty of movement, with frequent and long delays in initiation movement

Gait
0   Walks 10 m and returns in 15 sec or less
15   Walks 10 m and returns in 20 sec or less
30   Walks 10 m and returns in 25 sec or less
45   Walks 10 m and returns in 30 sec or less
60   Walks 10 m and returns in 35 sec or less
75   Walks 10 m and returns in 40 sec or less
100 Walks 10 m and returns in 41 sec or more

B. Use of Piribedil (Parenteral) in Parkinson's Disease and Other Extrapyramidal Disorders

Piribedil, 3 mg, was administered by slow intravenous injection to three patients with Parkinson's disease. Two of these patients had previously participated in the trial of oral piribedil. All three of the patients were receiving L-DOPA at the time of injection. All three had predominantly resting tremors. In one, there was also a large action innervation component. None of these patients had experienced AIMs on oral piribedil or on L-DOPA.

Piribedil was also administered to one patient with familial (essential) tremor and to one patient with the recessive form of dystonia musculorum deformans.

C. Use of Piribedil (Oral) in Extrapyramidal Disorders Other Than Parkinson's Disease

Piribedil (oral) was administered to two patients with familial (essential) tremor, two patients with Huntington's disease, and one patient with the recessive form of dystonia musculorum deformans. Starting doses and titration schedule were similar to those in patients with Parkinson's disease.

III. RESULTS

A. Use of Piribedil (Oral) in Parkinson's Disease

Three of the seven patients were judged by both the therapist and the blind observer to have a reduction in tremor on piribedil (Table 2). None of the other manifestations of Parkinson's disease (rigidity, bradykinesia, or gait disturbance) was judged to be improved. Four patients showed no improvement in any of the manifestations of Parkinson's disease (including tremor). One of these patients (CP) inadvertently discontinued L-DOPA during the first week of the study. There was a worsening in his condition, particularly rigidity and bradykinesia, and, to a lesser extent, tremor. After reinstitution of L-DOPA (with continuation of piribedil) the patient's condition returned to baseline. This patient showed an improvement in tremor on piribedil (maximum tolerated dose 160 mg/day).

Only one of the patients experienced AIMs on piribedil. This patient (HF) had previously experienced AIMs on L-DOPA. Two elderly patients (HF, SS) with mild organic mental syndromes (temporal disorientation, confusion, recent memory loss) with paranoid ideation but without grossly psychotic behavior or agitation became severely confused, temporally and spacially disoriented, psychotic, hallucinatory, and agitated on piribedil. The severe confusion, spacial disorientation, psychosis, hallucinations, and agitation subsided within 1 week of discontinuation of piribedil.

TABLE 2. *Piribedil (oral) in Parkinson's disease*

Patient	Age	Stage	Overall score[a]	Rigidity[a]	Tremor[a]	Brady-kinesia[a]	Gait[a]	AIM	Concomitant L-DOPA	Amount of piribedil (mg)	Duration of piribedil (weeks)
HF	71	5 5	355 355	100 100	80 80	75 75	100 100	Yes	0	280	4
SS	75	5 5	335 335	100 100	60 60	75 75	100 100	0	0	100	2
MB	62	5 5	260 295	90 80	30 40	40 75	100 100	0	0	240	3
JR	34	2 2	115 80	30 20	(50) (20)	30 40	15 0	0	0	200	9
SD	67	4 4	260 270	50 70	70 60	40 40	100 100	0	Yes	120	4
CP	61	3 3	145 115	50 50	(50) (10)	30 40	15 15	0	Yes	160	9
HM	46	3 3	140 115	50 50	(50) (10)	30 10	60 45	0	Yes	240	7

[a] Score is that of the "blind observer."

While three of the patients with stage 5 parkinsonism might be considered as having end-stage disease and thus incapable of further response to medication, it is of interest that two of these patients (SS and MB) showed a good response (advancing by more than one stage on the scale of Hoehn and Yahr) to L-DOPA administered in combination with a peripheral DOPA decarboxylase inhibitor (MK 486). One of these patients (MB) advanced by still another stage when amantadine was added to the combination of MK 486 and L-DOPA. This patient, who had been helpless and bedridden on 240 mg of piribedil a day, was now, while still obviously parkinsonian, able to arise from a chair unaided, and shave, dress, and feed himself. He was able to walk 10 m and return in 15 sec. All of the patients experienced nausea on piribedil, and this was the dose-limiting factor in five patients.

B. Use of Piribedil (Parenteral) in Parkinson's Disease and Other Extrapyramidal Disorders

1. *Parkinson's disease.* Two of the three parkinsonian patients and the one patient with essential tremor were judged to have a significant reduction in tremor—documented with moving pictures (Table 3). In one patient, SD, the tremor was bilateral, involved all four extremities, was worse on the left, and was present at rest as well as on action-innervation. Within 5 min of the injection of piribedil, there was a marked bilateral reduction in the tremor. The tremor of the left hand remained worse than the right. The tremor in the lower extremities was virtually gone. There was no change in bradykinesia or in rigidity. The effect lasted for 30 min. This patient's tremor had not responded to oral piribedil (maximum dose 120 mg/day). In another patient, SU, the tremor involved only the right hand and was present only

TABLE 3. *Piribedil (parenteral) in Parkinson's disease and other extrapyramidal disorders*

Patient	Diagnosis	Side of tremor	Score resting tremor	Score action tremor	AIM
SD	Parkinson's	Bilateral	70 40	70 40	0 0
JR	Parkinson's	Left	30 30	10 10	0 0
SU	Parkinson's	Right	20 0	0 0	0 0
JB	Familial tremor	Right	0 0	20 0	0 0
RR	Dystonia	Right	0 0	20[a] 0	0 0

[a] RR had a rhythmical movement resembling an action tremor.

at rest. It was moderate in amplitude and present most of the time, and was judged 3 on a scale of 4 in the Northwestern University Disability Scale — equivalent to a score of 20 on our Parkinson's Disease Evaluation Form. Within 5 min of injection of piribedil, the tremor disappeared and remained suppressed for $2\frac{1}{2}$ hr (score of "0" on our Evaluation Form). There was no change in rigidity or bradykinesia. This patient had never received piribedil orally. In a third patient, JR, parenteral piribedil resulted in no change in the resting tremor. This patient had previously failed to respond to oral piribedil, 200 mg/day.

2. *Other extrapyramidal disorders.* In JB, a patient with familial (essential) tremor, there was a marked diminution in the action-innervation tremor after parenteral piribedil. This patient had shown a good but less dramatic response to oral piribedil, 80 mg/day.

Piribedil was also administered to one patient with the recessive form of dystonia musculorum deformans with rhythmical movements resembling an action-innervation tremor. There was no response to piribedil.

All of the patients experienced nausea and vomiting within 5 min of the administration of piribedil. All experienced declines in supine blood pressure and all experienced bradycardia, and all experienced orthostatic hypotension upon standing. None became syncopal.

C. Use of Piribedil (Oral) in Extrapyramidal Disorders Other Than Parkinson's Disease

1. *Familial (essential) tremor.* Piribedil was administered to two patients with familial tremor, tremor most marked on action innervation. In one patient (JB), who was able to tolerate 80 mg of piribedil a day for 9 weeks, there was a decrease in the amplitude of his tremor. It was easier for the patient to shave and to dress himself and to hold a cup of coffee. The other patient was unable to tolerate 80 mg of piribedil a day, and discontinued the medication after 2 weeks. There had been no change in his tremor.

2. *Huntington's disease.* Piribedil was administered to two patients with Huntington's disease, characterized by abnormal involuntary movements and minimal mental changes. One of the patients noted a "calming effect" on 160 mg of piribedil (thought to be a "placebo effect"). No change (neither increase or decrease) of the AIMs was noted. The patient later discontinued the medication. The second patient was able to tolerate a maximum of 200 mg of piribedil a day before becoming severely nauseated and then psychotic. This cleared within 1 week. When observed by the physician-therapist 1 month after discontinuation of piribedil, the patient's mental status was judged to be unchanged from baseline; her AIMS were somewhat more frequent and her handwriting and ability to draw had worsened slightly. This was felt to represent progression of her underlying disease.

IV. DISCUSSION

Piribedil was moderately effective in relieving the tremor in three of seven patients with Parkinson's disease. The drug had no effect on the other symptoms of the disease (rigidity, bradykinesia, gait disturbance) in these patients. The drug had no effect on any of the symptoms (tremor, rigidity, bradykinesia, gait disturbance) in the four remaining patients. Relief of tremor, when it occurred, was more pronounced after parenteral administration. Abnormal involuntary movements occurred in only one patient, and in this patient were unassociated with relief of tremor. These findings are comparable to those of other investigators, Table 4 (24–28), and contrast with the more striking improvement noted in animal "parkinsonism": rats and monkeys with nigrostriatal lesions (20–22). In the monkey 3.0 mg/kg of piribedil results in relief of tremor for 3 to 4 hr and usually evokes abnormal involuntary movements. We were unable to use parenteral dosages of more than 0.05 mg/kg in man because of the development of orthostatic hypotension, nausea, and vomiting, effects not observed in the monkey.

TABLE 4. *Summary of other investigations of the use of piribedil (oral) in Parkinson's disease*

Series	Number of patients	Amount of piribedil (mg)	Number on L-DOPA	Number developing AIM	Comment
Shaw and Stern	17	120	7	0	5 patients worse (rigidity, tremor)
Calne et al.	15	20–240	15	11	4 patients improved (tremor)
McLellan et al.	9	200	5	1	3 patients improved (tremor)
Fieschi et al.	18	170–180	1	No comment	13 patients improved (tremor predominantly; also rigidity)
Emile and Chanelet	10	160–240	2	No comment	6 patients improved (tremor)

The relative failure of oral piribedil in human parkinsonism may be explained in one of two ways:

1. *Piribedil may be poorly absorbed from the gastrointestinal tract.* Although Sandler et al. were unable to demonstrate any effect of oral piribedil in relatively small quantities (120 mg/day) on the overall turnover of urinary dopamine as measured by gas chromatography with a sensitivity limit of 2 ng/ml in five parkinsonian patients (29), Campbell et al., in a later study utilizing thin-layer chromatography coupled with direct-inlet mass spectrometry, were able to show that, in seven parkinsonian patients receiving up to

240 mg/day of piribedil, metabolites could be detected in the urine as early as 30 min after oral administration. This would suggest that piribedil is rapidly absorbed from the gastrointestinal tract. The studies of Chase, showing a 57% reduction in mean cerebrospinal fluid homovanillic acid in eight patients receiving piribedil, also confirm the fact that the drug is absorbed and that it (or a metabolite) penetrates the central nervous system (31). The extent of CNS penetration of piribedil or its metabolite cannot be determined.

2. *A metabolite of piribedil may be the active agent.* Fuxe has shown that piribedil is inactive when given locally into the neostriatum and cannot counteract the nervous-induced release of dopamine *in vitro* from brain slices as can another dopaminergic receptor stimulating agent, apomorphine (32). Piribedil contains a methylenedioxybenzy group, and it has been shown that a major route of metabolism is cleavage of the methylene dioxyphenyl bridge to give a compound (identified as S 584), a catechol, bearing a close resemblance to dopamine (30,33). This compound, like dopamine, may be unable to penetrate the blood-brain barrier, and the antiparkinsonian effect of piribedil thus may be dependent on the ability of piribedil itself to avoid being metabolized in the periphery, penetrate the blood-brain barrier, and be converted to S 584. This may explain why agents, such as SKF 525, which slow the peripheral metabolism of piribedil, enhance the antiparkinsonian effect of the drug (32).

We believe that the discrepancies between the action of piribedil in man and in the monkey can best be explained by postulating that piribedil works through a metabolite and only after this metabolite has gained access to the central nervous system.

The differential effect of piribedil on parkinsonian tremor and not on rigidity and bradykinesia is consistent with the observation of others (11,12) that the mechanisms subserving tremor in Parkinson's disease may be different from those subserving rigidity and bradykinesia.

V. SUMMARY

Piribedil (ET 495), a dopaminergic receptor stimulating agent, when given orally in dosages of 2 to 4 mg/kg, was moderately effective in relieving the tremor in three of seven patients with Parkinson's disease. The drug had no effect on the other manifestations of the disease (rigidity, bradykinesia, gait disturbance) and was without effect in the other four patients. Abnormal involuntary movements were observed in only one patient. Relief of tremor, when it occurred, was more pronounced after parenteral (intravenous) administration in dosages of 0.05 mg/kg. These findings are in contrast to the more striking improvement in tremor in monkeys with nigrostriatal lesions after parenteral administration in dosages of 3 mg/kg, relief of tremor at these dosages being associated with the development of abnormal involuntary movements. Parenteral dosage in man was limited by the development of toxicity, chiefly vomiting, an effect not observed in the monkey. It is suggested that the effect of piribedil in relieving tremor and perhaps in producing abnormal involuntary movements is mediated through a metabolite, a catechol bearing a resemblance to dopamine.

ACKNOWLEDGMENTS

The authors wish to thank Dr. Menek Goldstein for his advice and counsel and Mrs. Joanne Nesbitte, R.N., for her help.

The authors also wish to express their thanks to M. Derome-Tremblay, Ph.D., and Les Laboratoires Servier, Neuilly-sur-Siene, France, for their generous support and assistance.

REFERENCES

1. Cotzias, G. C., Van Woert, M. H., and Schiffer, L. M. (1967): Aromatic amino acids and modifications of parkinsonism. *New England Journal of Medicine*, 276:374–379.
2. Cotzias, G. C., Papavasiliou, P. S., and Gellene, R. (1969): Modification of parkinsonism: Chronic treatment with L-DOPA. *New England Journal of Medicine*, 280:337–345.
3. Barbeau, A. (1969): L-DOPA therapy in Parkinson's disease. *Canadian Medical Association Journal*, 101:791–800.
4. Yahr, M. D., Duvoisin, R. C., and Schear, M. J. (1969): Treatment of parkinsonism with levodopa. *Archives of Neurology*, 21:343–354.
5. McDowell, F., Lee, J. E., and Swift, T. (1970): Treatment of Parkinson's syndrome with L-dihydroxyphenylalanine (levodopa). *Annals of Internal Medicine*, 72:29–35.
6. Calne, D. B., Stern, G. M., and Spiers, A. S. D. (1969): L-DOPA in idiopathic parkinsonism. *Lancet*, 2:973–976.
7. Mones, R. J. (1972): Parkinson's disease treated with L-DOPA: Three year follow-up report. *New York State Journal of Medicine*, 72:2749–2756.
8. Hunter, K. R., Shaw, K. M., and Laurence, D. R. (1973): Sustained levodopa therapy in parkinsonism. *Lancet*, 2:929–931.
9. Mones, R. J. (1973): An analysis of six patients with Parkinson's disease who have been unresponsive to L-DOPA therapy. *Journal of Neurology, Neurosurgery and Psychiatry*, 36:362–367.
10. Lieberman, A. N., Goodgold, A. L., and Goldstein, M. (1972): Treatment failures with levodopa in parkinsonism. *Neurology*, 22:1205–1210.
11. Abramsky, O., Carmon, A., and Lavy, S. (1971): Combined treatment of parkinsonian tremor with propranolol and levodopa. *Journal of Neurological Science*, 14:491–494.
12. Klawans, H., Ilahi, M. M., and Shenker, D. (1970): Theoretical implications of the use of L-DOPA in parkinsonism. *Acta Neurologica Scandinavica*, 46:409–441.
13. Andén, N. E., Rubenson, A., and Fuxe, K. (1967): Evidence for dopamine receptor stimulation by apomorphine. *Journal of Pharmacy and Pharmacology*, 19:627–629.
14. Roos, B. E. (1969): Decrease in homovanillic acid as evidence for dopamine receptor stimulation by apomorphine in the neostriation of the rat. *Journal of Pharmacy and Pharmacology*, 21:263–264.
15. Butcher, L. L., and Andén, B. L. (1969): Effects of apomorphine and amphetamine on schedule controlled behavior: Reversal of tetrabenzazine supression and dopaminergic correlates. *European Journal of Pharmacology*, 6:255–264.
16. Ungerstedt, U., Butcher, L. L., and Butcher, S. G. (1969): Direct chemical stimulation of dopaminergic mechanisms in the neostriatum of the rat. *Brain Research*, 14:461–471.

17. Cotzias, G. C., Papavasiliou, P. S., and Fehling, C. (1970): Similarities between neurologic effects of L-DOPA and of apomorphine. *New England Journal of Medicine,* 282:31–32.
18. Braham, J., Sarova-Pinhas, I., and Goldhammer, Y. (1970): Apomorphine in parkinsonian tremor. *British Medical Journal,* 3:768.
19. Duby, S. E., Cotzias, G. C., and Papavasiliou, P. S. (1972): Injected apomorphine and orally administered levodopa in parkinsonism. *Archives of Neurology,* 27:474–480.
20. Corrodi, H., Fuxe, K., and Ungerstedt, U. (1971): Evidence for a new type of dopamine receptor stimulating agent. *Journal of Pharmacy and Pharmacology,* 23:989–991.
21. Corrodi, H., Farnebo, L. O., and Fuxe, K. (1972): ET 495 and brain catecholamine mechanisms: Evidence for stimulation of dopamine receptors. *European Journal of Pharmacology,* 20:195–204.
22. Goldstein, M., Battista, A. F., and Onmoto, T. (1973): Tremor and involuntary movements in monkeys: Effects of L-DOPA and of a dopamine receptor stimulating agent. *Science,* 179:816–817.
23. Hoehn, M. H., and Yahr, M. D. (1967): Parkinsonism: Onset, progression and mortality. *Neurology,* 17:427–442.
24. Shaw, K. M., and Stern, G. M. (1972): Pirimidinyl piperonyl piperazine (ET 495) in the treatment of parkinsonism. Read before the Symposium International Trivastal, Monastir, Tunisia.
25. Calne, D. B., Testigo, J. V., and Reid, J. L. (1972): A double blind study of ET 495 in patients receiving levodopa. Read before the Symposium International Trivastal, Monastir, Tunisia.
26. McLellan, D. L., Chalmers, R. J., and Johnson, R. H. (1972): Clinical and physiological evaluation of the effects of ET 495 in patients with parkinsonism. Read before the Symposium International Trivastal, Monastir, Tunisia.
27. Fieschi, C., Nardini, M., and Sciannandrone, R. (1972): A single blind study upon the effects of trivastal in Parkinson's disease. Read before the Symposium International Trivastal, Monastir, Tunisia.
28. Emile, J., and Chanelet, J. (1972): Preliminary research on the use of piribedil in Parkinson's disease. Read before the Symposium International Trivastal, Monastir, Tunisia.
29. Sandler, M., and Ruthven, C. R. J. (1972): The effect of ET 495 on catecholamine metabolism. Read before the Symposium International Trivastal, Monastir, Tunisia.
30. Campbell, D. B., Taylor, A. R., and Jenner, P. (1972): Kinetics and metabolism of ET 495 in humans. Read before the Symposium International Trivastal, Monastir, Tunisia.
31. Chase, T. N. (1973): Central monoamine metabolism in man: Effects of putative dopamine receptor agonists and antagonists. *Archives of Neurology,* 29:349–351.
32. Fuxe, K., Corrodi, H., and Farnebo, L. O. (1972): On the neuropharmacology of ET 495. Read before the Symposium International Trivastal, Monastir, Tunisia.
33. Jenner, P., Taylor, A. R., and Campbell, D. B. (1972): The *in vitro* and *in vivo* metabolism of ET 495 in animals. Read before the Symposium International Trivastal, Monastir, Tunisia.

Neuropsychopharmacology of Monoamines and Their Regulatory Enzymes, edited by E. Usdin. Raven Press, New York © 1974.

Clinical Studies of Dopaminergic Mechanisms

Thomas N. Chase

Neurology Unit, National Institute of Mental Health, Bethesda, Maryland 20014

I. INTRODUCTION

The remarkable ability of L-DOPA to ameliorate parkinsonian signs has generally been attributed to a drug-induced increase in the availability of dopamine (DA) at striatal receptor sites. Since the nigrostriatal dopaminergic pathway degenerates in Parkinson's disease, there is some question as to what role surviving neurons comprising this system play in the decarboxylation of exogenous DOPA and in the release of the amine thus formed onto the postsynaptic DA receptors. Preclinical studies also raise the possibility that certain of the pharmacologic effects of L-DOPA may reflect alterations in serotonergic or noradrenergic function or be attributable to some DOPA metabolite other than DA (Ng, Chase, Colburn, and Kopin, 1972; Sandler, 1972). A comparison of the clinical effects of L-DOPA with those of a presumptive direct-acting DA receptor agonist provides one approach to the evaluation of these issues.

Recent preclinical evidence suggests that 1-(2″-pirimidyl)-4-piperonyl-piperazine (piribedil) acts to stimulate central dopaminergic receptors. Piribedil substantially diminishes DA turnover in rodent brain and causes rats with 6-hydroxydopamine-induced degeneration of one nigrostriatal DA pathway to rotate toward the unoperated side (Corrodi, Farnebo, Fuxe, Hamberger, and Ungerstedt, 1972). These actions closely resemble those of another suspected DA receptor agonist, apomorphine (Corrodi et al., 1972). Piribedil has also been found to mimic the effects of L-DOPA and apomorphine in reducing tremor and inducing repetitive, involuntary movements in monkeys with ventral tegmental lesions (Goldstein, Battista, Ohmoto, Anagnoste, and Fuxe, 1973). The effects of both piribedil and apomorphine on motor function are inhibited by DA receptor blocking agents such as haloperidol or pimozide (Goldstein et al., 1973).

II. HALOPERIDOL AND PIRIBEDIL EFFECTS ON DOPAMINE TURNOVER

Pharmacologic observations in the experimental animal suggest an inverse relationship between the activity of dopaminergic receptors and the rate of DA turnover (Carlsson, Kehr, Lindqvist, Magnusson, and Attack, 1972).

Neuroleptics that appear to block DA receptors characteristically accelerate the synthesis and release of this catecholamine (Corrodi, Fuxe, and Hökfelt, 1967; Cheramy, Besson, and Glowinski, 1970; Nybäck, Schubert, and Sedvall, 1970). Conversely, apomorphine and piribedil, which purportedly stimulate dopaminergic receptors, substantially diminish cerebral DA turnover (Andén, Rubenson, Fuxe, and Hökfelt, 1967; Nybäck et al., 1970; Corrodi et al., 1972). It has been suggested that these changes in DA metabolism reflect a compensatory feedback mechanism acting directly on impulse activity in presynaptic dopaminergic fibers (Corrodi et al., 1967; Nybäck et al., 1970; Bunney, Walters, Roth, and Aghajanian, 1973). Measurement of central DA turnover during the administration of piribedil might thus afford some insight into the drug's mechanism of action in man (Chase, 1973). In the present studies, DA turnover was estimated by the probenecid-induced accumulation of homovanillic acid (HVA) in lumbar cerebrospinal fluid (CSF). The rationale for this approach as well as a discussion of its assumptions and limitations have been presented elsewhere (Korf and Van Pragg, 1971; Bowers, 1972; Chase and Ng, 1972; Goodwin, Post, Dunner, and Gordon, 1973).

Haloperidol treatment of patients with various neurologic disorders lead to a 90% rise in the average probenecid-induced accumulation of HVA (Table 1). In another group of patients the administration of piribedil was associated with a 52% reduction in DA turnover as estimated by the oral probenecid loading technique (Table 1). Included in the haloperidol-treated group were five parkinsonian patients and five individuals with Huntington's chorea. Although no significant change in the HVA responses to probenecid occurred in the former group ($20 \pm 11\%$; $p > 0.05$), HVA increments were markedly increased by haloperidol in choreatic patients ($189 \pm 64\%$; $p < 0.05$). Similarly, piribedil treatment of 11 parkinsonian patients was associated with a $46 \pm 21\%$ reduction in HVA accumulations ($p \geq 0.05$), in contrast to a $61 \pm 4.5\%$ decline in the four nonparkinsonian patients ($p <$

TABLE 1. *Effect of haloperidol and piribedil on probenecid-induced accumulation of homovanillic acid in lumbar CSF*

Drug	Number of patients	Homovanillic acid Placebo	Treatment	Difference	p
Haloperidol	22	84 ± 16	160 ± 31	76 ± 24	<0.01
Piribedil	15	58 ± 11	27 ± 6	-30 ± 8	<0.01

Patients with various neurologic disorders were studied after receiving haloperidol (maximum dose 8 to 10 mg/day) for an average of about 2 weeks or piribedil (average maximum dose 227 mg/day) for about 3 weeks. The technique for administering probenecid as well as for the collection and assay of CSF are as previously described (Chase and Ng, 1972). Homovanillic acid values are the means ± SEM, expressed in nanograms per milliliter.

0.001). These differences presumably reflect the diminished functional capacity of the degenerating DA-containing neurons in parkinsonian patients.

III. PIRIBEDIL EFFECTS ON PARKINSON'S DISEASE

The antiparkinsonian efficacy and toxicity of orally administered piribedil was compared against an inert placebo in 16 patients (nine men, seven women, ages 43 to 77 years) with idiopathic parkinsonism (Chase, Woods, and Glaubiger, 1974). A single-blind design was employed, with the duration of active drug therapy averaging 38 days. Maximum daily dose levels of piribedil averaged 276 mg (range, 100 to 520 mg) given in four to six divided doses. Throughout the entire period of drug and placebo administration, four patients received a constant, low to moderate dose of a conventional, anticholinergic antiparkinsonian preparation. None of the other patients received centrally active drugs during or for at least 1 week prior to entering the study. Neurologic evaluations were performed at least twice weekly by two observers.

At maximum dose levels piribedil led to a statistically significant reduction in overall parkinsonian severity, averaging 30% and ranging from 14 to 67% for the individual patient (Fig. 1). For the entire group of 16 patients, each of the three cardinal parkinsonian signs improved significantly, although the individual responses varied from 0 to 71% improvement for akinesia and from 0 to 100% for rigidity and tremor. Tremor appeared somewhat more

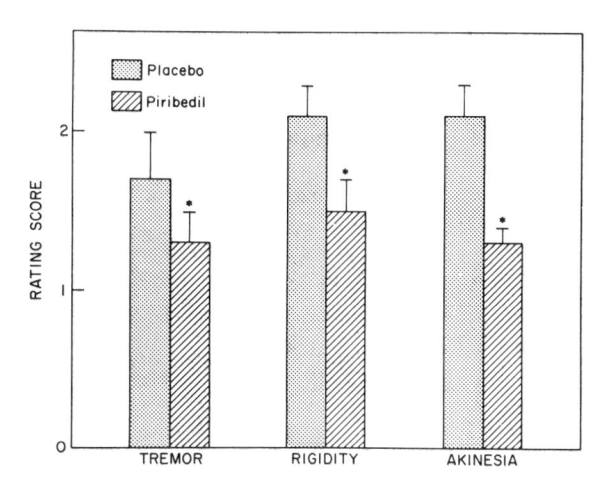

FIG. 1. Effect of piribedil on cardinal parkinsonian signs. Tremor, rigidity, and akinesia were each rated on a scale of 0 (absent) to 4 (very severe) during placebo administration and at maximum dose levels of piribedil (average dose 276 mg/day). Values are the means ±SEM for 16 patients. * $p < 0.001$ for difference from placebo score by t test for paired data.

resistant to piribedil therapy than the other two major signs of parkinsonism. Scores for tremor improved by 25% or more in only six individuals, while akinesia and rigidity scores improved by at least this amount in 11 patients. Adverse effects of piribedil included nausea and vomiting (seven patients), drowsiness (five patients), personality changes (five patients) and dyskinesias (eight patients). The clinical appearance of piribedil-induced dyskinesias closely resembled those observed with L-DOPA. Indeed, in any given patient the pattern of abnormal involuntary movements produced by piribedil was indistinguishable from that occurring with L-DOPA. There was no laboratory evidence of toxicity attributable to piribedil therapy.

IV. COMPARISON OF L-DOPA AND PIRIBEDIL

Twelve of the 16 patients given piribedil also received a comparable therapeutic trial of L-DOPA in combination with a peripheral decarboxylase inhibitor (carbidopa) either just before or shortly after receiving piribedil. In this group optimal dose levels of L-DOPA had nearly twice the overall antiparkinsonian efficacy of piribedil (51 ± 7 vs. $29 \pm 3\%$ improvement). Moreover, for each of the three cardinal parkinsonian signs, L-DOPA produced a significantly greater improvement than piribedil. Patients whose parkinsonian symptoms responded poorly to relatively high doses of L-DOPA (given alone or with carbidopa) also tended to respond poorly to piribedil. Indeed, there was a significant positive correlation between the amount or percent overall improvement on piribedil and the amount or percent overall improvement on L-DOPA (Fig. 2).

Patients included in the present study cannot be considered entirely

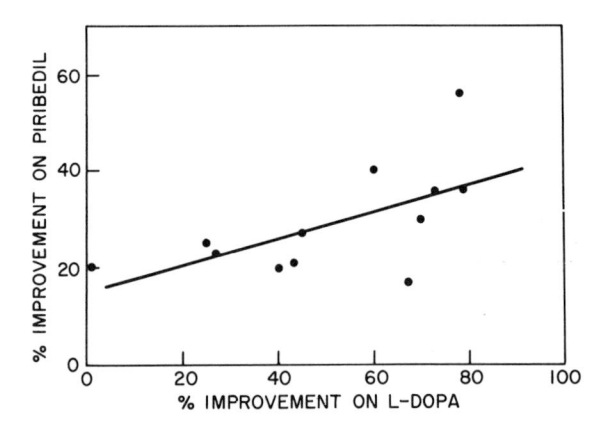

FIG. 2. Relationship between the response to piribedil and L-DOPA. There was a significant correlation between percent improvement in cardinal parkinsonian signs while receiving piribedil (average maximum dose 285 mg/day) and during optimal treatment with L-DOPA (average dose 1.4 g/day) together with carbidopa (75 to 100 mg/day) in these 12 patients ($r = 0.619$; $p < 0.05$).

representative of those suffering from Parkinson's disease. Fourteen had previously received therapeutic trials of L-DOPA. Results were clearly successful in only seven, while in four others L-DOPA had been withdrawn due to lack of efficacy or intolerable toxicity. L-DOPA was considered of limited usefulness in the three remaining patients due to adverse effects or relatively poor efficacy. A comparison of our results with piribedil against those obtained in a larger and possibly more typical group of L-DOPA-treated parkinsonian patients is shown in Table 2. The antiparkinsonian efficacy of L-DOPA again may be seen to exceed substantially that of piribedil, while the incidence and severity of dyskinesias were similar. Tremor tended to be the least responsive of the cardinal parkinsonian signs during treatment with either drug.

TABLE 2. *Comparative antiparkinsonian efficacy and toxicity of piribedil and L-DOPA*

Drug	Number of patients	Percent improvement	Percent dyskinesias
Piribedil	16	30 ± 4	50
L-DOPA + carbidopa	25	62 ± 5	80

Neurologic ratings were obtained in piribedil-treated patients after an average of 38 days of therapy (average maximum dose of 276 mg/day) or in patients who had been optimally treated with L-DOPA (average dose 1.4 g/day) together with carbidopa (75 to 100 mg/day) for at least 3 months.

V. CONCLUDING SPECULATIONS

Several possibilities arise to explain the relatively modest therapeutic efficacy of piribedil as compared with L-DOPA. In parkinsonian patients optimally treated with L-DOPA, the amount of overall improvement, but not the percent improvement, correlates with pretreatment severity of cardinal parkinsonian signs (Fig. 3). Since the number of residual nigrostriatal DA neurons capable of decarboxylating DOPA and releasing DA into the synaptic cleft probably diminishes as the disease progresses (Chase and Ng, 1972), one might expect that L-DOPA would be least effective in patients with advanced disease. Since our data fail to support this possibility (Fig. 3), it is conceivable that the nigrostriatal DA neurons are not obligatory participants in the mechanism of antiparkinsonian action of L-DOPA, or that a compensatory increase in postsynaptic receptor sensitivity attends the degeneration of presynaptic DA fibers. The latter possibility is indirectly supported by both preclinical (Corrodi et al., 1972; Schoenfeld and Uretsky, 1973) and clinical observations (Chase, Holden, and Brody, 1973). If denervation supersensitivity does occur, and if piribedil acts primarily as a DA receptor agonist, then piribedil might be expected to

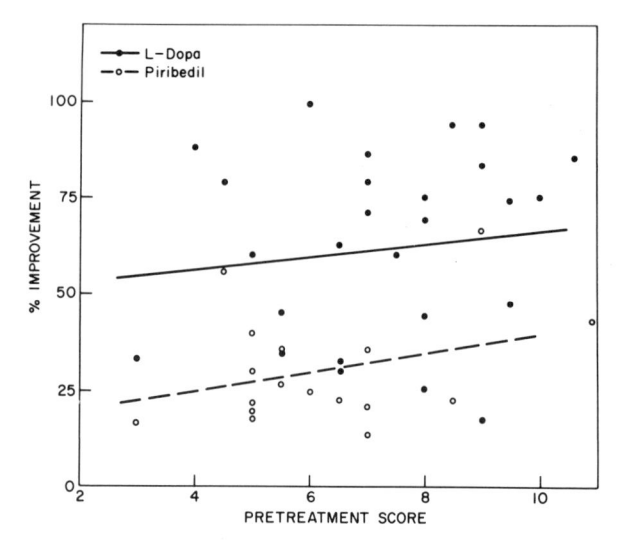

FIG. 3. Pretreatment parkinsonian severity and response to piribedil or L-DOPA. No correlation was found between pretreatment severity and percent improvement in 16 patients given piribedil (average maximum dose 276 mg/day; $r = 0.267$; $p > 0.05$) or in 25 patients receiving optimal dose levels of L-DOPA (average dose 1.4 g/day), together with carbidopa (75 to 100 mg/day; $r = 0.110$; $p > 0.05$).

be relatively more effective in patients with severe parkinsonism. This, however, does not appear to be the case (Fig. 3). As with DOPA, the pretreatment severity of parkinsonian signs correlated with the amount ($r = 0.626$) but not percent ($r = 0.267$) improvement on piribedil. Recent preclinical evidence suggesting that the mechanism of action of piribedil and apomorphine involves presynaptic events (Costall and Naylor, 1973; Goldstein et al., 1973) may provide a partial explanation for these clinical observations. The contribution of pharmacokinetic factors or the ability of piribedil as well as L-DOPA to influence monoaminergic systems other than those containing dopamine (Ng et al., 1972; Fuxe, Corrodi, Farnebo, Hamberger, and Ungerstedt, 1973) also cannot be excluded.

Although the antiparkinsonian efficacy of piribedil approximates that of conventional anticholinergic preparations (England and Schwab, 1961) more closely than that of L-DOPA, it would appear unlikely that this drug acts primarily as a central cholinolytic agent. Preclinical studies indicate that piribedil increases striatal acetylcholine levels (Ladinsky. Consolo. and Garattini, 1974). This effect is similar to that reported with chronic L-DOPA administration (Sethy and Van Woert, 1973) and the opposite to what occurs in rodents given atropinic agents (Consolo, Ladinsky, Peri, and Garattini, 1972). Moreover, although dyskinesias may appear in parkinsonian patients receiving anticholinergics (Fahn and David, 1973), they clearly occur more rarely than during treatment with either L-DOPA or piribedil. A close functional relationship undoubtedly exists between cholinergic and dopaminergic pathways in the basal ganglia. The role of these mutually interactive systems in the pathogenesis of Parkinson's disease and the mechanisms by which drugs such as piribedil or L-DOPA modify symptoms of this disorder remain to be fully elucidated.

ACKNOWLEDGMENTS

Piribedil was supplied by Servier Laboratories, Paris.

REFERENCES

Andén, N. E., Rubenson, K., Fuxe, K., and Hökfelt, T. (1967): Evidence for dopamine receptor stimulation by apomorphine. *Journal of Pharmacy and Pharmacology*, 19:627–629.
Bowers, M. B. (1972): Clinical measurements of central dopamine and 5-hydroxytryptamine metabolism: Reliability and interpretation of cerebrospinal fluid acid monoamine metabolite measures. *Neuropharmacology*, 11:101–111.
Bunney, B. S., Walters, J. R., Roth, R. H., and Aghajanian, G. K. (1973): Dopaminergic neurons: Effect of antipsychotic drugs and amphetamine on single cell activity. *Journal of Pharmacology and Experimental Therapeutics*, 185:560–571.
Carlsson, A., Kehr, W., Lindqvist, M., Magnusson, T., and Attack, C. V. (1972): Regulation of monoamine metabolism in the central nervous system. *Pharmacological Reviews*, 24:371–384.
Chase, T. N. (1973): Central monoamine metabolism in man. Effect of putative dopamine receptor agonists and antagonists. *Archives of Neurology*, 29:349–351.
Chase, T. N., Holden, E. M., and Brody, J. A. (1973): Levodopa-induced dyskinesias. Comparison in parkinsonism-dementia and amyotrophic lateral sclerosis. *Archives of Neurology*, 29:328–330.
Chase, T. N., and Ng, L. K. Y. (1972): Central monoamine metabolism in Parkinson's disease. *Archives of Neurology*, 27:486–492.
Chase, T. N., Woods, A. C., and Glaubiger, G. A. (1974): Parkinson disease treated with a suspected dopamine receptor agonist. *Archives of Neurology*, 30:383–386.
Cheramy, A., Besson, M. J., and Glowinski, J. (1970): Increased release of dopamine from striatal dopaminergic terminals in the rat after treatment with a neuroleptic: Thioproperazine. *European Journal of Pharmacology*, 10:206–214.
Consolo, S., Ladinsky, H., Peri, G., and Garattini, S. (1972): Effect of central stimulants and depressants on mouse brain acetylcholine and choline levels. *European Journal of Pharmacology*, 18:251–255.
Corrodi, H., Farnebo, L.-O., Fuxe, K., Hamberger, B., and Ungerstedt, U. (1972): ET 495 and brain catecholamine mechanisms: Evidence for stimulation of dopamine receptors. *European Journal of Pharmacology*, 20:195–204.
Corrodi, H., Fuxe, K., and Hökfelt, T. (1967): The effect of neuroleptics on the activity of central catecholamine neurons. *Life Sciences*, 6:767–774.
Costall, B., and Naylor, R. J. (1973): Neuropharmacologic studies on the site and mode of action of ET 495. In: *Advances in Neurology, Vol. 3: Progress in the Treatment of Parkinsonism*, edited by D. B. Calne, Raven Press, New York, pp. 281–293.
England, A. C., and Schwab, R. S. (1961): Medical Progress: Parkinson's syndrome. *New England Journal of Medicine*, 265:785–792.
Fahn, S., and David, E. (1973): Oral dyskinesias secondary to anticholinergic drugs. In: *Excerpta Medica International Congress Series No. 296. Tenth International Congress of Neurology, Barcelona, Spain.* Excerpta Medica Foundation, p. 43.
Fuxe, K., Corrodi, H., Farnebo, L.-O., Hamberger, B., and Ungerstedt, U. (1972): On the neuropharmacology of ET 495. Presented at the International Symposium Trivastal, Monastir, Tunisia, Nov. 27–Dec. 1.
Garattini, S. (1973): *Personal communication*.

Goldstein, M., Battista, A. F., Ohmoto, T., Anagnoste, B., and Fuxe, K. (1973): Tremor and involuntary movement in monkeys: Effect of L-DOPA and of a dopamine receptor stimulating agent. *Science,* 179:816–817.

Goodwin, F. K., Post, R. M., Dunner, D. L., and Gordon, E. K. (1973): Cerebrospinal fluid amine metabolites in affective illness: The probenecid technique. *American Journal of Psychiatry,* 130:73–79.

Korf, J., and Van Pragg, H. M. (1972): Amine metabolism in the human brain: Further evaluation of the probenecid test. *Brain Research,* 35:221–230.

Ladinsky, H., Consolo, S., and Garattini, S. (1974): Increase in striatal acetylcholine levels *in vivo* by piribedil, a new dopamine receptor stimulant. *Life Sciences,* 14:1251–1260.

Ng, L. K. Y., Chase, T. N., Colburn, R. W., and Kopin, I. J. (1972): L-DOPA in parkinsonism. A possible mechanism of action. *Neurology,* 22:688–696.

Nybäck, H., Schubert, J., and Sedvall, G. (1970): Effect of apomorphine and pimozide on synthesis and turnover of labeled catecholamines in mouse brain. *Journal of Pharmacy and Pharmacology,* 22:622–624.

Sandler, M. (1972): Catecholamine synthesis and metabolism in man: Clinical implications (with special reference to parkinsonism). In: *Catecholamines,* edited by H. Blaschko and E. Muscholl. Springer-Verlag, Berlin, pp. 845–899.

Schoenfeld, R. I., and Uretsky, N. J. (1973): Enhancement by 6-hydroxydopamine of the effects of DOPA upon the motor activity of rats. *Journal of Pharmacology and Experimental Therapeutics,* 186:616–624.

Sethy, V. H., and Van Woert, M. H. (1973): Effect of L-DOPA on brain acetylcholine and choline in rats. *Neuropharmacology,* 12:27–31.

Neuropsychopharmacology of Monoamines and Their Regulatory Enzymes, edited by E. Usdin. Raven Press, New York © 1974.

Open Discussion: Dopamine

Reporter: Earl Usdin

Lewander. Dr. Seiden asked several questions on the food given to the rats in the study on amphetamine anorexia; Dr. Lewander said that standard food pellets were available to the animals 7 hr/day. Since no food was available during the night, the animals were hungry at the start of the experiments in the morning. Dr. Lewander said that after high doses of the drug (>5 mg/kg, i.p.), the rats would not start eating until at least 2 or 3 hr after injection; even drug-tolerant animals were completely anorexic after the injection, but for a shorter time period than the nontolerant rats.

Wallach. Dr. Ellinwood inquired about the type of stereotopy produced by morphine; Dr. Wallach replied that he had not made this observation, but that Fog reported morphine stereotopy was similar to that produced by amphetamine. Dr. Way reported that chronic morphine in mice or rats produces repeated jumping behavior when the morphine is discontinued. He has observed that dopamine is more involved in this withdrawal behavior than NE, since there is always a sudden elevation of dopamine but not NE with jumping. Further, there seems to be a cholinergic dopaminergic involvement, because if acetylcholine is elevated, the sudden rise in dopamine is inhibited and the jumping behavior is blocked. Alpha-methyl-tyrosine will also decrease the jumping behavior.

In reply to Dr. Lewander's questions on the stereotopy resulting from LSD, Dr. Wallach said that Fog reported that LSD gives rise to a gnawing type of stereotopy rather than the head-twitching type. Similarly, rodents will gnaw after amphetamine or grab the wires of the cage so strongly that the cages may be picked up by lifting the animals by the tails. Dr. Lewander reported that he had measured the metabolic pattern of amphetamine in the urine of rats after giving benzodiazepines. The results showed that *para*-hydroxylation of amphetamine was inhibited by the benzodiazepines, which most probably causes increased brain concentrations of amphetamine with concomitant increased pharmacological effects. The potentiation of amphetamine effects by benzodiazepines reported by Dr. Wallach might well be explained by such a mechanism.

Dr. Seiden made an observation on stereotopy in general: when rats showing a good deal of gnawing stereotopy are placed in a box under scheduled control, the stereotyped behavior disappears. When asked by Dr. Wallach about the dose of amphetamine he administered in this work, Dr. Seiden stated that it was between 0.75 and 4 mg/kg. Dr. Wallach countered that the usual dose for obtaining stereotyped behavior in Sprague-Dawley rats is between 5 and 10 mg of amphetamine per kilogram. Dr. Seiden said that he had used even larger doses (> 4 mg/kg) and observed the disappearance of stereotopy. Dr. Perel said that he had observed continued stereotopy after rats were given 6 mg/kg of amphetamine.

Dr. McGaugh asked Dr. Wallach if stereotyped behavior was similar from species to species. Dr. Wallach replied that there are differences between species, but that stereotopy could be said to involve continuous, repetitive, abnormal behaviors. Although there are differences between species, within certain species, e.g. the cat, the same type of behavior is elicited with any precipitating agent.

Dr. Lewander said that his guinea pigs showed stereotopies after 20 mg/kg, i.p., of amphetamine as long as they were not disturbed by the observer or anything arousing in the environment. Dr. Wallach said that Randrup and Munkvad had reported this same phenomenon, but that it was not true of other species. Dr. Angrist noted that he had observed this in the dog.

Goodwin. Dr. McGaugh said that he found a very sizable increase in dopamine following diethyldithiocarbamate administration; this lasts for an hour or so and then settles to baseline. He wondered about the number of doses per day given to the subjects. Dr. Goodwin said that the biological half-life in man is 19 hr and that a single dose is given and plasma DBH followed. According to the literature, there is no consistent increase in dopamine, but there must be some effect on the dopamine system since HVA levels are affected.

Dr. Costa noted that Dr. Goodwin's data may be interpreted to indicate that fusaric acid blocks acid transport in CSF; such an action may explain the elevation of 5-HIAA and HVA in CSF. Fusaric acid is lipid-soluble, enters the brain readily, and has a relatively long half-life; it may be worth studying whether fusaric acid is a transport blocker. Dr. Goodwin said that some probenecid data in fusaric acid studies agree with this. However, some other data (increase of tryptophan and 5-HT) are not consistent with blockade of transport. Dr. Goodwin observed that fusaric acid had reached the clinic a lot sooner than most drugs, so that the amount of animal pharmacology on it is relatively sparse.

Dr. Schildkraut felt that the relatively slight (25%) decrease in MHPG in the fusaric acid experiments suggested that there was only minimal inhibition of DBH in the brain in contrast to the effect in the plasma. The findings are therefore difficult to interpret with respect to clinical efficacy, since a 25% decrease in NE synthesis might not be expected to produce much in terms of physiological, functional, or clinical effects. Dr. Goodwin said that he would fall back on Dr. Costa's comment that if there is a transport blockade effect involved here as well, there would be an explanation for the relatively small change in MHPG.

Dr. Stein asked for Dr. Goodwin's comments to the fact that in two out of the three cases in which a fusaric acid effect was noted there was an even greater increase in the psychosis rating following termination of fusaric acid. Dr. Goodwin said that in these two cases a psychotic episode was triggered and continued beyond the withdrawal of drug. He believes that a few other patients did get worse after they went off drug, and wondered whether this might be a kind of rebound phenomenon. Dr. Stein wondered if the drug might be exerting some direct (suppressant or "antipsychotic") effect other than enzyme inhibition in addition to the enzyme inhibition effect. When the drug is removed, the direct effect might be lost and only the enzyme inhibition left. Dr. Goodwin said that he had had similar speculations but had not been able to test this as yet.

Ellinwood. In reply to Dr. Hartmann's question on what DBH inhibitor had been used and with which species, Dr. Ellinwood said that he had used disulfiram in the cat. Interest in using disulfiram arose from the fact that it produces a psychosis in humans very similar to that produced by amphetamine. It should be remembered that disulfiram produces other metabolic and toxic effects, including a peripheral neuritis which had been discussed in reports on amphetamine and disulfiram experiments. To Dr. Wallach's question about indifferent electrode placement, Dr. Ellinwood replied that the indifferent electrodes were screwed into the back of the skull.

Meltzer. In reply to questioning by Dr. Janowsky, Dr. Meltzer said that he was trying to be conservative and not imply that the tract had anything to do with mental status, although he thought that is an interesting possibility. Dr. Jarvik wondered about the effects of amphetamine or apomorphine; Dr. Meltzer said they lowered prolactin levels in laboratory animals; no one has yet looked in man.

Angrist and Janowsky (combined discussion). Dr. Janowsky said that at the start of his studies Ritalin® was an FDA-approved drug given to patients prior to interviews to increase associations and emotional catharsis. Similarly, D- and L-amphetamine were accepted clinical drugs.

A member of the audience asked about the experimental design and Dr. Janowsky replied that the experiment was double-blind, with one of three drugs or a placebo given. The experimental room had a one-way mirror so that the subject could be watched with the least possible interaction from the injector. IV injections were given with the injector entering the room to give the injection; the rater did not know what drugs were being administered.

Dr. Moore stated that the controversy on differences in effects between D- and L-amphetamine really does not exist any longer.

Dr. Davis gave his opinion that it is hazardous to apply *in vitro* results to the clinic. Dr. Janowsky agreed and stated that it was interesting that various amphetamines increase psychotic symptoms.

Dr. Usdin wondered about the implication of the term "aberrant dopamine"; Dr. Janowsky said that he meant abnormally metabolized dopamine, possibly methylated. He wondered if some such abnormally metabolized dopamine product might have hallucinogenic properties and be involved in the etiology of schizophrenia.

McKenzie. In reply to a question by Dr. Baxter as to whether pimozide might be a more specific dopaminergic blocker than chlorpromazine, Dr. McKenzie said that he had not tried pimozide since none of the results he had seen indicated any advantage over chlorpromazine or haloperidol. Dr. Davis agreed that all three were essentially similar with regard to behavioral activity.

Dr. Goldstein reported that he had recently observed that apomorphine inhibits tyrosine hydroxylase activity in the striatum; this enzyme inhibition is particular strong in slices. Dr. Goldstein proposed that this might be the result of apomorphine penetrating into the neuron and could be involved in the decreased synthesis observed by Dr. McKenzie. Dr. McKenzie said that he had read Dr. Goldstein's paper in which Dr. Goldstein had inferred that since the effect could not be reversed with Haldol®, tyrosine hydroxylase was inhibited; he wondered why under these conditions apomorphine effects were reversible by chlorpromazine. Dr. Goldstein said that when higher doses of haloperidol were used, partial inhibition was obtained and it was possible to get reversal of this inhibition.

Roth. Dr. Goldstein mentioned that not only does a decrease in calcium result in an increase in dopamine formation, but cyclic AMP also strongly stimulates formation of dopamine. He has evidence that the two mechanisms are linked. Dr. Roth indicated that although removal of Ca^{2+} with EGTA and addition of cyclic AMP produce a similar increase in tyrosine hydroxylase activity, their mechanism of action differs.

Dr. Carlsson wondered whether in the homogenates the synaptosomes were intact or destroyed. Dr. Roth said that he had not used true homogenates, but that he had taken a high-speed supernatant for his enzyme kinetic studies. By using soluble enzyme in the absence of particles, he did not have to worry about problems of membrane binding. He observed in his experiments that he could completely reverse the effects of EGTA by adding calcium back to the medium. Similarly, he could reverse the effects of pretreatment of animals with γ-hydroxybutyrate by adding calcium to the *in vitro* medium, suggesting that calcium is involved in the effect. In reply to a question as to whether he thought haloperidol acted on the presynaptic membrane, Dr. Roth stated that experiments in his laboratory as well as in Dr. Aghajanian's laboratory indicate that haloperidol can act on both pre- and postsynaptic membranes.

Dr. McKenzie mentioned that the rate of transmitter release and the end-organ response are usually dependent upon the rate of stimulation. To a question as to whether it disturbed him that the rate of synthesis did not parallel the rate of stimulation, Dr. Roth replied that the dopamine neurons are quite different from other monoamine neurons. The other neurons seem to be able to modulate differences in a much better fashion. In dopaminergic neurons, when impulse flow is increased, there is a very nice correlation between increases in impulse flow and dopamine catabolism, that is, DOPAC concentrations. It is more difficult to make such comparison in terms of synthesis, however, since both an increase in impulse flow and a cessation of impulse flow result in an increase in dopamine synthesis.

Dr. Roth mentioned that if a fairly large electrode is used to stimulate the pathway, then, presumably as a result of interruption of a few of the dopaminergic fibers as the electrode is introduced, an increase in synthesis results; sham stimulation will give about a 20% increase in synthesis. By keeping the electrode at least 0.5 mm from the median forebrain bundle, Dr. Roth gets a better frequency dependence.

Dr. Goldstein reported that high concentrations of calcium result in the precipitation of TH, and that at some concentrations the precipitation is quite extensive. Dr. Roth indicated that he had observed similar effects of high Ca^{2+} on striatal TH.

In reply to a question of Dr. Breese's with regard to his studies on pargyline, Dr. Roth pointed out that his work was not with homogenates, but rather was an *in vivo* study. Two and a half hours after giving animals pargyline, a DOPA decarboxylase inhibitor is given, then the stimulation of synthesis of DOPA is followed, as an indication of tyrosine hydroxylase activity. It is observed that once the levels of dopamine are increased, synthesis shuts off very nicely. This is not a result of a direct effect of the MAO inhibitor on the enzyme, but related to the increase of dopamine. Dr. Breese inquired whether pargyline inhibited impulse flow in dopaminergic neurons. Dr. Roth said that although Dr. Aghajanian had found that MAO inhibitors block impulse flow in serotonergic neurons, Dr. Walters has observed that pargyline does not inhibit impulse flow in dopaminergic neurons found in the zona compacta.

Zigmond. Dr. Carlsson observed how similar Dr. Zigmond's results were to those reported by Markiewicz on the recovery in mice treated daily with small doses of reserpine. He felt that this might give some clue as to the mechanism involved, that it need not necessarily be an anatomic damage, that one could obtain the same adaptive response to a drug. Dr. Carlsson wondered whether it would be possible to distinguish between these effects on water and food intake and the akinesia, or whether the effect on water and food intake might be just an expression of the akinesia. Dr. Zigmond stated that there is a possibility that they are tied together during the early stages of recovery, but that, later, feeding deficits persisted after motor deficits appeared to have disappeared.

Moore. In reply to comments from the audience, Dr. Moore remarked that he did not think the dopamine effect on locomotor stimulation was necessarily due to the nigrostriatal projection, since other workers had observed that bilateral lesions of the nigrostriatal projection completely blocked amphetamine stereotopy but did not completely block amphetamine locomotor activity. Dr. Moore also stated that after 6-hydroxydopamine, maximum effect of DOPA is observed by the end of 10 days. After 6-hydroxydopamine, animals will show the L-DOPA or apomorphine-induced circling phenomenon for more than 60 days.

In reply to a question as to when peak sensitivity to various drugs was obtained, Dr. Moore said that this was dependent on what one was looking for and that he felt one deficit in his studies was a lack of measuring total activity.

Several questions were asked about U-14624. It was suggested that U-14624 was altering 5-HT metabolism. Dr. Moore discussed comparative results with U-14624 and FLA-63; both increase dopamine levels in whole brain.

The next series of questions concerned the effects of oral administration. Dr. Moore pointed out that when disulfiram or U-14624 were given intraperitoneally, both steroid and blood glucose levels rose very high, and he interpreted this as an indication of stress resulting from local irritation. He also stated that after oral administration, NE values decreased considerably, that the levels went down even lower than after α-methyl-p-tyrosine (even to 0.06 mg/g of whole brain). Further, if these mice have ^{14}C-tyrosine injected, there is a complete blockade of ^{14}C-NE formation. In earlier work, Dr. Moore had published that with oral administration of U-14624 depletion of NE was observed but no effect on behavior. The major advantages of administering the drug in the diet are that the animals do not have to be handled and they do not have to be intubated.

A member of the audience said that all of his animals had drug administered intraperitoneally. Behavioral depression is not observed in avoidance, but only in food-reinforcing tests. New cell stimulation is not observed. Furthermore, the 6-hydroxydopamine animals do not show a depression correlated with the reduction of NE.

Dr. Breese asked Dr. Moore, if the release of brain dopamine was responsible for amphetamine-induced motor activity, what function did he feel might be altered by amphetamine-induced release of NE stores in brain. Dr. Moore said that in his brain perfusion experiments, stimulation of the medial forebrain bundle causes release into the perfusate of NE and dopamine at about the same amount. Dr. Moore does not know whether the NE is coming from the lateral hypothalamus or from the septum. In contrast, amphetamine results in very little NE coming out.

Chase: In reply to a question as to whether removal of the area postrema would be effective in blocking nausea and vomiting produced by apomorphine or DOPA, Dr. Chase replied that in dogs ablation of the medullary trigger zone has been shown to prevent emesis induced by either of these drugs. Dr. Moore wondered if it might be possible to prepare a phenothiazine or a haloperidol-like compound which was charged so that it would still combine with the receptor but not get into the CNS, eliminating some undesirable effects. Dr. Chase replied that, while this might be theoretically possible, the problem is now handled in L-DOPA-treated patients by the coadministration of a peripherally acting decarboxylase inhibitor. Such drugs markedly diminish the gastrointestinal (G.I.) side effects of L-DOPA, while at the same time potentiating its antiparkinsonian efficacy. Moreover, G.I. effects of L-DOPA or piribedil are usually a problem only during the initial stages of treatment. Dyskinesias are the major adverse effect of long-term therapy, and no satisfactory means for handling this problem has yet been found. Dr. Chase felt that it would be important if someone could show that the receptors responsible for dyskinesias are different than those responsible for antiparkinsonian activity.

Dr. Moore's next question was whether the efficacy of L-DOPA could be attributed to the fact that it may be converted to dopamine in serotonin fibers and thereby act as a false transmitter. Dr. Chase replied that in rats, exogenous L-DOPA can be decarboxylated in serotonergic neurons and the dopamine released during nerve terminal depolarization. Changes in lumbar spinal fluid levels of 5-hydroxyindoleacetic acid during L-DOPA treatment suggest that L-DOPA may also enter serotonergic neurons in man. The critical question is whether serotonergic neurons merely serve as an alternative site for DOPA decarboxylation or whether their partial conversion to dopaminergic neurons affects extrapyramidal function by interferring with serotonergic transmission. Dr. Chase said that evidence against the latter possibility is that if either untreated or L-DOPA-treated parkinsonian patients are given parachlorophenylalanine, there is no modification of symptomatology. Dr. Chase expressed the opinion that it does appear possible that normally functioning dopamine neurons may not be necessary for L-DOPA to influence motor behavior. In 6-hydroxydopamine-treated rats and in patients with parkinsonism, where dopamine fibers degenerate, sufficient alternative sites remain for the decarboxylation of exogenous DOPA. Some of the dopamine formed may overflow into the extraneuronal space and diffuse onto dopamine receptors. Conceivably, normal transmission in the nigrostriatal system may be achieved simply by attaining sufficient levels of the amine in the synaptic cleft and that the neurally mediated, pulsatile release of dopamine may not be necessary. Supersensitivity of dopaminergic receptors may be important in conferring specificity to the DOPA response in parkinsonian patients by allowing dopamine to be most effective at denervated sites and least active at normally innervated dopamine receptors, where increased stimulation might cause unwanted clinical effects.

Dr. Goldstein commented that in some monkeys with about 80% depletion of dopamine, severe tremors were seen. Dr. Fuxe did some histofluorescence studies and found that some dopaminergic terminals were still present, fluorescing strongly. In the monkey studies, whenever a treatment resulted in the relief of tremors, involuntary movements developed almost invariably. Dr. Goldstein said that it was necessary to titrate dosage very closely to obtain a relief from tremor without producing involuntary movements. This suggests to him that the two effects are closely related to dopamine receptors.

Dr. Lieberman observed that patients who do not respond to L-DOPA usually do not develop dyskinesias.

Dr. Clark stated that he had a metabolite of piribedil available. Dr. Lieberman identified this compound as S-584. Dr. Chase said that S-584 is a catechol metabolite of piribedil and when given systemically might undergo rapid O-methylation before entering brain. Dr. Goldstein confirmed that it was not very effective except when administered intraventricularly.

In reply to a question as to why dopamine metabolites increased so much more in Huntington patients given haloperidol than in normals, Dr. Chase replied that the comparison was not with normals but with parkinsonian patients. In the case of the patients with Huntington's chorea, a normal compensatory increase in dopamine metabolism is observed when the receptor is blocked, since degenerative changes in this disease do not primarily involve the dopaminergic system. In parkinsonian patients the dopamine pathway degenerates and the ability of residual neurons to increase dopamine synthesis during haloperidol treatment appears quite limited.

Dr. Zigmond felt that there might be some support for the idea that dopamine is circulating within the extracellular space rather than being released directly onto the postsynaptic membrane. Referring to the results of Drs. Seiden and Carlsson, he noted that one does not have to go to brain-damaged animals to get an effect which is reversible by DOPA. For example, this may be done by giving reserpine, which blocks the capacity of both dopamine and serotonin neurons to release transmitters. Since the vesicles are impaired and an effect is still obtained from DOPA, he felt that the indications were that the DOPA was decarboxylated extraneuronally and then acted on receptors. He wondered if dopamine itself might really be acting as a neuro-hormone rather than as a synaptic transmitter. Dr. Chase said that others were wondering about this too. The ability of L-DOPA to reverse parkinsonian signs, even in patients with far advanced disease, suggests that virtually normal dopamine-mediated transmission may be restored despite a major loss of dopaminergic neurons. It thus appears that dopamine transmission may not occur according to the classical conception.

Dr. Lieberman described a patient who developed paranoid symptomatology after piribedil administration. Dr. Chase confirmed that he had occasionally observed a psychotic component to the behavioral changes occurring during L-DOPA or piribedil treatment.

Neuropsychopharmacology of Monoamines and Their Regulatory Enzymes, edited by E. Usdin. Raven Press, New York © 1974.

Discussion and Summary

Morris A. Lipton

Biological Sciences Research Center, Child Development Institute, University of North Carolina School of Medicine, Chapel Hill, North Carolina 27514

The papers presented in this volume are organized around three major topic areas: monoamine oxidase (MAO), monoaminergic enzymes other than MAO, and dopamine (DA). This grouping is deceptively simple because the scope of the papers includes consideration of basic chemistry of the enzymes involved in the synthesis and degradation of the monoamines, regulation of these enzymes, use of animal models, measurement of these enzymes in blood as potential diagnostic aids, and their pharmacological manipulation in the interests of therapeutics. Thus the collection encompasses and illustrates methods and results of basic biochemical investigation, utilization of this information in the study of brain behavior relations in subhuman species, and application of this knowledge to the diagnosis and treatment of serious human ills.

The workshop on which this volume is based is worth looking at in historical perspective. Somewhat more than a decade ago, the biochemical skeleton for the synthesis and degradation of the biogenic amines was eluciated. For norepinephrine (NE), the sequence of steps from tyrosine to DOPA to DA and finally to NE was worked out. Similarly, the steps from tryptophan to 5-hydroxytryptophan (5-HTP) to serotonin (5-HT) were demonstrated. Many of the characteristics of the enzymes involved in the synthesis were described. Inhibitory control of the rate of synthesis of the neurotransmitters was considered a consequence of end-product inhibition of the first and rate-limiting enzyme in the synthetic chain. Acceleration was by induction of new enzyme. Destruction of the transmitter released into the synaptic cleft was achieved in the case of NE by reuptake into the presynaptic neuron, O-methylation to a biologically inactive compound, or intracellular destruction by MAO. For 5-HT it was even simpler because O-methylation was not involved in the inactivation.

A decade ago the mode of action of phenothiazines at the molecular level was unknown, and at a clinical level the distressing tardive dyskinesia attendant upon its long-term use was not yet recognized. The utility of L-DOPA in the treatment of parkinsonism was not yet known, DA tracts in much of the central nervous system had not yet been discovered, and DA was considered important mainly as a step in the synthesis of NE. Chemical models of the psychoses considered mainly the acute psychotomimetics such as LSD, and animal models were almost inconceivable for uniquely human diseases.

A vast quantity of information has been obtained since that time and much of it is the subject of this volume. For example, the rate of synthesis of catecholamines now appears to be under much more complex control than had been thought previously. Even the measurement of rates of synthesis and turnover is very complex. Weiner, in a scholarly review, critically examines the various methods employed for the estimation of turnover of the biogenic amines. Two types of methods exist. A steady state method involves labeling stores and monitoring decay or measuring the rate of formation from a labeled precursor. Nonsteady state methods perturb the system by inhibiting synthesis or degradation and then monitor changes in tissue levels. All the methods have difficulties, and none provides an accurate measure of absolute turnover rates. The possibility must always be considered that overall turnover represents a composite of multiple compartmentalized pools, each of which has a considerably different turnover rate. Only animal work, with anatomically dissected neural pathways, can answer this question, but the question itself is frightening to those of us engaged in clinical research where only global procedures are thus far possible.

Molinoff agrees that tyrosine hydroxylase (TH) is usually the rate-limiting step in NE synthesis but questions whether this is always the case. He points out that a prolonged increase in nerve firing, generated pharmacologically or physiologically, leads to increases not only in TH but also in dopamine-*beta*-hydroxylase (DBH) and PNMT. Using sympathetic ganglia in organ culture, he was able to demonstrate that stable levels of DBH could be elevated by 40% in 24 hours by increasing the potassium concentration to depolarizing levels. Threefold increases in the rate of DBH synthesis can also be produced *in vivo* by reserpine treatment. Such increases are blocked by inhibitors of protein synthesis. These and other lines of evidence suggest that DBH levels are not static and that the enzyme is inducible. The question of whether DBH can be elevated when TH is not, or vice versa, is not yet answered. If TH can be elevated with DBH remaining unchanged, then DBH may indeed become rate-limiting for NE synthesis, but this remains to be demonstrated.

The long-term control of TH involves changes in the total number of enzyme molecules by induction, whereas the short-term control does not involve such a change. How then is it maintained? Carlsson and his group note that end-product inhibition, as a form of control, requires 10^{-5} M NE. Such enormous concentrations might be achieved locally, but the possibility must be entertained that totally different control mechanisms may also operate. Carlsson offers evidence that blocking DA receptors increases DA turnover, whereas stimulating DA receptors decreases it. In both the nigrostriatal system and the adrenal medulla, Carlsson finds that interruption of release of catecholamines, which causes an increase in levels, does not lead to a decrease in synthesis. He suggests that enzyme activity is under short-term control by a modulator and that the availability of this modulator is influenced by pre- or postsynaptic receptors. The nature of the modulator and of the mechanisms controlling its availability remain to be elucidated.

Costa's group also has been concerned with the short- and long-term regulation of TH. Costa recognizes three types of regulation: (1) end-product inhibition, (2) changes in the affinity constant of TH, and (3) induction of new enzyme. He points out that some regulation can be achieved by the competition for the pteridine cofactor by NE precursors such as DOPA or DA. The concentration of cofactor then must have a role in determining if a given concentration of NE precursor inhibits TH. Neuroleptics may alter the affinity constants of soluble TH, probably indirectly by increasing neuronal activity. Studies of induction of TH in the adrenal gland offer evidence that cyclic nucleotides may be involved.

Roth studied the effects of alterations in impulse flow on transmitter metabolism in central dopaminergic neurons. In NE and 5-HT neurons, increased impulse flow accelerates transmitter metabolism and turnover and decreased flow diminishes it. DA neurons are similar in that stimulation of those neurons originating in the substantia nigra and projecting to the neostriatum results in release of transmitter and transmitter metabolites. But interruption of flow by electrolytic lesions or by *gamma*-hydroxybutyrate results in a marked increase in the DA content of the terminals in the striatum and an unexpected rapid rise in TH. To study this phenomenon further, Roth used *gamma*-hydroxybutyrolactone, which reversibly inhibits impulse flow in dopaminergic neurons without affecting NE or 5-HT neurons. *In vivo*, in animals pretreated with a DOPA decarboxylase inhibitor, inhibition of impulses in the DA neuron increased the DA of the neostriatum by 200%. Studies with peripheral neurons suggest a mechanism (similar to that proposed by Carlsson) in which a presynaptic receptor may exist on the DA nerve terminal and may stimulate an increase in synthesis when impulse flow is interrupted and no DA is released into the cleft. The idea that a presynaptic receptor might regulate DA synthesis is novel and raises questions about the interaction of this type of regulation with endproduct inhibition of TH. Roth hypothesized that a presynaptic receptor could control TH activity by altering the permeability of the presynaptic membrane to substrates which might influence TH activity. Mandell had previously proposed that calcium might alter the kinetic properties of TH. Roth studied this in the striatum and found that removal of calcium by EGTA increased TH activity while calcium reversed this stimulation. The highly active TH formed by removal of calcium was insensitive to endproduct inhibition by DA. *In vivo* treatment of animals with *gamma*-hydroxybutyrolactone resulted in striatal TH similar to that formed when the enzyme is treated *in vitro* with EGTA. The lactone itself was not active *in vitro*. These new findings raise the interesting possibility that changes in calcium flux in the DA nerve terminal may be responsible for altering TH activity *in vivo*.

Regulation of tryptophan hydroxylase function has been studied for some years by Mandell. In his presentation he offers data dealing with the changes induced in tryptophan hydroxylase activity by drugs in regions of the brain rich in cell bodies and in synaptosomes. Earlier he had shown that parachlorphenylalanine (PCPA) diminished soluble enzyme in the midbrain in 2 days and that 13 to 18 days was required for these cell body changes to appear in synaptosomes. Similar slow changes were noted with lithium and morphine. Now he reports that high doses of methamphetamine or reserpine have a very short latency in altering the capacity of synaptosomal enzymes to convert tryptophan to 5-HT. The mechanism for this rapid effect is not clear. Mandell feels it may involve changes in the physical state of the enzyme which might affect its affinity for substrate or cofactor. He used tetrahydrobiopterin as a cofactor for tryptophan hydroxylase to demonstrate these rapid effects. This new cofactor yields a more sensitive assay. He has similar data showing that TH activity may also change rapidly after the administration of drugs.

The papers by Molinoff, Carlsson, Costa, Roth, and Mandell offer new concepts and mechanisms by which brain may respond to adaptive needs for more or less transmitter over very brief time intervals. Since it is very likely that rapid changes actually do occur, understanding the mechanisms of change promises to be very fruitful.

Four chapters deal with DBH. This enzyme is of interest not only because it is the last enzymatic step in the synthesis of NE, but also because it is extruded from the presynaptic neuron along with NE and then appears in the circulation where it may be assayed. It may therefore be a useful index of the degree of sympathetic activity. DBH has been purified and shown to be a copper-containing protein. The fact that homogenates of adrenergically innervated tissues show little DBH activity, whereas chemically fractionated homogenates do show such activity, suggests that endogenous inhibitors are present. Kirshner has isolated an inhibitor from bovine adrenal medulla and finds that it resembles glutathione, although it is not identical with it. Molinoff finds that the inhibitor is stable at $-10°C$ in intact organs but is less stable in homogenates. The inhibitor may also be partially organ-specific because treatments which destroy inhibitory activity in the heart do not do so in the spleen. The role of DBH inhibitors *in vivo* is not clear, but they are so potent *in vitro* that one cannot help but suspect that they may also exert a physiological role.

Goldstein, who has been concerned with the diagnostic utility of circulating DBH, has developed an immune assay. The results of the two assays do not always agree, perhaps because of peripheral inactivation of the enzyme shortly after release. The immune assay, which presumably measures both active and inactivated forms, may therefore reflect sympathetic activity more accurately. Goldstein finds abnormal levels of serum DBH in familial dysautonomia, torsion dystonia, and Down's syndrome. He was unsuccessful in using serum DBH to differentiate among patients with various psychiatric disorders, but he is now using this enzyme assay to investigate inheritance patterns in families with schizophrenia. Since Stein, using an enzyme assay, has reported low DBH in brains of schizophrenics, Goldstein suggests using the immune assay for further study of postmortem brains from schizophrenics to further validate Stein's findings.

Lovenberg reports remarkable stability in serum DBH in individuals over time, but as much as 100-fold variation among individuals in a large population. Siblings of low-level DBH subjects are also low in DBH, suggesting a genetic pattern. Marked elevations of serum DBH occur in pheochromocytoma and are reversed after surgical removal of the tumor. Slight elevations occur in Huntington's chorea, neuroblastoma, and torsion dystonia. Lovenberg feels that the utility of the enzyme assay as an index of sympathetic activity is still low. Maneuvers which markedly alter sympathetic activity *in vivo* yield highly variable results in serum DBH.

Levels of serum DBH correlate with hypertension in some laboratories but not in others. Schanberg, studying the relationship between serum DBH and arterial hypertension, selected low-DBH and high-DBH subjects and correlated these levels with urinary catecholamines and with arterial blood pressure. In general, he found that low DBH levels correlated with low urinary catecholamines and with low and stable blood pressure. In 50 patients he found good correlations between high serum DBH and primary fixed or labile hypertension. Since others, using new and sensitive assays, have shown that NE levels in blood are elevated in some forms of hypertension, Schanberg suggests that plasma DBH may be useful for the differential diagnosis of different forms of hypertension. It seems likely that further knowledge of the half-life of the serum enzyme as well as of the mechanisms of degradation will enhance the clinical utility of this assay. By the same token, Goldstein's immune assay may prove to be singularly useful.

Seven papers deal with the class of enzymes which oxidize monoamines. It is now clear MAO exists in multiple forms. Plasma MAO seems quite different from the family of enzymes found in platelets or tissues. It requires pyridoxal and copper as cofactors and also differs significantly from tissue MAO in its response to substrates and inhibitors. Plasma levels may be determined genetically but are altered in diabetes, thyrotoxicosis, cirrhosis, and congestive heart failure. Interest in plasma MAO is significantly less than in tissue and platelet MAO, and its role remains unclear.

Youdim presents evidence that a highly purified tissue MAO is a sulfhydryl-containing protein with flavin as a cofactor. Iron may play a role in its activity. Multiple forms of MAO have been unequivocally demonstrated *in vitro* and do not appear to be a product of procedural artifacts. Udenfriend demonstrated two enzymes immunochemically in beef brain, and Racker has physically separated two types. Youdim finds evidence for five forms of the enzyme in rat liver, basing his evidence on studies of substrate and inhibitor specificity, kinetic properties with specific substrates, and thermal stability. Eiduson solubilized mitochondrial rat brain MAO and used agarose column chromatography and Sephadex electrophoresis to separate fractions. Two major protein peaks were obtained. Fraction A has a much higher molecular weight than does B. Fraction B, when passed through a guanidine column, seems to disaggregate into a fraction C. On the other hand, studies on a urea column offer evidence that fraction B may aggregate to A.

The precise nature of the multiple forms of MAO in brain is not yet certain. Separate proteins may be involved or there may be a single protein whose properties are modified by attached phospholipids. The finding that at least some of these forms of MAO are interconvertible by aggregation and disaggregation techniques suggests that separate proteins may not be involved in all the separate enzymes.

Whatever the chemical nature of the relationship between the different forms may be, it is quite clear that they have different properties. Eiduson speculates that interconversion *in vivo* may be one form of regulating the turnover of biogenic amines because substrate specificity and affinity differ with the different enzymes. The multiple forms of MAO have been tentatively classified into two types: A and B. Type A is specific for the oxidation of norepinephrine and 5-HT but not for DA. Type B preferentially oxidizes phenylethylamines. Tyramine is a substrate for both types. Inhibitors have been found which are relatively specific for types A and B. Pharmacological differentiation of inhibitors *in vitro* can also be found *in vivo* and has been used ingeniously by Neff to examine the role of various neurotransmitters in common but poorly understood pharmacological syndromes. For example, he finds that a type A MAO inhibitor reverses the reserpine syndrome but a type B inhibitor does not. Since DA is oxidized preferentially by type B while NE and 5-HT are oxidized by type A, it follows that DA is not deeply involved in the reserpine syndrome. Sandler reports evidence for a deficit in type B MAO in migrainous subjects even during attack-free periods, and has also found that in double-blind trials, orally ingested phenylethylamine can trigger a migrainous attack. In his typically imaginative fashion, Sandler suggests that an increase in one of the specific forms of MAO may occur in depression, possibly leading to a selective deficit of 5-HT, tyramine, or some other monoamine. He speculates that it may someday be possible to define a specific monoamine hypothesis of depression.

The recognition that multiple forms of MAO exist in different tissues, and that each enzyme may have special affinity for and activity with a specific substrate, raises the possibility that the pharmaceutical industry may be able to develop more specific and less toxic MAO inhibitor drugs. Drs. Martin and Biel address themselves to the question of how a rational rather than empirical approach to the problem of the development of specific drugs can be approached by the medicinal chemist through the study of the enzyme, the receptor, the effector, and the competing processes. This approach has not yet paid off spectacularly but has great potential.

The significance of several features of MAO activity are still not understood. MAO activity increases with age and alters with the estrus cycle in females. Youdim offers several possible explanations for the *in vivo* physiological effects of steroids and seems to prefer the hypothesis that they affect the outer membrane of the mitochondria in which the MAO resides. He emphasizes that enzyme inhibitors have been found which inhibit MAO *in vitro* but not *in vivo,* and that perfusion techniques which maintain structural organ integrity may lead to useful information more directly than experiments using tissue extracts and fractions.

Despite the gaps in information on the nature and function of the different MAOs, interesting results in humans have nonetheless been obtained. Dr. Nies has studied genetic influences on MAO activity by examining platelet MAO activity in mono- and dizygotic twins, using benzylamine and tryptamine as substrates. He finds substantially less variability for this enzyme with both substrates in monozygotic twins than in dizygotic twins or controls. Using both substrates, Nies and his co-workers examined the platelets of schizophrenics compared with matched controls and found low MAO activity with tryptamine as the substrate but not with benzylamine as the substrate. On the other hand, in a few experiments, brains from schizophrenic patients did not differ from controls with either substrate. These preliminary results suggest that studies of enzyme activity in peripheral tissues of schizophrenic patients must be interpreted cautiously when it comes to inferring central nervous system events.

Murphy, who has been concerned with the meaning of MAO levels in the peripheral tissues of different clinical populations, offers the results of many careful experiments with platelet MAO. He finds that platelet MAO is electrophoretically homogeneous, and by a study of substrate activity and sensitivity to selective MAO inhibitors, concludes that it is a type B enzyme which does not readily oxidize 5-HT, the biogenic amine present in the highest concentration in the platelet—nor does it oxidize NE. It is more active with DA, tryptamine, and tyramine, and most active with nonphysiologic synthetic amines such as benzylamine.

Murphy has also studied MAO activity in brain and finds evidence for at least two and perhaps four types. Whatever the type, the concentration seems highest in the hypothalamus and lowest in the cerebral cortex and cerebellum. Since much of brain MAO is intraneuronal type A, and since only platelet type B MAO is readily accessible for study in patients, it is difficult to extrapolate to brain events the finding of low MAO activity in platelets of schizophrenics. Yet types A and B are somehow related, as Eiduson suggests, and the use of selective inhibitors may offer us some future insight into the status of MAO activity in the brains of psychiatric patients.

DA, long recognized as a precursor to NE, has emerged as a most popular subject for investigation. The large interest in DA relates to several events that occurred simultaneously: Carlsson's development of a chemical method for DA more than 10 years ago; the demonstration and mapping of dopaminergic tracts, initially in the caudate but more recently in the limbic system and cortex; the success of L-DOPA in parkinsonism; and the recognition that antipsychotic agents are DA receptor blocking agents. Hyperdopaminergic activity has been proposed as a mechanism in the pathogenesis of schizophrenia. This hypothesis arose from the convergence of four separate lines of research: (1) the DA receptor blocking effects of antipsychotic drugs manifested by their capacity to increase DA turnover; (2) the psychotomimetic effects of DOPA agonists such as D-amphetamine, L-amphetamine, L-DOPA, cocaine, apomorphine, and methylphenidate, all of which facilitate release of DA from central synapses or enhance DA receptor sensitivity (the amphetamine psychosis is of particular interest because it is clinically indistinguishable from paranoid schizophrenia); (3) the fact that these psychotomimetic agents in man produce stereotypy in animals; and (4) the success of blockade of amphetamine- or apomorphine-induced stereotypy in animals as a predictive test for antipsychotic drug efficacy. Why there should be excess DA activity in schizophrenia is not precisely formulated. Kety has suggested that a deficiency of DBH may lead to a relative excess of DA and a corresponding deficiency of NE. An antagonistic system, perhaps cholinergic, might be underactive, as suggested by Janowsky. A system such as GABA, which is under tonic DOPA inhibition, might be effective. And, finally, DA receptors might be hypersensitive.

In his presentation, Goldstein points out that the global dopaminergic hypothesis of schizophrenia has inconsistencies and that more precise hypotheses may be necessary. For example, although all drugs effective in the treatment of schizophrenia are DA blockers, there is no linear relationship between extrapyramidal side effects and clinical antipsychotic activity. Some drugs produce major extrapyramidal effects with little antipsychotic activity. Some do the opposite. Clozapine, an antipsychotic drug still relatively unknown in the United States, is reported to have no extrapyramidal effects. It has been suggested that clozapine blockade of DA receptors is surmountable whereas blockade by the other neuroleptics is not. Clearly, more clinical and basic work with clozapine is needed.

There are inconsistencies in animal work as well. For example, *alpha*-methyltyrosine (AMT) and reserpine antagonize stereotyped behavior induced by amphetamine and phenmetrazine, but not that induced by methylphenidate and L-DOPA. These and similar inconsistencies suggest not that the DA hypothesis is wrong, but rather that it needs refinement and testing by more sophisticated means. For example, neuroleptic activity might be related to blocking effects on DA neurons in the limbic system and cortex rather than in the striatum. This could occur because of differences in neuroleptic drug distribution in the brain, or possibly because of altered balance between DA activity and other transmitters such as 5-HT, NE, or acetylcholine.

Several chapters deal with the role of DA in normal and abnormal animal behavior. Cooper and Breese employed the technique of selective destruction of DA or NE tracts by the appropriate injection of 6-hydroxydopamine intracisternally or into selected portions of rat brain. Although the techniques are imperfect insofar as it is impossible to destroy DA fully without effecting some change in NE and vice versa, these investigators were nonetheless able to produce major changes in one system with only minor changes in another. In both adult and developing rats, they find that destruction of DA fibers results in major deficits in ingestive and operant behavior. They also find evidence that the behavioral depressant effects of AMT and reserpine are caused by disruption of DA rather than NE function in rat brain. Even amphetamine-induced locomotion seems to be mediated by DA rather than by NE. Moore also presents data supporting this view. So much of the behavior previously attributed to NE now has been found to be mediated by DA that questions arise about the role of NE. Possible functions in the cerebellum and in the control of the peripheral cardiovascular system are suggested. Clearly, the behavioral tests were far from exhaustive, and a specific role for NE may yet be found.

Seiden and Campbell studied the mutual interactions between catecholamines, drugs, and behavior using pharmacological, neurochemical, and behavioral techniques simultaneously. They note that not only do drugs which modify catecholamines alter behavior, but also that when behavior is the independent variable it alters catecholamine metabolism. They suggest that drug effects must be studied in the context of ongoing behavior and that chemical changes must be measured on a regional neuroanatomical basis.

Wallach reports on stereotyped behavior induced pharmacologically. He points out that the behavioral patterns of drug-induced stereotypy in animals are species-specific, and that doses of a given drug required for stereotypy can vary several hundredfold for different species. Stimulant drug-induced psychoses elicit a desynchronized EEG while perception-distorting psychotomimetics such as LSD, mescaline, and DOM produce a hypersynchronized EEG and different behavior patterns. Animal evidence that DA rather than NE is involved in amphetamine stereotypy is based on the finding that neither adrenergic blockers nor DBH inhibitors antagonize stereotypy, but that DA receptor blockers do. Involvement of the corpus striatum is suggested by the finding that lesions of this system prevent amphetamine and DOPA stereotypy. Cholinergic agents antagonize amphetamine stereotypy and anticholinergic agents potentiate it. Benzodiazepines potentiate amphetamine stereotypy. Wallach suggests that most stereotypy-producing agents have a DA link that may be modified by NE input or influenced by anticholinergic agents. Morphine stereotypy is an exception to this general rule.

Ellinwood, an experienced clinician who has studied drug-induced psychoses in man and also animal models of psychosis, makes the point that chronic rather than acute use of stimulants in humans produces conditions resembling paranoid schizophrenia. He suggests that hyperarousal sustained over months or even years permits the development of distorted cognitive processes and their neuronal substrates. Sustained hyperarousal also leads to the type of psychological deficits noted by Shakow many years ago. Acutely induced drug psychoses are inadequate in time to permit the establishment of fixed delusional systems. Ellinwood feels that temporal lobe epilepsy, another chronic condition, resembles amphetamine psychoses. Paranoid ideation, hallucinations, ideas of influence, and thought disorders are present in this condition. In both amphetamine psychoses and temporal lobe epilepsy, behavior frequently becomes stereotyped. Hyperarousal occurs in the functional psychoses, as shown by Kornetsky, as well as in temporal lobe epilepsy and the amphetamine psychoses.

Using cats chronically intoxicated with amphetamine, Ellinwood has studied the behavioral changes which lead ultimately to stereotypy. Simultaneously he records the EEG in olfactory lobe and finds increases in frequency and decreases in amplitude as hyperarousal ensues. He also finds that disulfiram, a DBH inhibitor, permits the development of stereotypy at half the usual doses of amphetamine, and thus offers further evidence that a relative deficiency of NE compared with DA is responsible for the stereotypy. Ellinwood argues that neurophysiological and psychological studies of arousal may provide heuristic and integrative functions to bridge the gap between molecular and behavioral hypotheses and between animal models and the human condition.

From 1953 to 1968 phenothiazines, antidepressants, antianxiety agents, lithium, and L-DOPA were discovered to be therapeutically useful. Indications for their use were based largely on symptom relief. Empirical pharmacotherapy far outstripped knowledge of etiology, pathogenesis, or cellular mechanism of drug action. Relatively simple and global hypotheses, for example, the catecholamine hypothesis of schizophrenia, were generated largely in an effort to understand the effects of drugs in either exacerbating or treating illness. Conceptually we were trying to complete a jigsaw puzzle in which many pieces were still missing.

This volume illustrates how dramatically the picture has changed in the past 5 years. As a result of the development of new techniques and of continued investigation of the mechanisms by which drugs alter behavior in animals and man, there has been a burgeoning of basic information. Evidence for a functional redundancy in the aminergic central nervous system is illustrated by the finding in animals that 80% of the neurons of the central sympathetic system must be destroyed before behavioral differences are noted. In humans, an approximately equal number of neurons of the nigrostriatal system must be destroyed before parkinsonian symptoms appear. Part of the apparent redundancy may be due to the enhanced sensitivity of receptors which follows neuronal destruction; considerable attention is now being paid to the nature of the postsynaptic receptor and its feedback regulation of the presynaptic neuron. Another part clearly seems to be related to the balance between systems of different chemical neurotransmitters. Thus, by now there is considerable evidence that, in addition to the DA system, serotonergic and cholinergic systems may also be involved in the control of fine voluntary movements and that imbalances in these systems may be responsible for the manifestations of Parkinson's disease. Both catecholaminergic and serotonergic systems are involved in the control of human mood, and cholinergic and dopaminergic systems are involved in the pathogenesis of schizophrenia. Curiously, as we learn that many of the functions previously attributed to NE systems actually involve DA, we are less certain of the functions of NE. Furthermore, it is increasingly evident that drugs we employ to affect one system directly or indirectly perturb other systems. Although the problems of studying neurotransmitter metabolism and function grow geometrically with the number of transmitter systems under simultaneous investigation, there is little choice but to investigate their interactions. Several papers deal with such studies and several others discuss the short- and long-term regulation of monoamine biosynthesis.

Another prominent feature of this volume is the number of papers dealing with animal models and with the effects on animal behavior of drugs frequently used by humans. It is a truism that the investigation of human disease moves much more rapidly when the illness can be produced and studied in animals. For too long it was assumed that mental illness is uniquely human. Although it probably is true that the complex human brain is required for elaboration of the almost infinite number of specific symptoms by which mental illness may manifest itself, there is no *a priori* reason why the biological substrates for such illness may not be the same for animals and man. These papers illustrate the use of such models, not only to test new drugs but also to study the brain mechanisms involved in normal and pathological behavior.

Two other aspects of these papers are worth noting. First, as neurologists become increasingly concerned with behavior and as psychiatrists become concerned with the contents and function of the black box which subsumes behavior, there seems to be a trend toward the reintegration of neurology and psychiatry into the neuropsychiatry of almost a century ago. Second, many of the papers presented by basic scientists deal with matters immediately relevant to clinical diagnosis and therapy, and an approximately equal number presented by clinicians deal with important basic science findings. To me this speaks well for the future: basic scientists have clearly demonstrated their eagerness to participate in the solution of human problems, and clinicians are being produced who are competent in basic science. Together they have developed new methods and data that will permit the generation and testing of new hypotheses and treatments. In doing so, we must guard against premature closure and must recognize the likelihood that many bits of information are still missing. The amino acids and polypeptides widely distributed in the brain which act in the central nervous system perhaps as modulators or transmitters may represent still newer pieces to be worked into the puzzle.

INDEX

A

Adenyl cyclase
  denervation and, 210
  dopamine sensitive, 301
Adrenergic receptor blockade
  stereotypy antagonism and, 249
Adrenergic vesicles
  constituents of, 99-100
  dopamine-β-hydroxylase of, 99-102
  isolation of, 99-102
Affective disorders
  dopamine function in, 261-277
Agroclavine
  dopamine and, 348-350
Amantadine
  circling behavior and, 409-410
Aminoindanes
  as MAO inhibitors, 45-47, 88
Aminotetralins
  as MAO inhibitors, 45-47
Amphetamine
  behavior and, 212-215, 300-336, 363
  brain catecholamines and, 222-225, 404-408
  circling behavior and, 408-413
  as dopamine agonist, 403, 408-413
  EEG and, 283-284, 291-294
  effect on NE level, 333-334
  epilepsy and, 281-282
  6-hydroxydopamine and, 363-364
  hypothermia and, 229-232
  prolactin levels and, 437
  as psychosis model, 221, 281-294, 300, 317-318
  serotonin release and, 221
  serotonin synthesis and, 180-181, 183-184, 226-229
  stereotypy and, 212-213, 242-256, 435-436, 440
  tolerance to, 221-236
  tyrosine hydroxylase and, 373-375
Amygdala seizures, 289-290
Anorexogenic effects
  and tolerance, 234-235
Apomorphine
  behavior and, 212-214, 245-246, 252, 339-341, 347-348, 408-413

catechol-O-methyl-transferase and, 340-343
dopamine and, 164, 339-340, 343-346, 423, 428
methylation of, 341-342
prolactin levels and, 437
tyrosine hydroxylase and, 373-375, 378, 438
in Parkinson's disease, 214, 415-416, 428
Aromatic amino acid decarboxylase
  inhibition of, 148-149
Audiogenic seizures
  amphetamine and, 289

B

Behavior
  amphetamines and, 212-215, 245-246, 330-336, 408-413
  apomorphine and, 212-214, 245-246, 339-341, 347-348, 408-413
  catecholamines and, 325, 330-337
  circling, 408-413
  L-DOPA and, 245-246, 326-328, 336
  dopamine and, 341-350, 451
  drug effects, 325
  6-hydroxydopamine and, 353-365
  ingestive, 354-355, 385-400
  operant, 328-333, 355-357, 435
  stereotyped, amphetamine and, 241-242, 245-246, 282, 291-292, 363, 435-436
  stereotyped, apomorphine and, 340-341, 347-348
  stereotyped, catecholamines and, 249-250
  stereotyped, cholinergic effects, 251-253
  stereotyped, dopamine and, 250-251, 408-409
  stereotyped, drug-induced, 241, 245-247, 435, 451
  stereotyped, histamine and, 253-254
  stereotyped, mechanisms of, 248-256
  stereotyped, species differences, 247, 436
  stereotyped, stimulant-induced, 241-256, 435-436

455

Benzylamine
  as MAO substrate, 32, 64, 74, 77
  plasma MAO and, 78
Brain damage
  recovery from, 385-400
γ-Butyrolactone (GBL)
  dopaminergic system and, 372-379
  tyrosine hydroxylase and, 376-379, 445

C

Calcium
  cyclic AMP and, 438
  tyrosine hydroxylase and, 375-378, 382, 438
Carbamylcholine
  cyclic AMP and, 170-171
Carbidopa
  antiparkinsonian efficiency and, 430
β-Carbolines
  as MAO inhibitors, 41-43
Catecholamine biosynthesis
  amphetamine and, 223-225
  genetic control of, 195-201
  regulation of, 102, 135
Catechol-O-methyl-transferase (COMT)
  apomorphine and, 340-350
  behavior and, 340
  dopamine and, 341-342, 344, 349-350
  inhibitors of, 339-350
  kinetics of, 190-192
  in psychiatric disorders, 193
  in red blood cells, 189-190
  soluble and insoluble, 189-193
  substrate specificities, 190
  temperature effects, 192
Caudate nucleus
  apomorphine and, 343-346
  dopamine synthesis in, 343-346
  MAO activity of, 16-17, 20-22
  tryptophan hydroxylase of, 181-182
Chlorpromazine (CPZ)
  apomorphine and, 347-348
  behavior and, 347-348
  dopamine metabolism and, 347-349
  dopamine receptors and, 373-375
  dopamine turnover and, 163-164
  prolactin levels and, 310, 312

Cholinergic pathways
  behavior and, 251-252, 335, 453
  prolactin and, 303
Clorgyline
  MAO phospholipid and, 14-15
  platelet MAO and, 75-76
  reserpine and, 55-56
  specificity of, 4, 13-14, 49, 52-55, 80
Clozapine
  action of, 215-216, 450
Cocaine
  effect on NE uptake, 207-208
  seizures and, 289-290
  stereotypy, 212, 245-247
Corpus striatum
  stereotypy and, 250-251
Cyclic AMP
  ACTH and, 170-171
  calcium and, 438-439
  carbamylcholine and, 170-171
  catecholamine synthesis and, 93
  denervation and, 210
  genetic factors and, 200-201
  hypothermia and, 209
  MAO and, 19
  stereotypy and, 254
  tyrosine hydroxylase and, 172-173, 209

D

Denervation supersensitivity
  DOPA and, 431
2-Deoxy-D-glucose
  6-hydroxydopamine and, 388-389
Deprenyl
  platelet MAO and, 75-76
  reserpine and, 55-56
  specificity of, 4, 13-14, 49, 52-55, 88
Depression
  MAO activity in, 59, 64-65, 78, 89-90
Development
  6-hydroxydopamine and, 361-363
Dextroamphetamine
  audiogenic seizures and, 289
  as preconvulsant, 289
Diethyldithiocarbamate
  dopamine and, 436
Dihydroxyphenylacetic acid (DOPAC)
  impulse flow and, 370-371

Dihydroxyphenylacetic acid (DOPAC) contd.,
  metabolism of, 370-371
Dihydroxyphenylserine
  antagonism of stereotypy, 249
Dimethoxyphenylethylamine
  prolactin and, 302-303
Disulfiram
  DBH inhibition and, 437
  EEG and, 286-289
  psychosis from, 276, 286, 437
L-DOPA
  in affective disorders, 265-268, 273-277
  behavior and, 212-213, 245-246, 326-337
  6-hydroxydopamine and, 363-364
  metabolism of, 372-382
  motor activity and, 396
  Parkinson's disease and, 427, 441
  piribedil and, 430-431
  prolactin, effect on, 302
  psychosis and, 300-301
  reserpine and, 326-328
  turnover of, 145, 148-149, 161-163
Dopamine
  affective disorders and, 261-277
  agonists of, 403, 408-413
  agroclavine and, 348-349
  amphetamine and, 223-225, 300, 317-318, 363-364, 403-405
  apomorphine and, 343-346
  behavior and, 246, 326-337, 339-340, 358-361, 383-400, 451
  depletion of, 223, 357-358, 399-400, 441-442
  hypothermia and, 231
  6-hydroxydopamine and, 358-361
  impulse flow and, 370-382
  MAO and, 49-50, 53-55, 88, 372
  metabolism, 339-350, 370-382, 438
  migraine and, 5-7
  neuronal activity and, 369-382
  Parkinson's disease and, 427-433, 450
  pools of, 403-404
  prolactin and, 299, 312
  psychosis and, 211-217, 272-277, 300

receptors, 250-251, 299, 312, 373-374, 376, 382, 415-424
regulation of, 369-382
release of, 403-404, 406-408
schizophrenia and, 299-300, 317, 322, 438, 450
stereotypy and, 250
in telencephalon, 391
tropolone and, 346-347
turnover of, 137-139, 146-147, 163-164, 208, 266, 272-273, 427
tyrosine hydroxylase and, 135-141, 372, 378-382
Dopamine-β-hydroxylase (DBH)
  adrenergic vesicles and, 99-102
  amphetamine and, 404-405
  assay of, 205-206
  brain levels of, 116
  of CSF, 207
  copper in, 97
  endogenous inhibitors of, 97-99, 446
  factors influencing, 93, 102, 205, 445-446
  fusaric acid and, 262, 269-272, 276-277
  induction of, 95-97, 102, 444-445
  inhibition, 276, 358
  psychosis and, 269-272, 275-276, 447
  radioimmunoassay, 106-107, 125, 260, 447
  of serum
    assays for, 106-107, 117, 126, 206
    cold pressor test, 107-109
    in Down's syndrome, 114-115, 126
    in familial dysautonomia, 109-111
    genetic factors in, 106
    hypertension and, 124-125, 129-133, 207, 447
    levels of, 121-124, 129-130, 206
    in mental disorders, 114-116, 126
    in pheochromocytoma, 207
    regulation of, 106, 205
    sympathetic activity and, 105-107, 121-127, 129
    in torsion dystonia, 111-113, 126
  in urinary catecholamines and, 130-132

Down's syndrome
  serum DBH and, 114-115, 126
Dysautonomia
  serum DBH in, 109-113
Dyskinesias
  COMT inhibitors in management of, 350
  in parkinsonian patients, 432, 442

E

EEG
  in amphetamine psychosis, 283-294
  effect of drugs on, 242-243, 451
  of schizophrenia, 282-283
EGTA
  and activation of tyrosine hydroxylase, 382, 438, 439, 445
Epilepsy
  amphetamine and, 281-294
  psychosis with, 281-282
Extrapyramidal side effects
  and efficacy, 215

F

Familial tremor
  DBH level and, 109-110, 126
  piribedil and, 421
Fusaric acid
  CSF metabolites and, 270-271, 436-437
  dopamine-β-hydroxylase and, 262, 269-270
  mania and, 271-272, 275-277

G

Genetic factors
  in amine metabolism, 195-201
  cyclic AMP and, 200-201
Gerovital H_3
  as MAO inhibitor, 88
Glial cells
  MAO and, 89, 91
Guanylcyclase
  in protein synthesis, 209

H

Haloperidol
  dopamine receptors and, 373-375
  dopamine turnover and, 427-429
  tyrosine hydroxylase and, 164-167, 438
Hippocampus
  MAO activity in, 65-66
Histamine
  behavior and, 253-254, 335
Homovanillic acid (HVA)
  amphetamine and, 224-225
  fusaric acid and, 270-271, 436
  haloperidol and, 428-429
  schizophrenia and, 301
Huntington's disease
  DBH levels and, 126
  dopamine metabolism and, 428, 442, 447
  piribedil in, 418, 421, 428
γ-Hydroxybutyrate (GHB)
  effect on DOPA, 370, 372, 378
6-Hydroxydopamine
  actions of, 353-354
  amphetamine and, 363-364
  behavior and, 251, 256, 353-365, 386-400, 440
  catecholamines and, 386-387
  cold stress and, 389
  dopamine and, 358-361, 396-397
  glucopenia and, 388-389
  growth and, 361-363
  hypovolemia and, 389-390
  ingestive behavior and, 390-393
  lesions in striatum, 408
  nigrostriatal system and, 386-400
  norepinephrine and, 358-361
  recovery from, 386-389
  temperature regulation and, 354, 363
5-Hydroxyindoleacetic acid (5-HIAA)
  amphetamines and, 226-229
  MAO and, 4-5
  phenmetrazine and, 232
  in urine, 5
4-Hydroxy-3-methoxyphenylglycol (HMPG)
  MAO and, 4-5
  in urine, 5
p-Hydroxynorephedrine
  accumulation of, 223, 229
Hypertension
  dopamine-β-hydroxylase, 124-125, 129-133

Hypothalamus
dopamine-β-hydroxylase in, 116
6-hydroxydopamine effects on, 355
lesions in, 386, 395-397
MAO activity of, 16-17, 20-22, 79-80
norepinephrine uptake and, 395-397
Hypothermia
amphetamine and, 229-232

I

Impulse flow
dopamine and, 369-382
transmitter metabolism and, 369-382, 445

L

Lead poisoning
symptoms in, 399
Lithium
tryptophan hydroxylase and, 179-181, 186-187

M

Methamphetamine
serotonin and, 180-181
stereotyped behavior and, 245-246, 254
Methiothepia
tyrosine hydroxylase and, 164-166
3-Methoxytyramine
and stereotypy, 246
α-Methyl-para-tyrosine
behavior and, 254, 328-335, 403, 435
L-DOPA and, 262, 329-335, 403-404
mania and, 268-269
Methylphenidate
psychosis and, 317-322
stereotypy and, 212, 245-246
α-Methyltryptamine
stereotypy and, 253
α-Methyltyrosine
antagonism of steretypy, 212-213, 249
behavior and, 356-360
dopamine and, 403-405

effect on prolactin, 302
food intake and, 395-396
6-hydroxydopamine and, 356-360, 395-396
Midbrain
tryptophan hydroxylase in, 181-182
Migraine
phenylthylamine and, 5-7
sulfate conjugation and, 7-9
tyramine and, 5-9
MK 485
decarboxylase inhibition and, 266, 420
Monoamine oxidases (MAO)
in adrenal, 17-18, 21-22
age and, 20-22, 91-92, 449
of brain, 16-17, 20-22, 29-35, 50-56, 79-80, 87
in caudate, 16-17, 20-22
control of, 11
depression and, 59-60, 64-65, 78, 89-90, 274
development of, 20-22
dopamine and, 372, 379
genetic control of, 71, 73
half-life of, 19, 87
of heart, 21-22
hormonal effects, 11, 16-20
of humans, 11, 59-62, 71-81
in hypothalamus, 16-17, 20-22
inhibition of, 147-148, 163
inhibitors of
clinical use of, 2, 13
and inhibition of tyrosine hydroxylase, 377
properties of, 12-15
specificity of, 37-47, 50, 52-53, 75-76
structure-activity of, 37-47
of liver, 12-15, 21-24, 80, 89
in menstrual cycle, 11, 16-18, 87
migraine and, 5-9
molecular weight, 87
multiple forms of, 1-2, 3-9, 11-15, 30-32, 448-450
inhibitors of, 4, 37-47, 49-50, 80
in vivo studies of, 3-9
phospholipid and, 3-4, 14-15
in pineal, 88-89
of plasma
inhibition of, 79
properties of, 76-77

459

Monoamine oxidases (MAO) contd.,
of plasma contd.,
substrates of, 77-79
of platelets, 2, 7, 59, 67-69, 72-76, 90, 301, 321-322
progesterone and, 17-18
properties of, 11-12, 24, 29-30, 50, 73, 76, 77-79, 447-448
purification of, 29-35, 50-51, 87-88, 90, 447
schizophrenia and, 1-2, 59-60, 71, 63-69, 90-91, 301, 321-322
steroids and, 19
substrates of, 35, 50, 59-69, 74-75, 77-80
in twins, 59-63, 73
type A, 4-5, 13, 30-32, 49-50, 68, 81, 89, 449-450
type B, 4-7, 13, 30-32, 49-50, 68, 81, 89, 449-450
in vivo studies, 3-9, 22-25, 163
Morphine
stereotypy and, 245, 254, 435
tryptophan hydroxylase and, 179-180

N

Neuronal activity
transmitter metabolism and, 369-382
Nigrostriatal system
dopamine release in, 137, 407, 431
electrolytic lesions and, 386-393
functional recovery in, 393-400
6-hydroxydopamine and, 386-393
stereotypy and, 250
Norephinephrine (NE)
affective disorders and, 261-277
amphetamine and, 223-225, 330-336, 403-405
behavior and, 318, 330-337
depletion of, 223, 358
6-hydroxydopamine and, 358-361, 386-387
hypothermia and, 229-232
MAO and, 53-55
neuronal activity and, 369-382
prolactin and, 302-303
stereotypy and, 249-250, 255
synthesis of, 207-208, 444
uptake of, 395-397

P

Parachlorophenylalanine (PCPA)
tryptophan hydroxylase and, 178
in Parkinson therapy, 441
Pargyline
dopamine and, 372, 379, 439
6-hydroxydopamine and, 356, 387
tyrosine hydroxylase and, 136
Pargyline analogues
as MAO inhibitors, 43-45
Parkinson's disease
L-DOPA and, 427
dopamine depletion and, 399
dopaminergic receptor stimulant and, 415-424
MAO activity in, 91
Pemoline
stereotypy and, 212
Phenmetrazine
catecholamines and, 232-236
5-HT and, 232-236
dopamine receptor and, 299-300
serum prolactin and, 309-310
sterotypy and, 212, 242-256
tolerance to, 234-236
Phenylacetic acid
MAO and, 4-5
in urine, 5
Phenylethanol N-methyltransferase (PNMT)
factors influencing, 197-198
genetic regulation of, 195-201
induction of, 95-96
neural control of, 196
substrates of, 205
Phenylethylamine
MAO and, 5-7, 32, 89
migraine and, 5-7
stereotypy and, 245-246
Pheochromocytoma
serum DBH and, 127, 207
Pimozide
behavior and, 212-215, 438
as dopamine blocker, 438
tyrosine hydroxylase and, 164-165
Piribedil
L-DOPA and, 430-431
dopamine turnover and, 427-429
dopaminergic receptors and, 415-424, 427
Huntington's disease and, 418

Piribedil contd.,
    HVA and, 428-429
    metabolism of, 423
    Parkinson's disease and, 415-424,
        429-433, 441
    stereotypy and, 245
    tremor and, 422-423, 429-430
Plasticity
    model of, 386
Probenecid
    in turnover studies, 149-150, 301,
        428
Prolactin
    dopamine and, 302-303
    factors influencing, 303-309, 437
    schizophrenia and, 304-312
Prolactin-inhibiting factor, 302
Prolixin enanthate
    prolactin levels and, 310
Prostaglandins
    migraine and, 6
Protein synthesis
    reserpine and, 96
Psychiatric disorders
    COMT and, 193
    dopamine and, 211-217, 300
    serum DBH and, 114-116, 126
Psychosis
    amphetamine model of, 281-295
    arousal levels in, 282
    dopamine and, 261-277
    drug-induced, 291-295, 452
    stimulant-induced, 241-242
Pteridin
    tyrosine hydroxylase and, 167,
        445
Pyrogallol
    COMT and, 340-341

R

Receptor
    supersensitivity of, 412
Reserpine
    behavior and, 212-213, 326-328,
        439
    depression and, 356
    L-DOPA and, 326-328
    dopamine-β-hydroxylase and, 95-96
    MAO inhibitors and, 55-56
    stereotypy and, 248-249
    tryptophan hydroxylase and, 182,
        185-186

tyrosine hydroxylase and, 165

S

Schizophrenia
    amphetamine-induced, 281-283,
        300, 317-322
    COMT and, 193
    L-DOPA and, 300-301, 322
    dopamine and, 299-300, 317-318,
        322, 438, 450
    EEG in, 282-283
    HIAA and, 301
    MAO and, 1-2, 59-60, 63-69
    prolactin and, 304-312
    serum DBH in, 116-117
    stimulant-induced, 281
Serotonin (5-HT)
    amphetamines and, 226-229
    behavior and, 253
    MAO inhibitors and, 52-55
    as MAO substrate, 32
    phenmetrazine and, 232
    synthesis of, 177-188, 208, 370,
        440
Stereospecificity
    amphetamine, 317-322
    for PNMT substrates, 255
Stereotopy
    antagonism of, 211-212
    dopaminergic mediation, 211
    drug-induced, 241-256
    as psychosis model, 255
Striatum
    DOPA in, 162
    dopamine in, 162, 346, 408
    dopamine-β-hydroxylase in, 116
    plasticity of, 386
    stereotypy and, 250, 408
    tyrsosine hydroxylase and, 165
Substantia nigra
    stereotypy and, 250
Sulfate conjugation
    migraine and, 7-9
    substrates for, 87

T

Thioridazine
    prolactin levels and, 310, 312
    stereotypy and, 214-215
Tolerance
    amphetamines and, 221-231

Tolerance contd.,
  central stimulants and, 221-236
  phenmetrazine and, 232-236
Torsion dystonia
  serum DBH in, 111-113, 126
Tranylcypromine
  as MAO inhibitors, 80
Trifluperazine
  prolactin levels and, 310
Triton X-100
  dopamine-β-hydroxylase and, 100-102
Trivastal
  tyrosine hydroxylase and, 373-375
Tropolone
  catechol-O-methyl transferase and, 340-342
  dopamine synthesis and, 343-347
Tryptamine
  as MAO substrate, 32, 64, 66
  metabolism of, 1-2
  stereotopy and, 245
Tryptophan
  amphetamine and, 226-232
  metabolism of, 178-188
Tryptophan hydroxylase
  amphetamines and, 180-181, 183-184
  cofactors for, 182-183
  lithium and, 179-181
  parachlorophenylalanine and, 178-179
  regulation of, 177-188, 446
  reserpine and, 182, 185-186
  in synaptosomes, 178-187
Tuberculum olfactorium
  stereotypy and, 251
Turnover rate
  of DOPA, 145, 148-149, 162-163
  of dopamine, 137-139, 162-163
  mathematical models for, 150-156
  methods for, 143-156, 444
  of monoamines, 143-156, 161-163, 444
Tyramine
  conjugation of, 7, 9
  migraine and, 5-9
  plasma MAO and, 77, 78
Tyrosine
  inhibited metabolism of, 343-346

Tyrosine hydroxylase
  adrenal, 197-198
  affinity constant of, 166-168
  apomorphine and, 373-375, 438
  bound and soluble, 165
  calcium and, 375-378, 382, 445
  cofactor and, 166-168
  cyclic nucleotides and, 168-173, 209-210
  dopamine receptors and, 378-380, 382
  drugs influencing, 373-379
  forms of, 376
  genetic factors and, 195-201
  induction of, 95-97, 102, 168-173, 373, 444-445
  inhibition of, 345
  neuroleptics and, 164-168, 172
  product inhibition of, 135-141, 161-163
  regulation of, 95, 102, 135-141, 161-173, 372-379, 444
  striatal dopamine and, 138-141
  transsynaptic induction of, 168-173

U

U-14,624
  and NE levels, 405, 440
  and 5-HT metabolism, 440
Urinary catecholamines
  plasma DBH and, 130-132